Biodegradation and Bioremediation: Pollution Control and Waste Management

Biodegradation and Bioremediation: Pollution Control and Waste Management

Editor: William Chang

www.callistoreference.com

Callisto Reference,
118-35 Queens Blvd., Suite 400,
Forest Hills, NY 11375, USA

Visit us on the World Wide Web at:
www.callistoreference.com

ISBN: 978-1-64116-084-1 (Hardback)

Trademark Notice: Registered trademark of products or corporate names are used only for explanation and identification without intent to infringe.

Cataloging-in-Publication Data

Biodegradation and bioremediation : pollution control and waste management / edited by William Chang.
    p. cm.
Includes bibliographical references and index.
ISBN 978-1-64116-084-1
1. Biodegradation. 2. Bioremediation. 3. Pollution prevention. 4. Refuse and refuse disposal.
5. Salvage (Waste, etc.). 6. Environmental protection. I. Chang, William.
QH530.5 .B56 2019
572--dc23

Table of Contents

Preface...IX

Chapter 1   **Elimination of Antibiotic Multi-Resistant *Salmonella* Typhimurium from Swine Wastewater by Microalgae-Induced Antibacterial Mechanisms** 1
Melissa Paola Mezzari, Jean Michel Prandini, Jalusa Deon Kich and Márcio Luís Busi da Silva

Chapter 2   **Optimization of Biological Pretreatment of Green Table Olive Processing Wastewaters Using *Aspergillus niger***...5
Lamia Ayed, Nadia Chammam, Nedra Asses and Moktar Hamdi

Chapter 3   **Extraction, Determination and Bioremediation of Heavy Metal Ions and Pesticide Residues from Lake Water** 15
Sandip R Sabale, Bhaskar V Tamhankar, Meena M Dongare and B S Mohite

Chapter 4   **Antibiotic Susceptibility and Heavy Metal Tolerance Pattern of *Serratia Marcescens* Isolated From Soil and Water** 23
Natasha Nageswaran, P.W. Ramteke, O.P. Verma and Avantika Pandey

Chapter 5   ***Allium* Chromosome Aberration Test for Evaluation Effect of Cleaning Municipal Water with Constructed Wetland (CW) in Sveti Tomaž, Slovenia**.........28
Peter Firbas and Tomaž Amon

Chapter 6   **Biological Treatment of Meat Processing Wastewater Using Lab-Scale Anaerobic-Aerobic/Anoxic Sequencing Batch Reactors Operated in Series**..........33
David Nzioka Mutua, Eliud Nyaga Mwaniki Njagi, George Orinda, Geoffry Obondi, Frank Kansiime, Joseph Kyambadde, John Omara, Robinson Odong and Hellen Butungi

Chapter 7   **Microbial Degradation of Gasoline in Soil: Comparison by Soil Type**.................39
D A Turner, J. Pichtel, Y Rodenas, J McKillip and J V Goodpaster

Chapter 8   **Biodegradation of Coir Geotextiles Attached Media in Aerobic Biological Wastewater Treatment**...46
Gopan Mukkulath and Santosh G. Thampi

Chapter 9   **Biosorption of Arsenic by Living and Dried Biomass of Fresh Water Microalgae - Potentials and Equilibrium Studies**...50
Sibi G

Chapter 10 **Biodegradation of Tertiary Butyl Mercaptan in Water** 58
R. Karthikeyan, S.L.L. Hutchinson and L. E. Erickson

Chapter 11 **Bioremediation Rate of Total Petroleum Hydrocarbons from Contaminated Water by *Pseudomonas aeruginosa* Case Study: Lake Albert, Uganda**.................64
Kiraye M, John W and Gabriel K

Chapter 12 **Patent Analysis on Bioremediation of Environmental Pollutants**..................................69
Shweta Saraswat

Chapter 13 **Studies on Biosorption of Chromium Ions from Wastewater Using Biomass of *Aspergillus niger* Species** 75
Korrapati Narasimhulu and Y. Pydi Setty

Chapter 14 **Modelling Biogas Fermentation from Anaerobic Digestion: Potato Starch Processing Wastewater Treated Within an Up flow Anaerobic Sludge Blanket** 79
Philip Antwi, Jianzheng Li, En Shi, Portia Opoku Boadi and Frederick Ayivi

Chapter 15 **Development of Molecular Identification of Nitrifying Bacteria in Water Bodies of East Kolkata Wetland, West Bengal** 88
Mousumi Saha, Agniswar Sarkar and Bidyut Bandhophadhyay

Chapter 16 **Effect of Pesticide (Chlorpyrifos) on Soil Microbial Flora and Pesticide Degradation by Strains Isolated from Contaminated Soil**...93
Hindumathy CK and Gayathri V

Chapter 17 **Bioremediation of Polluted Soil obtained from Tarai Bhavan Region of Uttrakhand, India** 99
Rajdeo Kumar, Ashish Chauhan, Nisha Yadav, Laxmi Rawat and Manish Kumar Goyal

Chapter 18 **Influence of Different Organic-Based Fertilizers on the Phytoremediating Potential of *Calopogonium mucunoides* Desv. from Crude Oil Polluted Soils** 105
M. B. Adewole and Y. I. Bulu

Chapter 19 ***In-situ* Biological Water Treatment Technologies for Environmental Remediation**..111
Mohamed Ateia, Chihiro Yoshimura and Mahmoud Nasr

Chapter 20 **Rhizoremediation of Contaminated Soils by Comparing Six Roots Species in Al-Wafra, State of Kuwait**..116
Danah Khazaal Al-Ameeri and Mohammad Al Sarawi

Chapter 21 **The Potential of Shea Nut Shells in Phytoremediation of Heavy Metals in Contaminated Soil Using Lettuce (*Lactuca sativa*) as a Test Crop**.......................................130
Quainoo AK, Konadu A and Kumi M

Chapter 22 **Statistical Methodology for Cadmium (Cd(II)) Removal from Wastewater by Different Plant Biomasses** 137
Alaa El Din Mahmoud and Manal Fawzy

Chapter 23 **Monitoring of Pollution Using Density, Biomass and Diversity Indices of Macrobenthos from Mangrove Ecosystem of Uran, Navi Mumbai, West Coast of India** 144
Prabhakar R. Pawar

Chapter 24 **A Novel Subspecies of *Staphylococcus aureus* from Sediments of Lanzhou Reach of the Yellow River Aerobically Reduces Hexavalent Chromium** 152
Zhang X, Krumholz LR, Zhengsheng Yu, Yong Chen, Pu Liu and Xiangkai Li

Chapter 25   **Stable Isotope Probing - A Tool for Assessing the Potential Activity and Stability of Hydrocarbonoclastic Communities in Contaminated Seawater** 162
Petra J Sheppard, Eric M Adetutu, Alexandra Young, Mike Manefield, Paul D Morrison and Andrew S Ball

Chapter 26   **Eco-friendly and Cost-effective Use of Rice Straw in the Form of Fixed Bed Column to Remove Water Pollutants** 170
Baljinder Singh, Vasundhara Thakur, Garima Bhatia, Deepika Verma and Kashmir Singh

Chapter 27   **Evaluation of the Efficiency of Duckweed (*Lemna minor* L.) as a Phytoremediation Agent in Wastewater Treatment in Kashmir Himalayas** 177
Irfana Showqi, Farooq Ahmad Lone and Javeed Iqbal Ahmad Bhat

Chapter 28   **The Applicability of Electrical Current Based Treatment for the Remediation of Different Types of Polluted Soils Contaminated by Organic Compounds** 181
E. C. Rada and I.A. Istrate

Chapter 29   **The Role of Microorganisms in Distillery Wastewater Treatment** 190
Terefe Tafese Bezuneh

Chapter 30   **Heavy Metal Removal from Wastewater Using Low Cost Adsorbents** 196
Ashutosh Tripathi and Manju Rawat Ranjan

Chapter 31   **Potential of Microbial Inoculated Water Hyacinth Amended Thermophilic Composting and Vermicomposting in Biodegradation of Agro-Industrial Waste** 201
Alkesh Patidar, Richa Gupta and Archana Tiwari

Chapter 32   **Chitosan– A Low Cost Adsorbent for Electroplating Waste Water Treatment** 207
Bhavani K, Roshan Ara Begum E, Selvakumar S and Shenbagarathai R

Permissions

List of Contributors

Index

Preface

The main aim of this book is to educate learners and enhance their research focus by presenting diverse topics covering this vast field. This is an advanced book which compiles significant studies by distinguished experts in the area of analysis. This book addresses successive solutions to the challenges arising in the area of application, along with it; the book provides scope for future developments.

Bioremediation is the process used to treat contaminated soil or water by encouraging biodegradation or natural decay mechanisms. This is usually done by promoting growth of microorganisms. It can also be done by introducing microbial cultures in polluted media or by injecting nutrients to stimulate growth of microbes naturally. Bioremediation is a good alternative to traditional and industrial cleaning processes because of its affordability, minimal invasiveness and ecological viability. Research in this field extends into the core areas of soil remediation therapies, lowering radioactivity, curing metal contamination, groundwater treatment and oil spills remediation. In this book, the various advancements in bioremediation and biodegradation are glanced at and their applications in pollution control and waste management are looked at in detail. This book aims to equip students and experts with the advanced topics and upcoming concepts in this area.

It was a great honour to edit this book, though there were challenges, as it involved a lot of communication and networking between me and the editorial team. However, the end result was this all-inclusive book covering diverse themes in the field.

Finally, it is important to acknowledge the efforts of the contributors for their excellent chapters, through which a wide variety of issues have been addressed. I would also like to thank my colleagues for their valuable feedback during the making of this book.

Editor

Elimination of Antibiotic Multi-Resistant *Salmonella* Typhimurium from Swine Wastewater by Microalgae-Induced Antibacterial Mechanisms

Melissa Paola Mezzari[1], Jean Michel Prandini[2], Jalusa Deon Kich[3] and Márcio Luís Busi da Silva[3*]

[1]Biotechnology and Sciences Program, West University of Santa Catarina, Videira, SC 89560-000, Brazil
[2]Department of Chemical Engineering, Federal University of Santa Catarina, Florianópolis, SC 88040-900, Brazil
[3]EMBRAPA Swine and Poultry, Concórdia, Brazil

Abstract

The effect of microalgae-based swine wastewater treatment on the removal of antibiotic multi-resistant *Salmonella enterica* serovar Typhimurium was investigated. Photobioreactors (PBRs) containing diluted swine digestate with and without microalgae *Scenedesmus* spp. were inoculated with *S.* Typhimurium (10^8 Colony Forming Units per milliliters - CFU mL^{-1}). Viable cells of *S.* Typhimurium were quantified over time by plate counts and qPCR amplification of the *Salmonella* invasion gene activator, *hilA*. In the absence of microalgae, *S.* Typhimurium concentrations increased 1.5 log cells mL^{-1} in 96 h. In the presence of microalgae, *S.* Typhimurium was completely eradicated within 48 h. In the PBRs with controlled pH (6.8 ± 0.8), concentration of *S.* Typhimurium remained constant (2.8 ± 0.2 log CFU mL^{-1}) throughout 96 h. Thus, natural increase in pH>10 due to photosynthesis was detrimental to the antibiotic multi-resistant bacteria survival. Phycoremediation holds promises as an alternative for wastewater treatment process for the elimination of the serious public health threatening antibiotic multi-resistant bacteria, thus effectively avoiding Salmonellosis outbreaks arising from animal farming activities.

Keywords: Antibiotic-resistant bacteria; *hilA* gene; *Salmonella enterica* serovar Typhimurium; *Scenedesmus* spp.; Swine wastewater

Introduction

Major concerns exist over the several invasive and antibiotic resistant organisms thriving in swine wastewaters and that are known to threaten human and animal health. Among them, *Salmonella enterica* serovar Typhimurium deserves special attention since it is the most prevalent antimicrobial resistant serovar in swines, and also frequently related to human infections and outbreaks [1].

Several physicochemical approaches are described to control the proliferation of pathogens, such as exposure to UV irradiation [2,3] use of strong oxidant radicals [2], pH increase [4], and selective membranes [5]. However, most if not all of these approaches are not economically feasible. Therefore, biological anaerobic digestion followed by stabilization ponds are the most common treatment option adopted in swine farming worldwide [3,6,7]. Pathogen elimination can occur under thermophilic conditions, but not under mesophilic conditions that prevail in most digesters [8].

Phycoremediation has been considered as an efficient tertiary treatment method to reduce organic compounds and nutrients from wastewaters [9-12]. Some microalgae produce a wide variety of antibacterial substances that can inhibit or kill pathogens [13,14]. These waterborne pathogens may also be sensitive to high oxygen concentrations and increased pH produced by photosynthesis. Nevertheless, to the best of author's knowledge, the mechanism in which phycoremediation reduce or even eliminate invase antibiotic multi-resistant bacteria from swine wastewaters has not been fully explored. This study demonstrates whether phycoremediation of swine wastewaters could effectively control the proliferation of antibiotic-resistant *Salmonella enterica* serovar Typhimurium. Ancillary objective include the evaluation of a pre-enrichment method for rapid quantification of S. Typhimurium from complex environmental samples.

Materials and Methods

Photobioreactor conditions

Laboratory scale photobioreactors (PBRs) were used to simulate phycoremediation of swine wastewater. Each PBR (3.5 L beakers) was filled with 3 L of diluted swine digestate (6%, v:v). Effluent physicochemical characteristics were (g L^{-1}): pH 7.9, total solids (3-8), total organic carbon (1.5-6.5), total inorganic carbon (0.8-1), total nitrogen (1.5-2), ammonia-N (0.9-1.5), phosphate-P (0.045-0.06). PBRs were subjected to mixotrophic conditions (12 h, light: dark) using red light emission diode light (PGL-RBC 2500, Parus) at 630 nm and 121.5 µmol m^{-2} s^{-1}, room temperature (22 ± 2°C), and continuous mixing. PBRs were inoculated with *Salmonella enterica* serovar Typhimurium (10^5 CFU mL^{-1}) and *Scenedesmus* spp. (30% v/v, 70 mg L^{-1} dry weight biomass), except the negative control without microalgae. To discern microalgae antibacterial effects from pH-derived photosynthesis, pH was adjusted daily in a set of PBRs by adding either 0.1 M HCl or 0.1 M NaOH. All experiments were conducted in duplicate.

Inocula source and identification

Microalgae was collected from a field scale facultative open pond used as tertiary treatment process downstream from a biodigestor at the Brazilian Agricultural Research Corporation (EMBRAPA) swine wastewater facility (Concórdia, SC, Brazil). *Scenedesmus* spp. was the dominant microalgae in the PBRs as described elsewhere [9,10]. The strain was isolated and deposited in the collection of photosynthetic microorganisms for Agroenergy Research at Embrapa (Brasília, DF, Brazil) under access number Embrapa LBA#31 (IAN193.096).

---

*Corresponding author: Márcio Luís Busi da Silva, EMBRAPA Swine and Poultry, Concórdia, Brazil, E-mail: marcio.busi@embrapa.br

Salmonella enterica serovar Typhimurium was collected from a local river stream (Concórdia, SC, Brazil), characterized as a multi-resistant bacteria [1] and stored at the Embrapa's microorganism bank collection under accession # 12301.

Analytical methods

Samples were analyzed twice a day for pH, temperature (pH–mV, Hanna Instruments, Inc.), dissolved oxygen (DO) (Lutron DO-5519). Microalgae were quantified by optical density at 570 nm (Hach DR/2000). A satisfactory correlation (r^2=0.97) between DW (gravimetric measuremets) and OD_{570} (mg-DW L^{-1}=536.2 × OD_{570nm} -36.89) was obtained. The specific growth rate μ (day^{-1}) was calculated by fitting the microalgae dry weight obtained from samples within the first 24 hours of microalgae growth to an exponential function of (lnX-lnX_0/t), where t was the time between the two measurements, and X and X_0 (mg-DW L^{-1}) were concentrations of biomass at t (24 hours) and t_0, respectively.

S. Typhimurium quantification – plate culture and qPCR

S. Typhimurium inoculum was collected from an overnight culture plate, sub-cultured into nutrient broth (50 ml) and incubated in a shaker (100 rpm, 37°C, 24 h). Density of cell suspension was assessed with McFarland turbidity standard No. 05 (1-2 × 10^8 CFU mL^{-1}). Samples from PBRs were diluted (10^0-10^{-5}) in saline media (0.85%) and spread onto Chromogenic XTL4 agar plates (100 μL) to enumerate S. Typhimurium colonies. Plates (duplicates from PBRs) were incubated overnight and results were reported as CFU mL^{-1}.

Samples from PBRs were collected daily and diluted ten times in buffered peptone water for incubation at 37°C for 6 h – pre enrichment step [15] prior DNA extraction (MoBio UltraClean Microbial DNA kit). Template genomic standard curves (10^{-1} to 10^{-10}) were also subjected to the pre-enrichment step. *hilA*, a *Salmonella* gene required for pathogenicity invasion, was quantified by qPCR using specific primers [16]. Each 20 μL qPCR reaction mixture contained 2 × qPCR-SYBR-Green mix (Ludwig Biotec, Brazil), 500 nM forward and reverse primers (Prodimol Biotecnologia®, Brazil), sterile DNAse-free water (Ludwig Biotec, Brazil) and 16 μg of DNA template. qPCR reactions were performed (7500 Applied Biosystems, The Netherlands) in a two-step thermal cycling procedure (95°C, 10 min; followed by 40 cycles of 95°C for 15 s and 62°C for 30 s). Quantitation of S. Typhimurium was performed by interpolation from DNA template standard curves.

Statistical analysis

Tukey's Pairwise Comparison test was used to determine significant difference between two data sets at 95% confidence level (p<0.05). All data were subjected to one-way analysis of variance (ANOVA) using OriginPro 8© OriginLab Corporation.

Results and Discussion

Scenedesmus spp. growth

Scenedesmus spp. are known for their environmental ubiquity and predominance in waste stabilization ponds and high-rate algal ponds [17]. In the present study, microalgae growth rates of 96.5 ± 1.7 and 82.3 ± 3.4 mg-DW L^{-1} d^{-1} were reached in PBRs with and without pH adjustments, respectively. Specific growth rates (μ) were also calculated during the exponential growth phase of *Scenedesmus* spp. The mean value of μ in the interval from 0 to 24 h for PBRs with and without pH adjustments were 0.97 and 0.92 day^{-1}, respectively. These values were within the higher growth rates reported for *Scenedesmus* sp. in swine wastewaters [10,17]. The biomass increase measured in both PBRs treatments were very similar (p<0.05), with maximum microalgal

concentration of 265 ± 3.9 mg-DW L^{-1} observed at 72 h (Figure 1). These experimental results are within typical values of maximal biomass production (0.2-1.0 g-DW L^{-1}) reported for *Scenedesmus* spp. in domestic and swine wastewaters from mixotrophic growth [17]. Interestingly, *Scenedesmus* spp. is a promising microalgal species that can thrive in wastewaters with great performance on biomass production and lipid accumulation than other high-lipid-content microalgae [18]. Although not the scope of the present investigation, the obtained *Scenedesmus* spp. growth performance and biomass production are comparable to other studies focused on microalgal biomass to bioenergy conversion [19,20]. Thus, for a swine wastewater treatment plant that uses tertiary phycoremediation treatment to remove pathogens and nutrients, microalgae biomass production could also aid in valuable biofuel products.

Phycoremediation effects on S. Typhimurium

Salmonella spp. are able to thrive in different environments, survive several weeks in dry environments and several months in water, thus proliferation of S. Typhimurium in swine wastewater digestate was

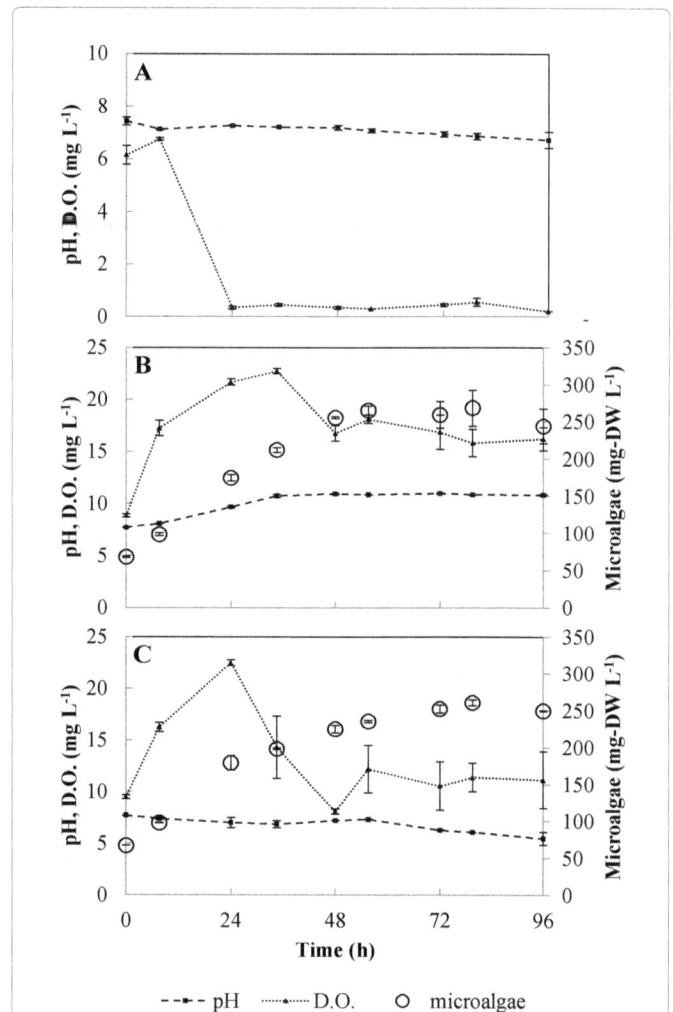

Figure 1: Dissolved oxygen (mg L^{-1}), pH and microalgae biomass concentration (mg-DW L^{-1}) profile in mixotrophic photobioreactors (PBRs) without microalgae (A), with microalgae (B) and with microalgae under pH control conditions (C). Data are presented as mean ± standard deviation (n=2).

somewhat expected and later confirmed in the biological reactors without microalgae (Figure 2). In these reactors, *S.* Typhimurium increased from $7.8 \pm 1.9 \times 10^2$ CFU mL^{-1} (t=0 h) to a maximum of $3.4 \pm 0.8 \times 10^4$ CFU mL^{-1} (t=72 h).

Removal of pathogens from phycoremediation in ponds are usually attributed to a combination of long retention times, UV irradiation, high temperature, increased dissolved oxygen (DO) concentrations and high pH [21]. Among these factors, elevated DO and pH levels, resultant from optimal microalgal photosynthetic activities, have greater effects on pathogen inhibition and inactivation [22]. While it is possible to artificially increase DO and pH by aeration and chemicals addition, respectively, these practices are usually economically unfeasible and can be challenging to dosage at field scale. In this regard, phycoremediation could be alternatively considered as a disinfection step in wastewater facility. S. Typhimurium was completely removed from PBRs within 48 hours (Figure 2). In pH adjusted PBRs, S. Typhimurium concentrations did not vary (p=0.12) throughout the entire experiment time frame and remained constant at $7.5 \pm 1.4 \times 10^2$ CFU mL^{-1}. Negligible difference in DO (p>0.05) concentrations were observed between PBRs treatments (Figure 1), thus discharging the probability of bacteria elimination by DO. The measured DO concentrations were above solubility levels (i.e., 8.3 mg L^{-1} at 1 atm and 25°C) (Figure 1). Oxygen saturation is normally found in ponds or photobioreactors due to high microalgae growth rate and associated photosynthetic activities [10,17,21]. The high pH values measured in the PBRs without pH adjustment seemed to play a major role on the inactivation of *S.* Typhimurium during *Scenedesmus* spp. growth.

The antibacterial properties of microalgae exudates should not be ruled out [14]. *Scenedesmus* sp. has been previously shown to act against Gram-positive (e.g., *Staphylococcus aureus*) and Gram-negative (e.g., *Escherichia coli* and *Pseudomonas aeruginosa*) bacteria, but it was not very effective against *Salmonella* sp. [14,23]. This fact associated with the lack of *S.* Typhimurium removal from the pH controlled PBRs suggests that microalgae metabolites had insignificant contribution on bacteria removal.

Sample preenrichment and qPCR fast approach

Colony count techniques consider CFU numbers, which can consist of more than one cell count and thus lead to data overestimation [24]. Therefore, to increase accuracy in bacterial concentration, qPCR quantification method was also performed in parallel. The plating method showed significantly higher cell concentrations (between 0.8 to 1.5 log differences) than qPCR (Figure 2). However, no differences (p>0.05) in *S.* Typhimurium growth ratio (X/X_0) over time was verified

between these two bacterial quantification methods.

Conclusions

Phycoremediation of swine wastewater digestate was very effective to produce valuable microalgae biomass and eliminate antibiotic multi-resistant *S.* Typhimurium within 48 h. Pathogen removal was linked to the inhibitory effects of high pH (as high as $\cong 11$) as a result of photosynthesis. Quantification of invasive *S.* Typhimurium in environmental complex samples was demonstrated by preenrichment followed by qPCR. Phycoremediation of swine wastewaters can be a promising treatment strategy to control the spread of antibiotic resistant bacteria in the environment. This can be particularly appealing when cogitating water reuse in animal farming and the risks associated with public health outbreaks from such practices.

Acknowledgements

The authors thank Remídio Vizzotto and Luiza Letícia Biesus from Embrapa Swine and Poultry for technical support. Authors thank financial support from Coordination for the Improvement of Higher Level or Education Personnel CAPES-EMBRAPA (#001/2011) and the Brazilian Agricultural Research Corporation – EMBRAPA (#02.12.08.004.00.05).

References

1. Palhares J, Kich J, Bessa M, Biesus L, Berno L, et al. (2014) Salmonella and antimicrobial resistance in an animal-based agriculture river system. Sci Total Environ 472: 654-661.

2. Macauley J, Qiang Z, Adams C, Surampalli R, Mormile M (2006) Disinfection of swine wastewater using chlorine, ultraviolet light and ozone. Water Res 40:2017-2026.

3. Jenkins M, Endale D, Fisher D, Adams M, Lowrance R, et al. (2012) Survival dynamics of fecal bacteria in ponds in agricultural watersheds of the Piedmont and Coastal Plain of Georgia. Water Res 46:176-186.

4. Viancelli A, Kunz A, Fongaro G, Kich J, Barardi C, et al. (2015) Pathogen Inactivation and the Chemical Removal of Phosphorus from Swine Wastewater. Water, Air, Soil Pollut 226: 263-272.

5. Masse L, Massé D, Pellerin Y (2007) The use of membranes for the treatment of manure: a critical literature review. Biosyst Eng 98: 371-380.

6. Da Silva M, Cantao M, Mezzari M, Ma J, Nossa C (2015) Assessment of bacterial and archaeal community structure in swine wastewater treatment processes. Microb Ecol 70: 77-87.

7. Hill V (2003) Prospects for pathogen reductions in livestock wastewaters: a review. Crit Rev Environ Sci Technol 33: 187-235.

8. Metcalf E, Eddy H (2003) Wastewater engineering: treatment and reuse. 4th edn. McGraw Hill, New York, United States of America.

9. Mezzari M, Da Silva M, Nicoloso R, Ibelli A, Bortoli M, et al. (2013) Assessment of N2O emission from a photobioreactor treating ammonia-rich swine wastewater digestate. Bioresour Technol 149: 327-332.

10. Prandini J, Da Silva M, Mezzari M, Pirolli M, Michelon W, et al. (2016) Enhancement of nutrient removal from swine wastewater digestate coupled to biogas purification by microalgae Scenedesmus spp. Bioresour Technol 202: 67-75.

11. Lowrey J, Brooks M, McGinn P (2014) Heterotrophic and mixotrophic cultivation of microalgae for biodiesel production in agricultural wastewaters and associated challenges. J Appl Phycol 27: 1-14.

12. Rahman A, Ellis J (2012) Bioremediation of Domestic Wastewater and Production of Bioproducts from Microalgae Using Waste Stabilization Ponds. J Bioremediation Biodegrad 3: e113.

13. Ghasemi Y, Moradian A, Mohagheghzadeh A, Shokravi S, Morowvat M (2007) Antifungal and antibacterial activity of the microalgae collected from paddy fields of Iran: Characterization of antimicrobial activity of Chroococcus dispersus. J Biol Sci 7: 904-910.

14. Guedes A, Barbosa C, Amaro H, Pereira C, Malcata F (2011) Microalgal and cyanobacterial cell extracts for use as natural antibacterial additives against

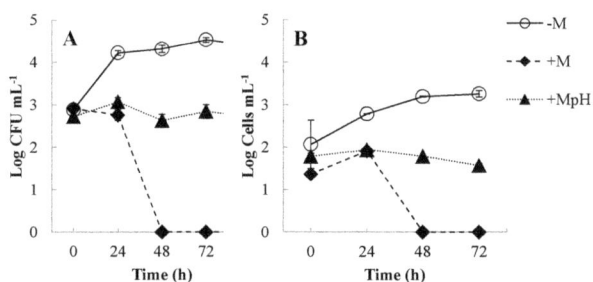

Figure 2: *Salmonella enterica* serovar Typhimurium concentration profile from PBRs without microalgae (-M), with microalgae (+M) and with microalgae and pH controlled conditions (+MpH). Bacteria were quantified by plate count (A) and qPCR (B). Error bars indicate standard deviation for replicates (n=4).

food pathogens. Int J Food Sci Technol 46: 862-870.

15. Malorny B, Löfström C, Wagner M, Krämer N, Hoorfar J (2008) Enumeration of Salmonella bacteria in food and feed samples by real-time PCR for quantitative microbial risk assessment. Appl Environ Microbiol 74: 1299-1304.

16. Brunelle B, Bearson S, Bearson B (2011) Salmonella enterica Serovar Typhimurium DT104 invasion is not enhanced by sub-inhibitory concentrations of the antibiotic florfenicol. J Vet Sci Technol 2: 1-4.

17. Wu Y, Hu H, Yu Y, Zhang T, Zhu S, et al. (2014) Microalgal species for sustainable biomass/lipid production using wastewater as resource: A review. Renew Sustain Energy Rev 33: 675-688.

18. Xin L, Hong-ying H, Jia Y (2010) Lipid accumulation and nutrient removal properties of a newly isolated freshwater microalga, Scenedesmus sp. LX1, growing in secondary effluent. N Biotechnol 27: 59-63.

19. Singh M, Reynolds D, Das K (2011) Microalgal system for treatment of effluent from poultry litter anaerobic digestion. Bioresour Technol 102: 10841-10848.

20. Shen Q, Jiang J, Chen L, Cheng L, Xu X, et al. (2015) Effect of carbon source on biomass growth and nutrients removal of Scenedesmus obliquus for wastewater advanced treatment and lipid production. Bioresour Technol 190: 257-263.

21. Butler E, Hung Y, Al Ahmad M, Yeh R, Liu R, et al. (2015) Oxidation pond for municipal wastewater treatment. Appl Water Sci pp: 1-21.

22. Gupta S, Ansari F, Shriwastav A, Sahoo N, Rawat I, et al. (2015) Dual role of Chlorella sorokiniana and Scenedesmus obliquus for comprehensive wastewater treatment and biomass production for biofuels. J Clean Prod 115: 255-264.

23. Ishaq A, Matias-Peralta H, Basri H, Muhammad M (2015) Antibacterial activity of freshwater microalga Scenedesmus sp. on foodborne pathogens Staphylococcus aureus and Salmonella sp. J Sci Technol 7: 858-871.

24. Krämer N, Löfström C, Vigre H, Hoorfar J, Bunge C, et al. (2011) A novel strategy to obtain quantitative data for modelling: combined enrichment and real-time PCR for enumeration of salmonellae from pig carcasses. Int J Food Microbiol 145: S86-95.

Optimization of Biological Pretreatment of Green Table Olive Processing Wastewaters Using *Aspergillus niger*

Lamia Ayed*, Nadia Chammam, Nedra Asses and Moktar Hamdi

Carthage University, National Institute of Applied Science and Technology (INSAT). Laboratory of Microbial Ecology and Technology, Department of Biological and Chemical Engineering, B.P .676,1080 Tunis, Tunisia.

Abstract

Aspergillus niger has been tested in a discontinuous process for the pretreatment of fresh green Table Olive Processing Waste waters (TOPW). The influence of seven factors like incubation time, initial pH, dilution, glucose, agitation, $(NH_4)_2SO_4$ and inoculum size on decolorization was screened by employing a fractional factorial design 2^{7-4}. Through factorial plots and analysis of variance, it was found that only glucose and agitation affected the decolorization. Further studies were carried out to adjust the level of determinant factors. Results showed that color removal (62%) was optimum at 150 rpm and 3 g/l glucose. HPLC and FTIR analysis of the treated TOPW suggested that the decolorization occurs through biosorption and biodegradation. Tannase and lignin peroxidase were detected in the cultures. However, only tannase is responsible for the color removal of the TOPW because lignin peroxidase is produced at a low level.

Keywords: Optimization; Decolorization; Fresh green table olive processing Wastewater; *Aspergillus niger;* Tannase

Introduction

In Tunisia, the table-olives production is highly important for the country's economy, constituting one of the major agro-industrial activities. There are three principal types of table olives: green, black and black through oxidation. The most common method for producing table olive is "Spanish-style" processing. The process is carried out through a series of steps: lye treatment (debittering), washing, brining, fermentation in brine, packaging and pasteurization [1]. Countries producing green table olive always have the problem of the disposal of the waste waters from debittering and washing steps.

These Table Olive Processing Waste waters (TOPW) contain large amounts of mineral and organic matter. The latter fraction contains a complex consortium of sugars and phenolic compounds, some nitrogenous compounds (especially amino acids), organic acids, tannins, pectins, carotenoids and oil residues [2]. It is therefore, obvious that these waste waters are highly polluting and are not simply treated by conventional methods.

Many strategies have been studied to reduce the environmental impact of these waste waters such as washing water re-use, reduction of washing waters and debittering with low-concentration lyes. Any of these approaches has resulted in meeting the needs [3].

The aerobic treatment of wastewaters was shown to be an encouraging way to reduce their toxicity and dark color [4-6]. The removal of color has been carried out by many fungal strains such as *Aspergillus* sp., which can decolorize [7-10] and / or biosorb recalcitrant compounds [11,12]. The biodegradation is due to their capacity to excrete various extracellular enzymes [13,14]. The efficiency of the biological treatment can be enhanced by carefully selecting the operational conditions that have been observed to influence the decolorization and enzymes' production [7,15,16].

No earlier publications concerning the parameters affecting decolorization of TOPW by *Aspergillus niger* and enzymes production has been cited in the literature. The present research reports the optimization of the decolorization of TOPW by *Aspergillus niger* and the produced enzymes were also investigated on the optimized medium.

Material and Methods

Used wastewaters

The wastewaters consisted of debittering and washing water resulting from green Table Olive Processing Waste waters (TOPW). Fresh TOPW has been stored at -20°C to avoid the auto-oxidation of phenolic compounds (Table 1).

The biological material and culture procedures

Aspergillus niger obtained from the University of Harokopio (Greece) was used for the present work [17]. The strain was maintained on potato-dextrose agar slants. Spore suspensions for inoculation were prepared from slants, after seven days of cultivation at 30°C, using Ringer's solution.

Characteristics	Fresh TOPW
pH	12.3 ± 0.4
TSS (g/l)	1.09 ± 0.13
CODs (g/l)	17.2 ± 3.2
Color (OD 390nm)	4.86 ± 0.33
Reducing sugars (g/l)	3.31 ± 0.18
TKN (g/l)	0.69 ± 0.05
TP (mg gallic acid/l)	506.9 ± 38

Data are reported as mean ± standard deviation of results carried out in triplicate
TS: total solids; TSS: total suspended solids; CODs: soluble chemical oxygen demand; TKN: total Kjeldahl nitrogen;TP: total phenolic compounds

Table 1: Main characteristics of used wastewaters in this work.

*Corresponding author: Lamia Ayed, National Institute of Applied Science and Technology (INSAT), Tunisia, E-mail: lamiaayed@yahoo.fr

The resulting suspension was centrifuged (6000 x g, 15 min) and the solids were resuspended in sterile distilled water. The spore concentration was measured by direct microscopic counts. A sample of five ml spore suspension was used as inoculum for 100 ml of substrate. 100 ml of the acidified TOPW (pH 5) were transferred in 500 ml Erlenmeyer flasks and sterilized by autoclaving. The medium was inoculated with the spore suspension and incubated under agitation condition at 30°C for four days. All cultures were used in triplicates.

Abiotic decolorization study

Spore suspensions of *Aspergillus niger* were transferred in 250 ml shake Elenmeyer flasks containing 50 ml of sabouroud brouth and incubated at 30°C for four days. After cultivation, the biomass was centrifuged at 5000 x g for 10 min, washed with sterile saline solution and used as inoculum. Autoclaved biomass was added into 250 ml Erlenmeyer flask containing 50 ml TOPW. The inoculated flasks were placed on rotary shaker (150 rpm) at 30°C for four days. Samples were withdrawn and centrifuged at 5000 x g for 10 min. The supernatants were read for the OD at 390 nm using a spectrophotometer (JENWAY 63200 UV/VIS) and analyzed with a Perkin–Elmer series 783 FTIR spectrophotometer.

Screening of process variables on TOPW decolorization by *Aspergillus niger*

Fractional factorial design 2^{7-4} was used to study the significance of seven factors (time, initial pH, dilution, glucose, agitation, $(NH_4)_2SO_4$ and inoculum size) on TOPW decolorization. The choice of factors was based on previous reports on *Aspergillus* growth [18,19], enzyme production [16] and the decolorization of wastewater by *Aspergillus niger* [8,20,21].

These factors were tested through eight experiments and each of which was carried out in triplicate. The fractional factorial design and responses measured during experiment were depicted in Table 2, with the low (-1) and high (+1) levels. The impact of each factor on response is evaluated by the determination of the coefficients relating to each of the seven factors:

$C_{xi} = 1/8 \ (\Sigma \ Aj \ x \ Xi)$

- Aj means either high (+) or low (-) level in experimental run i,

- Xi is the TOPW decolorization (% OD 390 removal),

Analytical methods

Growth was determined by Total Suspended Solid (TSS). TSS were obtained by filtration on Whatman filter (0.45 nm).The settled solid were then dried overnight at 105°C. Chemical analyses were carried out in triplicate according to standard methods [22].

Decolorization assay

The supernatants were adjusted to pH 5.0 using concentrated HCl and their absorbencies were measured at 390 nm against distilled water using a spectrophotometer (JENWAY 63200 UV/VIS).

Total phenolic compounds

Total Phenolic content (TP) was calculated in triplicate using the Folin-Ciocalteu's phenol reagent and requiring the addition of 200 µl Folin-Ciocalteu's phenol reagent in 3.6 ml diluted sample. After three minutes, 800 µl sodium carbonate (200 g/l) was supplemented. The mixture was heated at 100°C during one minute. After cooling, the optical density was measured at 750 nm [23].

Molecular mass distribution of polyphenolic compounds

Gel filtration on Sephadex G-50 was used to analyze the polyphenolic compounds present in different samples of TOPW. Three milliliters of sample were filtered and placed on a Sephadex coarse G-50 column (2.5 × 60 cm) previously equilibrated with NaNO_3 0.05 M containing 0.02% sodium azide at a flow rate of 0.6 ml/min. The effluent was collected on the basis of 3 ml per tube. The optical density of these fractions was measured at 280 nm. The column was calibrated with syringic acid (MM = 198 Da) and blue dextran (MM = 200 kDa).

HPLC analysis

Samples were acidified at pH 2 and extracted with ethyl acetate. The ethyl acetate extracts were identified and quantified by HPLC analysis. The HPLC chromatograph was performed on an Agilent Technologies apparatus (1200 series) composed of a VWD detector. Elutes were detected at 280 nm. The column was (4.6 x 250) mm (Shim-pack VP-ODS) and its temperature was maintained at 40°C. The flow rate was 0.5ml / min. The mobile phase used was 0.1% phosphoric acid in water (A) versus 70% acetonitrile in water (B) for a total running time of 40 min, and the gradient changed as follows: solvent B started at 20% and increased immediately to 50% in 30 min. After that, elution was conducted in the isocratic mode with 50% solvent B within 5 min. Finally, solvent B decreased to 20% until the end of running [6].

Fourier transform infra-red (FTIR) spectroscopy

A quantity of 1.5 mg of sample was compressed under vacuum with 250 mg of KBr. The pellets obtained were analysed with a Perkin–Elmer series 783 FTIR spectrophotometer (Nicolet Analytical Instruments, Madison, WI) covering a frequency range of 4000–400 cm⁻¹.

Enzyme assays

Supernatants from the final optimum decolorization medium were analyzed for tannase, lignin-peroxidase (LiP) and laccase

Experiments		Time (Days)	Glucose (g/l)	$(NH_4)_2SO_4$ (g/l)	pH	COD_0 (g/l)	Inoculum size	Agitation	
Low level		4	0	0	5	10	10^7	-	
High level		8	2	2	7	15	10^9	100rpm	% color removal
1	+	-	-	-	+	+	+	-	20 ± 2
2	+	+	-	-	-	-	+	+	32 ± 5
3	+	-	+	-	-	+	-	+	50 ± 1.6
4	+	+	+	-	+	-	-	-	48 ± 3
5	+	-	-	+	+	-	-	+	30 ± 2.4
6	+	+	-	+	-	+	-	-	19 ± 1.1
7	+	-	+	+	-	-	+	-	44 ± 4
8	+	+	+	+	+	+	+	+	53 ± 3.9

Table 2: 2^{7-4} fractional factorial design with observed response values for decolorization of TOPW using *Aspergillus niger*.

activities. Tannase activity was assayed by measuring the amount of hydrolyzed substrate. The method reported by Iibushi et al. [24] was based on estimation of decrease absorbance at 310 nm. A 4 ml solution containing 0.35% (w/v) tannic acid and 0.05 M citrate buffer, pH 5.5, was preincubated at 37°C to which 1 ml of the supernatant of the culture was added. The reaction mixture was maintained at 37°C. After t_1 and t_2 min, 40 µl then withdrawn and diluted 100 times with 90% (v/v) ethanol to stop the enzyme reaction. The absorbance at 310 nm was read and the difference between absorbencies of t_1 and t_2 was calculated. One IU of tannase activity was defined as the amount of enzyme which hydrolyzes 1 µmol ester bond in 1 min.

LiP activity was also determined colorimetrically with the method of Tien and Kirk [25] by measuring the absorbance increase at 310 nm (ε_{310} = 9300M^{-1} cm^{-1}). The reaction mixture (2 ml) containing 4 mM veratryl alcohol in 10 mM sodium tartarate buffer (pH 3) incubated with 100 µl of the culture fluid at 30°C. The reaction was initiated by addition of the suitable amount of 0.2mM H_2O_2. The blanks contained buffer in place of veratryl alcohol. One unit of LiP activity was defined as the amount of enzymes catalyzing the formation of 1µmol of veratraldehyde per minute under the assay conditions.

Laccase activity was determined by monitoring the oxidation of 2,2-azino-bis(3-ethylbenzothiazoline-6-sulfonic acid) at 420nm following the method of Palmieri et al. [26]. The reaction mixtures (1 ml) were prepared using 20 µl of culture broth, 880 µl of citrate-phosphate buffer (pH 3.0) and 100 µl 1 mM ABTS, and the temperature was adjusted to 30°C. The oxidation was followed by the increase in absorbance at 420nm (ε_{420} = 36,000M^{-1} cm^{-1}). The blanks received the buffer instead of ABTS. One unit of enzyme activity was defined as the amount of enzymes required to produce an absorbance increase of one unit per minute per milliliter of the reaction mixture.

Statistical analysis

All data presented are the average of triplicate measurements ± S.D. An analysis of variance (ANOVA) was conducted by employing performed (SPSS) version 16.0 software. The statistical software used to evaluate the experimental design results was Minitab 16. Effects were considered significant when the P value was less than 0.05.

Results and Discussion

The results of the present study establish that the decolorization of TOPW by Aspergillus niger is strongly affected by experimental conditions and substrate nutrient concentration. Optimization of the color removal was carried out in two steps. In the first step, the effect of seven factors (Time, dilution, glucose, (NH4)$_2$SO$_4$, inoculum size, initial pH and agitation) on TOPW decolorization was investigated using fractional factorial design 2^{7-4}. At the second time, levels of determinant factors were adjusted.

Effect of process variables on TOPW decolorization

Factors that influence the TOPW decolorization were evaluated by using factorial plots: main effect and the Pareto chart plot. ANOVA and P-value significant levels were used to check the significance of the effect on the color removal. Glucose had the greatest effect on decolorization, followed by agitation, initial COD, culture time, (NH4)$_2$SO$_4$, inoculum size and initial pH (Table 3). The main effects represent deviations of the average between the high and low levels for each factor. When the effect of a factor is positive, decolorization increases as the factor changes from low to high level. In contrast, if the effects are negative, a reduction in decolorization occurs for high level of the same factor.

Term	Effect	Coefficient (C_{xi})	t-value	P
Constant		40.125	24.06	<0.01
Time	5.5	2.7	1.65	0.138
Glucose	21	10.5	6.3	<0.01
(NH$_4$)$_2$SO$_4$	-2.75	-1.375	-0.82	0.434
Initial pH	-1.25	-0.25	-0.37	0.718
Initial COD	-5.5	-2.7	-1.65	0.138
Inoculum size	3.5	1.7	1.05	0.325
Agitation	11.25	5.625	3.37	0.01

Table 3: Estimated Effects and Coefficients for color removal (%).

From Figure 1, it is inferred that the larger the vertical line, the larger the change in TOPW decolorization when changing from level −1 to level +1. It should be pointed out that the statistical significance of a factor is directly related to the length of the vertical line. The effects of initial COD, (NH4)$_2$SO$_4$ and initial pH are negative, that is, a decrease of color removal is observed when the factor changes from low to high. For glucose, agitation, inoculum size and culture time, the opposite is true.

The increase in glucose concentration led to an increase of color removal. However, ammonium sulfate had a negative effect on the decolorization. It is probably because the TOPW itself contains a sufficient amount of N sources, which could be favorable to microbial growth and enzyme production.

The agitation is important to ensure the supply of nutrients and the formation of pellet. In fact, in the stationary cultures, Aspergillus formed filamentous mats on the surface of the growth medium, while in the cultures incubated with agitation, uniform pellets were formed. The enhanced efficiency in the decolorization could be due to the physiological state of the fungus as pellets and increased mass transfer between the cells and the medium.

With diluted TOPW, an improvement of decolorization was observed because of the dilution of phenolic compounds causing inhibition of Aspergillus niger. Moreover, an increase in the number of spores in the inoculum ensured the rapid proliferation of biomass as well as the enzyme synthesis, which consequently, accelerated decolorization. The initial pH played also an important role on the decolorization. It is reported that enzymatic activity and phenolic compound precipitation are pH depending. Assadia and Jahangirib [8] studied the effect of pH on the decolorization of textile wastewater using Aspergillus niger and they indicated that the maximum amount of decolorization and better fungal growth usually occurred at lower pH.

Pareto chart (Figure 2) illustrates the significance of each factor studied in the investigation. The horizontal bars show the calculated t-values, whereas the vertical line represents the table value of 2.306 for 95% level of significance. Furthermore, glucose and agitation exceed the reference line and are significant at the level of 0.05. Students't-values for factors namely dilution, time of culture, initial pH, inoculums size and (NH$_4$)$_2$SO$_4$ were found to be smaller than the table value indicating that these factors were not statistically significant.

Optimization of factors affecting TOPW decolorization

Glucose concentration and agitation were varied to find optimized conditions for TOMW decolorization. Glucose was supplemented at six levels of concentration varying from 1to 7g/l (Figure 3). The color removal was raised from 28% to 52% and decreased thereafter. The highest decolorization was obtained by the addition of 3 g/l glucose. An increase in glucose concentration above 3 g/l reduced the decolorization

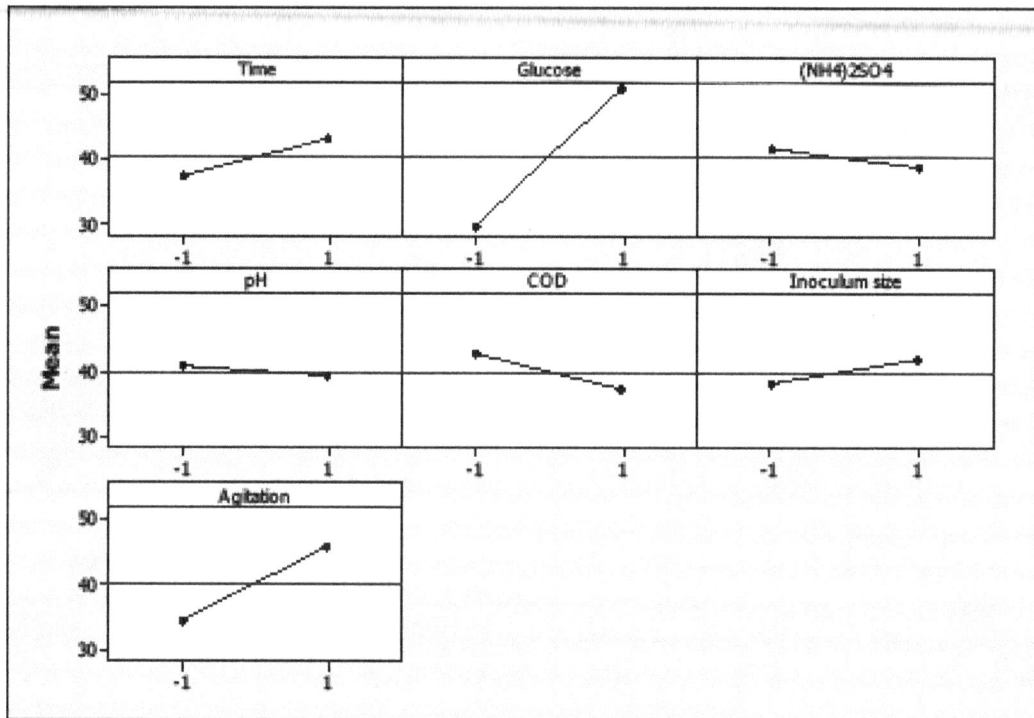

Figure 1: Main effects plot for color removal (%).

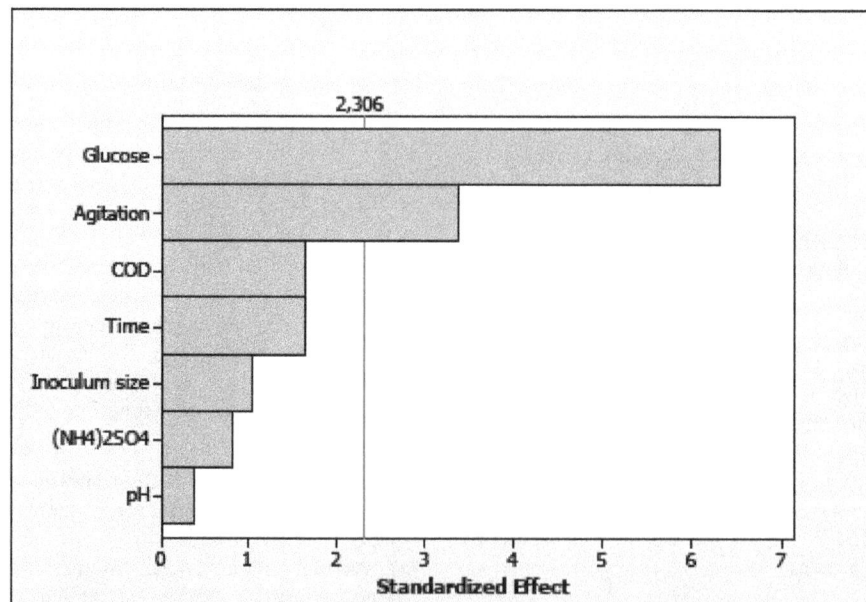

Figure 2: Pareto chart of standardized effects for TOPW decolorization.

because glucose is more easily used as a carbon source than phenolic compounds. Analogous results have been reported concerning the role of glucose on the decolorization of wastewater by *Aspergillus niger*. Assadia and Jahangirib [8] revealed that in the absence of glucose *Aspergillus niger* did not decolorize textile wastewater, and that glucose concentration affected the rate of decolorization.

Table 4 shows the effects of agitation rate (0 rpm to 200 rpm) on growth and decolorization. The results revealed that the optimal agitation rate was 150 rpm which produced a maximum decolorization. The agitation rate that was higher than 150 rpm ensued in low color removal, and this could be due to the fungal cell disturbances generated by shear stress. The agitation rate also determined the size of fungal pellet formed, and it seems that at 150 rpm, the medium size, rounded pellet promoted the decolorization. Purwanto et al. [27] found that there was significant correlation between the speed of agitation and the hyphal morphology, including its internal structures and the activity of enzyme production.

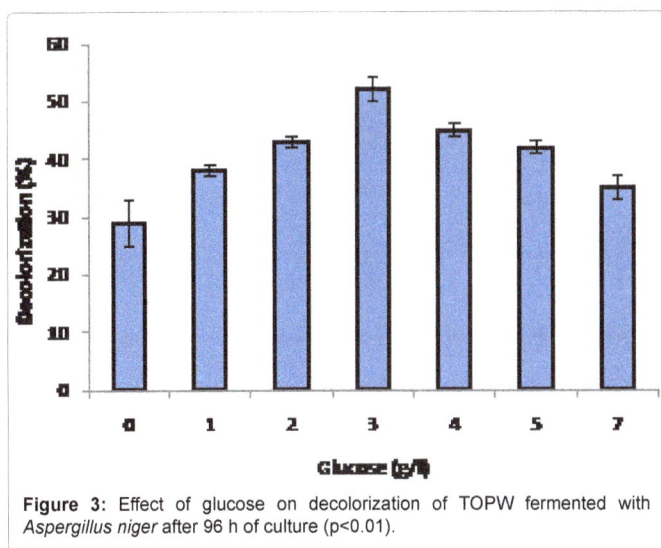

Figure 3: Effect of glucose on decolorization of TOPW fermented with *Aspergillus niger* after 96 h of culture (p<0.01).

Agitation speed (rpm)	Biomass (g/l)	Decolorization (%)
0	2.17±0.13	32 ± 1.5
100	4.18± 0.82	52± 2.3
150	3.93 ± 0.25	62 ±2.1
200	2.26±0.07	47±1
P-value	0.001	<0.001

Table 4: Effect of agitation on decolorization (%) of TOPW fermented with *Aspergillus niger* after 96 h of culture at 30°C.

Influence of optimal conditions on TOPW decolorization

The optimized factors previously set up were used for the study of the decolorization of TOPW by *Aspergillus niger*. The obtained results are presented in Table 5. Fungal biomass obtained during growth on TOPW peaked approximately 4.8 g/l after three days of culture and then remained constant. *Aspergillus niger* growth on TOPW decreased pH value from 5 to 3.22. The decrease of pH was due to the consumption of sugar content and to the generation of new organic and/or phenolic acids with low molecular masses from the polyphenolics with high molecular masses in the culture.

Most COD reductions occurred during the first three days. Biodegradation of TOPW coupled with *Aspergillus* growth induced 64%, 62 % and 48 % of COD, color and phenolic compounds removals respectively after four days of culture.This high removal efficiency could be attributed to both degradation and adsorption of phenolic compounds. In somewhat analogous studies, Hamdi et al. [28] reported that the decreased color intensity of the olive mill wastewater was possibly due to both the degradation of some phenolic compounds and the adsorption of the polyphenols and tannins onto the fungal mycelium. Moreover, Andleeb et al. [9] showed that the high percentage of the decolorization of dyes by *Aspergillus niger* is due not only to microbial biotransformation but also to biosorption.

Gel-filtration chromatography was used to analyze the polyphenolic compounds present in the liquid phase samples at the beginning and at the end of treatment (Figure 4). Results indicated that the molecular-mass distribution of polyphenol changes after four days of incubation. The high molecular-mass polyphenol fraction became progressively more populated at the expense of the low molecular- mass fraction.

Biodegradation was monitored also using FTIR Spectroscopy. The spectra of TOPW before and after biodegradation (Figure 5), indicating that noticeable qualitative changes have occurred during the biodegradation. The untreated TOPW spectrum (spectrum a) shows a high intensity for four bands: 3420, 1600, 1400 and 1076cm^{-1}. These characteristic bands confirm the presence of high content of phenols, alcohol and organic acids. The broad band centered on 3420 cm^{-1} is due to the hydrogen vibrations of alcoholic OH, phenol or carboxyl groups (COOH), but also to N–H vibrations of amides. Aliphatic compounds are also presented by a peak at 2929. The bands located at about 1600 are attributed to aromatic C=C vibrations, in addition to quinines, conjugated carboxyls and ketones. Signals around 1400 cm^{-1} are generated by –CH–, –CH2– and CH3 radicals.

The band at about 1070 cm^{-1} generally attributed to OH deformation and C–O stretching in phenolic OH, C–H deformation of CH2 and CH3 groups and COO– anti-symmetric stretching. The principal absorption bands in the FTIR spectra and their corresponding assignments based on the literature [29-31].

The Spectrum of treated TOPW (spectrum b), showed, a decrease in the signal of the –OH groups (3420 and 1076 cm^{-1}), another decrease between 1600, 1400 and 1076 cm^{-1} due to C- C in aromatic groups, aromatic ethers and polysaccharides. Moreover, the peak at 2929 cm^{-1} disappeared. It could correspond to biodegradation of aliphatic structures.

HPLC analysis of the final extract obtained from fresh TOPW at the beginning and at the end of incubation revealed the presence of several

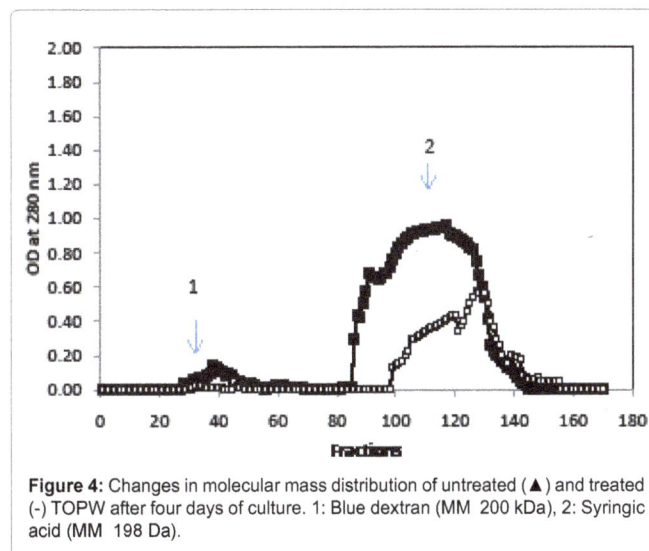

Figure 4: Changes in molecular mass distribution of untreated (▲) and treated (-) TOPW after four days of culture. 1: Blue dextran (MM 200 kDa), 2: Syringic acid (MM 198 Da).

Time (h)	0	24	48	72	96	P-value
TSS (g/l)	1.22±0.02	2.01 ± 0.32	3.77 ± 0.22	4.82 ± 0.18	4.84 ± 0.2	<0.001
pH	5.01 ± 0.2	4.590 ± 0.045	3.37 ± 0.049	3.12 ± 0.014	3.22 ± 0.035	0.002
COD (g/l)	13.7 ± 2	9.47 ± 0.58	6.25 ± 0.94	5.29 ± 0.28	4.75 ± 0.17	0.003
OD at 390 nm	2.4 ± 0.12	2.1 ± 0.01	1.39 ± 0.02	0.93 ± 0.06	0.91 ± 0.11	0.003
Polyphenols (mg gallic acid/l)	496 ± 10.3	426 ± 8	307 ± 11	262 ± 15	256 ± 9	<0.001

Table 5: Time course of biomass, pH, residual COD, residual OD$_{390}$ and polyphenols during the pretreatment of TOPW by *Aspergillus niger*.

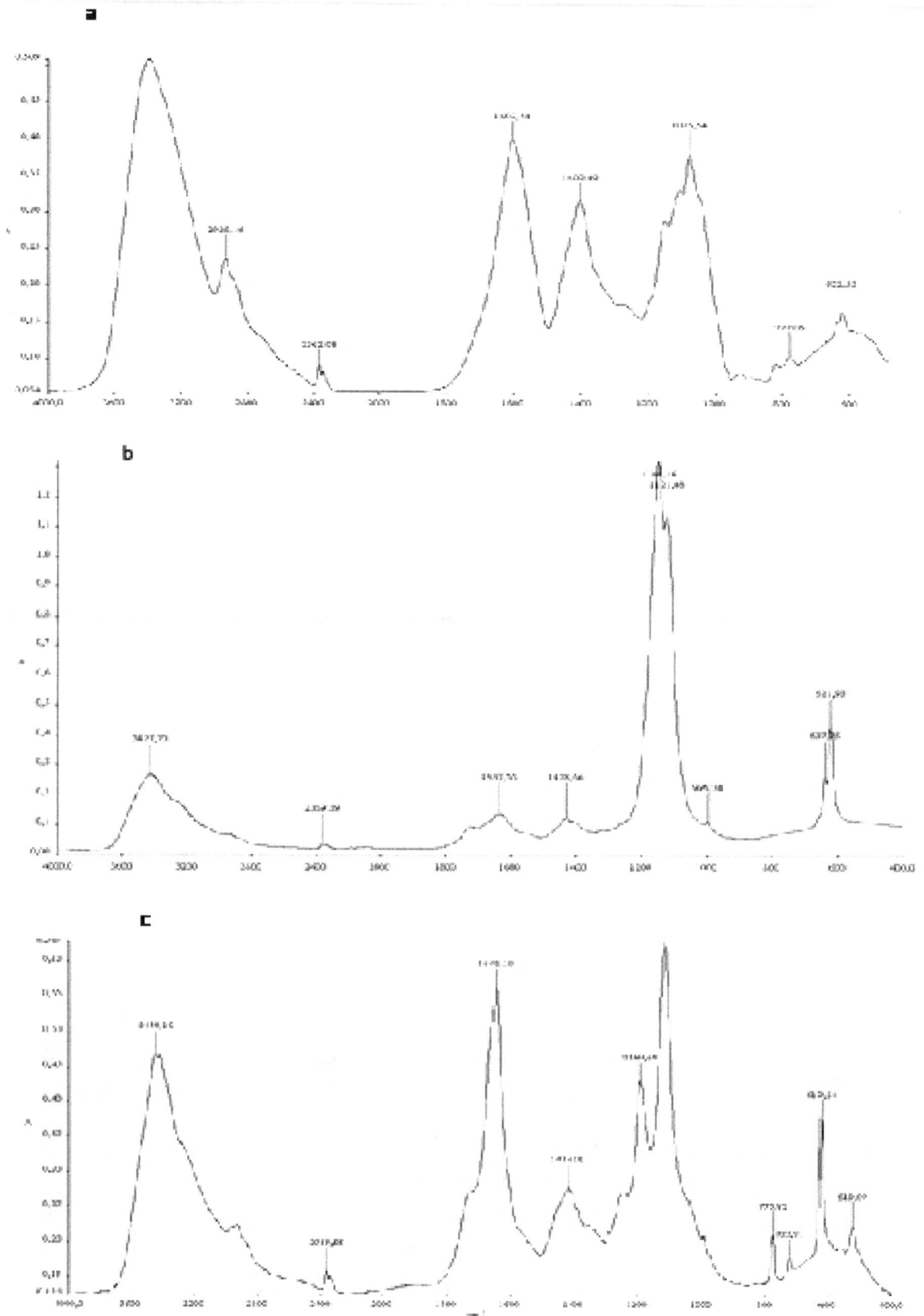

Figure 5: FTIR spectra of untreated TOPW (a); of TOPW after biodegradation by *Aspergillus niger* (b) and after biosorption (c).

simple phenol compounds as shown in Figure 6. Oleuropein, tyrosol and vanillic acid were found to be the most abundant components in the effluent. In addition to these compounds, important concentrations of hydroxytyrosol, hydroxybenzoic acid, benzoic acid caffeic acid, gallic acid and p-coumaric acid were also found. Treated fresh TOPW by *Aspergillus niger* showed a consequential removal in the concentration of total simple phenol's content and the emergence of new compounds resulting from the transformation of other phenolic compounds.

Adsorption of phenolic compounds by a heat-killed fungal biomass

Most of the time, decolorization involved both adsorption of the polyphenolic compounds in the cell and their enzymatic conversion into a less toxic form. To elucidate whether uptake of phenolic compounds were due to biodegradation or biosorption, samples of autoclaved biomass were suspended on TOPW under agitation. In presence of 1 g/l of autoclaved biomass, 7.21 % of the color was removed whereas with the same quantity of living biomass and after four days of culture, 17.33 % of the color was removed (unreported data).

The FTIR spectrum of heat inactived *Aspergillus niger* biomass before biosorption (Figure 7, spectrum a) shows peaks at the frequency level of 3400-3250 cm^{-1} representing -OH stretching of carboxylic groups and also representing stretching of -NH groups. Peaks due to bonded OH groups were observed by Kapoor and Viraraghavan at 3418 cm^{-1} [32]. The peak observed at 2926 was indicative of the C–H group [32]. The peaks at around 1644 cm^{-1} are caused by the bending of N–H groups of chitin on the cell wall structure of fungal pellets. The band present at 1546cm^{-1} indicated the presence of amide 2 and which was resulted from NH deformation.

The peaks at 1411, 1148 and 1044 cm^{-1} representing -CH$_3$ wagging (umbrella deformation), symmetric -SO$_3$ stretching and C-OH stretching vibrations, respectively, which were due to several functional groups present on the fungal cell walls. The peaks at 539cm^{-1} corresponding to C-O bending vibrations. The band at 470 cm^{-1} representing C-N-C scissoring is found in polypeptide structure. Similar results were reported for FTIR spectra of fungi [33-35].

Some differences were noticed in the characteristic frequencies of the functional groups in the FTIR spectrum of *Aspergillus niger* biomass after biosorption (Figure 7, spectrum b). In fact, the intensity of some peaks revealed significant variation after biosorption. The changes in the intensity of 3300 to 3600 cm^{-1} indicated that free OH and NH groups probably get bound with phenolic compounds. Moreover, the intensity of 1600 cm^{-1} in biomass belong N-acetyl glucosamine manifested significant variation after biosorption. The peak at 2359 cm^{-1} could be assigned to participation of C-H and C=O groups in biosorption process.

The comparative FTIR spectra of TOPW before and after biosorption (Figure 5, spectra a and c) exhibited significant differences. In fact, after biosorption, the structures absorbing at around 2929 and 1400 cm^{-1} decreased and those absorbing at around 1624, 1076 cm^{-1} rose. Peaks at 2929 cm^{-1} decreased reflecting a preferential aliphatic structures biosorption. In addition, the decrease of 1400 cm^{-1} explained the biosorption of phenolic structures. In contrast, the increase of the peak at 1624 cm^{-1} indicating the occurrence of oxidation reactions.

The comparative FTIR spectrums of TOPW treated in biotic and abiotic conditions showed that the living biomass of *Aspergillus niger* showed maximum decolorization. These results clearly indicated that the decolorization of TOPW was due to biological mechanisms and adsorption.

Study of enzymes responsible for TOPW decolorization

The enzymatic degradation was further explored in order to test the enzyme produced by *Aspergillus niger*. The detection of enzymes was investigated in the medium containing optimum conditions. The enzymatic activities are shown in Table 6.

Tannase activity was detected after 5 h of incubation (data not shown) and increased to a maximum up to 24h finished by decreasing by 48h, which may be due to the accumulation of the final product which hampers tannase production, or may be due to the accumulation of

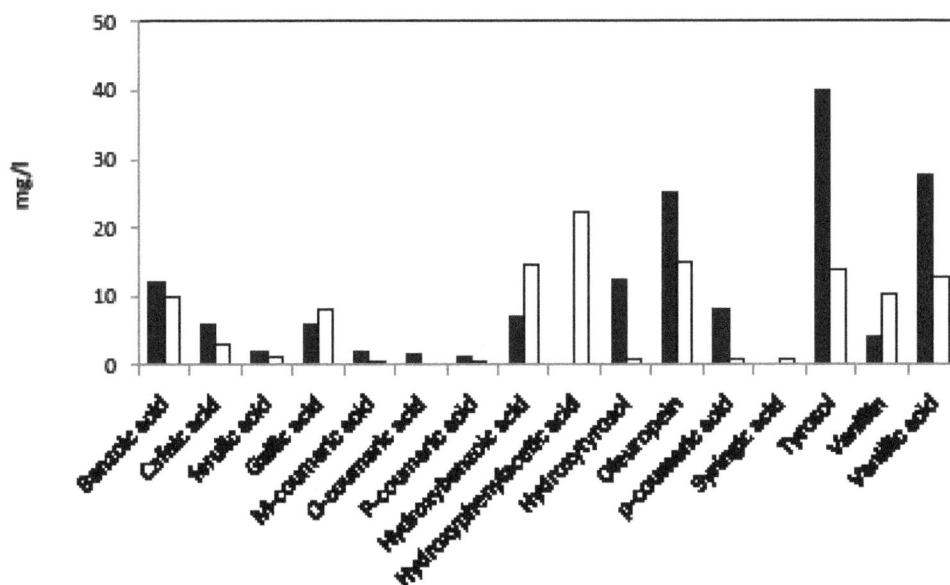

Figure 6: Main phenolic compounds present in untreated (■) and treated (□) fresh TOPW by *Aspergillus niger* after four days of culture.

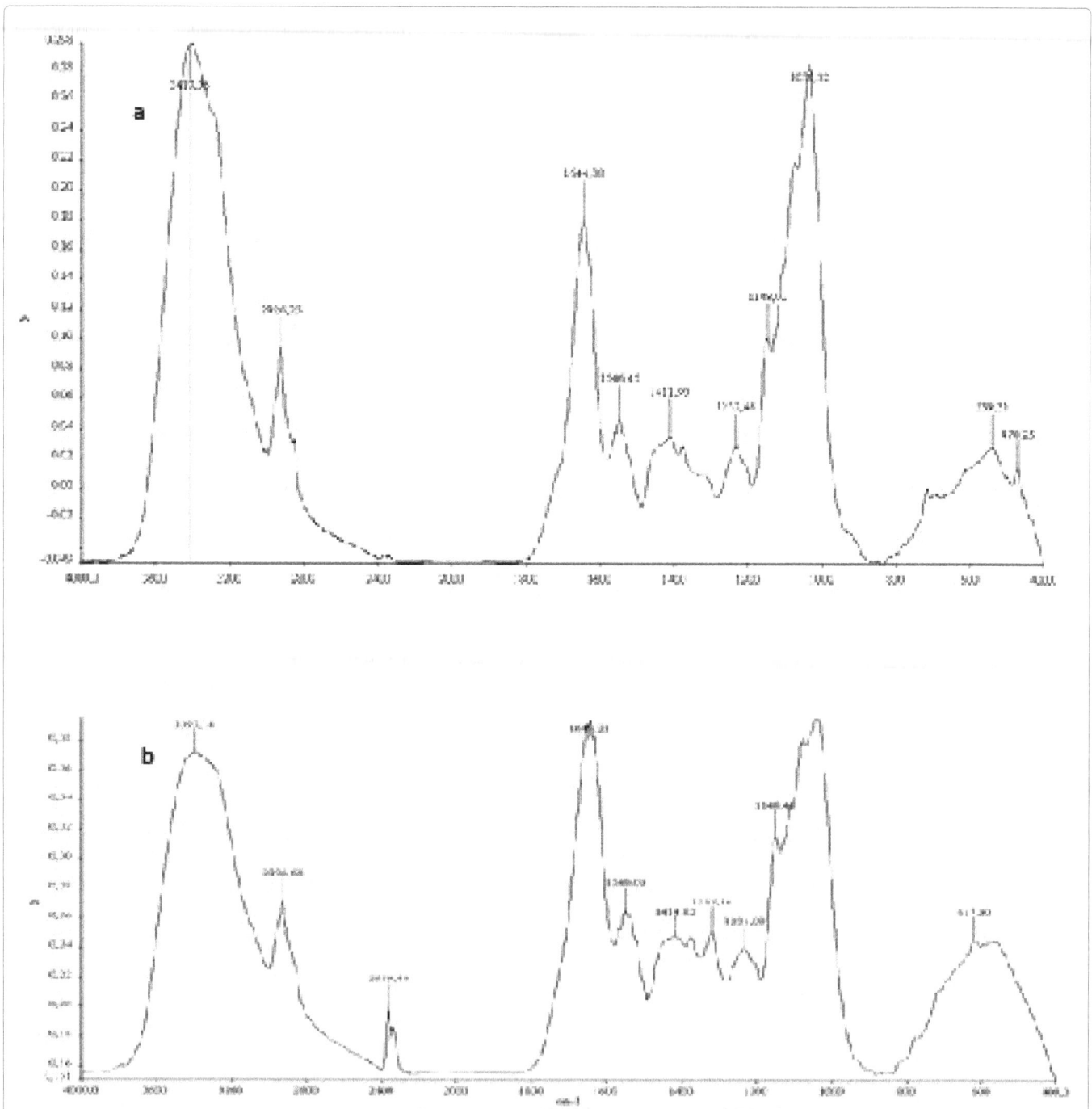

Figure 7: FTIR spectra of heat inactived *Aspergillus niger* before (a) and after biosorption of TOPW (b).

Time (h)	Tannase (U /ml)	LiP (U/l)	Laccase (U /l)
0	0	0	ND
24	0.89 ± 0.09	13.4 ± 0.02	ND
48	0.51 ± 0.07	11 ± 0.4	ND
72	0.3 ± 0.05	7.3 ± 0.1	ND
96	0.07 ± 0.01	0.6 ± 0.04	ND
120	0	0	ND

ND: not detected

Table 6: Quantitative estimation of tannase, lignin peroxydase and laccase produced by *Aspergillus niger* in TOPW (Initial pH 5, glucose 3 g/l; agitation speed 150 rpm, 10^7 spore /ml, COD 13 g/l,).

toxic metabolites secreted during fermentation. Maximum production reached 0.89 U/ml after 24 h of incubation. The LiP was also secreted but at low levels (13.4 U/l) although no laccase activity was detected. Aissam et al. [13] argued that the degradation of phenolic compounds of olive mill wastewater using *Aspergillus niger* HA37was due to the possible action of various enzymes, including those of laccase, tannase and the ligninolytic oxidative enzymes, and demonstrated that the growth of *Aspergillus niger* in diluted olive mill wastewater resulted in a 70% degradation of the phenolic compounds. However, tannase probably was the principal enzymatic system of color removal because *Aspergillus niger* had a low LiP activity and Sayadi and Ellouz [5] have demonstrated that low LiP activity did not efficiently decolorized OMW.

Conclusion

Biodegradation of table olive processing waste waters by *Aspergillus niger* led to a significant decrease in the color removal. To evaluate the importance of the various parameters on treatment efficiency, a fractional factorial design approach was followed. Of the seven parameters tested glucose addition and agitation affected decolorization at a statistically significant level. At high glucose concentration and high agitation rate than high loadings (over 0.3 % and 150rpm respectively), their effect on decolorization decreased. The average removal of COD, color and polyphenols under optimized conditions are 64, 62 and 48%, respectively. HPLC and FTIR study suggest their removals can be attributed to adsorption to biomass and degradation. The detection of important tannase activity (0.89 U/ml) and low activity of LiP (13.4U/l) in the culture indicated that tannase may play the most important role in the degradation of these wastewaters.

Acknowledgements

We gratefully acknowledge the financial support provided by the Tunisian Ministry of High Education. We wish to thank Dr. Katia Lasaridi and Dr. Adamantini Kyriacou (Harokopio University, Greece) for kindly supplying the fungal strain.

References

1. García-García P, Rejano Navarro L, Sánchez Gómez A H (2006) Trends in table olive production. Elaboration of table olives. Grasas Aceites 57: 86–94.

2. Parinos CS, Stalikas CD, Giannopoulos TS, Pilidis GA (2007) Chemical and physicochemical profile of wastewaters produced from the different stages of Spanish-style green olives processing. J Hazard Mater 145: 339-343.

3. Garrido Fernandez A, Fernandez Diez M, Adams N (1997) Table olives: Production and Processing. In: Chapman and Hall (ed). London, New York, Tokyo and Madras.

4. D'Annibale A, Crestini C, Vinciguerra V, Sermanni GG (1998) The biodegradation of recalcitrant effluents from an olive mill by a white-rot fungus. J Biotechnol 61: 209-218.

5. Sayadi S, Ellouz R (1992) Decolorization of olive mill waste waters by the white-rot fungus Phanerochaete chrysosporium: involvement of the lignin-degrading system. Appl Microbiol Biotech 37: 813-817.

6. Ayed L, Assas N, Sayadi S, Hamdi M (2005) Involvement of lignin peroxidase in the decolourization of black olive mill wastewaters by Geotrichum candidum. Lett Appl Microbiol 40: 7-11.

7. Pena Miranda M, Gonzalez Benito G , San Cristobal N, Heras Nieto C (1996) Colour elimination from molasses wastewater by Aspergillus niger. Bioresour Technol 57: 229-235.

8. Assadia M M, Jahangirib M R (2001) Textile wastewater treatment by Aspergillus niger. Desalination 141: l-6.

9. Andleeb S, Atiq N, Ali M I, Ur-Rehman F, Hameed A, et al.(2010) Biodegradation of anthraquinone dye by Aspergillus niger SA1 in self designed fluidized bed bioreactor. Iran J Environ Health Sci Eng 7: 371-376.

10. Manikandan N, Surumbar K S, Kumuthakalavalli R (2012) Decolorisation of textile dye effluent using fungal microflora isolated from spent mushroom substrate (SMS). J Microbiol Biotech Res 2: 57-62.

11. Fu – Ming Z, Jeremy SK, Kelvin NT (1999) Decolourisation of cotton bleaching effluent with wood rotting. Water Res 33: 919-928.

12. Rao JR, Viraraghavan T (2002) Biosorption of phenol from an aqueous solution by Aspergillus niger biomass. Bioresour Technol 85: 165-171.

13. Aissam H, Errachidi F, Penninckx M J, Merzouki M, Benlemlih M (2005) Production of tannase by Aspergillus niger HA37 growing on tannic acid and Olive Mill Waste Waters. World J Microbiol Biotechnol 21: 609-614.

14. Bhat TK, Makkar HP, Singh B (1997) Preliminary studies on tannin degradation by Aspergillus niger van Tieghem MTCC 2425. Lett Appl Microbiol 25: 22-23.

15. Lokeswari N, Jaya Raju K (2007) Tannase production by Aspergillus niger. E-J Chem 4: 192-198.

16. Kumar R, Sharma J, Singh R (2007) Production of tannase from Aspergillus ruber under solid-state fermentation using jamun (Syzygium cumini) leaves. Microbiol Res 162: 384-390.

17. Kotsou M, Kyriakou A, Lasaridi K, Pilidis G (2004) Integrated aerobic biological treatment and chemical oxidation with Fenton's reagent for the processing of green table olive wastewater. Process Biochem 39: 1653–1660.

18. Favela-Torres E, Cordova-Lopez J, Garda-Rivero M, Gutiérrez-Rojas M (1998) Kinetics of growth of Aspergillus niger during submerged, agar surface and solid state fermentations. Process Biochem 33: 103-107.

19. Bizukojc M., Ledakowicz S (2006) A kinetic model to predict biomass content for Aspergillus niger germinating spores in the submerged culture. Process Biochem 41:1063–1071.

20. Fadil K, Chahlaoui A, Ouahbi A, Zaida A, Borja R (2003) Aerobic biodegradation and detoxification of wastewaters from the olive oil industry. Int Biodeter Biodegr 51: 37 – 41

21. Singh SS, Dikshit AK (2010) Optimization of the parameters for decolourization by Aspergillus niger of anaerobically digested distillery spentwash pretreated with polyaluminium chloride. J Hazard Mater 176: 864-869.

22. Ijzerman MM, Falkinham JO, Reneau RB, Hagedorn C (1994) Field evaluation of two colorimetric coliphage detection methods. Appl Environ Microbiol 60: 826-830.

23. Catalano L, Franco I, De Nobili M, Leita L (1999) Polyphenols in olive mill waste waters and their depuration plant effluents: a comparaison of the Folin-Ciocalteau and HPLC methods. Agrochimica 43: 193-205.

24. Iibuchi S, Minoda Y, Yamada K (1967) Studies on tannin acyl hydrolase of microorganisms. Agri biol Chem 31: 513-518.

25. Tien M, Kirk TK (1984) Lignin-degrading enzyme from Phanerochaete chrysosporium: Purification, characterization, and catalytic properties of a unique H(2)O(2)-requiring oxygenase. Proc Natl Acad Sci U S A 81: 2280-2284.

26. Palmieri G, Giardina P, Bianco C, Scaloni A, Capasso A, et al. (1997) A novel white laccase from Pleurotus ostreatus. J Biol Chem 272: 31301-31307.

27. Purwanto L A, Ibrahim D, Sudrajat H (2009) Effect of Agitation Speed on Morphological Changes in Aspergillus niger Hyphae During Production of Tannase. World J Chem 4: 34-38.

28. Hamdi M, Khadir A, Garcia J L (1991) The use of Aspergillus niger for the bioconversion of olive mill waste-waters. Appl Microbiol Biotechnol 34: 828–831.

29. El Hajjouji H, Fakharedine N, Ait Baddi G, Winterton P, Bailly JR, et al. (2007) Treatment of olive mill waste-water by aerobic biodegradation: an analytical study using gel permeation chromatography, ultraviolet-visible and Fourier transform infrared spectroscopy. Bioresour Technol 98: 3513-3520.

30. El Hajjouji H, Bailly JR, Winterton P, Merlina G, Revel JC, et al. (2008) Chemical and spectroscopic analysis of olive mill waste water during a biological treatment. Bioresour Technol 99: 4958-4965.

31. Ait Baddi G, Hafidi M, Cegarra J, Alburquerque JA, Gonzálvez J, et al. (2004) Characterization of fulvic acids by elemental and spectroscopic (FTIR and 13C-NMR) analyses during composting of olive mill wastes plus straw. Bioresour Technol 93: 285-290.

32. Kapoor A, Viraraghavan T (1997) Heavy metal biosorption sites in Aspergillus niger. Bioresour Technol 6: 221–227.

33. Bayramoglu G, Gursel I, Tunali Y, Arica MY (2009) Biosorption of phenol and 2 chlorophenol by Funalia trogii pellets. Bioresour Technol 100: 2685-2691.

34. Mukhopadhyay M (2008) Role of surface properties during biosorption of copper by pretreated Aspergillus niger biomass. Colloïd Surface A. 329: 95–99

35. Guibal E, Roulph C, Le Cloirec P (1995) Infrared spectroscopic study of uranyl biosorption by fungal biomass and materials of biological origin. Environ Sci Technol 29: 2496-2503.

Extraction, Determination and Bioremediation of Heavy Metal Ions and Pesticide Residues from Lake Water

Sandip R Sabale[1]*, Bhaskar V Tamhankar[1], Meena M Dongare[3] and B S Mohite[2]

[1]P.G. Department of Chemistry, Jaysinpur College, Jaysingpur-416101, M.S., India
[2]Department of Chemistry, Shivaji University, Kolhapur-416004, M.S., India
[3]Department of Botany, Shivaji University, Kolhapur-416004, M.S., India

Abstract

The natural or manmade water bodies in and near the settlements play very important role in improvement and conservation of local environment but many of them are on the way of eutrophication. This problem needs an urgent and sincere attention of environmentalists. The pollution due to heavy metal ions, pesticides and fertilizers has been investigated and analyzed by using atomic absorption spectrometer and Liquid Chromatography Tandem Mass Spectrometry. The present study focuses on the determination of heavy metal ions from aquatic plants so as to evaluate its curative power as well as extraction method of pesticides from water and its determination using liquid chromatography tandem mass spectrometry. Nearly 11 pesticides have been found in ppb level. The curative power and remediation of pesticide residues and heavy metal ions using microbial culture has also been tried and reported.

Keywords: Heavy metal ions; Pesticides; Bioremediation; AAS; LC-MS/MS

Introduction

Pollution due to heavy metal ions and pesticides is dangerous as they tend to bioaccumulate and therefore toxic to the plant, human and animal health. Metal ions are accumulated in living beings when they are taken up and stored faster than they are metabolized or excreted. Heavy metal ions can enter a water supply through industrial and consumer waste. Lead, cadmium, chromium and mercury are the most potential heavy metal ions cause water pollution. Aquatic plants are well known for accumulating and in concentrating heavy metal ions [1,2]. Several workers have shown that constructed wetlands are very effective in removing heavy metal ions from polluted waste-waters [3]. Different wetland species differ, however, in their abilities to take-up and accumulate various trace elements in their tissues [4]. Laboratory studies of water hyacinth have demonstrated the potential use of this species in removing metals from polluted water [5,6].

Pesticides are the chemicals used in agriculture, in order to protect the crops from the attacks of pests, diseases and rodents. They are toxic and cause environmental contamination as well as generate public health problems. The analysis of pesticide residue in water is difficult, since these compounds occur at very low concentration level [7]. High Performance Liquid Chromatography (HPLC) is the method often used when the compounds to be analyzed are polar, nonvolatile and thermally labile [8]. The use of biodegradable pesticides in making HPLC is the favorite analytical technique, allowing large volumes of injection of aqueous samples [9]. Although, many authors reported the multiresidue pesticide analysis in water by HPLC [10] and Gas Chromatographic (GC) techniques [11], the use of hyphenated techniques like LC/MS and GC/MS are more easy, suitable and precise, and give accurate and confirmatory results [12,13].

Kolhapur, a prominent city of Southwestern Maharashtra (India) is rapidly emerging as a leading industrial and commercial centre. The development of the city has caused directly or indirectly, a number of water quality problems. In this city, there is a Wetland in the form of a beautiful Lake called as Rankala, which is man-made reservoir, situated in southwestern region of the city. The total catchment area is of 5.21 sq km. and is of 94.33 million cubic feet storage capacity. The maximum water depth is up to 88 feet [14]. In this lake, the *Eichhornia*

crassipes plant species were growing enormously up to 2008 [15], but in the year 2009 *Eichhornia* seems to be replaced by *Salvinia* occupying whole lake (Figure 1), which looks like a big foot ball ground. Because of rapid urbanization, increasing use of fertilizers and pesticides in the agricultural land around the lake subjected to great amount of ecological stress. The mixing of sewage and agricultural waste, results into the increase of organic matter, silt and nutrients like nitrogen, potassium and phosphorus in the lake. The lake is also suffering from abuse due to which soluble metal ions, detergents, fertilizers and solid wastes are accumulating. The large area and good sunlight, supports rapid growth of aquatic/eutrophic plants like *Salvinia*, *Eichhornia*, *hydrilla* and other macrophytes in the lake.

The microorganisms are the agents which consume many organic as well as inorganic compounds and degrade them. They are widely used for the degradation of organic compound like dyes, pesticides and other. The microorganism, *Proteus spp.* SUK 7 has been used by Patil et al. [16] for the decolorization of Reactive Blue 59. Several bacteria capable of dye decolourization have been reported [17-20]. The *Bacillus Thurigiensis* MOS-5 has been used for the biodegradation of malathion by Zeinat et al. [21]. Many other reports were also found for the biodegradation of pesticides by microorganisms or bacteria [22-25]. Microorganisms are also the agents used for the biosorption of the metals [26]. The mechanism of biosorption, bacterial biosorbants and the biosorption by the various microorganisms were described by Vijayraghavan et al. [27]. The Bacteria isolated from the various sources which were used for the biosorption of heavy metals are reported [28-30]. But there are no any attempts were found on the use of *Bacillus*

*Corresponding author:** Sandip R Sabale, P.G. Department of Chemistry, Jaysinpur College, Jaysingpur-416101, M.S., India
E-mail: sandip_ana@rediffmail.com

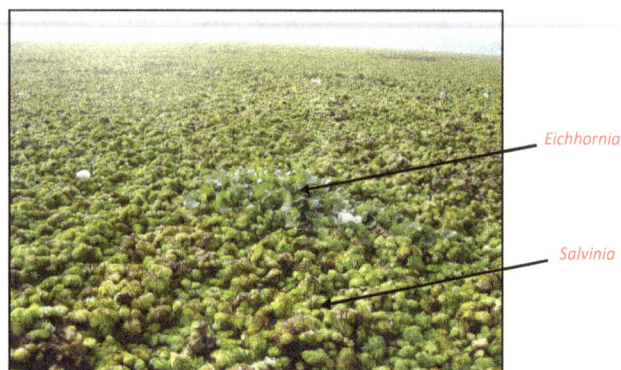

Figure 1: Figure shows the replacement of *Eichhornia* by *Salvinia*.

Thuringensis NCIM 2159 and *Proteus spp* SUK 7 for the biosorption of the metals and the removal of pesticide residue.

The investigation about the presence of heavy metal ions and the qualitative as well as quantitative assessment of pesticide residues present in water body (Rankala) is presented in this paper.

Experimental

Reagents and materials

Double distilled, deionized water has been used for washing the plant materials as well as for the determination of trace metals. All the reagents used are of analytical grade and are used without further purification.

Analytical reference standards of pesticides were obtained from Dr. Ehrenstorfer (Augsburg, Germany) and Accustandard (New Haren, USA) always with the purity certified >98%. All the solvents namely, dichloromethane, methanol and water were all HPLC grade. The anhydrous sodium sulphate was of analytical reagent grade and was purchased from the RFCL limited (New Delhi, India). It was activated by heating at 300°C for 4 hours before use and kept in oven at 60°C. HPLC grade water was collected from ultra pure water purification system (SG Water, Germany).

Instruments

The pH and conductance measurements were carried out using digital pH meter (LI-127, Elico, India) and digital conductivity meter (EQ-660 A, Equip-Tronics, India). The metal ion detection was carried out using Atomic Absorption Spectrometer, Perkin-Elmer, AAnalyst 300 series. A Perkin-Elmer 200 series HPLC system (Massachusetts, USA) hyphenated to an API 2000 mass spectrometer (Applied Biosystems, MDS Sciex, Canada) was used. A high speed homogenizer (Heidolph, Germany), low volume concentrator (Telemachanic, Magelies), high volume refrigerated centrifuge (Z 36 HK, HARMLE Labortechnik, GmbH), low volume refrigerated centrifuge (Z 233M K2, HARMLE Labortechnik, GmbH), were used for residue analysis.

Sampling

The plant samples of *Eichhornia*, *Salvinia* and *Hydrilla* were collected at random from different areas of lake covering all directions. The plant material was washed to remove periphyton, dust and sediment particles. The material was stored in polythene bags, at the same time the water samples around the plants were collected randomly and brought to laboratory. The temperature of water at the time of sampling was recorded. All individual plant samples were again washed with distilled deionized water in laboratory. These materials were sun dried for 10 days in separate containers and were heated in oven at 110°C for 12 hours. The dried samples were grounded using mortar till dry powder was formed. All the physical and chemical parameters of water were determined and the water was stored in refrigerator.

Sample preparation for AAS analysis of trace metals

The dried samples of *Eichhornia*, *Salvinia* and *Hydrilla* were weighed accurately and dissolved in HNO_3 and $HClO_4$ (in the ratio 3:1). The resulting mixtures were evaporated to dryness and extracted with distilled, deionized water. The solutions were heated to boiling and filtered. The volumes of the diluted sample were made to 100 mL each. 1.0 L water sample was heated to reduce the volume, acidified and total 100 mL volume was made. The metal ion concentrations in all the samples were analyzed by Atomic Absorption Spectrometer. For determination of unknown concentrations of all metals, the calibration charts for each element were used. All analyses were done in triplicate and one blank sample.

Extraction of pesticide residue from water sample

One liter water sample was filtered using vacuum and 100 g of sodium chloride was added. The salt was dissolved and the pesticides were extracted with 60 mL dichloromethane in a separating funnel. The organic layer was passed through sodium sulphate and collected in a 200 mL evaporation tube. The aqueous layer was then again extracted twice with two 60 mL fractions of dichloromethane in same manner. All the dichloromethane layers were then evaporated under a gentle stream of nitrogen in large volume concentrator up to 2 mL and the solution was transferred into 10 mL test tube. The mixture was evaporated to near dryness under a gentle stream of nitrogen in a low volume concentrator at 35°C. The residues were dissolved in a mixture of 1.0 mL 0.1% acetic acid (in water) and 1.0 mL methanol followed by vortexing. This solution was centrifuged at 10000 rpm for 5 min. The supernatant extract was filtered through 0.2 μm polyvinylidine fluoride (PVDF) membrane filter and then analyzed by LC-MS/MS.

Microorganism treatment for the removal of trace metals and pesticide residue

The microorganism *Bacillus thuringiensis* NCIM 2159 was obtained from MTTC Chandigarh (India) while the *Proteus spp*. SUK 7 was isolated by the known method [16]. These two bacteria were used in consortium.

Medium for the growth of microorganisms: 6.0 g/L Na_2HPO_4, 3.0 g/L K_2HPO_4, 2.0 g/L NH_4Cl, 5.0 g/L NaCl, 8.0 g/L Glucose, 0.1 g/L of $MgSO_4$ and 0.1 g/L of $MnSO_4$ were used as a nutritional medium for the growth of the microorganisms.

Microorganisms were grown in the minimal media composition as described above and 5% of them were transferred to water sample. Flask was kept on rotary shaker at 120 rpm at 30°C for 7 days. After 7 days, microbial cell mass was removed by centrifugation at 7000 rpm for 15 minutes. The supernatant obtained was used for further studies.

Liquid chromatography mass spectrometric method

The high performance liquid chromatography was performed using Perkin-Elmer 200 series liquid chromatography system equipped with series 200 Vaccum degasser, series 200 auto sampler, series 200 pump, and series 200 column oven. PrincetonSPHER C18 (150mm × 4.6 mm × 5 μm) column (Princeton, NJ, USA) was maintained at 30°C at the flow rate of 1.2 mL/min. The mobile phase was composed of (A)

Sr.No	Name of Pesticide	R.T.	Q1	DP	Scan Time	Q3	CE	CXP	Q3	CE	CXP
1.	Acephate	2.18	184	17.0	50	143	10.0	3.0	95	30.0	5.0
2.	Diazinon	7.73	305	24.0	20	169	27.0	4.0	153	26.0	3.0
3.	Dimethoate	4.93	230	18.0	40	199	12.0	5.0	125	27.0	2.0
4.	Omethoate	2.27	214	17.0	50	183	15.0	4.0	155	21.0	3.0
5.	Ethion	9.37	384.5	21.0	40	199	13..0	5.0	171	22.0	4.0
6.	Etrimphos	7.76	293	22.0	20	125	33.0	2.0	265	22.0	7.0
7.	Iprobenphos	7.36	289	10.0	20	205	13.0	5.0	91	24.0	5.0
8.	Malathion	6.61	331	15.0	20	127	18.0	3.0	285	10.0	4.0
9.	Methamidophos	2.09	142	65.0	50	94	18.0	5.0	125	18.0	2.0
10.	Monocrotophos	4.09	224	21.0	20	127	20.0	3.0	193	11.0	5.0
11.	Oxydemeton-methyl	4.09	247	17.0	20	169	18.0	4.0	105	18.0	2.0
12.	Phosalone	7.86	368	35.0	40	182	17.0	4.0	138	41.0	2.0
13.	Phosphomidon	5.19	300	50.0	20	174	25.0	8.0	227	18.0	10.0
14.	Profenophos	8.87	375	56.0	50	305	23.0	6.0	346	18.0	10.0
15.	Quinolphos	7.53	299	22.0	20	163	40.0	2.0	147	22.0	4.0
16.	Triazophos	6.69	314	20.0	20	162	23.0	4.0	97	46.0	2.0
17.	Atrazine	6.11	216	22.0	20	174	22.0	4.0	104	40.0	2.0
18.	Simazine	5.69	202	22.0	20	124	25.0	2.0	132	25.0	3.0
19.	Metalaxyl	6.02	280	19.0	20	220	27.0	2.0	248	16.0	4.0
20.	Carbaryl	5.69	202	15.0	20	145	11.0	3.0	127	37.0	2.0
21.	Carbofuran	5.52	222	19.0	20	165	16.0	4.0	123	27.0	2.0
22.	Carbosulphan	13.38	381	39.0	50	118	22.0	5.0	160	36.0	16.0
23.	Indoxacarb	8.86	527.5	67.0	40	293	20.0	8.0	249	18.0	9.0
24.	Methomyl	3.51	163	10.0	20	88	14.0	4.0	106	14.0	2.0
25.	Thiodicarb	5.60	355	20.0	50	88	23.0	4.0	108	21.0	2.0
26.	Fenarimole	6.94	331	29.0	40	268	32.0	7.0	81	50.0	3.0
27.	Bitertenol	7.86	338	9.0	40	269	13.0	7.0	99	20.0	4.0
28.	Flusilazole	7.11	316	40.0	40	247	24.0	7.0	165	36.0	4.0
29.	Hexaconazole	7.95	314	22.0	40	70	39.0	3.0	159	40.0	4.0
30.	Myclobutanil	6.69	289	30.0	40	70	37.0	4.0	125	36.0	2.0
31.	Penconazole	7.61	284	21.0	40	70	31.0	3.0	159	30.0	3.0
32.	Propiconazole	7.71	342	42.0	40	159	36.0	4.0	69	35.0	4.0
33.	Tebuconazole	7.53	308	37.0	40	70	42.0	3.0	125	43.0	2.0
34.	Tridimephon	6.69	294	19.0	40	197	20.0	5.0	225	17.0	6.0
35.	Tridimenol	6.86	296	10.0	40	70	21.0	2.0	227	13.0	6.0
36.	Difenconazole	8.11	406	30.0	50	251	34.0	6.0	337	22.0	10.0
37.	Carbendazim	5.19	192	21.0	20	160	24.0	4.0	132	40.0	3.0
38.	Thiophenate- methyl	5.44	343	30.0	20	151	25.0	3.0	311	16.0	8.0
39.	Acetamiprid	4.85	223	35.0	20	126	28.0	2.0	90	45.0	5.0
40.	Clothinidin	4.68	250	22.0	40	132	19.0	3.0	169	16.0	4.0
41.	Imidacloprid	3.60	256	45.0	40	209	22.0	6.0	175	21.0	5.0
42.	Thiacloprid	5.21	253	43.0	40	126	27.0	2.0	186	19.0	4.0
43.	Thiamethoxam	3.85	292	22.0	20	211	17.0	6.0	132	25.0	3.0
44.	Cymoxanil	5.02	199	27.0	50	128	12.0	6.0	111	25.0	2.0
45.	Dimethomorph	6.27	388	29.0	50	301	29.0	9.0	165	47.0	4.0
46.	Buprofezine	9.20	306	16.0	20	201	16.0	5.0	116	22.0	2.0
47.	Emamectin benzoate	11.12	886	45.0	50	158	51.0	3.0	126	64.0	2.0
48.	Spinosad A	10.83	732	73.0	50	142	41.0	3.0	98	98.0	4.0
49.	Spinosad D	12.23	784	59.0	50	142	43.0	3.0	98	98.0	4.0
50.	Difenthiuron	10.44	385	31.0	40	329	24.0	10.0	278	42.0	7.0
51.	Azoxystrobin	6.02	404	14.0	20	372	17.0	11.0	344	32.0	9.0
52.	Flufenoxuron	9.95	489	52.0	50	158	25.0	4.0	141	63.0	2.0
53.	Propargite	10.12	368	11.0	40	231	14.0	6.0	175	21.0	4.0
54.	Fenpyroximate	10.79	422	17.0	40	366	22.0	11.0	214	43.0	5.0
55.	Femoxadone	7.36	392	10.0	40	331	11.0	10.0	238	22.0	7.0
56.	Fenamidone	6.27	312	10.0	40	236	19.0	6.0	92	34.0	5.0
57.	Diflubenzuron	7.19	311	28.0	40	158	19.0	4.0	141	44.0	3.0
58.	Trifloxystrobin	7.95	408.6	14.0	20	186	20.0	4.0	206	21.0	6.0
59.	Pyraclostrobin	7.61	388	10.0	20	194	16.0	5.0	164	25.0	4.0

Table 1: LCMS/MS Pesticides MRM.

methanol/water (20:80 v/v) with 5.0 mM ammonium formate and (B) methanol/water (90:10 v/v) with 5.0 mM ammonium formate, gradient 0-1.0 min/90% A, 1-2 min/90%-0% A, 2-16 min/0%A, 16-17 min/0%-90% A, 17-23 min/90% A. Prior to use, the solvents were filtered through 0.22 μm filter with applied vacuum. 20 μL of the sample extract was injected.

The triple quadrupole mass spectrometer was equipped with Turbo Ion Spray Interface (ESI). The instrument was operated in positive ion electrospray mode with 118 transitions monitored during LC separation in the multiple reaction monitoring (MRM) modes. The MS parameters included ion spray voltage of 5500 V; nebulizer gas 30 psi; curtain gas 20 psi; heater gas 60 psi and the ion source temperature of 500°C.

Selection and tuning of the transitions as well as analyte dependent parameters, DP (Declustering Potential), CE (Collision Energy) and CXP (Cell Exit Potential) were performed by direct infusion of individual pesticide solution in 10 mM ammonium formate in methanol/water (1:1, v/v) at a concentration of 1.0 mg/L. Analyte MS/MS transitions, retention times and instrument conditions are presented in Table 1.

Stock solution of individual pesticides at 500 mg/kg was prepared by exact weight and solution in methanol for LC-MS/MS (gravimetric control was followed). The stock solution was stored at -10°C in a freezer. The intermediate stock standard mixture of 5 mg/L was prepared by mixing the appropriate quantities of the individual stock solutions followed by requisite volume makeup by methanol as required. A working standard mixture of 0.5 mg/L was prepared by diluting the intermediate stock solution, from which the calibration standards within the range 0.01-0.250 mg/L were prepared by serial dilution with methanol-water (1:1, v/v) for LC-MS/MS to plot the calibration curve, from which unknown concentration of pesticide residues from water sample was determined.

Results and Discussion

Determination of heavy metal ions from water and plant samples by AAS

The pollution in stagnant water bodies is mainly due to accumulation of plant nutrients, heavy metal ions and pesticide residues. Apart from actual chemical analysis the pollution level can be judged by observation of vegetative and animate life in the water. Both the plants and the animals in a polluted ecosystem show different modifications, indicating the types and extent of the pollution. Among these indicators plants perform better. The monstrous growth of aquatic plants indicates the eutrophication of water body. This indicates the excessive presence of plant nutrients in water. The heavy metal ion pollution is amongst most severe problems of water pollution as the toxic metals like lead, mercury, copper and cadmium directly heat the vital organs of the human beings.

The aquatic plants are bioaccumulators of heavy metal ions, so they are both indicator and corrective systems. The capacity and preference of accumulation of particular metal ion in plant body is a finger print i.e. for different aquatic plants preference of accumulation and the extent varies. For the present work the plants *Salvinia*, *Eichhornia* and *Hydrilla* were studied.

The analysis carried out with the help of AAS technique revealed that, Cd, Co, Cr, Cu, Fe, Mn, Ni, Pb and Zn ions are present in these plant samples. The data of concentrations in ppm with standard

deviation are presented in Table 2. Similar analysis carried out for water samples around plants reveals that, there is relatively less concentration of toxic metal ions in water. Thus the accumulation of these ions by the plants is quite evident.

The results show that the accumulation of Pb, Cd, Ni and Co is highest in *Salvinia* than *Eichhornia* and *Hydrilla*, while the accumulation of Cu, Mn, Zn, Cr and Fe is found to be higher in *Eichhornia*. The extent of concentration of metal ions in *Salvinia*, *Eichhornia* and *Hydrilla* are compared and the order is

Salvinia: Mg > Ca > Fe > Cu > Zn > Pb > Mn > Ni > Cr > Cd > Co

Eichhornia: Ca > Cu > Fe > Mg > Mn > Zn > Cr > Ni > Pb > Co > Cd

Hydrilla: Ca > Cu > Mg >Fe > Zn > Ni > Mn > Cr > Pb > Co > Cd

The present investigation shows that the plants, *Salvinia*, *Eichhornia* and *Hydrilla* are effective and inexpensive absorbents for many toxic metal ions. The plants can be used to purify water from industrial effluents. In the Rankala Lake, the sources of toxic metal ions like Pb, Cr must be dumping of batteries, effluents from the electroplating industries, the submersion of Ganesh idols, sewage, leather industry effluent and leaching of the residue due to rain water and effluent to the lake water. The enormous growth of these plants (Figure 1) indicates the nature and volume of pollution of Rankala Lake which shows the initiation of eutrophication. Figure 2 reveals that the *Salvinia* plant was found to be efficient for the absorption of Pb, Cd, Zn and Cr while *Eichhornia* was found efficient towards Zn and Cr ions.

S.No	Name of Pesticide	Concentration in water sample (μg/L)	After microorganism treatment (μg/L)	Percentage of degradation
1	Iprobenphos	23.9	0.0	100
2	Malathion	17.3	14.0	19.06
3	Propenophos	223	55.2	65.24
4	Quinolphos	91.2	37.2	59.21
5	Triazophos	24	5.76	76.0
6	Acetamiprid	4.59	3.22	29.84
7	Carbaryl	9.74	9.4	7.59
8	Hexaconazole	55.6	33.4	39.92
9	Carbendazim	37.1	0.0	100
10	Imidacloprid	32.0	26.70	16.56
11	Propargite	2400	456	81.0

Table 2: Determination of pesticides in water sample.

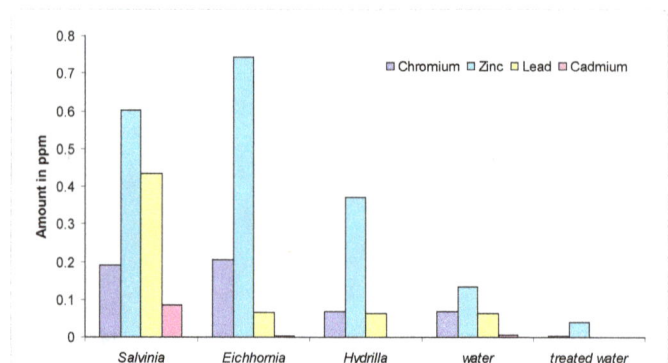

Figure 2: Comparative graph of amount of chromium, zinc, lead and chromium ions with respect to samples.

Determination of heavy metal ions from water sample treated with microorganism

The water sample was treated with the microorganisms as per the described procedure. The treated water sample was subjected to analyze the metal concentration by AAS. From the results it is found that there is a decrease in the concentration of metal ion which shows that these microorganisms are consuming the metals and playing very important role in the removal of the metal ions. It has been found that the concentration of metal ions in the microbial treated water is decreased, while the concentration of magnesium and manganese ions seems to be slightly increased after the treatment. This may be due to the metal ions entering from cultural media. The microorganisms, *Bacillus thuringiensis* NCIM 2159 and *Proteus spp.* SUK 7 were found to be effective for removal of toxic metal ions from water sample under study. It has been found that, at a given concentration of the metal ions the cobalt, nickel, lead and cadmium were removed completely while zinc, copper and chromium were partially removed (Table 3).

Remediation of copper is about 45% while that of chromium is about 90%.

Pesticide residue analysis from water by liquid chromatography tandem mass spectrometer

Figure 3 shows the chromatogram of standard 59 pesticides of 0.250 mg/L. The water sample was extracted as per the procedure and 20 µL of sample was run in LCMS/MS. The water sample after treatment with microorganisms was extracted by the same procedure and sample was run in LCMS/MS.

The results show that 11 pesticide residues are found to be present in the studied water sample. The amount of propargite found was very high. The results show that the Rankala Lake is contaminated with pesticides. Most of the pesticides enter the lake through water streams coming from the agricultural land. The pesticide residues are found in the order

Sr No	Metal ion	Salvinia	Eichhornia	Hydrilla	Water	Water Treated with Micro-organisms
1	Ca	2.051(0.003)	1.972(0.005)	1.176(0.003)	2.632(0.003)	1.203(0.004)
2	Mg	2.218(0.096)	1.173(0.005)	0.482(0.003)	2.725(0.002)	2.917(0.007)
3	Cr	0.189(0.005)	0.205(0.015)	0.067(0.012)	0.067(0.002)	0.004(0.00)
4	Mn	0.248(0.007)	1.02(0.005)	0.088(0.000)	0.122(0.001)	0.150(0.002)
5	Fe	0.966(0.006)	1.453(0.031)	0.466(0.008)	0.362(0.005)	0.168(0.005)
6	Co	0.065(0.001)	0.042(0.001)	0.043(0.003)	0.052(0.003)	0.000(0.001)
7	Cu	0.828(0.029)	1.655(0.012)	0.743(0.017)	2.179(0.020)	1.230(0.013)
8	Ni	0.211(0.003)	0.183(0.002)	0.109(0.012)	0.064(0.002)	0.000(0.001)
9	Zn	0.601(0.008)	0.744(0.004)	0.372(0.000)	0.133(0.001)	0.040(0.003)
10	Pb	0.434(0.001)	0.064(0.004)	0.061(0.004)	0.061(0.017)	0.000(0.000)
11	Cd	0.086(0.001)	0.003(0.001)	0.000(0.000)	0.005(0.000)	0.000(0.001)

Values in parenthesis indicate standard deviation

Table 3: Metal Analysis in *Salvinia, Eichhornia, Hydrilla* and water sample collected from Rankala Lake by Atomic Absorption Spectrometer (Perkin Elmer; AAnalyst 300) [concentration of metals in ppm].

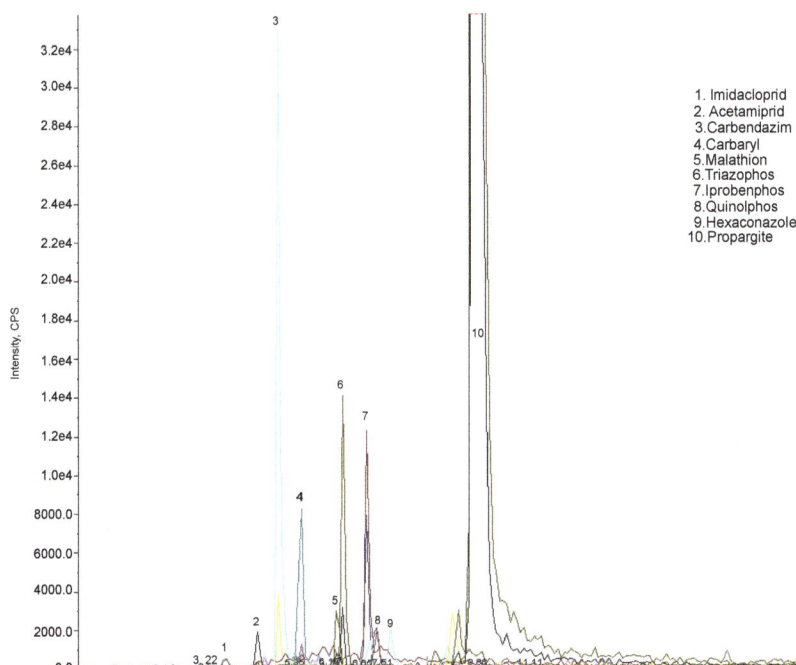

Figure 3: Liquid Chromatogram of standard solution containing 250 µg/L of each of the 59 pesticide selected for the experimentation.

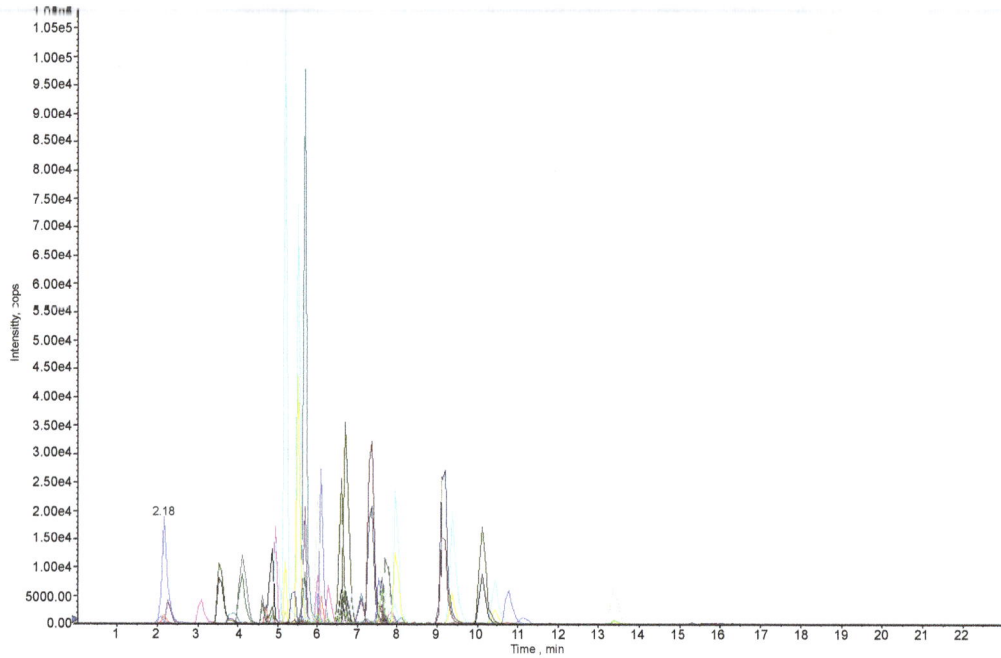

Figure 4: Liquid Chromatogram of water sample from Rankala Lake analyzed for pesticide residue.

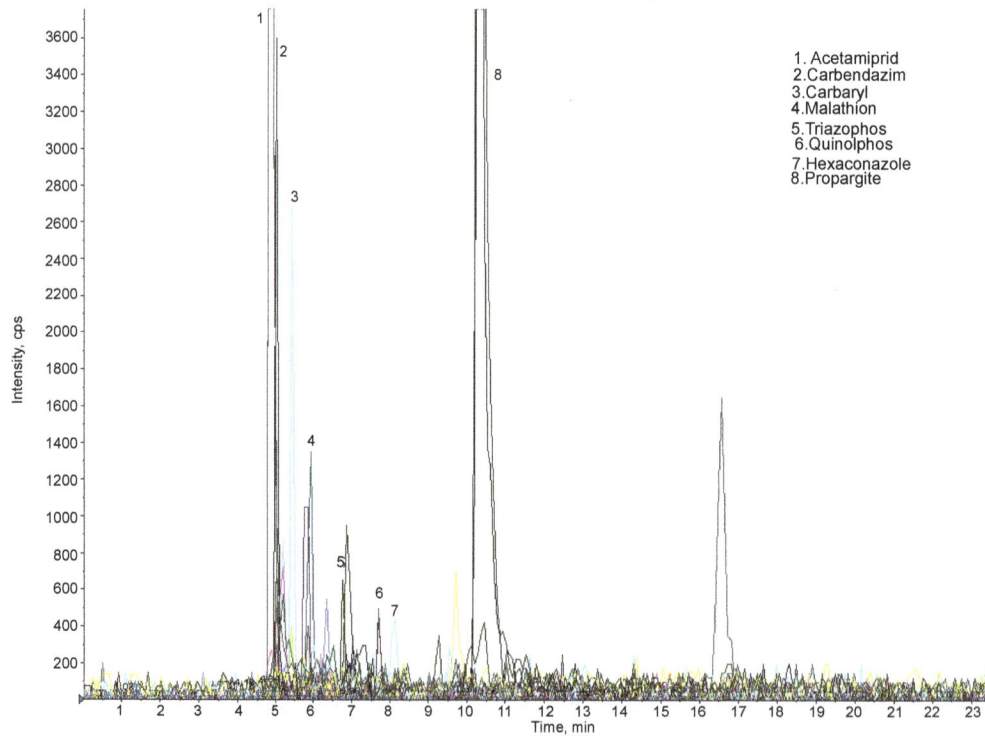

Figure 5: Chromatogram of water sample after treatment with microorganisms.

Propargite > Propenophos > Quinolphos > Hexaconazole > Carbendazim > Imidacloprid > Triazophos > Iprobenphos > Malathion > Carbaryl > Acetamiprid

The Figure 4 shows the chromatogram of water sample indicating the 10 peaks of different pesticides (except propenophos, low response hence there is no peak). Figure 5 shows the chromatogram of water

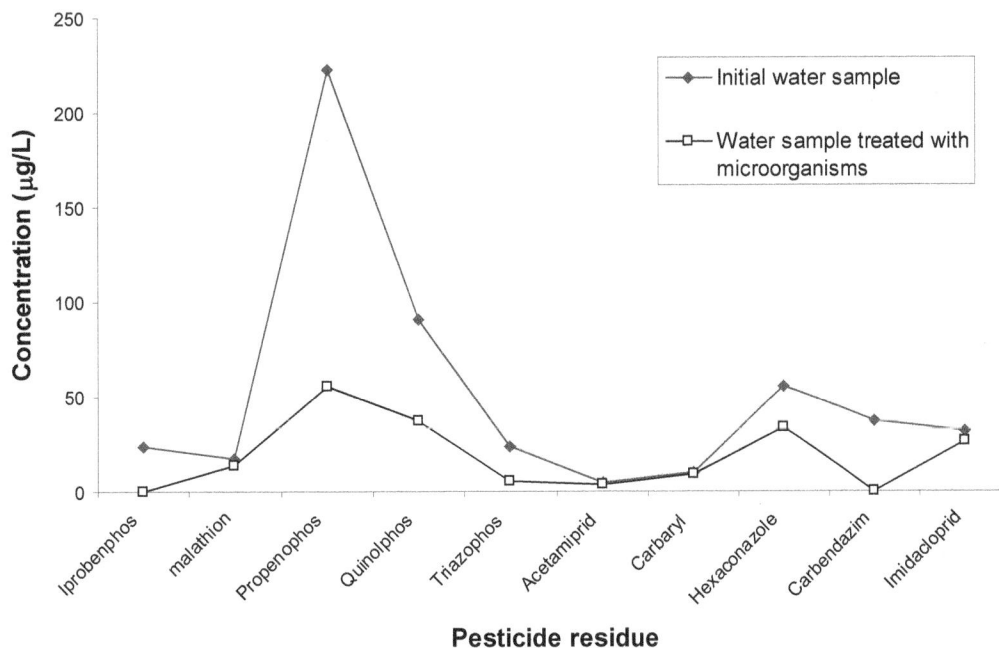

Figure 6: Graph of concentration of pesticides present in water sample before and after treatment with microorganism (except Propargite).

sample treated with microorganisms. The results show that, these organisms are efficient for the degradation of all the 11 pesticides. The amount and percentage of degradation of pesticides by the micro organisms are as shown in Table 2. The graph of concentration of pesticide residue in water sample and microbial treated water sample (Figure 6) shows the efficiency of microorganisms for the removal of pesticides residue from water, except malathion, carbaryl and imidacloprid.

Conclusions

The study reveals that, the pollution levels of stagnant water bodies in small cities have also reached to a severe stage. From the study it is concluded that, the lake water is polluted by pesticide residues entering through the irrigation of agricultural lands in vicinity. The residues of eleven pesticides have been found in the water and the level of propargite is very high. The lake water is also found to be contaminated by heavy metal ions like Pb, Cr, Zn and Cd probably entering from industrial waste, through drainage and battery industries in the vicinity. At the same time, the aquatic plants like *Eichhornia* and *Salvinia* appears to play a vital role in removing the heavy metal ions and correcting the system. But the plants should be carefully controlled and harvested to remove the toxic metal residues. The microorganisms like *Bacillus thuringiensis* NCIM 2159 and *Proteus spp*. SUK 7 are found effective in assimilation and degradation of many of the pesticide residues. However, further investigation regarding scaling up and commercial operation is necessary. The natural or manmade water bodies in and near the settlements play very important role in improvement and conservation of local environment, but many of them are on the way of eutrophication. This problem needs an urgent and sincere attention of environmentalists.

Acknowledgement

Author is thankful to Shanghai University for post doctoral fellowship and also thankful to Prof. Dr. Genxi Li, Dean, School of Life Sciences, Shanghai University, Shanghai, China, for valuable discussion and support during the work. Author is also thankful to Prof. K. J. Patil and Dr. R. R. Kumbhar for their valuable suggestions for this work.

References

1. Outridge PM, Noller BN (1991) Accumulation of toxic trace elements by freshwater vascular plants. Rev Environ Contam Toxicol 121: 1-63.

2. Zayed A, Gowthaman S, Terry N (1998) Phytoaccumulation of trace elements by wetland plants:I. Duck weed. J Environ Qual 27: 715-721.

3. Qian JH, Zayed A, Zhu YL, Yu M, Terry N (1999) Phytoaccumulation of trace elements by wetland plants III. Uptake and accumulation of ten trace elements by twelve plant species. J Environ Qual 28: 1448-1455.

4. Rai UN, Sinha S, Tripathi RD, Chandra P (1995) Wastewater treatability potential of some aquatic macrophytes: removal of heavy metals. Ecol Eng 5: 5-12.

5. Soltan ME, Rashed MN (2003) Laboratory study on the survival of water hyacinth under several conditions of heavy metal concentrations. Advances in Environmental Research 7: 321-334.

6. Mohamed AA, Amer HA, Shawky S, El-Thahawy M, Kandil AT (2009) Instrumental neutron activation analysis of water hyacinth as a bioindicator along the Nile river, Egypt. Journal of Radioanalyical and Nuclear Chemistry 279: 611-617.

7. Brondi SHG, Lanças FM (2005) Development and validation of a multi-residue analytical methodology to determine the presence of selected pesticides in water through liqid chromatography. J Braz Chem Soc 16: 650-653.

8. Liska I, Slobodnik J (1996) Comparison of gas and liquid chromatography for analyzing polar pesticides in water samples. J Chromatogr A 733: 235-258.

9. Grosser ZA, Ryan JF, Dang MW (1993) Environmental chromatographic methods and regulations in the United States of America. J Chromatogr A 642: 75-87.

10. Rebeiro ML, Poleze L, Draetta MS, Minelli EV, Del'Acqua A (1991) Mixed column application in pesticide residue analysis. J Braz Chem Soc 2: 102-104.

11. Munch JW (1995) Determination of Nitrogen- and Phosphorus- containing pesticide in water by Gas Chromatography with a Nitrogen-Phosphorus detector.

12. Tahara M, Kubota R, Nakazawa H, Tokunaga H, Nishimura T (2006) Analysis

of active Oxon forms of nine organophosphorus pesticides in water samples using gas chromatography with mass spectrometric detection. Journal of Health Science 52: 313-319.

13. Perez M, Alario J, Vazquez A, Villén J (2000) Pesticide Residue analysis by off-line SPE and on-line Reversed-phase LC-GC using the through-oven-transfer Adsorption/ Desorption Interface. Anal Chem 72: 846-852.

14. Rasool S, Harakishore K, Satyakala M, Suryanarayana MU (2003) Studies on the physico-chemical parameter of Rankala lake, Kolhapur, India. J Environ Prot 23: 961-963.

15. Sabale S, Jadhav V, Jadhav D, Mohite BS, Patil KJ (2010) Lake Contamination by accumulation of heavy metal ions in *Eichhornia crassipes*: A case study of Rankala Lake, Kolhapur (India). J Environ Sci Eng 52: 155-156.

16. Patil PS, Shedbalkar UU, Kalyani DC, Jadhav JP (2008) Biodegradation of Reactive Blue 59 by isolated bacterial consortium PMB11. J Ind Microbiol Biotechnol 35: 1181-1190.

17. Haug W, Schmidt A, Nörtemann B, Hempel DC, Stolz A, et al. (1991) Mineralization of the sulfonated azo dye Mordant Yellow 3 by a 6-aminonaphthalene-2-sulfonate degrading bacterial consortium. Appl Environ Microbiol 57: 3144-3149.

18. Itoh K, Kitade Y, Nakanishi M, Yatome C (2002) Decolorization of methyl red by a mixed culture of *Bacillus sp.* and *Pseudomonas stutzeri*. J Environ Sci Health A Tox Hazard Subst Environ Eng 37: 415-421.

19. Moosvi S, Kher X, Madamwar D (2007) Isolation characterization and decolorization of textile dyes by a mixed bacterial consortium JW-2. Dyes Pigm 74: 723-729.

20. Yatome C, Yamada S, Ogawa T, Matsui M (1993) Degradation of crystal violet by Nocardia coralline. Appl Microbiol Biotechnol 38: 565-569.

21. Zeinat kamal M, Nashwa AH, Fetyn A, Ibrahim MA, El-Nagdy S (2008) Biodegradation and detoxification of malathion by *bacillus Thuringiensis* MOS-5. Aust J Basic Appl Sci 2: 724-732.

22. Kanekar PP, Bhadbhade BJ, Deshpande NM, Sarnaik SS (2004) Biodegradation of Organophosphorus Pesticides. Proc Indian natn Sci Acad B70: 57-70.

23. de Pasquale C, Fodale R, Lo Piccolo L, Palazzolo E, Alonzo G, et al. (2009) Biodegradation of organophosphorus pesticides by soil bacteria. Geophysical Research Abstracts 11: EGU2009-3147.

24. Sarkar S, Satheshkumar A, Premkumar R (2009) Biodegradation of Dicofol by *Pseudomonas* strains isolated from tea *Rhizosphere microflora*. Int J Integr Biol 5: 164-166.

25. Kanazawa J (1987) Biodegradability of pesticides in water by microbes in activated sludge, soil and sediment. Environ Monit Assess 9: 57-70.

26. Nebera VP, Oo SKN (2007) Biosorption of metals from geotechnology solutions. Physicochemical Problems of Mineral Processing 41: 251-258.

27. Vijayaraghavan K, Yun YS (2008) Bacterial biosorbants and biosorption. Biotechnol Adv 26: 266-291.

28. Leung WC, Chua H, Lo W (2001) Biosorption of heavy metals by bacteria isolated from activated sludge. Appl Biochem Biotechnol 91-93: 171-184.

29. Masud Hosin S, Anantharaman N (2005) Studies on Copper (II) biosorption using *Thiobacillus ferroxidan*. Journal of the University of Chemical Technology and Metallurgy 40: 227-234.

30. De J, Ramaiah N, Vardanyan L (2008) Detoxification of toxic heavy metals by matrix bacteria highly resistant to mercury. Mar Biotechnol (NY) 10: 471-477.

4

Antibiotic Susceptibility and Heavy Metal Tolerance Pattern of *Serratia Marcescens* Isolated From Soil and Water

Natasha Nageswaran[1], P.W. Ramteke[2], O.P. Verma*[3] and Avantika Pandey[3]

[1]*Department of Microbiology and Fermentation Technology, Jacob School of Biotechnology & Bioengineering, SHIATS Allahabad-211007, Uttar Pradesh, India*
[2]*Department of Biological Sciences, Jacob School of Biotechnology & Bioengineering, SHIATS Allahabad-211007, Uttar Pradesh, India*
[3]*Department of Molecular & Cellular Engineering, Jacob School of Biotechnology & Bioengineering, SHIATS Allahabad-211007, Uttar Pradesh, India*

Abstract

The antibiotic and metal tolerance patterns of Serratia marcescens strains isolated from soil and water around the Sangam region of Allahabad were obtained. Using the standard minimum inhibitory concentration (MIC) for each antibiotic respectively, the Kirby-Bauer disc-diffusion method was used to obtain antibiotic resistance patterns of the *Serratia* strains and the MIC of the metals - chromium, cadmium, cobalt, copper, lead and nickel for each of the strains were also obtained. Plasmid curing was carried out for specific antibiotic and metal resistances to ascertain plasmid-borne transfer of resistance genes. Results obtained showed that Multi-drug resistant (MDR) strains of *Serratia* were resistant to certain metals as well suggesting specific metal-antibiotic resistant gene patterns in the different strains.

Keywords: Disc-diffusion; Minimum inhibitory concentration; Multi-drug resistance; Plasmid curing

Introduction

Microorganisms are ubiquitous in nature and are involved in almost all biological processes of life. With rapid urbanization and natural processes, heavy metals have been found in increasing proportions in microbial habitats. Metals have been known to play a major role either directly or indirectly in almost all metabolic processes, growth and development of microorganisms [1,2]. However, increasing concentrations of metals beyond tolerance levels have forced these organisms to adapt to various biological mechanisms to cope with this condition. Hence, microbes have developed mechanisms like metal efflux systems, complexation, reduction of metal ions or utilization of the metal as a terminal electron acceptor in anaerobic respiration to tolerate heavy metal accumulation [3]. Bacteria that are resistant to such heavy metals and have the ability to grow in high concentrations of these metals play an important role in their biological cycling which has great potential in bioremediation of poorly cultivable soil high in heavy metal content. Heavy metal tolerance has been observed in the *Enterobacteriaceae* member, *Serratia marcescens* and has been thought to be attributed to plasmid-borne resistant genes. 8 isolated strains of this microorganism have been used for heavy metal tolerance testing against various metal salts in order to identify specific strains that can be used for removal of particular metals from environments such as soil and water where they are present as pollutants.

Antibiotic susceptibility has also emerged as an ever increasing health hazard due to the indiscriminate use of antibiotics. This has led to severe complications in patients especially with gram negative bacterial infections as the number of drugs to combat this category of infections are limited. Multidrug resistance (MDR) can also be caused by another mechanism of accumulation of multiple antibiotic resistance genes each coding for a single antibiotic occurring on resistance (R) plasmids [4]. Multi Drug Resistance organisms are posing to be a huge threat in treatment procedures due to the presence of plasmid borne mobile resistance genes that can readily spread through bacterial populations and efflux systems to counter third and even fourth generation cephalosporin [5].

Serratia marcescens, formerly known as *Chromobacterium prodigiosin* [6] is a gram-negative facultative anaerobe [7] that has emerged in recent times as a nosocomial pathogen. It functions as an opportunistic organism in immunocompromised patients because of its invasive ability to adhere to hospital instrumentation such as catheters, endoscopes and intravenous tubing [8]. *Serratia marcescens* is one of the major nosocomial pathogen found associated with urinary and respiratory tract infections, endocarditis, osteomyelitis, septicemia, wound infections, eye infections (conjunctivitis) and meningitis [9]. *Serratia marcescens* is the only pathogenic species in its genus. The organism inhabits a wide variety of ecological niches and causes diseases in plant, vertebrate as well as invertebrate hosts [10].

Various metals such as cadmium, copper, chromium, cobalt, lead and nickel were used for metal tolerance tests against 8 strains of *Serratia marcescens* as well as antibiotic susceptibility test was conducted for the above mentioned strains using 14 different antibiotics. All the 8 strains were isolated from soil and water near the Sangam region (confluence of the Ganga and Yamuna rivers) of Allahabad.

All metals and antibiotics towards which the strains showed resistance (for metals at a particular concentration) were subjected to plasmid curing experiments to determine the likelihood of plasmid-borne resistance pattern and relationship between heavy metal and antibiotic resistance genes.

Materials and Methods

Preliminary tests for Identification

All 8 strains of *Serratia marcescens* were identified as Gram negative bacilli by the Gram Staining technique and produced the characteristic red pigment, Prodigiosin.

---

*Corresponding author: O.P. Verma, Department of Molecular & Cellular Engineering, Jacob School of Biotechnology & Bioengineering, SHIATS Allahabad-211007, Uttar Pradesh, India, E-mail: vermaop.aaidu@gmail.com

Biochemical tests were conducted for all the above according to Bergey's Manual of Determinative Bacteriology and all strains were confirmed as *Serratia marcescens*.

Antibiotic susceptibility test

The isolated strains were checked for antibiotic resistance with Ceftriaxone, Ciprofloxacin, Cloxacillin, Gentamicin, Amikacin, Cefuroxime, Netilmicin, Erythromycin, Ceftazidime, Chloramphenicol, Vancomycin, Ampicillin, Cefepime, and Imipenem [11].

The Kirby-Bauer disk diffusion method [12] was employed for antibiotic susceptibility of the 8 strains of *Serratia marcescens* [13]. The strains of *Serratia* were inoculated into 5 ml of Nutrient broth and incubated overnight at 37°C. These pure broth cultures were then swabbed on 20 ml of Mueller Hinton pre-made agar plates each. All 14 antibiotic disks were placed on each of the swabbed plates at appropriate distances from one another and the plates were then incubated at 37°C for another 24 hrs. Zones of inhibition were obtained by measuring the diameter across the centre of each zone in millimeters.

In addition to this, powdered form of other antibiotics were also used for antibiotic sensitivity tests such as Ampicillin, Ceftazidime, Cefepime, Cephalexin, Nalidixic acid, Neomycin Sulphate, Polymyxin, Tetracycline, Cephataxime, Streptomycin and Trimethoprim.

Bacterial isolation to obtain pure culture

Nutrient agar medium was prepared by adding sodium chloride (5.0 g), peptone (5.0 g), beef extract (3.0 g) and agar (15.0 g) to 1000 ml of distilled water and at pH 7.0 and autoclaved at 15 psi.

Isolates of *Serratia marcescens* were plated on nutrient agar and incubated at 37°C for 24 hours. The pure cultures obtained were inoculated into tubes containing 5 ml of sterile nutrient broth each and incubated overnight at 37°C.

Preparation of metal stock solutions

1. Metal salts were used to prepare stock solutions of 100 µg/ml and 1000 µg/ml (for further tests)

2. Weight of the salt required to make stock solutions is calculated [14] in the following Table 1.

Minimum inhibitory concentration (MIC) of heavy metals

The isolated bacterial cultures were checked for their respective MICs towards heavy metals ions such as chromium (Cr), cadmium (Cd), Lead (Pb), copper (Cu), cobalt (Co) and nickel (Ni). The isolated pure cultures were plated on NA medium containing 3 different concentrations of each of the metals- 25 µg/ml, 50 µg/ml and 100 µg/ml respectively. NA medium was prepared by the above mentioned method. 20 ml of NA was poured into each Petri plate and appropriate volume of metal stock solutions were pipetted into the plate. The molten NA was mixed with the metal solution by swirling the plate and was allowed to solidify.

The isolated cultures were streaked onto the NA medium containing metal salts using sterile loops and then incubated at 37°C for 24-48 hrs. The plates were checked for bacterial growth. The concentration of metal where there was no growth is observed as the MIC for that salt for that strain [14]. The isolates, for which growth was visible even at 100 µg/ml for a particular metal, were further grown at higher metal concentration ranges of 200 µg/ml, 400 µg/ml and 800 µg/ml respectively from a 1000 µg/ml metal stock solution. These were then incubated at 37°C for 24-48 hrs and checked for visible growth.

Plasmid curing

Each strain was checked for their resistance towards a particular antibiotic and metal. For metals, the highest concentration which showed growth of a particular strain was taken for plasmid curing. The antibiotics and metals that were used for curing were selected after MIC of metals and antibiotic resistance patterns were obtained as shown in the following Table 2. For Plasmid Curing, the protocol followed was in accordance to [15]. 200 µg/ml of acridine orange solution (mutagenic agent) was prepared in distilled water. 9 ml of peptone water and 10 ml of acridine orange solution were autoclaved separately. 9 ml of peptone water was mixed with 1 ml acridine orange solution. In another tube, 5 ml of peptone water was inoculated with *Serratia marcescens* culture at 37°C in the morning. After about 7-8 hrs, 0.1 ml of this broth was transferred to a fresh 10 ml solution of acridine orange and incubated at 37°C for 24 hrs under shaking conditions. Appropriate dilutions were plated on LB agar to obtain about 120 colonies. These colonies were picked aseptically and transferred on Antibiotic and metal infused plates of LB agar separately towards which the organism showed resistance earlier. The colonies which grew on the plate were those which still showed resistance and those which were cured, failed to grow on the plate since they were plasmid borne genes and were sensitive to the particular metal or antibiotic.

Results and Discussion

All the eight isolates of *Serratia marcescens* that were obtained from the soil and river water samples of the Sangam region (confluence of the Ganga and Yamuna rivers) at Allahabad produced the characteristic red pigment, prodigiosin. Their antibiotic susceptibility and tolerance

Serial No.	Metal	Salt used	Mol wt.	Atomic wt.	Weight of 10ml stock solution (mg)
1	Chromium	$K_2Cr_2O_7$	294	52	170
2	Cadmium	$CdCl_2$	183	112	49
3	Cobalt	$CoCl_2$	130	59	66
4	Copper	$CuSO_4$	160	64	75
5	Lead	$PbCl_2$	278	207	40
6	Nickel	$NiCl_2$	130	57	68

Table 1: Preparation of stock solution of heavy metal compounds.

Isolate No.	Antibiotics	Heavy Metals
1	Ampicillin, Ceftazidime, Cefepime, Cephalexin, Nalidixic Acid, Trimethoprim	Nickel, Lead, Chromium, Copper, Cobalt
2	Ceftazidime, Cefepime, Cephalexin, Nalidixic Acid, Streptomycin, Trimethoprim	Nickel
3	Ampicillin, Ceftazidime, Cefepime, Nalidixic Acid, Streptomycin, Trimethoprim	Nickel, Lead
4	Ampicillin, Ceftazidime, Cefepime, Cephalexin, Nalidixic Acid, Streptomycin, Trimethoprim	Nickel, Lead, Copper
5	Imipenem, Ampicillin, Ceftazidime, Cefepime, Nalidixic Acid, Streptomycin, Trimethoprim	Nickel, Lead, Copper
6	Ceftazidime, Cefepime, Cephalexin, Nalidixic Acid, Trimethoprim	Nickel, Lead, Copper, Cobalt, Chromium
7	Ceftazidime, Cefepime, Cephalexin, Nalidixic Acid, Streptomycin, Trimethoprim	Nickel
8	Ceftazidime, Cefepime, Nalidixic Acid, Streptomycin, Trimethoprim	Nickel, Lead, Copper, Cobalt

Table 2: Selection of antibiotics and heavy metals for Plasmid Curing.

to heavy metals are shown in Table 3 & 4 respectively. All strains were sensitive towards Ceftriaxone, Chloramphenicol, Vancomycin, Netilmicin, Erythromycin, Ciprofloxacin, Cefuroxime, Amikacin, Gentamicin, Cloxacillin, Neomycin sulphate, Polymyxin, Tetracycline, and Cephataxime. All strains were sensitive towards Imipenem, a β-lactam Carbapenem [16] except for Strain 5. Imipenem is a high end broad spectrum drug and is used to combat very severe gram positive and gram negative infections especially nosocomial infections.

The isolates were further tested for heavy metal tolerance against six heavy metal compounds in order to study their ability to grow in the presence of increasing concentrations of respective heavy metals as shown in Figure 1. Cadmium proved to be toxic to all the isolates of *Serratia* even at low concentrations of 25 μg/ml. Isolate one grew in Chromium and Cobalt at 25 μg/ml but was inhibited at 50 μg/ml respectively. It grew at higher concentrations of 100 μg/ml for Copper, Lead and Nickel. Isolate 2 did not grow in any of the metals except Nickel at 50 μg/ml and was inhibited at 100 μg/ml. Isolate 3 showed visible growth for Lead and Nickel at 50 μg/ml and 25 μg/ml respectively but did not grow in the other metal compounds. Isolate 4 and 5 showed identical tolerances to 50 μg/ml of Copper and Nickel and 100 μg/ml of Lead. Both did not grow on Chromium and Cobalt. Isolate 6 tolerated 100 μg/ml of Copper and Lead, 50 μg/ml of Nickel and 25 μg/ml of Chromium and Cobalt. Isolate 7 could only tolerate 25 μg/ml of Nickel. Isolate 8 was able to tolerate 50 μg/ml of Copper, Lead and Nickel and 25 μg/ml of cobalt.

All the isolates which grew at and below 100 μg/ml concentration of a particular metal were further tested at higher concentrations of 200 μg/ml, 400 μg/ml and 800 μg/ml of the respective metals in order to establish the concentration at which the concerned isolate was inhibited for that particular metal. Only Isolate 1 grew at 400 μg/ml concentration of Nickel and Isolates 4 and 5 were able to tolerate up to 200 μg/ml concentration of Lead. Plasmid curing studies showed variable results for all the strains of *Serratia marcescens* revealing that they behaved differently in their plasmid borne resistant gene patterns.

Isolate 1 showed about 90% curing for Nickel alone. Isolate 3 showed about 69% curing for Lead. Isolate 6 revealed 76% curing for Copper, 66% curing for Chromium and 83% curing for Lead. The rest of the isolates did not give significant curing results. The above results have been graphically represented in Figure 2.

Isolate 1 showed moderate levels of curing for ampicillin and ceftazidime and absolute curing for cefepime and cephalexin. Isolate 2 showed significantly high levels of curing for ceftazidime, cefepime, cephalexin and streptomycin and 60% curing for trimethoprim. Isolate 3 showed moderate levels of curing for ampicillin, ceftazidime, and nalidixic acid and 90% for streptomycin. Isolate 4 showed very high levels of curing for nearly all antibiotics. Results obtained showed 100% curing for ampicillin, cephalexin, nalidixic acid and streptomycin. It also showed 87% curing for ceftazidime, 91% for cefepime and 76% for trimethoprim. Isolate 5 showed very high levels of curing for ampicillin, ceftazidime, cefepime, nalidixic acid, streptomycin and trimethoprim and moderate levels of curing for imipenem. Isolate 6 showed reasonable levels of curing for ceftazidime and cefepime and 100% curing for cephalexin. Isolate 7 moderate levels of curing for ceftazidime and cefepime and high levels of curing for cephalexin. Isolate 8 showed reasonable levels of curing for only streptomycin and trimethoprim (Figure 3).

For each strain, percentage curing was carried out numerically towards the different antibiotics and metals, and those which showed

curing beyond 50% were said to have been significantly cured as seen in Table 4.

Environments containing elevated concentrations of heavy metals potentially give rise to heavy metal tolerant organisms [17]. The area around the Sangam region of Allahabad, Uttar Pradesh, India has shown a large variety of microorganisms out of which *Serratia marcescens* has shown significant antibiotic and heavy metal resistant patterns. In order to survive these metal stressed conditions, microorganisms have evolved to tolerate the uptake of heavy metal ions [18]. Different strains of *Serratia* have shown varied resistant patterns suggestive of plasmid-borne transfer of resistant genes or pump effluxing systems. However, it has been suggested that increasing antibiotic resistance of Gram-negative bacteria is mainly due to mobile genes on plasmids that can readily travel through bacterial populations [5]. The studies conducted

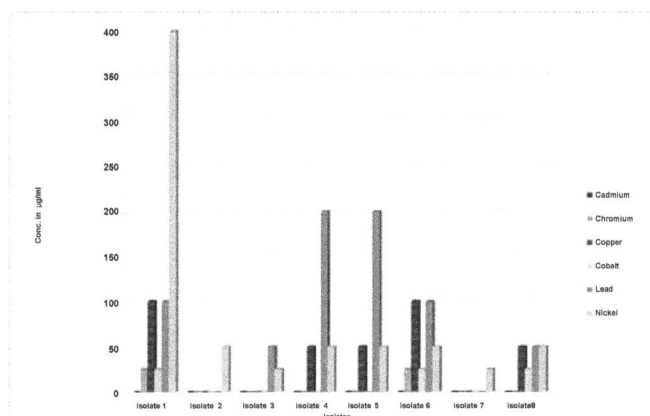

Figure 1: Graph showing metal tolerance patterns of 8 isolates of *Serratia marcescens*.

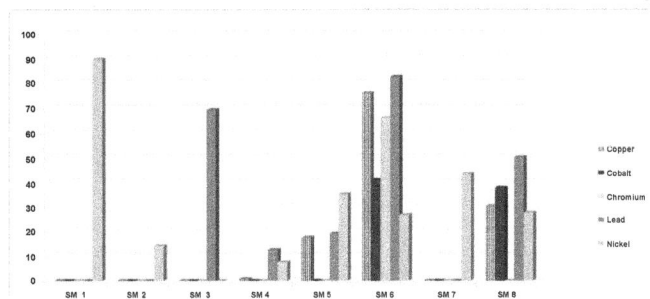

Figure 2: Graph showing plasmid curing pattern of 8 isolates of *Serratia marcescens* towards copper, cobalt, chromium, lead and nickel.

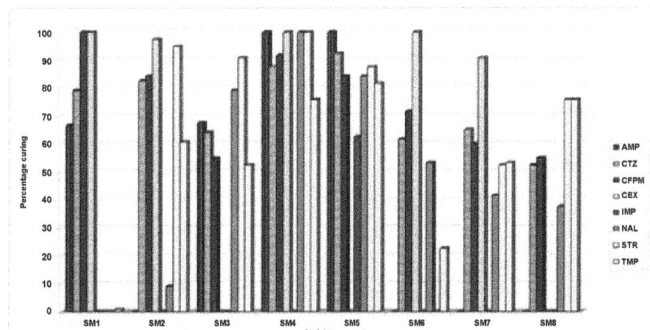

Figure 3: Graph showing plasmid curing pattern of 8 isolates of *Serratia marcescens* towards ampicillin, ceftazidime, cefepime, cephalexin, imipenem, nalidixic acid, streptomycin and trimethoprim.

Name of Antibiotic	Conc. (mcg)	Isolate 1	Isolate 2	Isolate 3	Isolate 4	Isolate 5	Isolate 6	Isolate 7	Isolate 8
Ceftriaxone	30	S	S	S	S	S	S	S	S
Chloramphenicol	30	S	S	S	S	S	S	S	S
Vancomycin	30	S	S	S	S	S	S	S	S
Netilmicin	30	S	S	S	S	S	S	S	S
Erythromycin	15	S	S	S	S	S	S	S	S
Ciprofloxacin	5	S	S	S	S	S	S	S	S
Cefuroxime	30	S	S	S	S	S	S	S	S
Amikacin	30	S	S	S	S	S	S	S	S
Gentamicin	10	S	S	S	S	S	S	S	S
Cloxacillin	1	S	S	S	S	S	S	S	S
Ampicillin	10	R	S	R	R	R	S	S	R
Ceftazidime	30	R	R	R	R	R	R	R	R
Cefepime	30	R	R	R	R	R	R	R	R
Cephalexin	30	R	R	S	R	S	R	R	S
Nalidixic Acid	30	R	R	R	R	R	R	R	R
Imipenem	30	S	S	S	S	R	S	S	S
Neomycin Sulphate	30	S	S	S	S	S	S	S	S
Polymyxin	30	S	S	S	S	S	S	S	S
Tetracycline	30	S	S	S	S	S	S	S	S
Cephataxime	30	S	S	S	S	S	S	S	S
Streptomycin	10	S	R	R	R	R	S	R	R
Trimethoprim	23.75	R	R	R	R	R	R	R	R

*'R' resistant
*'S' sensitive (for powder form antibiotics, since no zone is available)

Table 3: Antibiotic Susceptibility Patterns of the 8 isolates of *Serratia marcescens*.

Isolate No.	Cured Metals and Antibiotics
1	Ni Ampicillin, Ceftazidime, Cefepime, Cephalexin
2	- Ceftazidime, Cefepime, Cephalexin, Streptomycin, Trimethoprim
3	Pb Ampicillin, Ceftazidime, Nalidixic Acid, Streptomycin, Trimethoprim
4	- Ampicillin, Ceftazidime, Cefepime, Cephalexin, Nalidixic Acid, Streptomycin, Trimethoprim
5	- Ampicillin, Ceftazidime, Cefepime, Nalidixic Acid, Streptomycin, Trimethoprim, Imipenem
6	Cu. Cr, Pb Ceftazidime, Cefepime, Cephalexin
7	- Ceftazidime, Cefepime, Cephalexin
8	- Streptomycin, Trimethoprim

Table 4: Antibiotics and metals showing % curing more than 50%.

for *Serratia* show evidence of resistant genes in accordance to [4]. All *Serratia* strains that were tolerant to metals were also tolerant towards several antibiotics also seen by [19] who studied bacteria isolated from drinking water that contained organisms that were metal as well as antibiotic resistant. This has significance suggesting that the metal and antibiotic resistances are a function.

Conclusion

The above studies have shown the dual properties of the bacterial genome in rendering the metal tolerant organisms antibiotic resistant as well. Further research in closely examining this interesting nature of the *Serratia* genome will enable us to find out the exact mechanism of these two shared properties and their close association with one another. The metal tolerant nature of *Serratia marcescens* has tremendous potential in the bioremediation of heavy metal accumulation in soil and water and also in the treatment of sewage and toxic wastes.

References

1. Beveridge TJ, Doyle RJ (1989) Metal ions and Bacteria. Wiley, New York.

2. Bergey DH, Breed RS (1994) Biochemical Identification of Enterobacteriaceae. Bergey's Manual of Determinative Bacteriology 1: 416-417.

3. Ehrlich HL, Brierley CL (1990) Microbial Mineral Recovery. McGraw-Hill, NewYork.

4. Nikaido H (2009) Multidrug resistance in Bacteria. Annu Rev Biochem 78: 119-146.

5. Kumarasamy KK, Toleman MA, Walsh TR, Bagaria J, Butt F, et al. (2010) Emergence of a new antibiotic resistance mechanism in India, Pakistan, and the UK: a molecular, biological, and epidemiological study. Lancet Infect Dis 10: 597-602.

6. Hejazi A, Falkiner FR (1997) Serratia marcescens. J Med Microbiol 46: 903-912.

7. Patton TG, Katz S, Sobieski RJ, Crupper SS (2001) Genotyping of clinical Serratia marcescens isolates: a comparison of PCR-based methods. FEMS Microbiol Lett 194: 19-25.

8. Marty KB, Williams CL, Guynn LJ, Benedik MJ, Blanke SR (2002) Characterization of a cytotoxic factor in culture filtrates of Serratia marcescens. Infect Immun 70: 1121-1128.

9. Yah SC, Eghafona NO, Forbi JC (2008) Plasmid-borne antibiotics resistant markers of Serratia marcescens: an increased prevalence in HIV/AIDS patients. Scientific Research and Essay 3: 028-034.

10. Coulthurst SJ, Williamson NR, Harris AK, Spring DR, Salmond GP (2006) Metabolic and regulatory engineering of Serratia marcescens: mimicking phage-mediated horizontal acquisition of antibiotic biosynthesis and quorum-sensing capacities. Microbiology 152: 1899-1911.

11. Loureiro MM, de Moraes BA, Quadra MRR, Pinheiro GS, Asensi MD (2002) Study of multi-drug resistant microorganisms isolated from blood cultures of hospitalized newborns in Rio De Janeiro City, Brazil. Braz J Microbiol 33: 73-78.

12. Bauer AW, Kirby WMM, Sherris JC, Turck M (1966) Antibiotic susceptibility testing by a standardized single disc method. Am J Clin Pathol 45: 493-591.

13. Schoenknecht FD (1973) The Kirby-Bauer Technique in Clinical Medicine and Its Application to Carbenicillin. The Journal of Infectious Diseases 127: S111-115.

14. Narasimhulu K, Sreenivasa Rao PS, Vinod AV (2010) Isolation and identification of bacterial strains and study of their resistance to heavy metals and antibiotics. J Microbial Biochem Technol 2: 074-076.

15. Tewari S, Ramteke PW, Garg SK (2003) Isolation and R-plasmid studies in enterotoxigenic E.coli (ETEC). Geobios 30: 13-16.

16. Yong D, Lee K, Yum JH, Shin HB, Rossolini GM, et al. (2002) Imipenem-EDTA disk method for differentiation of metallo-beta-lactamase-producing clinical isolates of Pseudomonas spp. and Acinetobacter spp. J Clin Microbiol 10: 3798-3801.

17. Clausen CA (2000) Isolating metal-tolerant bacteria capable of removing copper, chromium, and arsenic from treated wood. Waste Manag Res 18: 264-268.

18. Nies DH, Silver S (1995) Ion efflux systems involved in bacterial metal resistances. J Ind Microbiol Biotechnol 14: 186-199.

19. Calomiris JJ, Armstrong JL, Seidler RJ (1984) Association of metal tolerance with multiple antibiotic resistance of bacteria isolated from drinking water. Appl Environ Microbiol 47: 1238-1242.

5

Allium Chromosome Aberration Test for Evaluation Effect of Cleaning Municipal Water with Constructed Wetland (CW) in Sveti Tomaž, Slovenia

Peter Firbas[1]* and Tomaž Amon[2]

[1]Private Researcher, SICRIS Id. No.11784, Private Laboratory for Plant Cytogenetics, Ljubljanska c. 74. SI-1230 Domžale, Slovenia
[2]Researcher, Gorazdova ul. 3. SI–1000 Ljubljana, Slovenia

Abstract

We show the effectiveness of the communal waste water cleaning plant of the type "LIMNOWET® Constructed Wetlands (CW)". The tests were done with the *Allium* metaphase test and show the degree of genotoxicity, by observing the aberrations of the metaphasic chromosomes of the plant *Allium cepa* L. that are evoked by genotoxic substances in the polluted water. The CW plant reduced the degree of genotoxicity 29.0% to 3.5% (the Fisher's Exact test: $p=9.2e^{-13}<0,00001$). Therefore CW is a very effective ecoremediation technique, since it removes up to 96% of waste.

Keywords: Constructed wetland (CW) LIMNOWET; Allium metaphase test; Chromosome damage

Introduction

Urban and agricultural waste can add significant amounts of contaminants to surface water and sediments. The resulting water pollution presents a serious problem for the health of the biota and humans that interact with these aquatic ecosystems [1].

The standardized and sensitive testing method "*Allium* test" is widely used for testing the quality of drinking water and environmental water pollution [2,3]. This officially accepted method has been in use for more than forty years. It is especially useful for testing the possibility contaminated water streams like rivers [4-6], rain and snow [7], earth [8], waste water monitoring [9], pesticides like atrazin [10], benzo(a) piren [11], pharmaceutical effluents [12], outflows from hospitals [13], including possible radioactive wastes [14]. *Allium* test shows a good correlation with other plant [15] and animal tests [16,17]. The test procedures described here show the effective degree of combined genotoxic activity of the pollutants on the metaphase chromosomes. Complex interactions may occur *in vivo* because component pharmacokinetics increases the unpredictability of pharmacodynamic outcomes [18].

There are six bioassays used in this technique: *Allium* and *Vicia* root tip chromosome break, *Tradescantia* chromosome break, *Tradescantia* micronucleus, *Tradescantia*-stamen-hair mutation and *Arabidopsis*-mutation bioassays were establish from four plant systems that are currently in use for detecting the genotoxicity of environmental agents [15]. The *Allium* root tip chromosome aberration assay has been adopted by the International Program on Plant Bioassay (IPPB), for monitoring or testing environmental pollutants [19]. Higher plants are recognized as excellent genetic models to detect environmental mutagens, and are therefore, frequently used in monitoring studies [20].

Constructed Wetlands (CW) imitate the self-cleaning ability of nature for the treatment of polluted waters. In general, CWs operate without machine power and need no electricity to run. This saves construction, maintenance and operation costs. The system consists of several successive beds filled with substrate (e.g. stone pebbles), and at bottom isolated with plastic foil. The flow of water through the bed substrate follows gravitational gradient. There is also added dry substrate on the bedtop, which reduces unpleasant odors and drives away insects. Perkins and Hunter [21] report the findings of an investigation of Faecal Coliform (FC) bacteria and Faecal Streptococci (FS) removal in four small, parallel *Typha*-dominated, surface flow reed beds. They constitute the tertiary phase of treatment at the Crow Edge sewage treatment works near Holmfirth in Yorkshire. Reduction in concentration was observed between inflow and outflow wastewater for both indicator bacteria, giving mean bed removal efficiency values of approximately 85-94%. The water is treated with the help of microorganisms, wetland plants, as well as physical and chemical processes.

Onion (*Allium cepa* L.) is very suitable for genotoxic studies. Let us list some of its advantages: (i) The root growth dynamics is very sensitive to the pollutants; (ii) The mitotic phases are very clear in the onion; (iii) It has a stable chromosome number; (iv) Diversity in the chromosome morphology; (v) Stable karyotype; (vi) Clear and fast response to the genotoxic substances; (vii) Spontaneous chromosomal damages occur rarely. Therefore, this test has become well established for the determination of the genotoxic substances in various environments. In this study, we describe various chromosome and chromatide damages in the root meristeme cells of the onion (*Allium cepa* L.), which serve as biomarkers for the different types of environmental pollution. With *Allium* test, one can obtain both the effective degree of cleaning the waste water and the impact factor of the so cleaned waste water on the stream into which it flows.

Materials and Methods

Sample example

Influent and effluent water-Constructed wetlands (CW).

Onion preparation for the test

Small onion (*Allium cepa* L.) bulbs of the same size 16-18 mm,

*Corresponding author: Peter Firbas, Private Researcher, SICRIS Id. No.11784, Private Laboratory for Plant Cytogenetics, Ljubljanska c. 74. SI-1230 Domžale, Slovenia, E-mail: peter.firbas@gmail.com

weighing about 3-3.5 g, aged maximum 6 months were denuded by removing the loose outer scales and scraped, so that the root primordia were immersed into the tested liquids.

Experimental procedure

The exposure time of small onion (*Allium cepa* L.) bulbs in each experiment was 72 hours at 22°C and protected against direct sunlight. In order to eliminate the influence of daylight rhythms, the plants were exposed to constant artificial light of middle intensity. In an alternative version of our Allium metaphase test, the five onion seeds were placed directly in the experimental water containers. The water sample under investigation was divided into three portions, which were successively applied to the onion roots in 24 hour periods. So each 24 hours, the roots obtained a fresh bath of the sample solution. After 72 hours, the samples were removed from water bath. The macroscopic and microscopic tissue morphology investigation followed.

Chromosome preparations

The squash technique for onion root as described [7,22], was used for the chromosome investigation. Chromosome samples were taken from root meristems containing actively growing cells. The developing roots with bulbs were pretreated with 0.1% water solution of colchicine for 3 hours at 21°C. After washing in distilled water for 20 min, the terminal developing roots of 2 mm length were fixed for 1 h in methanol: propionic acid mixture (3:1 or 1:1). Then they were macerated and stained in order to obtain a cellular suspension. This sample was stained with 0.5% aceto-carmine for 4-5 min at 60°C without hydrolysis, and squashed in aceto-carmine [23]. For observation was used optical microscope Olympus-BX 41, with the photo system PM 10 SP, typical magnifications used were 400 X and 1000 X.

Macroscopic parameters

After 72 hours growth in the test solution, we measured the root length and noted related parameters as the shape of the roots, number, color and turgescence [2,3]. Toxicity and genotoxicity measurements were performed with 10 ppm concentration of the test MMS chemicals as Positive Control (PC).

Microscopical parameters

To study the damage of chromosome and chromatid (breaks), we used the treatment with colchicines. Onion *Allium cepa* L. has 16 monocentric chromosomes (2n=16), with basic number x=8 (Figure 1). The possible aberrations seen at metaphase are: (i) single break chromatide, (ii) double break chromatide, (iii) gap break and (iv) centromere break (Figure 2).

Level genotoxicity

This is the percentage between all the metaphase cells and the cells with their chromosomes damaged. The total number studied was 200 metaphase cells.

Parallel control test

Integral parts of *Allium* test are the so called negative and positive control. Negative control shows the degree of toxicity in unexposed onions, and serves as control of the test efficiency. Positive control is used with known material, which normally induces a high degree of toxicity, and is necessary for controlling the test response. In other words-nearer the results of tested samples to negative control, the better the quality of water. On the other hand-farther the values from

the results of negative control and nearer to the positive control–this points to poorer water quality.

Physical and chemical parameters

The physical and chemical properties of the effluent sample were determined in accordance with standard analytical methods [24]. Suspended substances in the sample were determined by the standard (SIST ISO 11923), COD with standard (SIST ISO 6060), and BOD with the standard (SIST EN 1899-1).

Statistic calculation

Statistically established significant differences among the investigated samples are confirmed by the statistical calculation of paired data analysis using the two-way Fisher's exact test, which gives the p value property between pairs of data [25]. These pairs (investigated samples) are either different (statistically significant) or the same (statistically insignificant), and tell us what the risk is. The most common values are 0.05, followed by 0.01, 0.001 and 0,0001. With regard to these limits, we can also speak about 5%, 1%, 0.1% and 0.01% levels of significant results. Whichever of these levels we choose depends a great deal on the nature of the data and the problem addressed by the basic assumption. Statistical significance should not be the only determining factor when evaluating the results.

Results

We used the Allium metaphase test (Allium M test), which

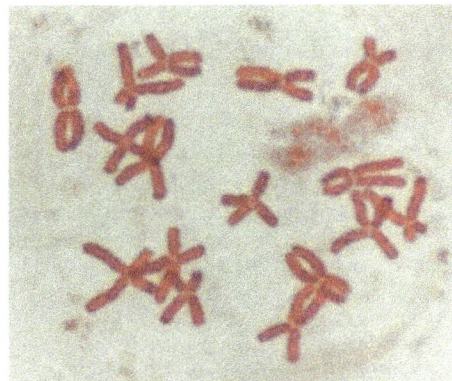

Figure 1: Diploid metaphase chromosome from the root cells of the onion (*Allium cepa* L.), containing 2n of 16 (2n=16).

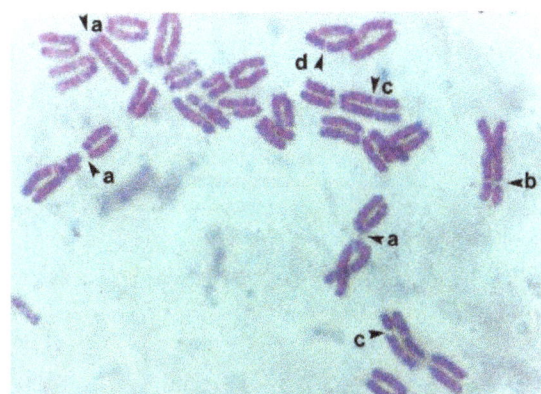

Figure 2: Damaged chromosomes: 4a–break in centromere, 4b–double break chromatide, 4c–single break chromatide, 4d–gap chromatide.

indicates partially damaged chromosomes or chromatids. The results of the Allium M test are shown in tables 1-3, and figures 1-5, and the physical and chemical parameters are displayed in the table 4.

Constructed wetland (CW) described here is located in village Sveti Tomaž in Slovenske gorice (46° 29'N; 16° 4'E); Ev. No. 13/56-08/P (Limnos, Slovenia). CW follows the system Limnowet° [26]. Its GPS coordinates are YX: 583011 149381. The village Sveti Tomaž has 254 inhabitants. It is surrounded by fields, wineyards and forests [27].

CW was built in the year 2000. It is intended for cleaning of household waste water and effluent of urban-municipal activities. Load CW is 250 PE (population units), with an area of 700 m². The system consists of two stage beveled settler, filtration bed, treatment beds and polishing bed. The slope of the shaft is 1%, the substrate is composed of different grades of sand. Filtration bed is planted with bent-grass (*Carex*) and rush (*Juncus*), treatment bed the shaft with reds (*Phragmites*), and rush (*Juncus*) polishing bed [26]. If necessary, the system can end with an open lagoon for multipurpose use for purified water (irrigation or watering of green areas, aqua culture), or as a landscape element. The sludge from mechanical treatment is composted in the composting bed, which is basically similar to CW beds [26].

General toxicity–root growth inhibition and malformation

The cleaned water (Sample II) shows longer roots and lesser general toxicity than the uncleaned waste water (Sample I), p<0.01. The latter also induces longer roots than the positive control (Sample IV). The cleaned water induces equal root length as the negative control (Sample III).

Level genotoxicity–induction of chromosome aberration in root meristems cell

The cleaned outflowing water (Sample II) is significantly less genotoxic than the inflowing polluted water (p=9.2e⁻¹³< 0,00001). So the level of genotoxicity decreases from 29.5% to 3.5%. This is confirmed both by PC and NC. The inflow water (Sample I) is not statistically significantly different from PC (p=1>0.05) but is significantly different by NC (p=7.9e⁻¹⁴< 0.00001). The outflowing water not statistically significantly different with NC (p=0.7887>0.005), but is significantly different by NC (p=7.9e⁻¹⁴<0.00001).

Physical and chemical parameters

CW effectively reduces the concentration suspended solids for 65.52/71.43%, markedly reducing the COD (chemical need when oxygen) to 84.2/85% and BOD$_5$ (biochemical oxygen consumption in five days) for 94.0/95.8%.

Discussion

Our testing has shown the high efficiency of the CW plant in Sveti Tomaž. Constructed Wetlands (CW) reproduce the self-cleaning ability of nature. They coexist nicely with environment, improve the appearance of degraded areas and provide new biotopes for plants and animals. Even most of the pharmaceutical substances (e.g. antibiotics and other drugs) get decomposed there.

Waste waters are typically very mitodepressive. That is clearly shown in the root growth inhibition of the experimental plant *Allium cepa* L. [2,3]. Mutagen parameters are represented by the damages of chromosomes like single or multiple chromatid breaks, gap damages, circular centric or acentric chromatids. Further abnormalities that appear are centric or acentric fragments, dicentric chromosomes and circular chromosomes [7,22,28-30].

The cell is called aberrant, if at least one chromosome gets damaged [31]. The onion *Allium cepa* has 16 (2n=16) chromosomes-several of them can get damaged. There is also possible to observe several types of damage on a single chromosome. Typically one to two chromosomes become damaged, sometimes 3 to 7 or 8, in extreme cases, even 12 or all. This corresponds to the degree of pollution of the tested water. In addition, multiple damages on a single chromosome point to serious genotoxicity [31]. Effluent wastewater has a complex mixture of inorganic and organic substances (*chemical complex mixture*), and is highly mutagenic. Typically at least 4-10 chromosomes (sometimes even all) get damaged in such a water (Figure 5D). The cleaned water samples from the CW show a much lesser degree of genotoxicity–typically one chromosome, rarely two are damaged.

The discharge of wastewater into natural water courses, ponds and wetlands has been an ancient practice. Urbanization led to the development of more engineered solutions for domestic sewage, as well as industrial effluents. This technical development temporarily lessened the economical importance of the earlier approaches that were closer to nature [32]. However, in the last years, there has increased interest in simpler, more natural methods for wastewater treatment including Macrophyte Treatment System like CWs [32]. Constructed Wetlands (CWs) imitate the self-cleaning ability of nature. They reduce the concentrations of nitrogen, phosphorous and other toxic substances. Their removal efficiency up to 96%; they are cheap to build and maintain, and fit nice into the local environment.

In the period between 1960-1970 in Slovenia, there was a common practice of melioration and regulation of small streams. So, many of them were biologically destroyed. In addition, there was paid very little attention to the waste water cleaning. However, from 1990 on the situation has improved very much. The projects resulting in ecoremediations and revitalizations were undertaken. This was accompanied by the testing also with the Allium method, which helps to monitor the general water quality. Until 2011, there was made more than 1230 Allium tests on 390 localities [31], such trends increasing,

Example	Number of identify metaphase cell	Average metaphase cells with chromosome damage	Average level genotoxicity (%)	Average length of root (mm)	Percentage (%) length root of negative control
Wastewater; influent in CW	200	58 ± 6.0	29.0* ± 3.0	17 ± 3.5	41.4
Cleaning water effluent out CW	200	7 ± 2.0	3.5* ± 1.0	42 ± 3.5	100.2
Negative control-NC (tap water filtered with R.O.-reverse osmosis	200	6 ± 2.0	3.0 ± 1.0	41 ± 2.0	100.0
Positive control-PC (10 ppm Methan Methilsulphonate–MMS 4016 Sigma)	200	61 ± 5.0	31.0 ± 2.5	12 ± 3.0	29.2

Table 1: The average length of the roots and cytological effects of investigated samples and both controls of the test plants of *Allium cepa* L. Cytological effects–investigation of genotoxicity level and Average root length of test plant *Allium cepa* L.–investigation of general toxicity. From each of 5 bulbs in a series of 5, one root tip is taken for each of 5 slides. From each slide, 200 metaphase cells are scored. The degree of general toxicity of the analysed samples (5 bulbs per sample) was assessment from the mean root lengths expressed as a percentage of the mean root length of the negative control.

since there are many CWs and other waste water treatment plants in Slovenia

In our further work, we intend to expand our testing methods with the so called micronucleus test (MN) [33], on the water animals,

Samples	NC	PC	Inflow	outflow
NC				
PC	7.9e^{-14}**			
inflow	1.7e^{-13}**	0.8269*		
outflow	0.7887*	7.3e^{-11}**	9.2e^{-13}**	

Table 2: Comparison of frequency of aberrant cells among controls (NC–negative controls, PC–positive controls) and effluents samples (two by two tables-Fisher´s Exact Test which gives the p value property between pairs of data.); * p>0.05 samples is not characterized as being statistically significant; **p<0,00001 samples is characterized as statistically significant.

Samples	NC	PC	Inflow	outflow
NC				
PC	0.01**			
inflow	0.01**	0.88*		
outflow	1*	0.02**	0.01**	

Table 3: Comparison of root length among controls (NC–negative controls, PC–positive controls) and effluents samples (two by two tables-Fisher´s Exact Test which gives the p value property between pairs of data.); * p>0.05 samples is not characterized as being statistically significant; **p<0.05 samples is characterized as statistically significant.

Figure 3: The length of the roots of the test plants of *Allium cepa* L. treated in a sample of urban waste-municipal water.

Figure 4: The length of the roots of the test plants of *Allium cepa* L. treated in a sample of the cleaned water.

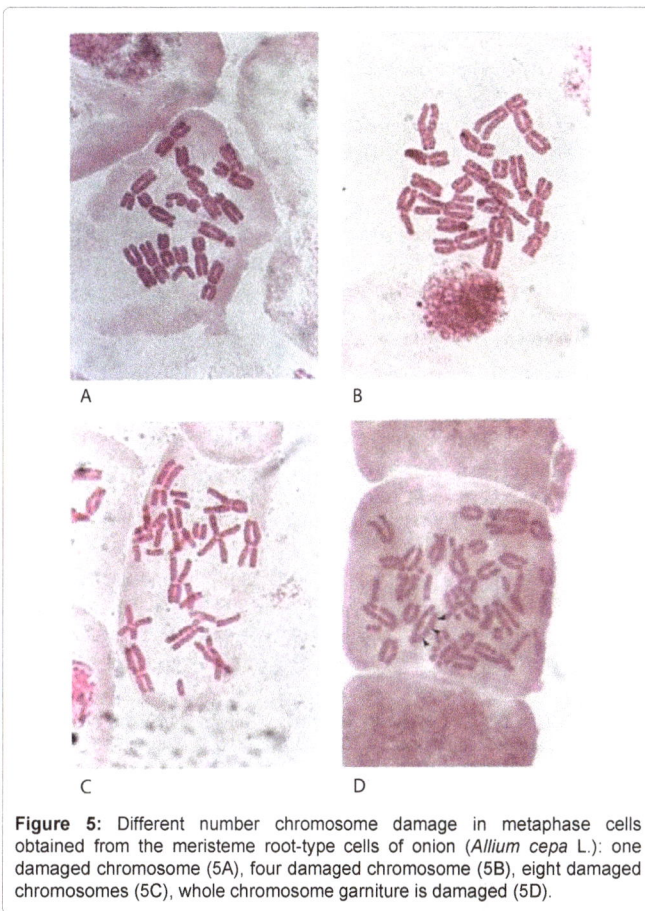

Figure 5: Different number chromosome damage in metaphase cells obtained from the meristeme root-type cells of onion (*Allium cepa* L.): one damaged chromosome (5A), four damaged chromosome (5B), eight damaged chromosomes (5C), whole chromosome garniture is damaged (5D).

Parameter	Wastewater (inflow) July/October	Cleaning water (outflow) July/October	Effect CW (%) July/October
Suspended solids (mg/l)	35/29	10/10	71.43/65.52
COD (mg/l)	200/190	30/30	85.0/84.2
BOD$_5$ (mg/l)	70/50	3/3	95.8/94.0

Table 4: Some physical-chemical measurements of inflow waste water and outflow cleaning water CW in Sveti Tomaž. Wastewater inflow and cleaning outflow water samples taken common twice, once in July and once in October.

especially when testing larger water volumes like lakes. There are more than 200 lakes in Slovenia and more than 1300 water habitats [34]. They are sensitive ecosystems, where one needs to monitor the quality of water, in order to keep the Slovene waters clean and healthy.

Introduction of genotoxicity research in the environment protection policies is of great importance, since it enables us to understand the impacts and consequences of genotoxic substances present in water. The goal of our research is to give an immediate and important contribution to preserving the health of the most precious life source– water. Due to our lack of knowledge and carelessness, we have already polluted some water sources; therefore, it is our obligation to correct our mistakes. We should be aware that as regards genotoxicity, there are no safe "Maximum Permissible Concentrations" (MPC), which would ensure a good and reliable quality of water.

References

1. Egito LCM, Medeiros MG, De Medeiros SRB, Agnes-Lima LF (2007) Cytogenetic and genotoxic potential of surface water from the Pitimbu river, northeastern/RN Brasil. Genet Mol Biol 30: 1-15.

2. Fiskesjö G (1985) The *Allium* test as a standard in environmental monitoring. Hereditas 102: 99-112.

3. Rank J (2003) The method of *Allium* anaphase-telophase chromosome aberration assay. Ekologija 1: 38-42.

4. Majer BJ, Grummt T, Ulh M, Knasmuller S (2005) Use of plant bioassays for the detection of genotoxins in the aquatic environment. Acta Hydrochimica Et Hydrobiologica 33: 45-55.

5. Vujosević M, Andelković S, Savić G, Blagojević J (2008) Genotoxicity screening of the river Rasina in Serbia using the *Allium* anaphase-telophase test. Environ Monit Assess 147: 75-81.

6. Barbério A, Barros L, Voltolini JC, Mello ML (2009) Evaluation of the cytotoxic and genotoxic potential of water from the River Paraíba do Sul, in Brazil, with the *Allium* cepa L. test. Braz J Biol 09: 837-842.

7. al-Sabti K (1989) *Allium* test for air and water borne pollution control. Cytobios 58: 71-78.

8. Watanabe T, Hirayama T (2001) Genotoxicity of soil. J Health Sci 47: 433-438.

9. Fiskesjö G (1993) The *Allium* test–A potential standard protocol of assessment of environmental toxicity. In: Gorsuch JW, Dwyer FJ, Ingorsoll CG, et al. (Eds) Environmental toxicology and risk assessment. American Society for Testing and Materials, Philadelphya, USA.

10. Bolle P, Mastrangelo S, Tucci P, Evandri MG (2004) Clastogenicity of atrazine assessed with the *Allium* cepa test. Environ Mol Mutagen 43: 137-141.

11. Cabaravdic M (2010) Induction of chromosome aberrations in the *Allium* cepa test system caused by the exposure of cells to benzo(a) pyrene. Med Arh 64: 215-218.

12. Abu NE, Mba KC (2011) Mutagenicity testing of phamarceutical effluents on *Allium* cepa root tip meristems. J Toxicol Environ Health Sci 3: 44-51.

13. Bagatini MD, Vasconcelos TG, Laughinghouse HD 4th, Martins AF, Tedesco SB (2009) Biomonitoring hospital effluents by the *Allium* cepa L. test. Bull Environ Contam Toxicol 82: 590-592.

14. Evseeva TI, MaÄstrenko TA, Geras'kin SA, Belykh ES (2006) [The influence of cadmium and potassium on the level of cytogenetic effects induced by thorium-232 in *Allium* cepa root meristem]. Tsitol Genet 40: 50-58.

15. Ma TH, Cabrera GL, Owens E (2005) Genotoxic agents detected by plant bioassays. Rev Environ Health 20: 1-13.

16. Torres de Lemos C, Milan Rödel P, Regina Terra N, Cristina D'Avila de Oliveira N, Erdtmann B (2007) River water genotoxicity evaluation using micronucleus assay in fish erythrocytes. Ecotoxicol Environ Saf 66: 391-401.

17. Bolognesi C, Hayashi M (2011) Micronucleus assay in aquatic animals. Mutagenesis 26: 205-213.

18. Martin FL (2007) Complex mixtures that may contain mutagenic and/or genotoxic components: a need to assess *in vivo* target-site effect(s) associated with *in vitro*-positive(s). Chemosphere 69: 841-848.

19. Ma TH (1999) The international program on plant bioassays and the report of the follow-up study after the hands-on workshop in China. Mutat Res 426: 103-106.

20. Leme DM, Marin-Morales MA (2009) *Allium* cepa test in environmental monitoring: a review on its application. Mutat Res 682: 71-81.

21. Parkins J, Hunter C (2000) Removal of enteric in a surface flow constructed wetland in Yorkshire, England. Water Res 34:1941-1947.

22. Al-Sabti K, Kurelec B (1985) Chromosomal aberrations in onion (*Allium* cepa) induced by water chlorination by-products. Bull Environ Contam Toxicol 34: 80-88.

23. Firbas P, Al-Sabti K (1995) Cytosistematic studies on the Charophyta in Slovenia. Archive biological doctrine 47: 45-58.

24. Standard Methods (2005) Standard methods for the examination of water and wastewaters. (21 Edn), American Water Works Association (AWWA), Water Pollution Control Federation (WPCF) and American Public Health Association (APHA), Washington DC, USA.

25. Agresti A (1992) A Survey of exact inference for contegency tables statistical. Science 7: 131-153.

26. Vrhovšek D, Vovk-Korže A (2007) Ecoremediations. University of Maribor, Faculty of Arts, International Center for ERM, Maribor and Limnos doo, Ljubljana, Slovenia.

27. KLS (1995) Local lexicon Slovenia. DZS Ljubljana, Slovenia.

28. Kumar P, Panneerselvan N (2007) Cytogenetic studies of food preservative in *Allium* cepa root meristem cells. Facta Universitatis Series: Medicine and Biology 14: 60-63.

29. Ragunathan I, Panneerselvam N (2007) Antimutagenic potential of curcumin on chromosomal aberrations in *Allium* cepa. J Zhejiang Univ Sci B 8: 470-475.

30. Panneerselman N, Palanikumar L, Gopinathan S (2012) Chromosomal aberration by Glycidol in *Allium cepa* L. root meristem cell. Int J Pharm Sci Res 3: 300-304.

31. Firbas P (2011) Level of chemicals in the environment and cytogenetic damage. Ekslibris, Ljubljana, Slovenia.

32. Evans GM, Furlong JC (2011) Environmental biotechnology: Theory and application. (2nd Edn), Wiley-Blackwell, UK.

33. Al-Sabti K (1991) Handbook of genotoxic effects and fish chromosomes. J Stefan Institute, Ljubljana, Slovenia.

34. Firbas P (2001) All Slovenian lake: Lexicon Slovenian standing water. DZS Ljubljana, Slovenia.

6

Biological Treatment of Meat Processing Wastewater Using Lab-Scale Anaerobic-Aerobic/Anoxic Sequencing Batch Reactors Operated in Series

David Nzioka Mutua[1]*, Eliud Nyaga Mwaniki Njagi[1], George Orinda[1], Geoffry Obondi[1], Frank Kansiime[2], Joseph Kyambadde[3], John Omara[3], Robinson Odong[3] and Hellen Butungi[3]

[1]Department of Biochemistry and Biotechnology, School of Pure and Applied Science, Kenyatta University, PO Box 43844, 00100, Nairobi, Kenya
[2]Department of Biochemistry, Institute of Environment and Natural Resources, Makerere University, PO Box 7062, Kampala, Uganda
[3]Department of Biochemistry, Faculty of Science, Makerere University, PO Box 7062, Kampala, Uganda

Abstract

In the eastern Africa sub-region, many industries discharge untreated effluents to nearby water resources, thereby polluting the environment. This is because the technologies applicable for wastewater treatment are expensive for these small-medium sized companies with low profit margins. Slaughterhouses belong to this category of industrial setup. The objective of this study was to investigate treatment of meat processing wastewater using anaerobic–aerobic/anoxic Sequencing Batch Reactors (SBRs) operated in series. Reactors were operated for one year using meat processing wastewater. Hydraulic retention time was 2 days for the anaerobic SBR, and 1 day for the aerobic/anoxic SBR while the organic loading was 12.8 kg COD/m³/day. In the anaerobic SBR, removal efficiencies for total and soluble chemical oxygen demand (TCOD and SCOD), total suspended solids (TSS) and turbidity were 79, 76, 79, and 70%, respectively, with effluent mean concentrations of 3554 ± 58 mg/L, 762 ± 3 mg/L, 2307 ± 21, and 2800 ± 9 FAU. Conductivity, ammonia-nitrogen, ortho-phosphates and total phosphorus concentrations increased by 38, 80, 81 and 71%. Pollutant removal efficiencies in the aerobic/anoxic SBR were 98, 96, 97, 89, 74, 97, 91, 90, and 86% for TCOD, SCOD, BOD, TSS, turbidity, ammonium nitrogen (NH_4^+–N), total nitrogen (TN), orthophosphorus (o-PO_4^{3-}–P), and total phosphorus (TP), respectively. Except TKN (35 ± 4 mg/L) and o-PO_4^{3-}–P (8 ± 1 mg/L), all other parameters in the aerobically treated effluent met national discharge standards. Thus, abattoir effluent can be treated using anaerobic–aerobic/anoxic SBR system.

Keywords: Biological treatment; City Abattoir; Meat processing; Sequencing batch reactors (SBR); Wastewater

Introduction

In the eastern Africa sub-region, facilities for the treatment of domestic and industrial effluents are either inefficient or non-existent thus leading to discharge of high contents of organic matter and nutrients (nitrogen and phosphorus) into nearby surface waters [1,2]. A typical example is City Abattoir (Kampala, Uganda), which on average discharges 400 m³/day of highly recalcitrant untreated effluent into Inner Murchison Bay of Lake Victoria causing oxygen depletion [3,4], eutrophication [5], health complications [1,6] and global warming [7]. As a result, there is pressure to meet environmental discharge standards in line with Sustainable Development Goal (SDG) number 6.

Activated sludge-based sequencing batch reactor (SBR) is an efficient system for treating organic-rich effluents [8]. The SBR technology is well documented in laboratory [5,9], pilot-scale [10] and full-scale studies [11,12]. The bioreactor can be anaerobic, aerobic or anoxic [3]. Each phase has four sequential steps: feed, react, settle and draw [9]. Transitioning between anaerobic, aerobic and anoxic conditions triggers the use of different electron donors and acceptors, thus promoting the transformation and thereby removal of carbon, nitrogen and phosphorus [8].

Biological nitrogen removal involves nitrification: a two-step aerobic oxidation of ammonia to nitrate [12]. Ammonia oxidizing bacteria such as *Nitrosospira*, *Nitrosomonas*, *Nitrosolobus* and *Nitrosococcus* oxidizes ammonia to nitrite and then, nitrite oxidizing bacteria such as *Nitrobacter*, *Nitrospina* and *Nitrococcus* oxidizes nitrite to nitrate [3,4]. This process is usually combined with denitrification, during which nitrate/nitrite is reduced to nitrogen-N_2 by anoxic heterotrophic bacteria like Bacillus and some fungi [12]. During denitrification, electrons from organic carbon are transferred to nitrate instead of oxygen to create a proton motive force for ATP generation

[8]. This process is inhibited by dissolved oxygen which represses nitrate reduction enzyme [13].

Simultaneous nitrification and denitrification can occur in a single reactor under low dissolved oxygen concentration [14]. This is due to the occurrence of aerobic/anoxic micro zones in the floc or within the bioreactor and also due to the presence of new types of microorganisms [12]. The environmental factors affecting nitrification include: pH, un-ionized ammonia, un-ionized nitrous acid, reduced sulphur components and metals [13].

Biological phosphorus removal (BPR) depends is dependent on excessive uptake of phosphorus by phosphorus-accumulating organisms (PAOs). This is achieved by alternating anaerobic, aerobic and anoxic conditions (those containing nitrate, not oxygen, as an electron acceptor) [8]. Under anaerobic conditions, organic matter is mineralized into volatile fatty acids (VFAs) which are taken up by PAOs and stored as poly-β-hydroxyalkaonates (PHAs) [14]. The energy to assimilate the VFAs is derived from breakdown of polyphospates (PPs) bonds, releasing phosphorus into the solution [10]. During the aerobic/anoxic phase, the stored PHA is oxidized to release energy for the PAOs to take up phosphorus from solution and form intracellular

*Corresponding author: David Nzioka Mutua, Department of Biochemistry and Biotechnology, School of Pure and Applied Sciences, Kenyatta University, PO Box 43844-00100, Nairobi, Kenya, E-mail: mutua.david@ku.ac.ke

PPs for cellular growth [3]. Oxygen acts as an electron acceptor [10]. Since phosphorus release in the anaerobic phase is less than its uptake under aerobic/anoxic conditions, net phosphorus removal from mixed liquor is achieved. Phosphorus is removed from the bioreactor by wasting excess phosphorus-rich sludge [15]. BPR is affected by pH, sludge retention time, excessive aeration, nitrate, nitrite, temperature and carbon source [13].

In combined denitrification and BPR systems, carbon availability is the limiting factor. Denitrifiers and PAOs are in competition for the available carbon. Both processes are disturbed by this competition hence a fine balance should be struck on the length of the aerobic and anoxic phase [10].

The configuration and operation of SBR depends on the type of the wastewater and the treatment objectives [3]. SBR can be anaerobic, aerobic or anoxic [8]. Most of the SBRs have one or two of these phases with only a few three-phase combination [3,16]. Besides, the bio-system performance efficiency depends on the sequential arrangement of these phases, the duration of each phase, hydraulic retention time, sludge retention time, organic loading and environmental factors [8,17,18]. The objective of this study was to investigate biological treatment of City Abattoir meat processing wastewater using anaerobic–aerobic/anoxic SBRs operated in series. To evaluate the treatment performance of the system, removal performances for carbon, nitrogen and phosphorus were analyzed.

Materials and Methods

Model reactors

Two reactors made of plastic, each with a total volume of 250 L and a working liquid volume of 200 L were set up at Makerere University and operated at room temperature. The anaerobic and aerobic SBR were seeded with anaerobic and aerobic sludge obtained from a brewery wastewater treatment Plant at Port bell, Luzira, Uganda. The initial concentration of mixed liquor volatile suspended solids (MLVSS) was approximately 10,000 mg/L. The reactors were batch-fed periodically with raw wastewater collected from City Abattoir, and the organic loading rate progressively increased by augmenting the volume of wastewater fed to the systems. After steady-state conditions were obtained (3 months), the reactors were operated sequentially (Figure 1), with each reactor operational cycle consisting of the feed, reaction, settling and, draw phases. The 24 hr operating cycle consisted of the following periods: (a) filling, 0.30 hours; (b) reaction, 41 hours; and (c) decanting, 0.30 hours for the anaerobic reactor, and (a) filling, 0.25 hours; (b) reaction, 17 hr; (c) settling, 6.5 hours and (d) decanting, 0.25 hours for the aerobic reactor. At the end of each cycle, 100 litres of the supernatant was decanted, followed by feeding of an equal amount of wastewater. The system operated at a nominal Sludge Retention Time (SRT) of 5 days and a total Hydraulic Retention Time (HRT) of 2 days and 1 day for the anaerobic and aerobic/anoxic SBR, respectively. The organic loading was 12.8 kg COD/m^3/day, during the study period. The properties of wastewater used during experimental studies are shown in Table 1.

Analytical procedure

Physical water quality variables (pH, electrical conductivity and temperature) were measured *in situ* twice a week using portable WTW (Wissenchaftlich Technishe Werkstatten) microprocessor probes and meters. Chemical parameters such as ammonium-nitrogen (NH_4–N), nitrite-nitrogen (NO_2–N), nitrate-nitrogen (NO_3–N), total kjeldahl nitrogen (TKN), ortho-phosphate (o-PO_4^{3-}–P), total phosphorus (TP),

biochemical oxygen demand (BOD_5), soluble chemical oxygen demand (SCOD), total chemical oxygen demand (TCOD), turbidity, and solids content (TSS) were analyzed according to standard methods.

Statistical analysis

Data analysis was done using ANOVA to compare the treatment performance of the SBRs based on parameter removal efficiency determined from influent and treated effluent mean ± standard error values.

Results

Treatment of abattoir effluent in anaerobic sequencing batch reactor

Figures 2-4 and Table 1 show the characteristics of inflow and outflow wastewater, together with pollutant removal efficiencies attained during anaerobic sequencing batch reactor treatment. The

Parameter	Anaerobic SBR		
	Inflow conc.	Outflow conc.	Removal efficiency (%)
TCOD	15812 ± 241	3554 ± 58	-79
SCOD	3176 ± 100	762 ± 3	-76
BOD_5	13659 ± 67	1869 ± 27	-86
TKN	1022 ± 139	400 ± 30	-61
NH_4–N	58 ± 9	288 ± 7	+80
NO_2–N	NIL	NIL	-
NO_3–N	NIL	NIL	-
TP	61 ± 8	129 ± 1	+71
o-PO_4^{3-}–P	16 ± 1	82 ± 1	+81
Turbidity	9335 ± 130	2800 ± 9	-70
TS	10760 ± 300	2307 ± 21	-79
pH	6.57 ± 0.12	6.56 ± 0.03	-0.2
EC	1.86 ± 0.2	3.02 ± 0.01	+38
Temperature	23.53 ± 0.1	25.7 ± 0.2	+9

Concentrations of TCOD, BOD_5, TKN, NO_2-N NO_3-N, TP, o-PO_4^{3-}, Turbidity and TS are expressed in mg/L; Turbidity, EC, and temperature are expressed in (FAU), (ms/cm) and (°C), respectively; - signifies reduction, + signifies increment.

Table 1: Mean ± standard error values of the physiochemical parameters determined for the raw wastewater, and anaerobic SBR effluent (n=8).

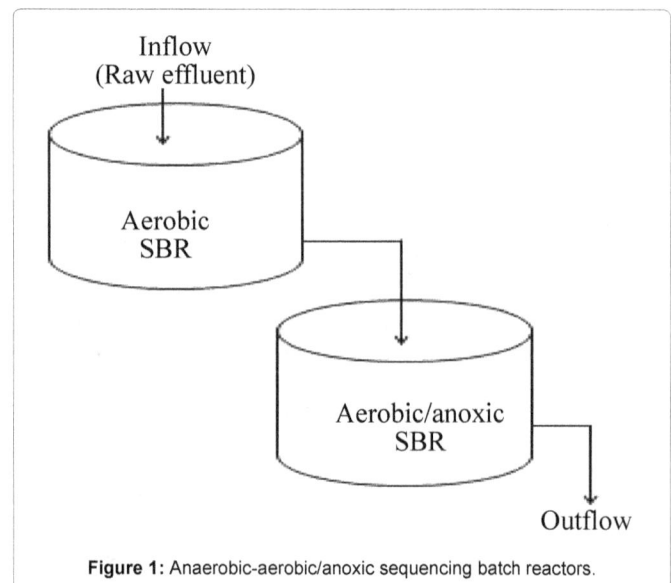

Figure 1: Anaerobic-aerobic/anoxic sequencing batch reactors.

Carbon removal efficiency

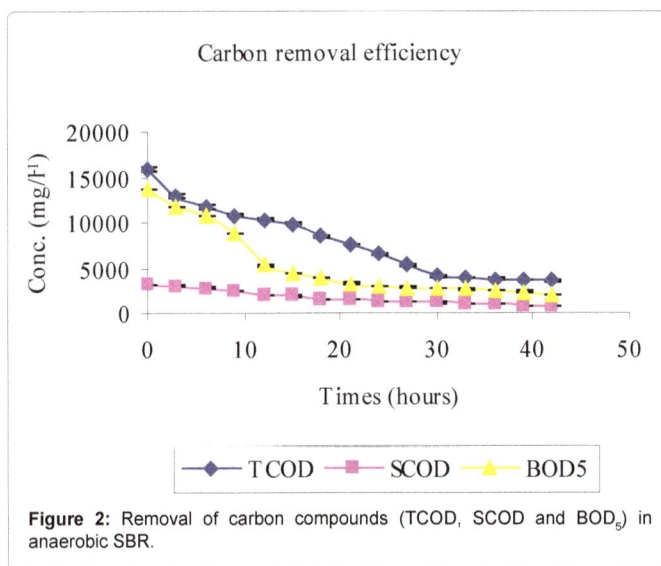

Figure 2: Removal of carbon compounds (TCOD, SCOD and BOD$_5$) in anaerobic SBR.

Nitrogen removal efficiency

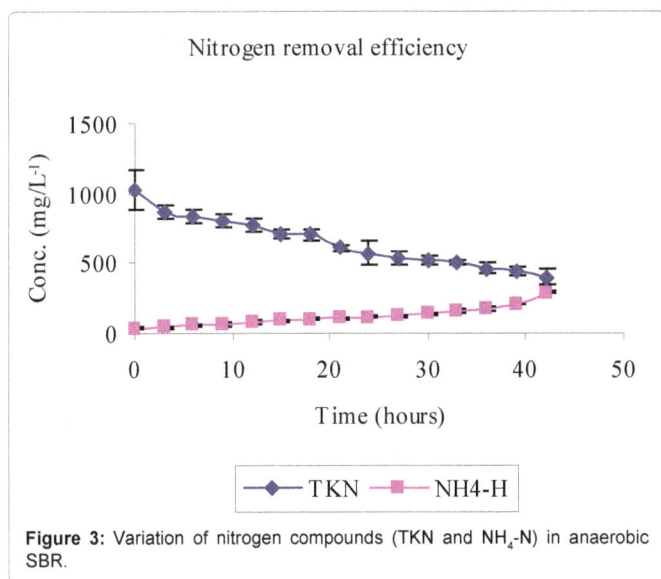

Figure 3: Variation of nitrogen compounds (TKN and NH$_4$-N) in anaerobic SBR.

influent concentrations of TCOD, SCOD, BOD$_5$, TKN, NH$_4$-N, TP, o-PO$_4^{3-}$, turbidity, TS, pH, EC and temperature were 15812 ± 241 mg/L, 3176 ± 100 mg/L, 13659 ± 67 mg/L, 1022 ± 139 mg/L, 58 ± 9 mg/L, 61 ± 8 mg/L, 16 ± 1 mg/L, 9335 ± 130 FAU, 10760 ± 300 mg/L, 6.57 ± 0.12, 1.86 ± 0.2 ms/cm³ and 23.53 ± 0.1°C, respectively.

The removal efficiencies for TCOD, SCOD, BOD$_5$, TKN, turbidity and TSS were 79, 76, 86, 61, 70 and 79% respectively, with effluent mean concentrations of 3554 ± 58 mg/L, 762 ± 3 mg/L, 1869 ± 27 mg/L, 400 ± 30 mg/L, 2800 ± 9 FAU and 2307 ± 21 mg/L, respectively. Comparably, NH$_4$-N, TP, o-PO$_4^{3-}$, pH, EC and temperature increased by 80, 71, 81, 0.2, and 38% registering an effluent concentration of 288 ± 7 mg/L, 129 ± 1 mg/L, 82 ± 1 mg/L, 6.56 ± 0.03, 3.02 ± 0.01 ms/cm and 25.7 ± 0.2°C, respectively.

Treatment of abattoir effluent in aerobic/anoxic sequencing batch reactor

Figures 5-7 and Table 2 show the inflow and outflow parameters of wastewater, together with pollutant removal efficiencies attained during aerobic/anoxic sequencing batch reactor treatment. The

influent concentrations of TCOD, SCOD, BOD$_5$, TKN, NH$_4$-N, NO$_2$-N, NO$_3$-N, TP, o-PO$_4^{3-}$, turbidity, TSS, pH, EC and temperature were 321 ± 75 mg/L, 923 ± 12 mg/L, 1210 ± 32 mg/L, 383 ± 20 mg/L, 233 ± 7 mg/L, 0 mg/L, 0 mg/L, 81 ± 1 mg/L, 67 ± 5 mg/L, 2762 ± 50 FAU, 1350 ± 47 mg/L, 0.91 ± 0.1 mg/L 6.98 ± 0.04, 2.91 ± 0.17 mS/cm and 23.84 ± 0.11°C, respectively.

In the aerobic-anoxic phase, TCOD, SCOD, BOD$_5$, TKN, NH$_4$-N, TP, o-PO$_4^{3-}$, turbidity, TSS and temperature removal efficiencies were 98, 96, 97, 91, 97, 86, 90, 74, 89 and 14% respectively, with effluent mean concentrations of 80 ± 5 mg/L, 31 ± 10 mg/L, 54 ± 12 mg/L, 35 ± 4 mg/L, 8 ± 1 mg/L, 18 ± 1 mg/L, 8 ± 1 mg/L, 738 ± 9 FAU, 254 ± 12 mg/L and 22.04 ± 0.02 ± 0.1°C, respectively. Comparably, NO$_2^-$, NO$_3^-$ and DO which had increased in aerobic phase by 115, 184 and 94% decreased in anoxic phase by 100, 98 and 93% to register an effluent concentration of 0.00 ± 0, 16 ± 8 and 1 ± 3 mg/L, respectively. During this phase pH, EC and temperature varied from 6.71, 1.64 ms/cm³,

Parameter	Aerobic/anoxic SBR		
	Inflow conc.	Outflow conc.	Percentage change
TCOD	3554 ± 58	80 ± 5	-98
SCOD	762 ± 3	31 ± 10	-96
BOD$_5$	1869 ± 27	54 ± 12	-97
TKN	400 ± 30	35 ± 4	-91
NH$_4$–N	288 ± 7	8 ± 1	-97
NO$_2$–N	NIL	.00 ± 0	-100
NO$_3$–N	NIL	16 ± 8	+16
TP	129 ± 1	18 ± 1	-86
o-PO$_4^{3-}$–P	82 ± 1	8 ± 1	-90
Turbidity	2800 ± 9	738 ± 9	-74
TS	2307 ± 21	254 ± 12	-89
pH	6.56 ± 0.03	7.00 ± 0.0	+6
EC	3.02 ± 0.01	1.64 ± 0.01	-46
Temperature	25.7 ± 0.2	22.04 ± 0.02	-14
DO	0.91 ± 0.1	1 ± 3	+9

Concentrations of TCOD, BOD$_5$, TKN, NO$_2$-N, NO$_3$-N, TP, o-PO$_4^{3-}$, and TS are expressed in mg/L; Turbidity, EC, and temperature are expressed in (FAU), (ms/cm³) and (°C), respectively. - signifies reduction, + signifies increment.

Table 2: Mean ± standard error values of the different parameters determined for the raw wastewater, and aerobic/anoxic SBR effluent (n=8).

Phosphorus variation

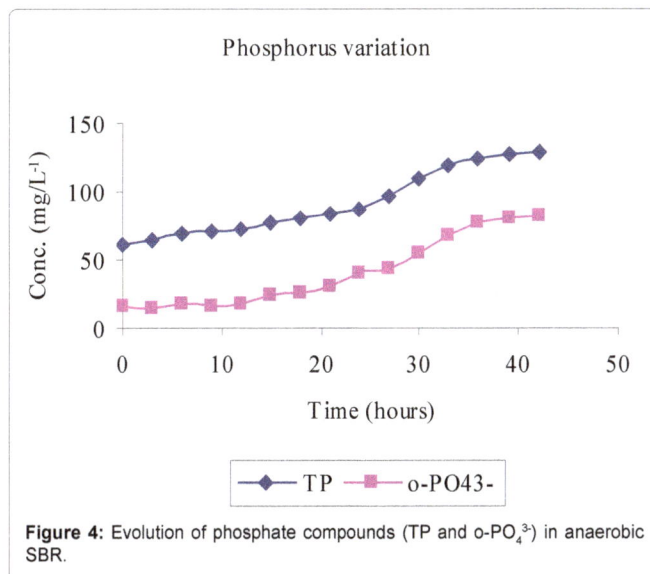

Figure 4: Evolution of phosphate compounds (TP and o-PO$_4^{3-}$) in anaerobic SBR.

and 22.04°C to 7.64, 7.71 ms/cm³ and 25.32°C to register an effluent concentration of 7.00 ± 0.0, 1.61 ± 0.01 ms/cm³ and 22.01 ± 0.02°C, respectively.

Discussion

This study has shown that meat processing effluent can be treated biologically using anaerobic–aerobic/anoxic SBRs operated in sequence. The high Carbon (TCOD, SCOD and BOD₅) levels in the influent (Table 1, Figure 2) were mainly due to the fact that the abattoir effluent employed in this study was of high organic strength [2]. Cellulose, which mainly originates from animal feed residues, is a major component of abattoir effluents which contributes significantly to COD and suspended solids [19]. Slaughterhouse wastewater is also rich in proteins originating from blood which has a TCOD of 375, 000 mg/L⁻¹ [1].

A reduction in SCOD in Figure 2 was due to microbial activity while total COD, TSS and turbidity reductions were due to microbial activity, solids settlement and floatation [10,14]. Anaerobic fill phase was marked by high TKN reduction (Figure 3) due to settling of the blood [20]. Heterotrophic ammonification further decreased organic nitrogen in a process that also produces CO_2 and HCO_3^- [14]. The

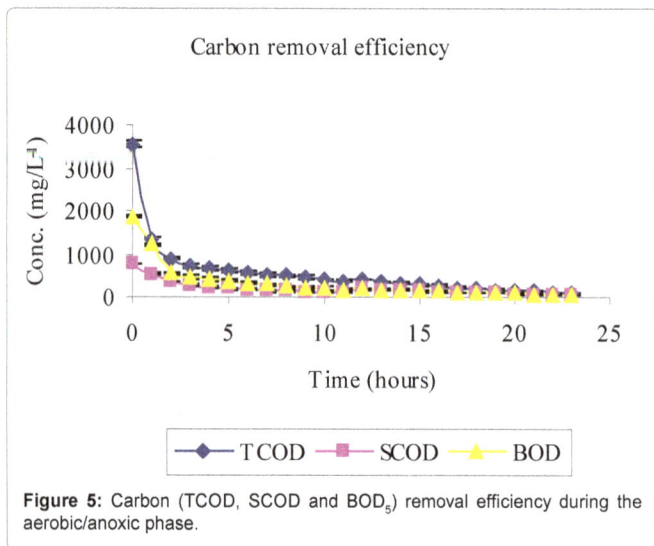

Figure 5: Carbon (TCOD, SCOD and BOD₅) removal efficiency during the aerobic/anoxic phase.

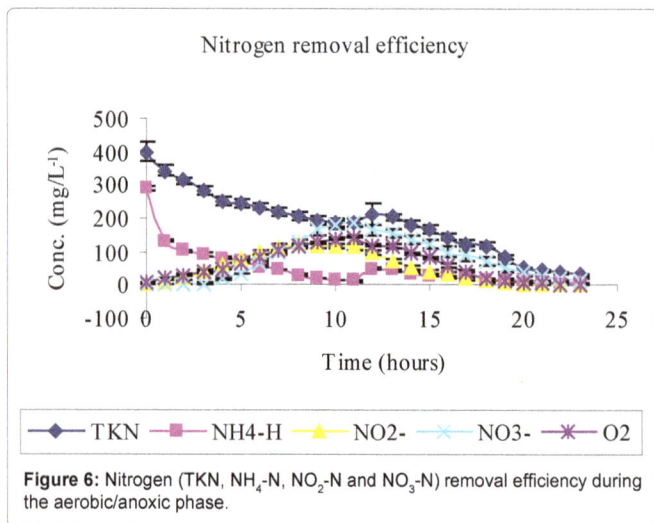

Figure 6: Nitrogen (TKN, NH₄-N, NO₂-N and NO₃-N) removal efficiency during the aerobic/anoxic phase.

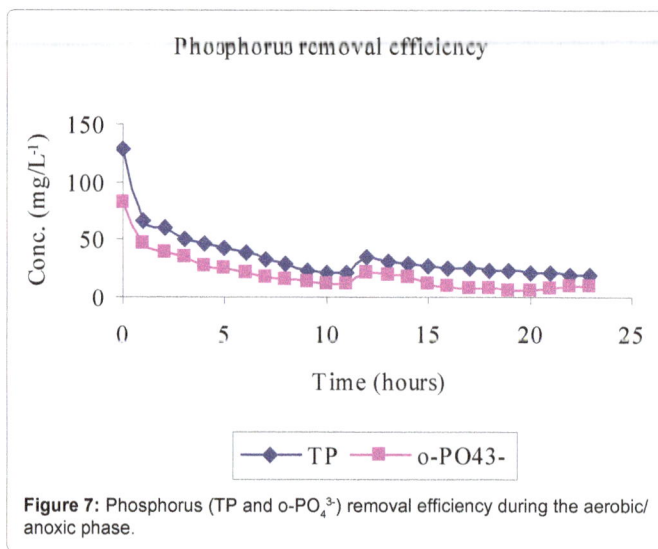

Figure 7: Phosphorus (TP and o-PO_4^{3-}) removal efficiency during the aerobic/anoxic phase.

HCO_3^- production increased the system alkalinity, and thus pH, while ammonification raised the pH and EC [21].

Lipids in the wastewater were anaerobically hydrolysed to long chain fatty acids (LCFA) and glycerol which are subsequently acidified to volatile fatty acids (VFAs) [14]. Phosphate accumulating organisms (PAOs; mainly *Candidatus Accumulibacter phosphatis*) preferentially assimilate VFA across their cell membranes for the synthesis of intracellular carbon/energy reserves of poly β hydroxyalkanoate (PHA) [3]. The energy required for VFA assimilation is provided by the hydrolysis of high-energy intracellular polyphosphate bonds [10]. This resulted in the release of orthophosphorus (Figure 4) and increased temperature. The early stages of anaerobic phase were characterised by low phosphorus release because only VFA in the influent was available. However, as more VFA was made available by bacterial fermentation, more polyphosphate bonds were cleaved to supply the additional energy demands for the VFA uptake hence more phosphorus was released.

Total phosphorus content was found to be higher than that of orthophosphate in the study carried out. This was mainly due to the fact that orthophosphate represents the reactive fraction of phosphorus which is biologically available, while dissolved organic and inorganic phosphorus are generally not biologically available [15].

Compared to the influent raw wastewater, the treatment efficiencies obtained for this reactor were high but still do not meet national discharge standards (COD, 100 mg/L; TSS, 100 mg/L; Turbidity, 300 NTU/FAU; NH₄-N, 10 mg/L, TN, 10 mg/L; ortho-P, 5 mg/L and total-P, 10 mg/L; [22]). Thus, the anaerobic effluent necessitated further processing in the aerobic SBR to reduce pollutant concentrations.

At the initial stages of aeration, there was a conspicuous delay in the first occurrence of both NO₂⁻ and NO₃⁻ when the DO was below 0.91 ± 0.1 mg/L⁻¹. Although effective nitrification has been reported in systems with residual oxygen as low as 0.5 mg/L⁻¹, DO concentrations below 1.5 ml/L limit the nitrification process [4,5]. Moreover, the nitrification is inhibited at pH below 6.8 [23] which existed at this point. Autotrophic nitrifiers first use NH₄-N for cell synthesis, the NH₄-N that is left after cell synthesis is then removed via nitrification [24].

Aeration caused loss of protons as a result of CO_2 stripping creating a localized high pH. The high pH caused re-distribution of ammonium ions to the more volatile ammonia form [25,26]. Heterotrophic assimilation and volatilization can thus account for the high carbon and nitrogen treatment efficiency observed in Figures 5 and 6, respectively.

Nitrification started once the DO and pH were in the applicable range of 6.9. Ammonia is oxidized to NO_2^- which is subsequently oxidized to NO_3^- concentration [3]. This explains the higher concentrations of NO_3^- than NO_2^- at the end of aerobic phase (Figure 6).

In the aerobic phase, the organic matter serves as the electron donor while the oxygen serves as the electron acceptor [8]. Heterotrophs have a higher affinity for NH_4-N and oxygen than autrotrophs [12]. PAOs, being heterotrophs, preferentially take up oxygen and hydrolyse the high-energy PHA phosphoanhydride bonds for energy to grow and assimilate phosphate (released in the anaerobic zone and additional phosphate present in the influent wastewater) in amounts that are much greater than biosynthetic needs. The phosphates are converted into intracellular polyphosphate stores [10]. This caused a decrease in phosphate levels in the bulk liquid (Figure 7). Since PHA oxidation releases 24 to 36 times more energy in the aerobic zone than is used to store polyhydroxybutyrate (PHB) in anaerobic zone, phosphorus uptake is significantly more than the phosphorus release [15].

In the anoxic phase, oxidized nitrogen compounds were reduced to dinitrogen gas [12]. Denitrifiers use organic carbon as energy source and NO_2^--NO_3^- as electron sink [8]. However, the preceding nitrification phase is known to consume reducing power [3] hence sufficient organic carbon must be provided for proper denitrification [10]. Obaja et al. [27] and Pedros et al. [28] suggest the use of wastewater as a source of carbon.

Anaerobic effluent (equivalent to 10% of the aerobic reactor operational volume) was used to supply organic carbon required for denitrification during the settling phase. This volume was experimentally found to be sufficient for complete denitrification of oxidized forms of nitrogen within the system. The introduction of NH$_4$-N rich anaerobic effluent accounted for the spike in nitrogen content at the onset of denitrification (Figure 6). Denitrification results in a rise in alkalinity of the system [13], with corresponding increase in pH. According to Azhdarpoor et al. [3] assimilative and dissimilative carbon utilization by denitrifying and other bacteria is responsible for further carbon reduction in Figure 5.

Phosphorus removal in the aerobic/anoxic SBR was above the recommended discharge limits, with outflow concentration of 8 ± 1 and 18 ± 1 mg/L registered for o-PO_4^{3-}-P and TP, respectively possibly due to its release during the anoxic settlement phase of the SBR system [10,13], and hence minimal phosphorus removal through sludge wasting [29].

Conclusions

Meat processing wastewater can be efficiently treated using anaerobic–aerobic/anoxic SBRs operated in series. Except TKN (35 ± 4 mg/L) and o-PO_4^{3-}-P (8 ± 1 mg/L), all other parameters (TCOD, SCOD, BOD, TSS, turbidity, NH_4^+-N, and TP) in the treated effluent met national discharge standards.

Acknowledgements

This study received financial support from the Swedish International Development Co-operation Agency (Sida)/Department of Research Co-operation (SAREC) under the East African Regional Programme and Research Network for Biotechnology Policy Development (BIO-EARN).

Conflict of Interest

The authors declare no conflict of interest.

References

1. Baskar M, Sukumaran B (2015) Effective Method of Treating Wastewater from Meat Processing Industry Using Sequencing Batch Reactor. International Research Journal of Engineering and Technology 2: 27.

2. Kundu P, Debsarkar A, Mukherjee S (2013) Treatment of Slaughter House Wastewater in a Sequencing Batch Reactor: Performance Evaluation and Biodegradation Kinetics. BioMed Research International 2013: 1-11.

3. Azhdarpoor A, Mohammadi P, Dehghani M (2016) Simultaneous removal of nutrients in a novel anaerobic–anoxic/aerobic sequencing reactor: removal of nutrients in a novel reactor. International Journal of Environmental Science and Technology 13: 543-550.

4. Kyambadde J (2005) Optimizing Processes for Biological Nitrogen Removal in Nakivubo Wetland, Uganda. PhD Thesis. Department of Biotechnology, Royal Institute of Technology, Stockholm, Sweden, pp: 1-62.

5. Abubakar S, Latiff AA, Lawal IM, Jagaba AH (2016) Aerobic treatment of kitchen wastewater using sequence batch reactor (SBR) and reuse for irrigation landscape purposes. American Journal of Engineering Research 5: 23-31.

6. Dhanalakshmi D, Maleeka BSF, Rajesh G (2016) Biodegradation and bioremediation of food industry waste effluents. International Journal of Advanced Research in Science, Engineering and Technology 3: 1195-1197.

7. Yang X, Wu X, Hao H, He Z (2008) Mechanisms and assessment of water eutrophication. Journal of Zhejiang University Science 9: 197-209.

8. Singhal N, Perez-Garcia O (2016) Degrading Organic Micropollutants: The Next Challenge in the Evolution of Biological Wastewater Treatment Processes. Frontiers in Environmental Science 4: 1-5.

9. Lam KY, Zytner RG, Chang S (2016) Treatment of High Strength Vegetable Processing Wastewater with a Sequencing Batch Reactor. Journal on Agricultural Engineering 2: 30-33.

10. Lochmatter S, Maillard J, Holliger C (2014) Nitrogen Removal over Nitrite by Aeration Control in Aerobic Granular Sludge Sequencing Batch Reactors. International Journal of Environmental Research and Public Health 11: 6955-6978.

11. Eslami H, Hematabadi PT, Ghelmani SV, Vaziri AS, Derakhshan Z (2015) The Performance of Advanced Sequencing Batch Reactor in Wastewater Treatment Plant to Remove Organic Materials and Linear Alkyl Benzene Sulfonates. Health Science 7: 33-34.

12. Fernandes H, Jungles MK, Hoffmann H, Antonio RV, Costa RHR (2013) Full-scale sequencing batch reactor (SBR) for domestic wastewater: Performance and diversity of microbial communities. Bioresource Technology 132: 262-268.

13. Puig S, Corominas LI, Balaguer MD, and Colprim J (2007) Biological nutrient removal by applying SBR technology in small wastewater treatment plants: carbon source and C/N/P ratio effects. Water sci technol. 55: 135-141.

14. Aponte-Morales VE, Tong S, Ergas SJ (2016) Nitrogen Removal from Anaerobically Digested Swine Waste Centrate Using a Laboratory-Scale Chabazite-Sequencing Batch Reactor. Environmental Engineering Science 33: 324-332.

15. Gebremariam SY, Beutel MW, Christian D, Hess TF (2011) Research Advances and Challenges in the Microbiology of Enhanced Biological Phosphorus Removal - A Critical Review. Water Environment Research 83: 195-219.

16. Kuśmierczak J, Anielak P, Rajski L (2012) Long-term cultivation of an aerobic granular activated sludge. Electronic Journal of Polish Agricultural Universities 15: 6.

17. Danial O, Salim MR, Salmiati (2016) Nutrient removal of grey water from wet market using sequencing batch reactor. Malaysian Journal of Analytical Sciences 20: 142-148.

18. Milia S, Malloci E, Carucci A (2016) Aerobic granulation with petrochemical wastewater in a sequencing batch reactor under different operating conditions. Desalination and Water Treatment, pp: 1-10.

19. Mittal G (2007) Regulations Related to Land-application of Abattoir Wastewater and Residues. Agricultural Engineering International: the CIGR E-journal, Invited Overview 10: 1-37.

20. Mane SS, Munavalli GR (2012) Sequential Batch Reactor- Application to Wastewater –A Review Proceeding of International Conference SWRDM-2012, Maharastra, India, pp: 1-8.

21. Merzouki M, Bernet N, Delgenes J, Benlemlih M (2005) Effect of pre-fermentation on denitrifying phosphorus removal in slaughterhouse wastewater. Bio-resource Technology 96: 1317-1322.

22. http://www.gov.ug/content/national-environment-management-authority-nema

23. Hayatsu M, Tago K, Saito M (2008) Various players in the nitrogen cycle: Diversity and functions of the microorganisms involved in nitrification and denitrification. Soil Science and Plant Nutrition 54: 33-45.

24. Szatkowska B (2007) Performance and control of biofilm systems with partial nitritation and anammox for supernatant treatment. PhD Thesis, p. 1035. ISBN 978-91-7178-729-3.

25. Paredes D, Kuschk P, Mbwette T, Stange F, Müller R, et al. (2007) New Aspects of Microbial Nitrogen Transformations in the Context of Wastewater Treatment – A Review. Engineering in Life Sciences 7: 13-25.

26. Kim J, Zuo Y, Regan J, Logan B (2008) Analysis of Ammonia Loss Mechanisms in Microbial Fuel Cells Treating Animal Wastewater. Biotechnology and Bioengineering 99: 112-1127.

27. Obaja D, Macé S, Mata-Alvarez J (2005) Biological nutrient removal by a sequencing batch reactor (SBR) using an internal organic carbon source in digested piggery wastewater. Bioresource Technology 96: 7-14.

28. Pedros PB, Onnis-Hayden A, Tyler C (2008) Investigation of nitrification and nitrogen removal from centrate in a submerged attached growth bioreactor. Water Environment Research 80: 222-228.

29. Haiming Z, Xiwu L, Abualhail S, Jing S, Qian G (2014) Enrichment of PAO and dPAO responsible for phosphorus removal at low temperature. Environment Protection Engineering 40: 67-80.

7

Microbial Degradation of Gasoline in Soil: Comparison by Soil Type

D A Turner[1], J. Pichtel[2], Y Rodenas,[3,2] J McKillip[3] and J V Goodpaster[1]*

[1]*Department of Chemistry and Chemical Biology, Forensic and Investigative Sciences Program, Indiana University Purdue University Indianapolis (IUPUI), Indianapolis, USA*
[2]*Department of Natural Resources and Environmental Management, Ball State University, Muncie, USA*
[3]*Department of Biology, Ball State University, Muncie, USA*

Abstract

During the investigation of a suspicious fire, debris is often collected from the scene and analyzed for residues of ignitable liquids (e.g., gasoline). In cases where the debris is contaminated with soil, it is known that heterotrophic soil microorganisms can alter the chemical composition of the ignitable liquid residue over time. The effects of soil type and season upon this phenomenon are not known, however. Hence, soil collected from locations under three different uses (residential, agricultural, brownfield) were spiked with gasoline and microbial degradation was monitored for 30 days. The soils were also chemically and biologically characterized. Gas chromatographic profiles showed that residential soil was most active and brownfield soil least active for the microbial degradation of gasoline. The brownfield soil possessed relatively high (497 mg/kg) concentrations of Pb, which may have affected bacterial activity. Predominant viable bacterial populations enumerated using real-time reverse transcriptase polymerase chain reaction (RT-PCR) included members of the *Alcaligenes, Acinetobacter, Arthrobacter, Bacillus, Flavobacterium*, and *Pseudomonas* genera. Principal Components Analysis (PCA) was found effective in elucidating trends of microbial degradation among the different soil types and seasons. The results of this study demonstrate the necessity of prompt analysis of forensic evidence for proper identification of possible ignitable liquids.

Keywords: Soil chemistry; Bacteria; Microbial degradation; Ignitable liquids

Introduction

The number of incendiary fires in the U.S. averages approximately 210,300 every year, which comprises about 13% of the total of all reported fires, according to FEMA's Topical Fire Report Series[1]. On an annual basis, incendiary fires claim 375 lives, injure over one thousand people, and cause approximately $1 billion in direct property damage [1]. In many cases, the arsonist uses an ignitable liquid to accelerate the fire. Gasoline is the most commonly used ignitable liquid as it is readily accessible, inexpensive and ignites easily [2]. Gasoline and other ignitable liquids are classified according to the American Society of Testing and Materials (ASTM) guidelines by their boiling point range and chemical composition [3]. In practice, a forensic chemist will use various extraction methods coupled with gas chromatography/mass spectrometry (GC/MS) to determine if an Ignitable Liquid Residue (ILR) is present in a fire debris sample. The ILR will then be classified according to ASTM guidelines [2,4].

Media rich in organic matter such as soil provides a rich source of carbon and typically contains substantial quantities of active bacterial biomass. Since ignitable liquids are composed of a range of hydrocarbons, they may be suitable as a carbon substrate by bacteria. Such transformations are problematic for fire debris analysis as samples are often stored for many weeks at room temperature before they are analyzed due to case backlog and lack of cold storage. As a result, selective loss of hydrocarbon species due to bacterial metabolism can occur, making the identification and classification of ignitable liquid residues difficult or even impossible. For example, five specific C_3-alkylbenzenes (3-ethyltoluene, 4-ethyltoluene, 1,3,5-trimethylbenzene, 2-ethyltoluene and 1,2,4-trimethylbenzene)must be identified in a sample in order to determine if residues of gasoline are present. Furthermore, because these compounds also occur in other materials, they must be present in relative amounts that are similar to that of a gasoline standard [3]. Among the serious consequences of microbial degradation are the selective losses of some of these compounds and/or changes in the ratios of these compounds in a gasoline sample.

Several factors affect bacterial numbers and activities in soil including soil type and season. Chemical and physical characteristics of soils including pH, nitrogen level and phosphorus content will vary, as do soil physical properties (e.g., texture). In turn, varying populations of bacteria may impact the degree of microbial degradation observed in fire debris samples containing soil.

Previous work has demonstrated that bacteria readily degrade normal alkanes (e.g. decane) and lesser substituted alkyl benzenes (e.g., toluene, ethyl benzene, propyl benzene) while more highly substituted alkyl benzenes (e.g., 1,2,4-trimethylbenzene) and highly branched alkanes are more resistant to microbial attack [4-6]. While treatment of hydrocarbon-contaminated soils by bacteria is a well-known phenomenon in the environmental engineering community [7-17], microbial processes are not wellunderstood in forensic science. This phenomenon likely varieswith soil type and over different seasons as soil chemical properties, temperature and moisture status may impact heterotrophic bacteria.

The overall objectives of this study were to assess the degradation of a common ignitable liquid (i.e., 87 octane gasoline) in soil as affected by soil type. The focus of this paper will be upon the effect of soil type, to include: (1) analysis of GC/MSdata from gasoline added to three different soils; (2) identification and quantification of bacterial populations present in the study soils; and (3) semi-quantification

---

***Corresponding author:** JV Goodpaster, Department of Chemistry and Chemical Biology, Forensic and Investigative Sciences Program, Indiana University Purdue University Indianapolis (IUPUI), Indianapolis, USA; E-mail: jvgoodpa@iupui.edu

of organic and inorganic compounds present in the study soils by Principal Component Analysis (PCA).

Materials and Methods

Soil chemical analyses

Soil material was obtained from an agricultural field (Pella clay), a residential property (Miamian sandy clay), and a brown field site (Urban land/Wawaka-Miami complex clay) in central Indiana. Soil material was collected from the surface 0-20 cm of each site using a stainless steel sampling probe. The soil was composited in the field, and air-dried and sieved (< 2 mm mesh) in the laboratory.

Particle size distribution of the soils was determined using the hydrometer method [18]. Total organic carbon (TOC) and total nitrogen (N) were analyzed on a Perkin Elmer Series II CHNS/O Analyzer 2400 (Shelton, CT). Acetanilide was the standard used. Soil pH was determined using a 1:2 (w:v) solids: deionized water slurry with an AB15 Accumet pH meter.

Soil nitrate (NO_3) concentrations were measured using Szechrome reagents [19] in a BioteK Power Wave XS2 micro assay system. Soil ammonium (NH_4) concentrations were determined by the method of Sims et al. which uses a modified indophenol blue technique [20]. The method was adapted for the BioteK Power Wave system. Soil extractable P was determined by the Bray-1 method [21]. Soil K was extracted with neutral 1.0 M ammonium acetate and analyzed using atomic emission spectrophotometry (Perkin Elmer A Analyst 2000). Extractable metal (Cd, Cr, Fe, Zn, Pb) concentrations were determined by extraction with 5 mM DTPA (diethylene triamine penta acetic acid) with 10 mM $CaCl_2$, pH adjusted to 7.3. Briefly, the method involved mechanical shaking (120 osc./min. for 2 h) of 5 g soil with 25 ml of 5 mM DTPA in acid-washed Nalgene® bottles. The suspension was filtered through Whatman No. 2 filter paper and analyzed for Cd, Cr, Fe, Zn and Pb using flame atomic absorption spectrophotometry (Perkin Elmer A Analyst 2000). For the above analyses, there were four replicates of each sample.

Soil Microbiological Analyses

Populations of total culturable bacteria were determined in each soil type using the standard plate count [22] on Plate Count Agar (Teknova, Hollister, CA).Soil borne actinomycetes were enumerated on Actinomycete Isolation Agar (Sigma-Aldrich, St. Louis, MO) and yeasts and molds were quantified using Sabouraud Dextrose agar (Fisher Scientific, Waltham, MA). Six replicates of each soil type colony counts were averaged following 48h incubation of all inoculated plates. Colony counts were assessed using exponential and log transformations via Sigma Stat 3.5 (Point Richmond, CA). Control and experimental groups were compared using a one-tailed Student's t test, and different media combinations were compared using one-way ANOVA (Minitab 16, State College, PA), followed by Student-Neuman-Keuls post hoc analysis and two-factor factorial analysis using SAS (SAS Institute, Cary, NC). Data were considered significantly different at $P< 0.05$.

For the genetic identification of bacteria, DNA was obtained from 3-5 g soil samples using a commercial system (MoBIO, Solana Beach, CA) and quantified spectrophotometrically. Real-time PCR was carried out in a SmartCyclerII (Cepheid,Sunnyvale, CA). Extracted DNA (1 µg) was added to real-time SYBR Green™ Supermix (Quanta Biosciences, Gaithersburg, MD); a no-template contamination control was analyzed for each sample/primer set, as well as positive control specimens consisting of genomic DNA from ATCC (Manassas, VA) type strains

or other reference strains of *Acinetobacter, Alcaligenes, Arthrobacter, Bacillus, Burkholderia,* and *Flavobacterium* (Table 3). All PCR primers were designed with the software analyses options available through the National Center for Biotechnology Information Basic Local Alignment Search Tool (NCBI BLAST) (www.ncbi.nlm.nih.gov/BLAST), which allows for sequences to be screened for nonspecific annealing frequencies and non-target homology determination. Internal standard primer targets in each case were the highly conserved prokaryotic gyrase subunit B gene, *gyrB* [23]. Each primer pair was tested on all non-target strains to ensure appropriate specificity and eliminate the appearance of false-positive amplification signal. Cycling conditions were 10 min. at 95°C, followed by 40 three-stepcycles of 15 s at 95°C , 1 min. at 55°C and 1 min at 72°C, with fluorescence acquisition monitored at the end of each cycle.In order to sensitively enumerate viable cell density, reverse transcriptase PCR was also subsequently performed on whole RNA extracted from 5 g soil of each treatment type using Trizol (Invitrogen, Grand Island, NY). RNA was standardized to 1 µg following DNAse-I treatment and subjected to cDNA synthesis and amplification using the qScript™ One-Step SYBR® Green qRT-PCR Kit (Quantas Biosciences) and genus-specific primers (Table 3). Viable cell densities were ascertained using the calculations described below for DNA targets and compared by soil type.

Standard curves to determine number of copies of target genomes (and mRNA)for each bacterial genus were constructed using quantified bacterial templates obtained from each reference strain 1:10 serially diluted in nuclease-free water to 10^{-6} (each diluted in triplicate) and subjected to amplification as described above. Bacterial template concentrations were converted to amplicon (PCR product) copies by multiplying the mean grams of DNA purified for each set of extraction replicates by 6.02×10^{23} and dividing that product by the product of the respective amplicon length in base pairs × 650 Daltons. Resulting plots depict the number of amplicon copies as a function of respective cycle threshold (Ct) values.

Microbial Degradation Studies

For each soil type, eightsample time points were prepared in triplicate by spiking 20 µL of commercial unleaded gasoline (87 octane) onto ~100 g soil in aclean, unused, but non-sterile quart-size paint can (i.e., real fire debris samples would not be collected in sterile paint cans). The samples were sealed and stored for 0, 2, 4, 7, 11, 15, 22, and 30 days. On each specified day, the samples were extracted using passive headspace adsorption-elution (a popular and widespread extraction technique for fire debris) [24]. In this method, one third (~7 × 9 mm²) of a charcoal strip (Albrayco Technologies, Cromwell, CT) was placed in each can and suspended in the headspace on a pre-baked (at 85°C) paper clip using nylon string. The re-sealed cans were heated at 85°C for 4 h. After cooling,the charcoal strips were removed and extracted with 400 µL of pentane with vortexing for ~1 min. Samples were then analyzed by GC-MS (Agilent 6890 GC with an Agilent 5975 MSD) using a standard method for fire debris analysis, which includes a 1 µL injection volume, 20:1 split ratio, inlet temperature of 250°C, flow rate of 1 mL/min (helium), a DB-5 30 m × 0.25 mm × 0.25 µm column, initial column temperature of 40°C held for 2 min, temperature ramp of 20°C/min, final column temperature of 280°C held for 3 min, solvent delay of 2 min, MS scan of 40-300 m/z, MS quad temperature of 150°C and an MS source temperature of 230°C [4-6].

Data Analysis

Each analyte (Table 2) was identified based on comparison to the retention time and mass spectrum of an authentic standard. A

Bacterial Genus	Agricultural	Residential	Brownfield
Acinetobacter	4.32×10^{16}	3.12×10^{16}	1.31×10^{17}
Alcaligenes	7.29×10^{19}	2.37×10^{20}	2.26×10^{18}
Arthrobacter	4.135×10^{10}	1.636×10^{13}	1.396×10^{12}
Bacillus	4.06×10^{14}	5.12×10^{14}	3.36×10^{14}
Flavobacterium	1.8×10^{16}	4.53×10^{10}	4.21×10^{11}
Pseudomonas	1.624×10^{14}	1.682×10^{14}	2.56×10^{15}

Table 1: qPCR-based determination of genome copies per gram of soil from each bacterial genus in this study. Values are the mean value from triplicate samples analyzed using SYBR Green-based standard curves as described in Materials and Methods.

Media	Agricultural	Residential	Brownfield
Plate Count Agar	5.02×10^5 (1.2×10^2)	2.54×10^5 (4.5×10^2)	6.86×10^5(3.2×10^2)
ACT	4.62×10^6 (8.2×10^2)	7.1×10^5 (2.2×10^2)	1.38×10^6(3.2×10^2)
SDA	2.28×10^5(1.93×10^2)	4.48×10^4(8.94×10^2)	1.24×10^5(7.54×10^2)

Table 2: Microbiological plating-based results.
Values shown represent mean colony counts of eight replicates, which were not significantly ($p > 0.05$) different across the sampling times (Fall, Winter, Spring, & Summer). Plate count agar for total chemoheterotrophic bacteria; ACT = actinomycete agar, for soilborne actinomycetes; SDA = Sabouraud dextrose agar for total molds and yeasts. Standard deviations are in parentheses.

comparison was also made to the National Institute of Standards and Technology (NIST) mass spectral database. The peak areas from the summed extracted ion profiles (alkane: m/z 57, 71, 85, 99; aromatic: m/z 91, 120; benzaldehyde: m/z 77, 106) were exported into Microsoft Excel from the Xcalibur data analysis software (Thermo Scientific, Hanover Park, IL). Extracted ion profiles are employed in fire debris analysis in order to filter out interfering signals that could otherwise impede the classification of the ignitable liquid. The peak areas of the compounds listed in Table 1were normalized and then auto scaled. Normalization corrects for differences in overall instrument response and auto scaling allows the variance for each variable to be weighted equally [25]. XLSTAT (Addin Soft), an add-in for Microsoft Excel, was used to run Principal Component Analysis (PCA) on the autoscaled data for each soil type. PCA is a data reduction technique that allows for the visualization of samples in a two-dimensional plot despite the fact that the samples are described by many variables. For example, PCA has been used to discern differences in the relative chemical composition of samples that underwent evaporation versus microbial degradation [6].

Results and Discussion

Soil chemical analyses

The results of the soil chemical analyses are summarized in Table 2. All soil samples contained high percentages of clay (29.8 to 53.9%). Soil textures ranged from sandy clay to clay. These textures are typical for much of the northern two-thirds of the state of Indiana, which is overlain by substantial deposits of till from the Wisconsin glacial epoch [26].

Soil pH ranged from 6.3 (residential) to 6.6 (agricultural and brownfield). Total soil nitrogen (N) ranged from 0.23 mg/kg (residential and brownfield) to 0.44 mg/kg (agricultural) (Table 1). The brownfield soil had the lowest level of nitrate (NO_3) at 26.3 mg/kg, whereas the agricultural soil had the highest level (60.9 mg/kg). Soil ammonium (NH_4) levels were similar across soil type, ranging from 1.9 mg/kg (brownfield) to 3.0 mg/kg (residential). Soil TOC was similar across treatments with values ranging from 0.9 to 1.0 %. The residential soil

contained high levels of phosphorous (P) and potassium (K) due to treatment with commercial fertilizers.

Levels of extractable Cd, Cr, Fe and Zn were all within range for non-contaminated soils (Table 1). However, extractable Pb levels in the brownfield soil measured 497 mg/kg. An upper limit for Pb content of a normal soil is approximately 70 mg/kg [27]. The levels of Pb in soils that are toxic to soil microorganisms and plants are a function of species, Pb concentration and soil factors (e.g., pH, fertility status, presence of other toxins); thus, threshold toxicity levels will vary. Soil Pb levels considered toxic to biota have ranged from 100 to several thousand mg/kg [28,29]. In addition to the high Pb concentration, the brown field soil had a "massive" structure (*i.e.*, it was highly compacted and did not allow water to infiltrate or to percolate). Such a structure is detrimental to plant growth and may inhibit microbial growth and activity, due to lack of both moisture and O_2. The other two soils had a granular structure, which is much more conducive to water (and air) movement.

Soil microbiological analyses

Populations of recoverable aerobic chemoheterotrophic bacteria were analyzed from each soil treatment using Standard Plate Count on Plate Count Agar. Table 2 reveals the average density was not significantly different ($p > 0.05$) across soil types at 3.8×10^5 cfu/g. Likewise, detectable actinomycetes remained stable as well (average = 6.9×10^5 cfu/g, and were not significantly different for any soil treatment ($p > 0.05$). Total recoverable fungal counts similarly revealed no significant differences among treated soils(1.1×10^5 cfu/g). However, since culture-based methods reveal only a subset of a microbial population in any given sample, PCR was used to determine comparative levels of selected bacterial genera across soil types in order to deduce what roles if any each genus may play in degradation of the ignitable liquid used here.

rDNA-based PCR detection was used to quantify total genome equivalents for representative bacterial genera demonstrating a previous history in the literature of chemical adulterant metabolism in soil environments [30-34]. Specifically, *Acinetobacter, Alcaligenes, Arthrobacter, Bacillus, Flavobacterium*, and *Pseudomonas* genera were

Strain designation	5'→3' primer sequences[1]	Reference
Acinetobacter calcoaceticus 346[2]	TAC GCA GGG TAA TGA ATC AA TCC GTG TCT CAG TAC CAG TG	Chang et al., 2005
Alcaligenes faecalis subsp. *faecalis* ATCC 8750[3]	CAT CCC GCG GTG TAT GAT GAA TCT GAC ATA CTC TAG CTC GG	Phung et al., 2012
Arthrobacter globiformis 607[2]	GTC GCG TCT GCT GTG AAA GC TTT AGC CTT GCG GCC GTA CT	Crocker et al., 2000
Bacillus cereus ATCC 14579[3]	AGA GTT TGA TCC TGG CTC AG TAC GGC TAC CTT GTT ACG ACT T	Bavykin et al., 2004
Flavobacterium capsulatum 315[2]	TAC TCG CAG AAT AAG CAC CG GTA TCT AAG TTC CCG AAG GC	GenBank Accession M59296
Pseudomonas fluorescens 13525	GGTCTGAGAGGATGATCAGT TTAGCTCCACCTCGCGGC	Widmer et al., 1998

[1]Top sequence given for each species = forward primer; bottom sequence = reverse
[2]Reference strain obtained from Presque Isle Cultures, Erie, PA
[3]Reference strain obtained from American Type Culture Collection, Manassas, VA

Table 3: ATCC reference strains and PCR primers used in the rDNA-based quantification aspect of this study.

Bacterial Genus	Agricultural	Residential	Brownfield
Acinetobacter	5.30×10^1	1.16×10^2	1.27×10^0
Alcaligenes	1.43×10^7	5.56×10^{13}	8.23×10^4
Arthrobacter	6.84×10^7	5.32×10^{12}	4.00×10^8
Bacillus	1.80×10^{13}	3.06×10^{17}	2.61×10^{10}
Flavobacterium	3.35×10^{25}	7.68×10^{11}	1.72×10^3
Pseudomonas	None detected	3.00×10^5	2.37×10^5

Table 4: qRT-PCR based determination of specific mRNA transcript copies per gram of soil to ascertain viable cell densities from each bacterial genus in this study according to treated soil type. Values are the mean value from triplicate samples analyzed using SYBR Green-based standard curves as described in Materials and Methods.

analyzed here by qPCR using genus-specific PCR primers and standard curves generated with ATCC type strain DNA (Table 3). The calculated total genome equivalent of each bacterial genus is shown in Table 3. These qPCR results reveal that in all soil treatments, *Alcaligenes* spp. was consistently detectable at significantly higher levels ($p < 0.05$) than any other genus. *A. faecalis* has been reported to degrade the chlorinated insecticide endosulfan, found routinely in many soil types, water, and as residue on foods due to its widespread use in, and rapid transport through, the natural environment [35]. Other species of this genus have been documented to degrade PCB [36,37], whereas *A. faecalis*, the same species detected in our study, has been reported to metabolize the organo-chlorine insecticide endosulfan [38]. However, studies such as the ones cited are frequently performed in sterile soil spiked with known strains of interest rather than assessing naturally-occurring strains as was done herein. Nevertheless, such work reinforces the notion that the success of any bioremediation approach is directly related to the quality and diversity of competing species inhabiting the soil community, among other factors. Our selection of these six genera was based on dominant populations seen in soils of this region from past work in our laboratory. The possibility exists that levels of this and perhaps other species of *Alcaligenes* are present in the soils analyzed in our study due to some effect by a ubiquitous chlorinated derivative or similar chemical adulterant. Calculated genome copies of each bacterial group using rDNA-specific PCR revealed densities many orders of magnitude above recovered bacterial densities on plate count agar. This is attributed to the fact that our DNA-based PCR is detecting template copies from both viable bacterial targets and from dead cells accumulated in the soil biomass. However, these numbers are nevertheless still quite revealing on relative levels of each genus across soil types, and will be even more interesting in further examination of these soils when compared seasonally.

It is well known that DNA-based PCR detects both dead and viable bacteria, however, so in order to more accurately ascertain levels of only viable bacterial genera playing an active role in biodegradation, we targeted mRNA and quantified only bacteria actively transcribing their respective rDNA genes (Table 3) [39]. *Alcaligenes* appears at the highest viable density in residential soil, and the lowest in Brownfield soil. Moreover, in the DNA-based PCR detection assays, *Alcaligenes* spp. was among the most prevalent genera, suggesting that these bacteria do represent a substantial member of the microbial community in all soil types analyzed here. However, even in the more industrially contaminated soil type, 80,000 copies/g of *Alcaligenes* were detected suggesting these bacteria persist using an as-yet-undetermined physiological mechanism in the presence of lower NO_3 and higher Pb levels, and/or in the presence of chlorinated adulterants as noted above, an area of future research interest.

Based on DNA PCR results (Table 1), residential soil exhibited the largest variation in levels of bacterial genera under study compared to

other soil types, as well as the highest levels of bacteria overall. However, qRT-PCR results (Table 4) speak to a slightly different proportion of viable bacteria. *Pseudomonas* spp. were detected at the lowest levels in residential soil (3.00×10^5/g) while the spore-forming *Bacillus* spp and *Alcaligenes* were at the highest viable density (3.06×10^{17} and 5.56×10^{13}, respectively).

Microbial degradation studies

Microbial degradation of gasoline was observed in the residential, agricultural, and brownfield soils (Figure 1 and 2). In all three soil types, n-alkanes were degraded in a similar fashion in that degradation is almost complete after 7 d. In fact, no peaks remained in the chromatograms by 15 d except those attributed to volatile aldehydes that are present in the headspace of all soil samples (Figure 1). In contrast, we noted differences in the ratios of the C_3-alkylbenzenes depending upon soil type (Figure 2). For example, all profiles appear nearly identical on day 0, but on day 2 propylbenzene (peak 1) is significantly reduced in

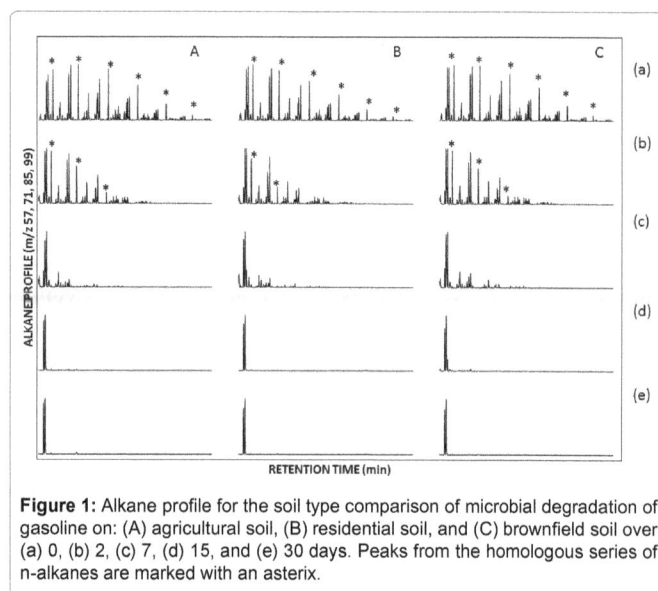

Figure 1: Alkane profile for the soil type comparison of microbial degradation of gasoline on: (A) agricultural soil, (B) residential soil, and (C) brownfield soil over (a) 0, (b) 2, (c) 7, (d) 15, and (e) 30 days. Peaks from the homologous series of n-alkanes are marked with an asterix.

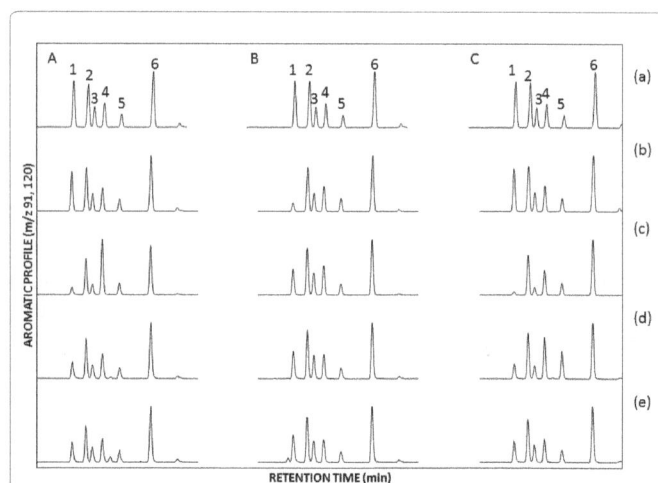

Figure 2: Aromatic profile for the soil type comparison of microbial degradation of gasoline on: (A) agricultural soil, (B) residential soil, and (C) brownfield soil over (a) 0, (b) 2, (c) 7, (d) 15, and (e) 30 days. Peaks: (1) propylbenzene, (2) 3-ethyltoluene, (3) 4-ethyltoluene, (4) 1,3,5-trimethylbenzene, (5) 2- ethyltoluene, and (6) 1,2,4-trimethylbenzene.

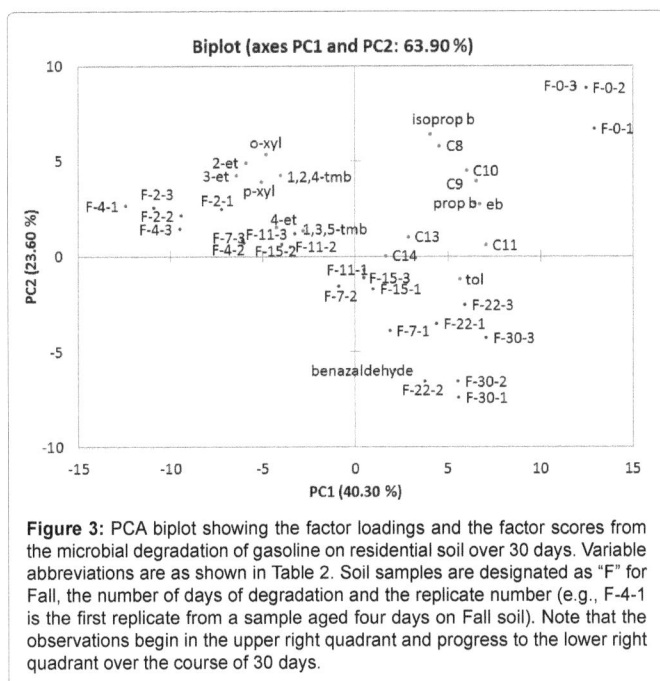

Figure 3: PCA biplot showing the factor loadings and the factor scores from the microbial degradation of gasoline on residential soil over 30 days. Variable abbreviations are as shown in Table 2. Soil samples are designated as "F" for Fall, the number of days of degradation and the replicate number (e.g., F-4-1 is the first replicate from a sample aged four days on Fall soil). Note that the observations begin in the upper right quadrant and progress to the lower right quadrant over the course of 30 days.

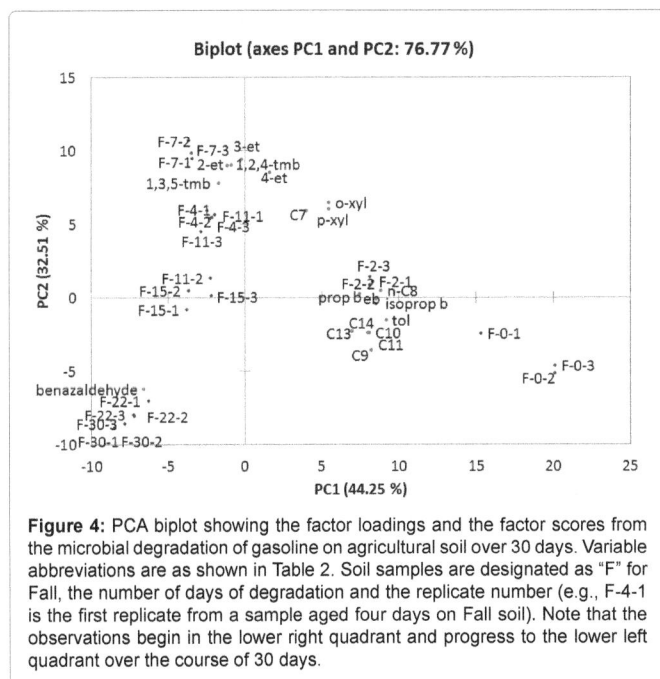

Figure 4: PCA biplot showing the factor loadings and the factor scores from the microbial degradation of gasoline on agricultural soil over 30 days. Variable abbreviations are as shown in Table 2. Soil samples are designated as "F" for Fall, the number of days of degradation and the replicate number (e.g., F-4-1 is the first replicate from a sample aged four days on Fall soil). Note that the observations begin in the lower right quadrant and progress to the lower left quadrant over the course of 30 days.

in residential or industrial soil. This peak has been tentatively identified as a bicyclic hydrocarbon based on the mass spectral library. Gasoline contains a variety of hydrocarbons and could very likely contain this compound in trace amounts which would not be detected in a fresh or even slightly degraded sample. Additionally, the variability of degradation might explain why this minor peak is not detected in the significantly degraded residential and industrial soil samples.

These trends may be the result of the higher levels of nutrients in the residential and agricultural soils - both soils contained higher NO_3, NH_4 and K concentrations compared with the brownfield soil (Table 1). Furthermore, the residential soil contained more than twice the extractable P compared with the brownfield soil (154 versus 74 mg/kg, respectively).Another factor may be the Pb concentration in the brownfield soil (497 mg/kg), which may have impairedthe activity of heterotrophic bacteria. A number of researchers have determined a direct relationship between concentrations of soil Pb and microbial activity in soil [40-42]. Shi et al. found that soil Pb decreased microbial activities and led to accumulation of soil organic carbon [43]. Furthermore, Pb was found to pose a greater stress to soil microbes than did other heavy metals. Application of Pb at concentrations of >500 mg kg^{-1} caused an immediate and significant decline in microbial biomass [41].

Zeng et al. [40] state that soil Pb concentrations >500 mg Pb/kg may be a "critical concentration", causing a significant decline in soil microbial activity. Therefore, nutrient content is most likely responsible for significant degradation in the residential and agricultural soil sampling, while high lead content is likely responsible for the reduction in microbial degradation of gasoline in the brownfield soil sampling.

Principal Components Analysis (PCA) was then used to calculate a set of latent variables that would better explain the overall variance in the data set. Figures 3-5 shows the factor loadings and the factor scores (known as a "bi-plot") from PCA of the fall sampling from each soil type. This biplot shows both the projections of the data in the new factor space and the projections of the original variables. By examining the relative locations of the observations and variables over time, we can determine what variables are more important for a given data point.

For example, in the residential soil at day 0, the major contributors to the relative composition of the gasoline samples are the normal alkanes and the mono-substituted alkyl benzenes (Figure 3). As degradation proceeds, however, the major contributors to the relative composition of the gasoline samples are the xylenes, ethyltoluenes, and trimethyl benzenes. By days 22 and 30, all of these compounds are completely degraded and the only compound strongly associated with the gasoline samples is benzaldehyde, which is a suspected degradation product of toluene [44].

The overall degradation rate appears to be slower in the agricultural soil (Figure 4) as the n-alkanes and the mono-substituted alkyl benzenes are still major contributors in the day 2 samples. Recall that in the residential soil the major contributors in the day 2 samples were the xylenes, ethyltoluenes, and trimethyl benzenes. However, in the end the agricultural soil samples are still significantly degraded by 22 days when the gasoline samples are only associated with benzaldehyde.

The brownfield soil (Figure 5) follows a similar trend to the agricultural soil except the day 22 and 30 samples are largely associated with toluene as well as benzaldehyde. Thisindicates that the brownfield soil was less active than the other soil samples, as toluene is the most abundant compound in gasoline and is one of the first compounds to be significantly decomposed by soil bacteria.

comparison to 3-ethyltoluene (peak 2) in the residential soil whereas in the agricultural and brownfield soils only minimal reduction is apparent. By 30 d the gasoline in residential and agricultural soils experienced the greatest microbial degradation while the gasoline in the brownfield soil experienced the least.It is important to note that in all chromatograms, the data is normalized to the most abundant peak in the chromatogram so that overall changes can be observed relative to the day 0 samples due to the lack of an appropriate internal standard. Therefore it may appear that some peaks are increasing in abundance, due to the changes in ratios of these compounds over time. Also, a minor peak appears between peaks 4 and 5 in the chromatograms after 15 and 30 days on agricultural soil that does not appear to be present

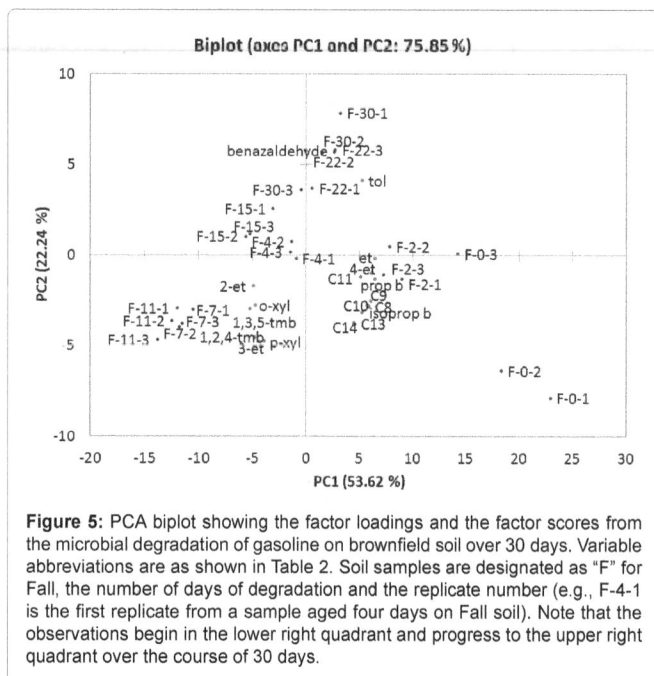

Figure 5: PCA biplot showing the factor loadings and the factor scores from the microbial degradation of gasoline on brownfield soil over 30 days. Variable abbreviations are as shown in Table 2. Soil samples are designated as "F" for Fall, the number of days of degradation and the replicate number (e.g., F-4-1 is the first replicate from a sample aged four days on Fall soil). Note that the observations begin in the lower right quadrant and progress to the upper right quadrant over the course of 30 days.

Conclusions

Our results demonstrate the necessity for prompt analysis of forensic evidence from fire scenes in order to properly identify ignitable liquids. In this study, microbial degradation was apparent in all soil samples. However, ignitable liquid residues in the brownfield soil suffered the least while ILR in the residential soil suffered the most. It was expected that the residential soil would suffer the most degradation as the bacteria counts were highest in this soil type. It was not anticipated that the brownfield soil would show the least microbial degradation, although, the increased levels of heavy metals in this soil (lead in particular) likely had a toxic effect on the soil bacteria, as suggested by the RNA-based quantification experiments. *Alcaligenes* and *Pseudomonas* spp., widely recognized as universal soil borne microbes capable of degrading many chemical adulterants, were not major metabolic participants in this study, although future work that specifically detects activity of genes encoding known factors for biodegradation would reveal the potential for these genera in these soils.

Acknowledgement

The authors would like to thank Mark Ahonen and Kathy Boone of the Microanalysis Unit of the Indiana State Police Laboratory for the idea behind this project and for their continued feedback on this work. Financial support for this work originated from the Research Support Funds Grant at IUPUI and the National Institute of Justice, Office of Justice Programs, U.S. Department of Justice (Award No. 2010-DN-BX-K036). The opinions, findings, and conclusions or recommendations expressed in this publication are those of the authors and do not necessarily reflect those of the U.S. Department of Justice.

Reference

1. Intentionally Set Fires, in Topical Fire Report Series, U.S.F. Administration, Federal Emergency Management Agency (FEMA), U.S. Department of Homeland Security, USA.

2. Stauffer, E., J.A. Dolan, and R. Newman, Fire Debris Analysis 2008, Burlington: Elsevier Inc.

3. Standard Test Method for Ignitable Liquid Residues in Extracts from Fire Debris Samples by Gas Chromatography-Mass Spectrometry. American Society for Testing and Materials 14: 1-11.

4. Turner DA, Goodpaster JV (2009) The effects of microbial degradation on ignitable liquids. Anal Bioanal Chem 394: 363-371.

5. Turner DA, Goodpaster JV (2011) The effect of microbial degradation on the chromatographic profiles of tiki torch fuel, lamp oil, and turpentine. J Forensic Sci 56: 984-987.

6. Turner DA, Goodpaster JV (2012) Comparing the effects of weathering and microbial degradation on gasoline using principal components analysis. J Forensic Sci 57: 64-69.

7. Crawford RL, DL (1996) Crawford, eds. Bioremediation: Principles and Applications. Biotechnology Research Series, ed. J. Lynch, Press Syndicate of the University of Cambridge, Great Britain.

8. Heipieper HJ (2007) Bioremediation of Soils Contaminated with Aromatic Compounds. Earth and Environmental Sciences, ed. NS Series, Springer: Dordrecht, The Netherlands.

9. Hernandez BS, et al. (1997) Terpene-utilizing isolates and their relevance to enhanced biotransformation of polychlorinated biphenyls in soil. Biodegradation 8: 153-158.

10. Huang H, S Larter (2005) eds. Biodegradation of petroleum in subsurface geological reservoirs. Petroleum Microbiology, ed. B Ollivier, M Magot, ASM: Washington, USA.

11. Magot M (2005) Petroleum Microbiology. Indigenous microbial communities in oil fields, ed. B. Ollivier and M. Magot, ASM: Washington DC, USA.

12. Mars AE, Gorissen JP, van den Beld I, Eggink G (2001) Bioconversion of limonene to increased concentrations of perillic acid by Pseudomonas putida GS1 in a fed-batch reactor. Appl Microbiol Biotechnol 56: 101-107.

13. McLoughlin E, Rhodes AH, Owen SM, Semple KT (2009) Biogenic volatile organic compounds as a potential stimulator for organic contaminant degradation by soil microorganisms. Environ Pollut 157: 86-94.

14. Misra G, SG Pavlostathis (1997) Biodegradation kinetics of monoterpenes in liquid and soil-slurry systems. Appl Microbiol Biotechnol 47: 572-577.

15. Sasek V, JA Glaser, P Baveye (2003) The Utilization of Bioremediation to Reduce Soil Contamination. Problems and Solutions. Earth and Environmental Sciences, ed. N.S. Series, Kluwer Academic Publishers: Dordrecht, The Netherlands.

16. Singh A, OP Ward (2004) Biodegradation and Bioremediation. Soil Biology, ed. A Varma, Springer-Verlag: Berlin, Germany.

17. Van Agteren MH, Keuning S, DB Janssen (1998) Handbook on Biodegradation and Biological Treatment of Hazardous Organic Compounds. Kluwer Academic Publishers, The Netherlands.

18. Day PR (1965) Particle Fractionation and Particle-Size Analysis, in Methods of Soil Analysis, Part 1, C.A. Black, Ed, American Society of Agronomy: Madison, USA.

19. Szechrome Reagents (2009) Technical Data Sheet 239, Polysciences Inc, Warrington, USA.

20. Sims GK, TR Ellsworth, RL Mulvaney (1995) Microscale determination of inorganic nitrogen in water and soil extracts. Communications in Soil Science and Plant Analysis 26: 303-316.

21. Olsen SR, Sommers LE (1982) Phosphorus, in Methods of Soil Analysis, Part 2, A.L. Page, R.L. Miller, D.R. Keeney. American Society of Agronomy: USA.

22. Cappuccino JG, N Sherman (1998) Experiment 53: Microbial populations in soil: Enumeration, in Microbiology: A Laboratory Manual, Addison Wesley Longman: New York, USA.

23. Watanabe K (2001) Microorganisms relevant to bioremediation. Curr Opin Biotechnol 12: 237-241.

24. Standard Practice for Separation and Concentration of Flammable or Combustible Liquid Residues from Fire Debris Samples by Passive Headspace Concentration (1995) Annual Book of ASTM Standards 14.02: 897. USA.

25. Brereton RG (2007) Applied Chemometrics for Scientists. John Wiley & Sons Ltd, West Sussex, England.

26. Service NRC (2012) Soil Survey of Delaware County, Indiana, USA.

27. Pichtel J (2007) Fundamentals of Site Remediation. 2nd ed, Boca Raton, FL: ABC Groups, USA.

28. Kabata-Pendias A (2001) Trace Elements in Soil and Plants. Third ed, Boca Raton, FL: CRC Press, USA.

29. Pichtel J, Salt CA (1998) Vegetative Growth and Trace Metal Accumulation on Metalliferous Wastes. Journal of Environmental Quality 27: 618-624.

30. Vanbroekhoven K, Ryngaert A, Wattiau P, Mot R, Springael D (2004) Acinetobacter diversity in environmental samples assessed by 16S rRNA gene PCR-DGGE fingerprinting. FEMS Microbiol Ecol 50: 37-50.

31. Menn FM, Easter JP, Sayler GS (2008) Genetically Engineered Microorganisms and Bioremediation, in Biotechnology: Environmental Processes, Wiley, USA.

32. Westerberg K, Elväng AM, Stackebrandt E, Jansson JK (2000) Arthrobacter chlorophenolicus sp. nov., a new species capable of degrading high concentrations of 4-chlorophenol. Int J Syst Evol Microbiol 50 Pt 6: 2083-2092.

33. O'Loughlin EJ, Sims GK, Traina SJ (1999) Biodegradation of 2-methyl, 2-ethyl, and 2-hydroxypyridine by an Arthrobacter sp. isolated from subsurface sediment. Biodegradation 10: 93-104.

34. McSpadden Gardener BB (2004) Ecology of Bacillus and Paenibacillus spp. in Agricultural Systems. Phytopathology 94: 1252-1258.

35. Kong L, Zhu S, Zhu L, Xie H, Wei K, et al. (2013) Colonization of Alcaligenes faecalis strain JBW4 in natural soils and its detoxification of endosulfan. Appl Microbiol Biotechnol .

36. Haluska L, Barancíková G, Baláz S, Dercová K, Vrana B, et al. (1995) Degradation of PCB in different soils by inoculated Alcaligenes xylosoxidans. Sci Total Environ 175: 275-285.

37. Aráoz B, Viale AA (2004) Microbial dehalogenation of polychlorinated biphenyls in aerobic conditions. Rev Argent Microbiol 36: 47-51.

38. Kong L, Zhu S, Zhu L, Xie H, Wei K, et al. (2013) Colonization of Alcaligenes faecalis strain JBW4 in natural soils and its detoxification of endosulfan. Appl Microbiol Biotechnol .

39. Widmer F, Seidler RJ, Gillevet PM, Watrud LS, Di Giovanni GD (1998) A highly selective PCR protocol for detecting 16S rRNA genes of the genus Pseudomonas (sensu stricto) in environmental samples. Appl Environ Microbiol 64: 2545-2553.

40. Zeng LS, Liao M, Chen CL, Huang CY (2006) Effects of lead contamination on soil microbial activity and rice physiological indices in soil-Pb-rice (Oryza sativa L.) system. Chemosphere 65: 567-574.

41. Liao M, Chen CL, Zeng LS, Huang CY (2007) Influence of lead acetate on soil microbial biomass and community structure in two different soils with the growth of Chinese cabbage (Brassica chinensis). Chemosphere 66: 1197-1205.

42. Konopka A, Zakharova T, Bischoff M, Oliver L, Nakatsu C, et al. (1999) Microbial biomass and activity in lead-contaminated soil Appl Environ Microbiol 65: 2256-2259.

43. Shi W, et al. (2002) Long-term effects of chromium and lead upon the activity of soil microbial communities. Applied Soil Ecology 21: 169-177.

44. Farrell RL, Rhodes PL, Aislabie J (2003) Toluene-degrading Antarctic Pseudomonas strains from fuel-contaminated soil. Biochem Biophys Res Commun 312: 235-240.

8

Biodegradation of Coir Geotextiles Attached Media in Aerobic Biological Wastewater Treatment

Gopan Mukkulath* and Santosh G. Thampi

Department of civil engineering, National Institute of Technology Calicut, Calicut-673 601, Kerala, India

Abstract

Coir is a hard and tough organic fibre, rich in cellulose and lignin, and possesses high specific area and wetting ability - factors which are important for bacterial adhesion in fixed film processes. Few studies have been reported on the durability of coir geotextiles, when used in Civil Engineering applications such as slope protection works. However, its durability when used as the media in biofilters, for wastewater treatment extensively has not been studied yet. Durability of coir geotextiles has been assessed by keeping it in wet condition and measuring the tensile strength of the fibres, at different times. In the present study, coir geotextiles that were used as the attachment media in a biofilter, for treatment of organic rich wastewater was tested for durability. The units were operated as aerobic, gravity flow filters. Different wastewater parameters were employed in each of these runs. Tensile strength of the fibres was determined after each run. Results show that loss of tensile strength (or degradation) is influenced by the characteristics of the effluent wastewater to the filter, due to the growth of microorganisms which depends upon the concentration of substrates in the wastewater. Permeability studies shows reduction in permeability, is influenced by the organic loading rate on the filters and the duration of filter operation. SEM studies and E-DAX analysis shows the biodegradation of the fibres constituents, influenced by the substrate concentration.

Keywords: Biofilters; Coir geotextiles; Permittivity; Tensile strength

Introduction

Coir fibre contains more lignin than all other natural fibres, such as jute, flax, linen, cotton etc. It has a lignin content of 45.84%, which makes it as the strongest of all known natural fibres. As other natural fibres like jute, sisal etc. has much less lignin content, they are degraded faster than coir. Studies conducted under similar environments, jute geotextile is degraded within one year whereas a coir geotextile degraded after 8-10 years for slope protection works [1]. The feasibility of using of coir geotextiles, made out of natural material, as the medium in filters for treatment of wastewater has not been studied extensively yet. Filters employing coir geotextile media exhibit high organic and nutrient removal rates, compared to filters using plastic or sand as the medium [2]. The major drawback of using coir geotextiles media in biofilters may be its degradation and subsequent clogging of the filter. Media degradation may be attributed to microorganisms feeding on it, rather than on the substrates in the effluent wastewater [3].

Subaida has conducted experimental studies on tensile strength and pull out studies on various types of coir geotextiles. Reported works pertaining to the durability of coir geotextiles are limited in number. Schurholz [4] found that coir geotexlies retained 20% of their original tensile strength, after one year in incubator tests with highly fertile soil. It was also observed that coir geotextiles exhibited no damage, even after being put in a shower room and kept wet for 167 days in conditions to simulate flooding. Sarsby et al. [5], reported about 20% strength reduction of coir ropes embedded in a soil with pH 8.7 after 10 months. Reduction in strength of the coir ropes was attributed to the alkaline environment, rather than the moisture present in the soil. Rao and Balan [6], conducted studies using coir yarns of retted and unretted types embedded in different soil media, and water at a temperature of 30 ± 1°C and a relative humidity of 90 ± 1 %. The study showed that major strength loss takes place in 4 to 8 months. The loss of strength is more for sand with some organic content, compared to clay with high organic content owing to less organic content in sand. In the present study, tests were conducted on coir geotextiles that were used as the attachment media in a biofilter for treatment of organic rich wastewater, to assess its durability.

Material and Methods

Geotextiles are relatively thick planar sheets which may be woven, non-woven or knitted. These are capable of transmitting fluids across or in-plane or in both ways, and retain suspended particles. Woven geotextiles are manufactured by interlacing yarns, usually at right angles, whereas non-woven geotextiles are made by mechanical, heat or chemical bonding of directional or randomly oriented yarns [1]. The coir geotextiles used for the study is supplied from Central Coir Research Institute in Alleppey, Kerala India. Salient properties of commonly used coir geotextiles are listed (Table 1). Coir mesh matting of different mesh sizes is most established coir geotextiles. Mesh matting having different specifications is available under quality code numbers H2M1 to H2M10. These qualities represent coir textiles of different mesh sizes ranging from "1/8 to 1". Another classification is based on the place of manufacture. For example: MMA1 stands for Mesh Matting Anjengo Grade 1. Woven coir geotextiles (H2M2) was used as the media in the experimental setup, and these were later subjected to tests for assessing its durability.

The experimental setup of the biofilter was made of a PVC pipe of diameter 20cm and length 90cm, with the coir geotextile media packed inside; the packing density is estimated at 71.13 kg/m^3 and porosity is 84.4%. Wastewater prepared in the laboratory was used, so that the influent characteristics would remain constant throughout the filter operation [7,8]. The filter was operated as an aerobic, down flow filter

*Corresponding author: Gopan Mukkulath, Department of civil engineering, National Institute of Technology Calicut, India, E-mail: gopan@nitc.ac.in

Designation	Type of warp yarn	Appro. Scorage of warp yarn	Ends per dm	Type of weft yarn	Picks per dm	Mass kg/m 2	Mesh opening mm x mm
MMA1(H2M1)	Anjengo	14	9	Vycome	8	0.650	9x12
MMA2(H2M 4)	Anjengo	12	19	Aratory	11	1.400	1x2
MMA3(H2M8)	Anjengo	12	11	Aratory	7	0.700	7x10
MMA4(H2M9)	Anjengo	11	13	Aratory	7	0.900	5x10
MMA5(H2M10)	Anjengo	11	18	Anjengo	9	1.300	4X10
MMR1(H2M3)	Aratory	15	14	Aratory	14	0.875	2x2
MMV1(H2M5)	Vycome	13	9	Vycome	8	0.740	9x9
MMV2(H2M6)	Vycome	12	4.6	Vycome	4	0.400	20x20
MMB1(H2M2)	Beach	9	8	Beach	7	0.700	10x10
MMY1(H2M7)	Beypore	-	4	Beypore	6	1.250	15x15

Table 1: Properties of Coir Geotextiles [6]

with continuous flow operating conditions. Six runs of duration 50 ± 5 days were performed with organic rich wastewaters having different characteristics, maintaining a constant hydraulic loading rate of 200 l/m²/day [9].

Bacterial pure culture for Fecal Coliforms was prepared in the laboratory .Commercially available M 7 hr FC Agar was used for the preparation of the media for fecal coliforms. A biological contaminated water sample was inoculated with the prepared media, for initiation of bacteriological growth. The fecal coliform media was kept for incubation for 7-9 hrs at 41.5°C. Fecal Coliforms appeared as bright yellow colonies. After incubation, the colonies developed on the media were transferred using nickel alloy wire loops, and they were cultured again for the development of pure culture. These pure cultures are preserved in nutrient agar slants. The pure culture stored in nutrient agar media is streaked into the broth media. This media is incubated for the development of biological organisms. After incubation, these organisms were transferred to the filter media using nickel alloy wire loops and kept for 15 days, for maximum biological organism growth [10].

Tensile strength

The tensile strength of the coir geotextiles used in the biofilter was determined, following the procedure outlined IS: 13162 (Part 5) [11] and ASTM D 4595 [12]. This standard IS: 13162 (Part 5) [11] prescribes an index test method for determination of the tensile properties of geotextiles and related products, using a wide width strip. A test specimen is held across its entire width, in the jaws of a tensile testing machine operated at a given rate of strain, applying a longitudinal force to the test specimen until the specimen ruptures. The tensile properties of the test specimen are calculated from machine scales, dials, autographic recording charts, or an interfaced computer. A Constant Rate of Extension (CRE) tensile testing machine, as specified in IS 1969 [13] was used. Jaws were sufficiently wide to hold the entire width of the specimen, and with appropriate means to limit slippage or damage. Compressive grips were used for the gripping of coir geotextiles.

The test specimen was 200mm in width, which contained at least 5 nos of selected Coir woven Geotextile elements with the width, and at least one row of the nodes or cross-members excluding the nodes or cross-members by which the test specimen was held in the jaws (Figure 1). The preload of 1% of the maximum load or 350 N (whichever is less) is prescribed for extension at preload. The test specimen is 200mm wide with at least 5 complete tensile elements; the gauge length is 20cm. Strain was applied at a rate of 20 ± 5% per minute. The tensile strength of coir geotextiles is calculated by the formula given by Eq. (1)

$$a_f = F_t \, x \, C \tag{1}$$

$a_f = tensile \ strength \ in \ kN$

$F_t = observed \ \max imum \ Load \ in \ kN$

$$C = \frac{N_m}{N_s} \tag{2}$$

$N_m = the \ \min imum \ no \ of \ tensile \ elements \ in \ 1m \ width \ of \ Coir \ Geotextile$

$N_S = the \ no \ of \ tensile \ elements \ within \ the \ test \ specimen (Coir \ Geotextile)$

Permeability

Permeability, k of Geotextiles indicates the rate at which the flow of liquid occurs under a differential pressure. Permittivity, Φ is the term used to denote volumetric flow rate of water in the normal direction per unit cross-sectional area and per unit head difference. The relationship between permeability and permittivity can be expressed as (Eq. 3) [14].

$$\varphi = \frac{k}{d} \tag{3}$$

$k = permeability \ in \ m \, / \, s$

$d = thickness \ of \ geotextiles \ used \ in \ m$

$\varphi = permittivity \ in \ s^{-1}$

Permeability tests were carried out using a constant head permeameter. The specimens were as specified in IS: 14324-1995 [14] with a diameter of 73 mm; a constant head of 50cm was maintained. The quantity of flow was measured versus time. The constant head test was used when the flow rate of water through the geotextile is so large that it is difficult to obtain readings of head change versus time, in the falling head test. Specified stress was applied on the fabric. Flow per unit area was recorded and permittivity was calculated using the following Eqs. (3) and Eqs. (4) [14].

$$\varphi = \frac{Q}{hAt} \tag{4}$$

$Q = quantity \ of \ flow \ in \ mm^3$

$h = head \ of \ water \ in \ mm$

$A = the \ cross \ sec tional \ area \ in \ mm^2$

$t = time \ of \ flow \ in \ sec onds$

$\varphi = permittivity \ in \ s^{-1}$

Results and Discussion

Tensile strength

The stress-strain curve obtained is presented (Figure 2). From the tensile strength values obtained for the six types of wastewaters (expressed in terms of the COD values), the percentage reduction in tensile strength is determined. Results are presented in Table 2 and Figure 3.

The strength of the fibre is characterized by the observed maximum load, beyond which will cause coir fibre to break. The observed maximum load decreases with increase in COD values of the influent to the filters. The reduction in observed maximum load with higher organic content in the influent shows the loss of strength in fibres. The slope of the stress -strain curve for the geotextiles shows a descending trend, with increase in COD values. The percentage reduction in strength increases with the increase in organic content of the influent. The influent organics causes an increase in microbial population, as per Monods Kinetics; increase in microbial concentration depends upon the influent substrate concentration, these microbes feeds on

Figure 1: Test setup for determining Tensile Strength of Coir geotextiles

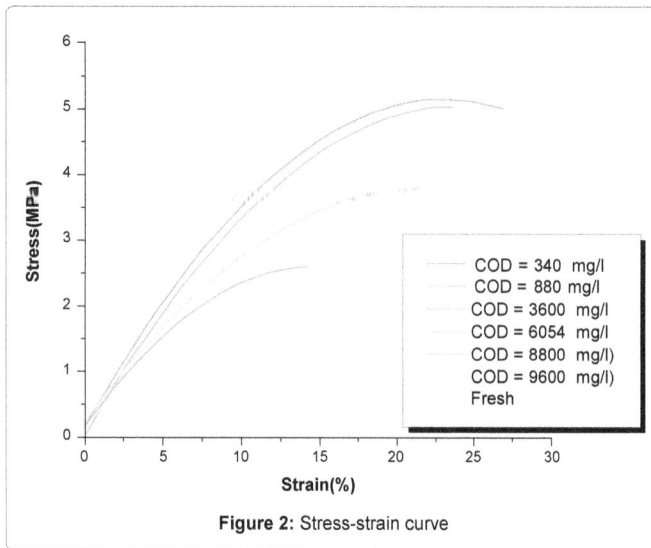

Figure 2: Stress-strain curve

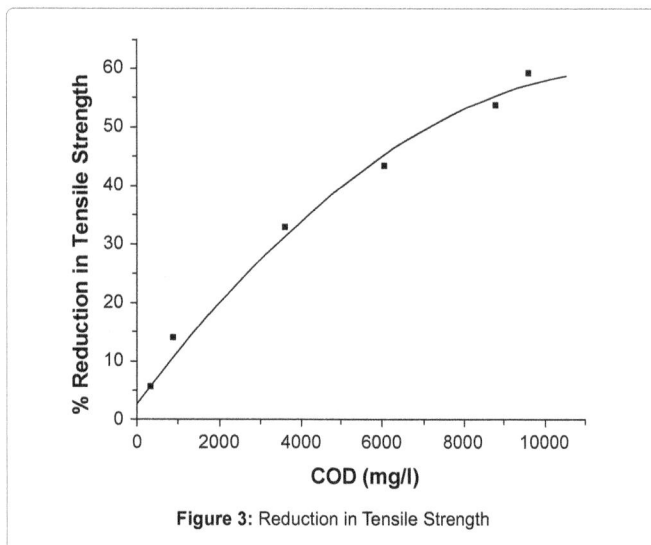

Figure 3: Reduction in Tensile Strength

COD of the Influent (mg/l)	pH of the Influent	Observed Maximum Load (kN)	Tensile Strength (kN)	Reduction in Tensile Strength (%)
0		0.108.5	1.845	0.00
340	5.9	0.1023	1.739	5.71
880	8.1	0.0933	1.586	14.00
3600	6.5	0.0729	1.238	32.87
6054	6.65	0.0615	1.044	43.37
8800	7.4	0.0503	0.855	53.68
9600	6.5	0.0442	0.752	59.24

Table 2: Reduction in Tensile Strength with COD

these media which results degradation. Subsequently the fibres become brittle, leading to lower tensile strength.

The pH value of the influent is different in the six runs, and hence it can most likely that it also influences the tensile strength. The study shows that the degradation or reduction in tensile strength of the coir geotextiles was at minimum for the range 6.5- 8.0. The deviation from the neutral ranges makes the influent acidic/ alkaline, which also makes fibres of the coir geotextiles brittle, which leads to reduction in tensile strength [15].

Permeability tests

As evident from the Permittivity vs. COD curve (Figure 4) the reduction in permittivity, with respect to influent COD values, shows a second order decay pattern. As in the case of tensile strength, the permittivity decreases with the increase in the organics content of the influent. The suspended solids in the influent gets trapped in the pores of the coir geotextiles; this decreases porosity, subsequently the permittivity through the coir geotextile media of the biofilter. Another reason for the decrease in permittivity is the increase in microbial population due to the higher substrate concentration of the influent to the biofilter [15].

SEM studies

Scanning Electron Microscope (SEM) images of the fresh as well as used media were obtained, using SEM-Hitachi SU6600. The fractured fibres were closely observed under the Scanning Electron Microscope (SEM).

Direct evidence of fibre cracking and pore matrix degradation, during exposure to wastewater media was gathered using Scanning Electron Microscopy (SEM). (Figure 5) (Figure6). EDS (Energy Dispersive X-ray Spectroscopy) or EDAX analysis is a technique used for identifying the elemental composition of the specimen, or an area of interest thereof. The EDAX analysis system works as an integrated feature of a scanning electron microscope (SEM), and cannot operate on its own without the latter. EDAX analysis was conducted on fresh and used geotextiles. Studies show that there is significant reduction of about 42% in silicon content in used coir geotextiles, than the fresh coir geotextiles. As silicon is the component which imparts strength to the coir fibres,results from EDAX analysis show that there is considerable degradation of the fibres in the filter media.

Conclusion

This study on coir geotextiles media (coir geotextiles H2M2) degradation was primarily intended to obtain some insight into the influence of substrate concentration in the influent wastewater, to a biofilter on the attached media. From the results of this study, the following conclusions can be drawn.

1. Fibres from the fresh and used media exhibit identical stress-strain behaviour.

2. Reduction in tensile strength of coir geotextile media fibres increases with the substrate concentration in the influent wastewater. This could be because growth of microorganisms is influenced by availability of substrates. However, these microorganisms cause media degradation.

3. Influent pH also influences the tensile strength of the coir geotextile media.

4. Permittivity is also affected by the concentration of substrates in the influent. Suspended solids in the influent may be intercepted in the pores of the coir geotextile medium, thereby restricting flow of the influent through the filter.

5. Images from a scanning electron microscope complemented by visual observations clearly show that the coir fibres shrink, and become brittle. Suspended solids adsorbed onto the surface of the coir fibres, can be seen in the SEM images. This reduces the medium permittivity.

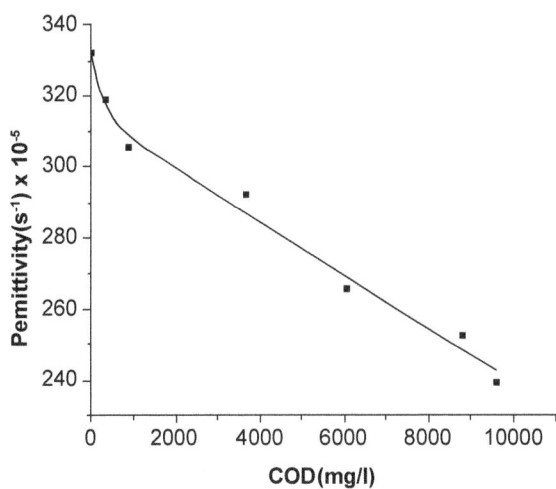

Figure 4: Reduction in Permittivity

Figure 5: Fresh woven coir geotextiles

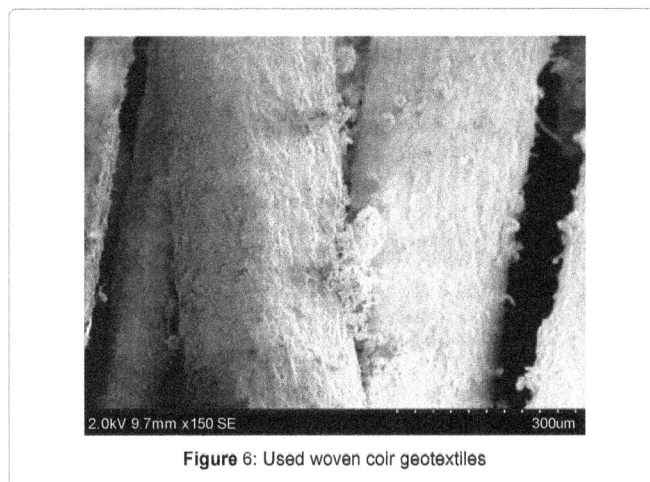

Figure 6: Used woven coir geotextiles

Acknowledgments

The authors wish to express their deep sense of gratitude to the Kerala Sate Council for Science Technology and Environment (KSCSTE), for providing financial assistance under the Ecology and Environment Research Scheme to conduct this study.

References

1. http://www.ncrmi.org/publications.htm

2. Praveen A, Sreelakshmy PB, Gopan M (2008) Coir geotextile-packed conduits for the removal of biodegradable matter from wastewater. Current Science 95: 655-658.

3. Gopan Mukkulath,Basil Cherian, Santosh G Thampi (2009) Novel method for onsite treatment of low volume organic rich wastewater by continuous flow biofilters. The young water talents symposium organized by Nanyang Technological University (NTU), Suntec Singapore.

4. Schurholz H (1991) Use of woven coir geotextiles in Europe, Coir, 35: 18-25.

5. Sarsby RW, Ali M, DelAluis R, Khoffot JH, Medongall JM (1992) Low cost soil reinforcement for developing countries. Proceedings of International Conference on Non-woven, the Textile Institute, North India Section, 297-310.

6. Rao GV, Balan K (2000) Coir geotextiles-emerging trends. The kerala state coir corporation Ltd, Kerala, India.

7. Nemerow NL (1978) Industrial water pollution. Addison -Wesley publishing company, Massachusetts, United States.

8. Bhatia SC (2002) Handbook of Industrial Pollution & Control: Air, Water, Wastes, & Pollution Control in Chemical Process and Allied Industries. CBS Publishers & Distributors, New Delhi.

9. http://idea.library.drexel.edu/handle/1860/158

10. Eaton AD, Franson MAH (2005) Standard Methods for the Examination of Water & Wastewater. American Public Health Association,Washington DC.

11. IS 13162: Part 5 (1992) Determination of tensile properties using a wide width strip. Bureau of Indian Standards, New Delhi, India.

12. ASTM D-4595 (2009) ASTM D4595 - 11 Standard Test Method for Tensile Properties of Geotextiles by the Wide-Width Strip Method. ASTM International, West Conshohocken, USA.

13. IS 1969 (1985) Methods for determination of breaking load and elongation of woven textile fabrics, Bureau of Indian Standards, New Delhi, India.

14. Bureau of Indian Standards (1995) Geotextiles - Methods of test for determination of water permeability-permittivity (IS 14324). Paper Back, New Delhi, India.

15. Metcalf Eddy (1993) Wastewater engineering treatment disposal and reuse. (4thedn) Tata McGraw-Hill Publishing Company Limited, New Delhi.

9

Biosorption of Arsenic by Living and Dried Biomass of Fresh Water Microalgae - Potentials and Equilibrium Studies

Sibi G*

Department of Biotechnology, Indian Academy Degree College, Centre for Research and Post Graduate Studies, Bangalore, India

Abstract

Biosorption of arsenic (III) and (V) from aqueous solutions by living and dried biomass of fresh water microalgae was investigated. A total of five microalgae namely *Chlorella, Oscillatioria, Scenedesmus, Spirogyra* and *Pandorina* were identified as arsenic tolerant and used for further studies. Varying conditions of pH, temperature, biomass and contact time were studied for As (III) and As (V) adsorption properties of living and dried biomass of microalgae. Significant biosorption of arsenic was found at pH 4, 32°C and 0.8 g/l levels. Both living and dried biomass were further studied for As (III) and (V) removal from the aqueous solutions containing 30 mg/l under the same experimental conditions. Removal rate of As (III) was higher than the As (V) by the microalgae under the experimental conditions and dried biomass showed higher biosorption rate and faster kinetic than the living ones.

Keywords: Biosorption; Arsenic removal; Microalgae; *Scenedesmus, Pandorina*

Introduction

Aquatic ecosystems are mainly affected by heavy metals and represent a potential risk to the health of humans and ecosystems [1]. Heavy metal ions in the environment are biomagnified in the food chain and are accumulated in tissues. Although low concentrations of some heavy metals are metabolically important to many living organisms, at higher levels they can potentially be toxic [2].

Arsenic (As) is a component of many industrial raw materials, products and wastes. Elevated levels of arsenic in drinking water have been implicated in human diseases and mortality [3]. Chronic exposure to arsenic causes neurological and haematological toxicity [4]. Arsenic impacts the major organs and is a potential carcinogen [5-7]. The most common arsenic species observed in the environment are the trivalent form arsenite As (III) and pentavalent form arsenate As (V) in which As (III) is more toxic than As (V) [8]. Because arsenic readily changes valence state and reacts to form species with varying toxicity and mobility, effective treatment of arsenic can be challenging. Arsenic treatment technologies require peroxidation step to form As (V) from As (III) [9] but the cost and secondary product formation during other conventional methods reduce the practices [10]. Use of biological processes provides a means for cost-effective removal of metals for the treatment of metal contaminated waters.

Microbes are ideal candidates to decrease the heavy metal ion concentration from ppm to ppb levels [11]. Microalgae are known to sequestrate heavy metals due to their cell wall constituents which act as binding sites for metals [12-17]. Bioaccumulation of metals by algae may create a feasible method for remediating water contaminated with metals [18,19]. It is well established that several marine and fresh water algae are able to take up various heavy metals selectively from aqueous media and to accumulate these metals within their cells [20-22]. Microalgae have been shown to accumulate arsenic and could potentially remediate through adsorption and biotransformation of inorganic arsenic [23-25].

In this study, the emphasis has been laid to know the efficiency of fresh water algal species in removing the arsenic from aqueous solutions. The research was to simultaneously determine the differences in arsenate and arsenite sorption capacities of living and dried algal biomass.

Materials and Methods

Sample collection and identification

Algal samples were collected from Bangalore fresh water habitats (13°04'N, 77°58'E) and washed several times with tap water and then with deionized water before analysed by using microscope. The family and genus identified with reference to the biology of algae [26]. The identified algae were Chlorophyceae (*Botryococcus, Chlamydomonas, Chlorella, Gonium, Pandorina, Scenedesmus, Spirogyra, Volvox*), Cyanophyceae (*Oscillatoria, Spirulina*) and Euglenophyceae (*Phacus*).

Chemicals

All chemicals used in this study were of analytical reagent grade. Stock solutions of 100 mg/L concentration each of As (III) and As (V) were prepared by dissolving $NaAsO_2$ and $Na_2HAsO_4.7H_2O$ in deionized water respectively. Solutions for adsorption and metal analyses were prepared by appropriate dilution of the freshly prepared stock solution.

Cultivation

The identified algae were cultivated in appropriate media BG 11 media (*Oscillatoria*), Bold 1NV media (*Spirogyra*), modified Bold 3N media (*Pandorina, Volvox, Phacus*), Proteose medium (*Scenedesmus, Gonium, Chlorella*), Soil extract medium (*Chlamydomonas*) and soil water media (*Spirulina*) [27,28].

Growth inhibition studies

Growth inhibition bioassays for the influence of As (III) and As (V) at a concentration of 10 mg/L on algal isolates (10 mg/l) were studied. The final cell density and biomass after arsenic exposure for a period of 15 days were compared to controls in dilution water (media). Samples

*Corresponding author: Sibi G, Indian Academy Centre for Research and Post Graduate Studies, India, E-mail: gsibii@gmail.com

from each flask was taken and fixed with Lugol's iodine solution to measure the cell density using haemocytometer. The mean number of cells produced at each concentration after exposing period was expressed as percentage growth reduction with respect to control. The algal biomass was determined by the spectrophotometric transmission of algal suspension (Schimadzu UV-2600) at 550 nm. From the results, arsenic tolerant microalgae were chosen and used for biosorption studies.

Arsenic biosorption studies

Living biomass: Cells in the exponential growth phase were obtained by centrifuging at 2500 rpm for 7 minutes and inoculated into flasks with fresh media at a concentration of 1×10^4 cells/ml. The flasks were shaken by hand and randomly placed in a growth cabinet ($27 \pm 1°C$, 12:12 h light/dark cycle, Philips TL 40W cool white fluorescent lighting, 140 μmol photons/m^2/s). The quantity of biomass used for the biosorption studies varied from 1- 2×10^5 cells/ml.

Non-living biomass: The algal biomass was sun-dried and then dried in oven at 50°C for 24 h. The dried algal biomass was shredded, ground in a mortar and an average size of 500-600 μm was used for biosorption experiments at a concentration of 0.5 g/l.

Arsenic concentration: Both living and non-living algal biomass were inoculated separately into 250 ml conical flasks containing 100 ml of media supplemented with As (III) and As (V) with initial concentrations ranging from 10 to 50 mg/l. The flasks were kept under illumination at 2500 lux for 12 hr light - dark cycle at $24 \pm 2°C$. Metal-free and algae-free blanks were used as control groups. Test flasks were rotated and shaken twice daily to ensure sufficient gas exchange. Separation of biomass from metal bearing solution after appropriate incubation time was achieved through centrifugation at 10,000 rpm for 10 minutes.

pH study: Measurements for the effects of varied pH on Ar (III) and (V) biosorption performance were chosen by adding 0.1M NaOH or 0.1M HCl to achieve pH values of 2.0 through 10.0.

Temperature study: Measurements for the effects of increased temperature on arsenic adsorption performance by the algal isolates were studied at temperature between 23.0 to $35.0 \pm 2°C$.

Biomass study: Various biomass concentrations of living and non-living algal cells in the range of 0.2, 0.4, 0.6, 0.8 and 1 g/l were used to study the adsorption of arsenic.

Apparatus

All measurements were carried out on Agilent 280FS AA spectrometer (Agilent Technologies, USA). The organic phase was aspirated directly without elution into the flame and the signals related to As (III) and As (V) was recorded at 193.7 and 197.2 nm (lamp current 10mA). The data were analyzed by using Spectra AA software.

Biosorption kinetics studies

The metal biosorption (q) by the algae and bioremoval efficiency (R) were calculated by the following formulae [29,30].

$$q = \frac{(C_i - C_f)V}{M}$$

$$R(\%) = \frac{(C_i - C_f)}{C_i} \times 100$$

Where q = metal adsorption (mg/g); M = dry mass of a lgae (g); V = volume of i nitial metal ion solution (L); R = bioremoval efficiency

(%); C_i= initial concentration of metal in aqueous solution (mg/l); C_f= final concentration of metal in aqueous solution (mg/l).

Biosorption equilibrium models: The biosorption equilibrium isotherm was obtained by the Freundlich model (Equation 1) and the Langmuir model (Equation 2) respectively [31].

$$q = K_f C_e^{1/n} \tag{1}$$

Where, K_f and n are the distribution coefficient and a correction factor, respectively and C_e is equilibrium concentration of heavy metal (mg/l). By plotting the linear form of Eq. (1), $\ln q = 1/n \ln C_e + \ln K_f$, the slope is the value of 1/n and the intercept is equal to $\ln K_f$.

$$q_{eq} = \frac{q_{max}bC_e}{1 + bC_e} \tag{2}$$

The linear form of Langmuir is

$$C_{eq}/q_{eq} = 1/q_{max}b + C_{eq}/q_{max} \tag{3}$$

Where q_{max} is the Langmuir constant (mg/g) reflecting the maximum adsorption capacity of the metal ion per unit weight of biomass to form a complete monolayer on the surface bound at high C_{eq}. Langmuir constant b(l/mg) represents a ratio of adsorption rate constant to desorption rate constant, which also gives an indication of the affinity of the metal for binding sites on the biosorbent. q_{max} and b can be determined from the linear form of Langmuir equation (3) by plotting C_{eq}/q_{eq} vs. C_{eq}.

The specific metal biosorption q was calculated using the following equation

$$q_e (mg/g) = \left[\frac{C_i - C_e}{M}\right] XV \tag{4}$$

Where q_e is the specific metal biosorption (mg metal/g biomass), V is the volume of metal solution (l), C_i and C_e are the initial and equilibrium concentration of metal (mg /l), respectively, and m is the dry weight of the biomass (g).

Arsenic biosorption potential

Based on the arsenic biosorption equilibriums (mg/g), optimum conditions for pH, temperature and biomass were determined and further studied to determine arsenic biosorption potential of microalgae based on contact time. The living and non-living algal biomass (0.8 g/l) was suspended in 100 ml of Ar (III) and (V) solution (30 mg/l) in 250 ml flasks containing a pH of 4.0. The cell/metal suspensions were shaken (150 rpm) at 32°C. Samples were taken from the solutions at different time intervals from 0-60 hours and analysed for metal biosorption.

Results and Discussion

Growth pattern of isolated microalgae in the presence of arsenic has revealed that the degree of growth inhibition by arsenic varied widely between the isolates. The growth rate of *Spirogyra, Volvox, Phacus, Gonium* and *Chlamydomonas* was inhibited by As (III and V) in 5 days at 10 mg/l concentration. *Oscillatoria* and *Spirogyra* reached stationary phase in 8-10 days followed by decline phase. The growth rate of *Pandorina* was high up to 11 days where as *Chlorella* and *Scenedesmus* had exponential phase up to 13 days. These five genera have been considered as arsenic tolerant microalgae for further studies. The arsenic tolerant microalgae were grown in the presence As (III) and As (V) with a initial concentrations of 10-50 mg/l during which increased concentration of arsenic in the growth medium has led to deceased adsorption rate. The environmental setups had varying pH, temperature and biomass which produced variable results irrespective of the species studied.

AS(III)	pH	Living					Dried				
		Chlorella	Oscillatoria	Scenedesmus	Spirogyra	Pandorina	Chlorella	Oscillatoria	Scenedesmus	Spirogyra	Pandorina
10 mg/l	2.0	4	3	5	4	4	4	4	5	5	4
	4.0	7	6	7	6	7	7	6	7	6	6
	6.0	7	6	7	6	7	7	5	7	6	6
	8.0	5	4	5	4	5	5	4	5	5	5
	10.0	4	3	4	3	4	4	3	4	3	5
20 mg/l	2.0	8	6	9	8	7	9	8	9	9	9
	4.0	11	9	13	12	12	12	11	14	13	15
	6.0	10	9	12	12	14	11	10	14	13	14
	8.0	8	5	8	6	6	7	4	8	7	7
	10.0	5	3	5	3	3	4	2	5	3	3
30 mg/l	2.0	10	9	11	8	8	9	9	10	9	8
	4.0	20	18	23	20	21	21	18	22	20	21
	6.0	19	17	22	20	20	21	18	21	18	19
	8.0	9	7	8	9	8	9	8	8	9	9
	10.0	5	3	5	4	4	5	4	5	4	4
40 mg/l	2.0	10	9	12	11	12	12	9	13	11	11
	4.0	17	16	20	18	19	17	15	19	18	17
	6.0	13	11	14	12	12	14	8	13	13	12
	8.0	11	7	9	8	9	9	7	8	9	8
	10.0	9	6	8	7	7	9	6	7	6	6
50 mg/l	2.0	12	7	11	9	9	12	8	10	9	9
	4.0	14	10	15	13	12	14	11	14	12	12
	6.0	11	8	12	10	9	10	7	11	9	9
	8.0	10	7	9	9	8	9	7	9	8	8
	10.0	9	8	7	8	7	8	6	8	8	8

Table 1: Biosorption of AS (III) (mg/g) under varying pH.

AS(V)	pH	Living					Dried				
		Chlorella	Oscillatoria	Scenedesmus	Spirogyra	Pandorina	Chlorella	Oscillatoria	Scenedesmus	Spirogyra	Pandorina
10 mg/l	2.0	5	4	5	4	4	5	3	5	4	5
	4.0	6	4	6	6	6	7	5	7	6	7
	6.0	6	4	5	6	5	6	4	6	5	6
	8.0	4	3	4	4	4	3	2	4	4	4
	10.0	3	1	2	2	2	3	2	3	3	3
20 mg/l	2.0	9	7	9	8	8	10	8	11	9	10
	4.0	12	10	13	11	11	13	12	14	12	13
	6.0	11	9	12	10	10	11	12	12	10	10
	8.0	9	7	9	8	8	9	8	9	8	8
	10.0	6	4	6	5	6	7	4	6	6	5
30 mg/l	2.0	13	11	14	12	12	14	12	17	15	16
	4.0	19	16	20	17	18	20	18	21	19	20
	6.0	17	14	17	13	15	17	12	16	15	14
	8.0	11	9	10	9	9	11	9	11	10	9
	10.0	7	5	7	6	6	7	5	6	6	6
40 mg/l	2.0	11	9	12	11	12	12	9	13	11	11
	4.0	14	12	16	14	15	15	13	18	15	15
	6.0	12	10	13	11	12	12	9	15	13	12
	8.0	9	7	9	7	7	10	8	10	9	9
	10.0	8	6	7	7	6	7	7	8	7	7
50 mg/l	2.0	15	11	15	13	14	16	10	16	14	14
	4.0	19	15	21	18	18	20	14	21	17	17
	6.0	16	13	17	15	16	15	12	17	13	14
	8.0	15	11	13	12	11	13	10	13	10	10
	10.0	9	8	10	9	10	10	9	11	9	9

Table 2: Biosorption of AS (V) (mg/g) under varying pH.

The pH of the solution has played a key role in the biosorption of arsenite and arsenate by the microalgae. Aqueous solutions containing As (III) and As (V) were prepared with varying pH ranging from 2.0 to 10.0. Arsenic sorption decreased with increasing pH and the maximum arsenic removal occurred at pH 4.0 for both living and dried biomass. The highest As (III) uptake (Q) of 23 mg/g was recorded with living biomass of *Scenedesmus* at a initial concentration of 30 mg/l followed by *Pandorina* (21 mg/g). Maximum uptake of 22 mg/g As (III) was found with the dried biomass of *Scenedesmus* which was followed by *Chlorella* and *Pandorina* (Table 1). The metal sorption of As (V) was highest in living biomass of *Scenedesmus* (20 mg/g) and the trend was reflected in dried biomass of *Chlorella* and *Pandorina* at 30 mg/g (Table 2)

Various temperatures in the range of 23°C - 35°C were used to study the metal uptake by the algal isolates. The results indicated that 32°C was found optimum in which maximum adsorption has taken place and there were no significant changes in the metal uptake at temperature below 29°C and above 32°C. The living biomass of *Scenedesmus* and *Chlorella* has adsorbed maximum As (III) from the aqueous solution followed by *Spirogyra* and *Pandorina* at 30 mg/l initial concentrations. An uptake of 25 mg/g was recorded with dried biomass of *Scenedesmus* followed by *Chlorella* and *Pandorina* (24 mg/g) under similar conditions (Table 3). The uptake levels of As (V) were comparatively lower than As (III) with a maximum sorption of 16 mg/g by living biomass of *Scenedesmus* and *Oscillatoria*. The dried biomass of *Scenedesmus* and *Chlorella* has up taken 19 and 17 mg/g of As (V) at 30 mg/l levels (Table 4).

The biomass dosage was varied from to 0.2 to 1.0 g/l. Increased biomass has increased the arsenic sorption and the maximum uptake of arsenic was obtained at a biomass concentration of 0.8 g/l. At 30 mg/l initial concentration of As (III), maximum uptake was seen with living

biomass of *Pandorina* (27 mg/g) followed by *Scenedesmus* (26 mg/g) whereas the dried biomass of *Scenedesmus* (28 mg/g) has recorded the highest AS (III) uptake over *Pandorina* (26 mg/g) (Table 5). The trend from temperature variations was repeated in biomass dosage in As (V) sorption with living cells of *Oscillatoria* and *Pandorina* has exhibited maximum uptake of 18 mg/g which relatively lower than As (III) uptake under similar conditions. The dried biomass of *Pandorina* has maximum As (V) sorption (17 mg/g) followed by *Oscillatoria* (16 mg/g) (Table 6).

The optimum pH, temperature and contact time obtained from equilibrium isotherm experiments were used to determine the arsenic biosorption of both living and non-living algal biomass based on time dependent manner. The time dependent biosorption of arsenic by living and dried biomass, samples equivalent to 0.8g/l were placed in solutions of As (III) and As (V) adjusted to pH 4.0 and 32°C temperature. The results of arsenic sorption as a function of time under the controlled conditions have been depicted in Figure 1-4. The kinetic studies revealed that arsenic biosorption increases with time and recorded highest at 36 hrs contact time. However the soption rate decreased progressively and became almost constant during the 60 hrs contact with arsenic. *Scenedesmus* has showed fast kinetic of As (III) binding and was adsorbed appreciable quantities of As (III) (20 mg/g) during the first 24 -36 hrs from the solution. The second highest uptake of As (III) was observed with *Pandorina* (21 mg/g) followed by *Chlorella* and there was no significant changes in the metal uptake after 36 hrs of contact time by the isolates. Similarly highest As (V) was recorded with *Scenedesmus* with a metal uptake of 20 mg/g followed by *Chlorella* and *Pandorina* after 36 hrs of contact time. *Oscillatoria* and *Spirulina* have recorded moderate arsenic uptake throughout the study.

Algae are known for their capability to accumulate heavy metals

AS(III)	Temp	Living					Dried				
		Chlorella	Oscillatoria	Scenedesmus	Spirogyra	Pandorina	Chlorella	Oscillatoria	Scenedesmus	Spirogyra	Pandorina
10 mg/l	23°C	4	2	3	2	3	3	3	4	2	3
	26°C	4	3	5	4	4	4	4	4	3	3
	29°C	7	5	7	5	6	6	5	7	5	5
	32°C	7	6	6	6	7	6	6	6	5	6
	35°C	6	5	6	5	5	5	4	7	4	5
20 mg/l	23°C	7	5	7	6	7	8	7	10	7	8
	26°C	9	8	10	9	10	11	9	12	9	10
	29°C	11	11	14	12	13	15	12	16	13	14
	32°C	12	12	15	13	14	15	12	17	14	14
	35°C	9	9	12	10	11	13	11	14	11	11
30 mg/l	23°C	10	9	11	10	11	14	10	12	11	12
	26°C	12	11	15	13	13	17	13	18	14	15
	29°C	20	17	21	18	20	24	21	25	22	23
	32°C	21	17	21	20	20	24	23	25	23	24
	35°C	16	13	18	16	17	21	18	22	19	19
40 mg/l	23°C	11	9	12	10	10	11	10	12	11	11
	26°C	14	19	13	11	13	13	11	15	12	12
	29°C	16	12	15	14	14	17	13	18	12	14
	32°C	18	13	17	15	15	19	14	18	14	15
	35°C	16	12	17	14	15	18	12	17	12	14
50 mg/l	23°C	13	9	11	11	10	14	10	12	12	12
	26°C	15	10	14	13	12	14	10	16	17	16
	29°C	17	14	28	16	17	21	19	22	20	20
	32°C	20	17	21	18	20	24	21	25	22	23
	35°C	16	25	20	17	18	21	19	23	20	20

Table 3: Biosorption of AS (III) (mg/g) under varying temperature.

AS(V)	Temp	Living					Dried				
		Chlorella	Oscillatoria	Scenedesmus	Spirogyra	Pandorina	Chlorella	Oscillatoria	Scenedesmus	Spirogyra	Pandorina
10 mg/l	23°C	6	5	6	5	6	5	4	6	5	5
	26°C	6	6	7	7	6	6	5	5	7	6
	29°C	7	7	7	6	6	6	5	6	6	7
	32°C	8	6	8	6	6	7	5	7	6	5
	35°C	7	7	8	6	5	7	5	6	6	5
20 mg/l	23°C	8	6	8	7	7	8	7	8	8	7
	26°C	8	7	9	7	7	9	7	9	8	8
	29°C	10	11	12	9	10	10	9	9	10	10
	32°C	11	13	13	11	12	11	11	11	11	11
	35°C	11	11	12	10	10	11	10	10	10	10
30 mg/l	23°C	12	11	11	11	12	12	10	12	11	11
	26°C	13	11	13	13	14	14	12	16	13	13
	29°C	14	12	15	14	13	14	10	17	14	12
	32°C	15	16	16	14	15	17	15	19	16	16
	35°C	14	12	17	16	15	14	13	18	15	16
40 mg/l	23°C	8	6	8	6	7	8	6	8	7	7
	26°C	10	8	11	8	9	9	8	11	9	9
	29°C	13	11	13	10	11	12	10	14	10	12
	32°C	16	11	15	13	14	17	11	17	12	14
	35°C	14	10	13	11	11	14	9	14	10	11
50 mg/l	23°C	9	6	9	8	8	8	6	8	7	9
	26°C	12	9	13	10	11	11	8	11	10	12
	29°C	16	13	17	14	16	17	11	17	15	17
	32°C	18	13	10	16	17	10	14	20	17	20
	35°C	15	11	16	13	14	16	13	15	13	15

Table 4: Biosorption of AS (V) (mg/g) under varying temperatures.

AS(III)	Biomass g/l	Living					Dried				
		Chlorella	Oscillatoria	Scenedesmus	Spirogyra	Pandorina	Chlorella	Oscillatoria	Scenedesmus	Spirogyra	Pandorina
10 mg/l	0.2 g	4	3	5	4	4	5	3	4	5	4
	0.4 g	9	7	10	9	9	11	8	11	10	11
	0.6 g	15	12	16	15	15	16	13	14	13	15
	0.8 g	22	18	23	21	21	22	19	21	17	19
	1.0 g	19	16	21	18	17	17	16	19	19	20
20 mg/l	0.2 g	5	4	6	6	5	6	4	6	5	6
	0.4 g	12	9	12	11	11	13	10	15	13	15
	0.6 g	17	15	18	17	18	18	15	17	16	17
	0.8 g	26	23	29	26	27	28	25	27	23	25
	1.0 g	20	19	24	21	21	22	20	23	22	23
30 mg/l	0.2 g	5	5	5	5	5	5	5	4	5	4
	0.4 g	15	11	16	12	13	13	9	14	11	12
	0.6 g	18	15	18	16	17	18	14	19	15	15
	0.8 g	23	24	26	23	27	25	24	28	25	26
	1.0 g	20	20	21	19	22	22	20	23	20	21
40 mg/l	0.2 g	6	5	7	7	6	6	4	6	6	6
	0.4 g	16	11	17	16	18	16	13	17	15	15
	0.6 g	20	19	21	21	22	22	20	23	21	21
	0.8 g	28	26	32	29	30	30	28	31	26	29
	1.0 g	23	22	27	24	26	27	25	29	25	27
50 mg/l	0.2 g	7	6	7	7	7	7	5	6	6	7
	0.4 g	16	13	18	17	17	17	13	17	16	16
	0.6 g	21	20	23	21	23	23	17	23	21	22
	0.8 g	28	27	33	29	30	31	28	31	27	30
	1.0 g	24	23	29	25	26	28	25	29	26	28

Table 5: Biosorption of AS (III) (mg/g) under varying biomass concentrations.

AS(V)	Biomass g/l	Living					Dried				
		Chlorella	Oscillatoria	Scenedesmus	Spirogyra	Pandorina	Chlorella	Oscillatoria	Scenedesmus	Spirogyra	Pandorina
10 mg/l	0.2 g	2	2	2	1	2	2	1	3	1	3
	0.4 g	4	2	4	3	4	4	2	3	3	4
	0.6 g	5	3	6	3	5	5	3	6	4	5
	0.8 g	5	4	7	5	7	6	4	7	5	6
	1.0 g	6	5	6	4	7	6	4	6	5	6
20 mg/l	0.2 g	3	2	3	2	3	3	2	3	2	2
	0.4 g	5	3	5	3	4	4	3	5	3	5
	0.6 g	6	6	7	6	6	6	6	7	6	6
	0.8 g	9	9	10	8	10	10	9	11	9	11
	1.0 g	9	9	10	9	11	9	9	10	9	11
30 mg/l	0.2 g	7	5	8	6	7	8	6	8	5	7
	0.4 g	10	8	12	8	11	10	9	11	8	11
	0.6 g	12	10	15	11	15	11	11	13	11	14
	0.8 g	14	18	16	16	18	13	16	15	15	17
	1.0 g	13	17	14	15	18	13	16	13	13	16
40 mg/l	0.2 g	11	11	12	11	11	12	11	12	11	10
	0.4 g	16	15	18	16	16	14	15	19	14	15
	0.6 g	21	16	24	21	25	21	16	22	22	23
	0.8 g	25	18	27	21	28	22	18	25	23	26
	1.0 g	24	19	26	20	26	20	18	24	22	25
50 mg/l	0.2 g	15	14	17	15	15	16	13	15	14	15
	0.4 g	26	23	26	22	23	25	23	24	22	22
	0.6 g	34	30	34	31	33	31	31	34	30	30
	0.8 g	36	32	36	34	36	33	33	35	33	33
	1.0 g	34	32	33	33	33	32	31	30	31	32

Table 6: Biosorption of AS (V) (mg/g) under varying biomass concentrations.

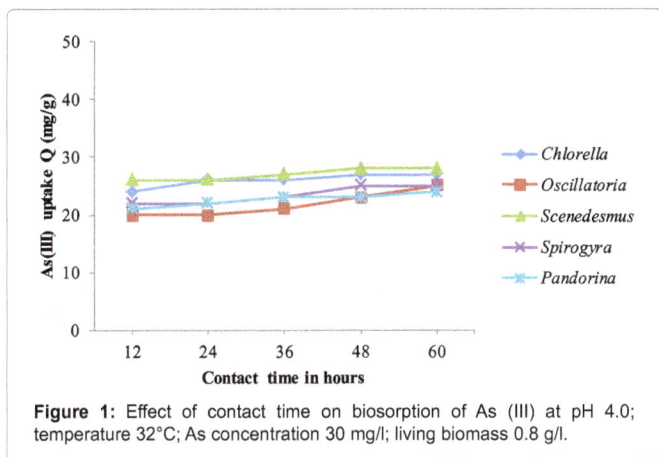

Figure 1: Effect of contact time on biosorption of As (III) at pH 4.0; temperature 32°C; As concentration 30 mg/l; living biomass 0.8 g/l.

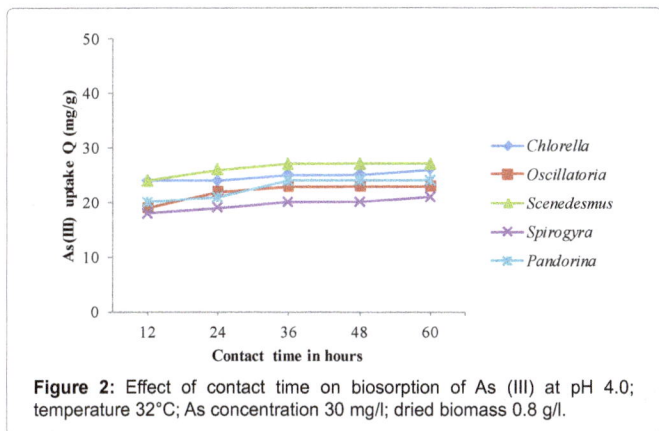

Figure 2: Effect of contact time on biosorption of As (III) at pH 4.0; temperature 32°C; As concentration 30 mg/l; dried biomass 0.8 g/l.

as they are required as essential nutrients [32] and have been explored for metal removal [33]. The level of metal removal by a microalga depends on biomass concentrations, pH and contact time [34-36]. In this experiment, arsenic sorption values of dried biomass were high in comparison with living biomass. This could be due to the larger surface area due to the destruction of cell membranes during the dried biomass preparation. Non-living microbial cells have lower sensitivity to toxic metal ions concentrations over living cells which offer to use them at adverse operating conditions [37]. The metal sorption capacity was drastically reduced at pH values above 6.0, which could be due to the formation of insoluble metal hydroxides. Further, it was noted that the arsenic uptake was lowered at pH below 4.0. The temperature of the solution could influence the metal biosorption of living cells [37] and culture temperatures have profound effects on the chemical composition of the algal cells. The biosorption of As increased with temperatures from 23°C-32°C and the results indicated that elevated temperatures tend to increase the biosorptive properties of isolated microalgae with an optimum temperature of 32°C. However the variations in uptake of As (III) and As (V) based on temperature need to be investigated. It was observed that higher biomass levels have reduced the adsorption amount which was influenced by the formation of aggregates at higher concentrations that has ultimately resulted in reduced biosorption area [38].

The accumulation of heavy metals in algae involves and rapid uptake initially followed by slower uptake [37,39-41]. The same trend was observed in the experiments as the metal sorption rate was higher in first 36 hours followed by significantly slower uptake in the next hours. Further, As (III) and As (V) sorption was more at initial time (0-36 hrs) followed by almost constant after 36 hrs of contact time.

Based on the pH studies, As (III) was effectively removed by both

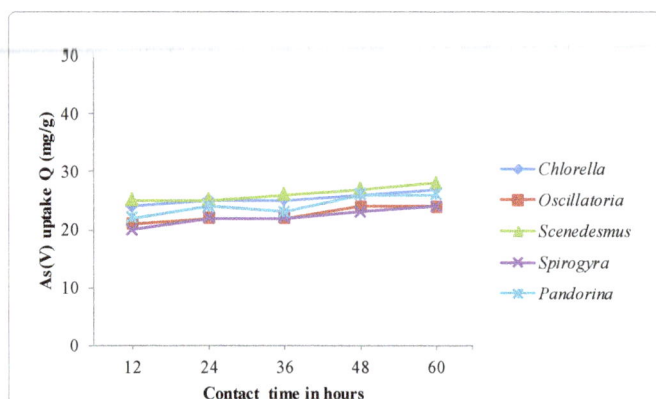

Figure 3: Effect of contact time on biosorption of As (V) at pH 4.0; temperature 32°C; As concentration 30 mg/l; living biomass 0.8 g/l.

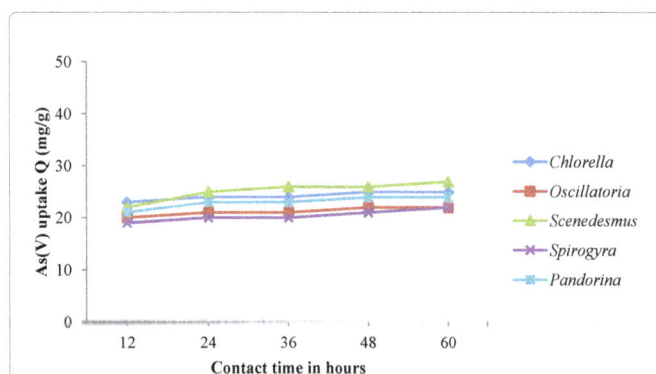

Figure 4: Effect of contact time on biosorption of As (V) at pH 4.0; temperature 32°C; As concentration 30 mg/l; dried biomass 0.8 g/l.

living and dried biomass of *Scenedesmus* and *Pandorina* whereas As (V) was removed by *Scenedesmus* and *Chlorella*. Significant levels of As (III) was removed by the living biomass of Spirogyra at varying pH. Under varying temperatures, *Scenedesmus, Chlorella* and *Pandorina* were efficient in removing As (III) and As (V) from the solutions. In general, *Scenedesmus, Pandorina* and *Chlorella* were found as potential arsenic removers from the aqueous solutions under varying environmental conditions. Heavy metal removal capacities of *Scenedesmus* was reported in earlier studies [42-45]. Arsenite and arsneate tolerance levels of *Chlorella* were studied earlier [23,24,46,47]. *Spirogyra* was efficient in As (III) removal than As (V) and the previous studies reported the arsenic removal by *Spirogyra* [48]. Samal et al. [49] has revealed the arsenic removing capacity of *Oscillatoria* in their studies and in this study dried biomass of *Oscillatoria* was significant at 30 mg/l of As (V).

In this study, removal rate of As (III) was higher than the As (V) by the microalgae under the experimental conditions. The trivalent arsenite is more toxic than the pentavalent arsenate and this study reveals the potential use of microalgae in removing the toxic As (III). The dried biomass was found effective for arsenic sorption than the living biomass. This study emphasizes that microalgae are ideal candidates to be further exploited for the removal of other heavy metal ions.

References

1. Rai PK (2010) Phytoremediation of heavy metals in a tropical impoundment of industrial region. Environ Monit Assess 165: 529-537.

2. Phillips DJH (1995) The chemistries and environmental fates of trace metals and organochlorines in aquatic ecosystems. Mar Pollut Bull. 31: 193-200.

3. Wang S, Zhao X (2009) On the potential of biological treatment for arsenic contaminated soils and groundwater. J Environ Manage 90: 2367-2376.

4. Byron WR, Bierbower GW, Brouwer JB, Hansen WH (1967) Pathologic changes in rats and dogs from two-year feeding of sodium arsenite or sodium arsenate. Toxicol Appl Pharmacol 10: 132-147.

5. Naqvi SM, Vaishnavi C, Singh H (1994) Toxicity and metabolism of arsenic in vertebrates. Arsenic in the environment, Part II: Human Health and Ecosystem effects, Edited by Jeome O. Nriagu. Pp: 55-91.

6. Morton W, Starr G, Pohl D, Stoner J, Wagner S, et al. (1976) Skin cancer and water arsenic in Lane County, Oregon. Cancer 37: 2523-2532.

7. Duker AA, Carranza EJ, Hale M (2005) Arsenic geochemistry and health. Environ Int 31: 631-641.

8. Cullen WR, Reimer KJ (1989) Arsenic speciation in the environment. Chem. Rev. 89: 713-764.

9. Kim MJ, Nriagu J (2000) Oxidation of arsenite in groundwater using ozone and oxygen. Sci Total Environ 247: 71-79.

10. Wang S, Mulligan CN (2006) Natural attenuation processes for remediation of arsenic contaminated soils and groundwater. J Hazard Mater 138: 459-470.

11. Wang J, Chen C (2006) Biosorption of heavy metals by Saccharomyces cerevisiae: a review. Biotechnol Adv 24: 427-451.

12. Gong R, Ding Y, Liu H, Chen Q, Liu Z (2005) Lead biosorption and desorption by intact and pretreated spirulina maxima biomass. Chemosphere 58: 125-130.

13. Davis TA, Volesky B, Mucci A (2003) A review of the biochemistry of heavy metal biosorption by brown algae. Water Res 37: 4311-4330.

14. Sheng PX, Ting YP, Chen JP, Hong L (2004) Sorption of lead, copper, cadmium, zinc, and nickel by marine algal biomass: characterization of biosorptive capacity and investigation of mechanisms. J Colloid Interface Sci 275: 131-141.

15. Schiewer S, Volesky B (2000) Biosorption Processes for Heavy Metal Removal in: D.R. Lovely (Ed.), Environmental Microbe Metal Interactions, ASM Press, Washington, DC, 2000, Pp. 329-362.

16. Yu Q, Matheickal JT, Kaewsarn P (1999) Heavy metal uptake capacities of common marine macro-algal biomass, Water Res. 33: 1534-1537.

17. Holan ZR, Volesky B (1994) Biosorption of lead and nickel by biomass of marine algae. Biotechnol Bioeng 43: 1001-1009.

18. Darnal DW, Greene B, Henzl MT, Hosea JM, McPherson RA, et al. (1986) Selective recovery of gold and other metal ions from an algal biomass. Environ Sci Technol 20: 206-208.

19. Nakajima A, Horikoshi T, Sakagushi T (1981) Recovery of varnlum by immobilized microorganisms. Eur. J. Appl. Microbiol. Biotechnol. 16: 88-91.

20. Afkar A, Ababna H, Fathi AA (2010) Toxicological response of the green alga Chlorella vulgaris to some heavy metals. Am. J. Environ. Sci. 6: 230-237.

21. Kumar D, Gaur JP (2011) Metal biosorption by two cyanobacterial mats in relation to pH, biomass concentration, pretreatment and reuse. Bioresour Technol 102: 2529-2535.

22. Chen CY, Chang HW, Kao PC, Pan JL, Chang JS (2012) Biosorption of cadmium by CO(2)-fixing microalga Scenedesmus obliquus CNW-N. Bioresour Technol 105: 74-80.

23. Levy JL, Stauber JL, Adams M, Maher W, Kirby JK, et al. (2005) Toxicity, biotransformation, and mode of action of arsenic in two freshwater microalgae (Chlorella sp. and Monoraphidium arcuatum). Environmental Toxicology and Chemistry. 24: 2630-2639.

24. Maeda S, Nakashima S, Takeshita T, Higashi S (1985) Bioaccumulation of arsenic by freshwater algae and the application to the removal of inorganic arsenic from an aqueous phase. Part II. By Chlorella vulgaris isolated from arsenic-polluted environment. Separ Sci Technol 20: 153-161.

25. Giddings JM, Eddlemon GK (1977) The effects of microcosm size and substrate type on aquatic microcosm behaviour and arsenic transport. Arch. Environ. Contam. Toxicol. 6: 491-505.

26. Round FE (1973) The Biology of the Algae (2nd ed). Edward Arnold (Publishers) Ltd, UK, Pp. 1-23.

27. Anderson RA (2005) Algal culture techniques. 1st edition. Elsevier Academic press. California, USA.

28. Stanier RY, Kunisawa R, Mandel M, Cohen-Bazire G (1971) Purification and properties of unicellular blue-green algae (order Chroococcales). Bacteriol Rev 35: 171-205.

29. Volesky B (1992) Removal of heavy metals by biosorption. In: Ladisch MR, Bose A (eds). Harnessing biotechnology for the 21stCentury.Amer. Chem. Soc., Washington DC, Pp. 462-466.

30. Zhang L, Zhao L, Yu Y, Chen C (1998) Removal of Pb2+ from aqueous solution by non-living Rhizopus nigricans. Water Res. 32: 1437-1444.

31. Volesky B (1990) Removal and recovery of heavy metals by biosorption. In: Volesky, B. (Ed.), Biosorption of Heavy Metals. CRC Press, Boca Raton, Fla. Pp. 7-43.

32. Gadd GM (1990) Heavy metal accumulation by bacteria and other microorganisms. Cell Mol Life Sci 46: 834-840.

33. Wilde EW, Benemann JR (1993) Bioremoval of heavy metals by the use of microalgae. Biotechnol Adv 11: 781-812.

34. Aksu Z, Donmez G (2006) Binary biosorption of cadmium(II) and nickel(II) onto dried Chlorella vulgaris: Co-ion effect on mono-component isotherm parameters. Process Biochem. 41: 860-868.

35. Solisio C, Lodi A, Soletto D, Converti A (2008) Cadmium biosorption on Spirulina platensis biomass. Bioresour Technol 99: 5933-5937.

36. Tang YZ, Gin KYH, Aziz MA (2002) Equilibrium model for cadmium adsorption by green algae in a batch reactor. J Environ Eng 128: 304-312.

37. Arica MY, Tuzun I, Yalc E, Ince O, Bayramoglu G (2005) Utilisation of native, heat and acid-treated microalgae Chlamydomonas reinhardtii preparations for biosorption of CrI) ions, Process Biochemistry, 40: 2351-2358.

38. Ahuja P, Gupta R, Saxena RK (1999) Zn2+ biosorption by Oscillatoria anguistissima, Process Biochemistry, 34: 77-85.

39. Hamdy AA (2000) Biosorption of heavy metals by marine algae. Curr Microbiol 41: 232-238.

40. Vinoj Kumar V, Kaladharan P (2006) Biosorption of metals from contaminated water using seaweed, Current Science, 90: 1263-1267.

41. Bates SS, Tessier A, Campbell PGC, Buffle J (1982) Zinc adsorption and transport by Chlamydomonas variabilis and Scenedesmus subspicatus (Chlorophyceae) grown in semi-continuous culture. J. Phycol. 18: 521-529.

42. Jahan K, Mosto P, Mattson C, Frey M, Derchak L (2006) Microbial removal of arsenic. Aater, air and Soli ollution: Focus 6: 71-82.

43. Monteiro CM, Castro PML, Malcata FX (2009) Use of the microalga Scenedesmus obliquus to remove cadmium cations from aqueous solutions. World J Microbiol Biotechnol. 25: 1573-1578.

44. Ajayan KV, Selvaraju M, Thirugnanamoorthy K (2011) Growth and heavy metals accumulation potential of microalgae grown in sewage wastewater and petrochemical effluents. Pak J Biol Sci 14: 805-811.

45. Kizilkaya B, Turkera G, Akgulb R, Doganc F (2012) Comparative Study of Biosorption of Heavy Metals Using Living Green Algae Scenedesmus quadricauda and Neochloris pseudoalveolaris: Equilibrium and Kinetics. Journal of Dispersion Science and Technology. 33: 410-419.

46. Jiang Y, Purchase D, Jones H, Garelick H (2011) Effects of arsenate (AS5+) on growth and production of glutathione (GSH) and phytochelatins (PCS) in Chlorella vulgaris. Int J Phytoremediation 13: 834-844.

47. Rao PH, Kumar RR, Raghavan BG, Subramanian VV, Sivasubramanian V (2011) Is phycovolatilization of heavy metals a probable (or possible) physiological phenomenon? An in situ pilot-scale study at a leather-processing chemical industry. Water Environ Res. 83: 291-297.

48. Kumar JI, Oommen C (2012) Removal of heavy metals by biosorption using freshwater alga Spirogyra hyalina. J Environ Biol 33: 27-31.

49. Samal AC, Bhar G, Santra SC (2004) Biological process of arsenic removal using selected microalgae. Indian J Exp Biol 42: 522-528.

10

Biodegradation of Tertiary Butyl Mercaptan in Water

R. Karthikeyan[1]*, S.L.L. Hutchinson[2] and L. E. Erickson[3]

[1]*Biological and Agricultural Engineering, Texas A&M University, College Station, TX 77845-2117, USA*

[2]*Biological and Agricultural Engineering, Kansas State University, Manhattan, KS 66502, USA*

[3]*Chemical Engineering, Kansas State University, Manhattan, KS 66502, USA*

Abstract

Tertiary butyl mercaptan (TBM) belongs to the alkyl mercaptan family and possesses a characteristic odor. Tertiary butyl mercaptan (TBM) can enter aquatic environments through anthropogenic activities as well as the natural processes. Undefined microbial cultures from different soils along with a pure culture were used to study the biodegradation of TBM in water under aerobic conditions. There were about 17% losses in gas phase TBM concentrations attributed to abiotic losses over the period of 14 days. Environmental microbial consortium from sandy soils with low organic matter content and significantly lower heterotrophs resulted in the lowest biodegradation, only slightly higher than abiotic losses. Microbial cultures isolated from soils with previous contamination history resulted in higher degradation rates. In general, biodegradation of TBM in water followed first-order kinetics. The first-order kinetic constant ranged from 0.002 to 0.005 h^{-1}. TBM was partly degraded to two significant intermediate products and partly mineralized to CO_2 in water with mixed culture isolated from a petroleum contaminated soil. The half-life of TBM in water with this mixed culture was only six days. A Gram-ve bacterium isolated from a grey-water bioprocessor, *Alcaligines faecalis* subsp. *phenolicus* subsp. nov, was able to mineralize 50% of TBM within four days under laboratory conditions. The degradation rate was slightly increased with the addition of tertiary butyl alcohol while slightly inhibited with the addition of phenol.

Keywords: Malodorants; Remediation; Sulfur compounds; Water quality

Introduction

Potential human health problems related to sulfur compounds such as hydrogen sulfide and mercaptans may occur due to deterioration of sewer and wastewater networks [1-4]. Mercaptans are primarily produced from anaerobic decomposition of proteins and are grouped under Volatile Sulfur Compounds (VSCs). Mercaptans possess a characteristic odor. Tertiary butyl mercaptan (TBM) belongs to the alkyl mercaptan family and has been listed as hazardous under Occupational Safety & Health Administration (OSHA) regulations [5,6]. Exposure to TBM may cause dizziness, eye and skin irritation, nausea, and other allergic reactions [7]. Commercial uses of TBM include natural gas odorizing agent for leak detection and starting material in the manufacture of several agricultural chemicals. Physical and chemical properties of TBM are listed in Table 1 [7].

Based on a thorough literature review, Kalainesan et al. [5] reported that there were no studies on the degradation characteristics of TBM. They studied the rate and extent of disappearance of TBM in six different soils under aerobic conditions and reported that 99% of TBM disappeared in 33 days. This was attributed to biodegradation and/or chemical oxidation. The estimated first-order degradation rate constant ranged from 0.044 to 0.14 day^{-1}. Tertiary butyl mercaptan (TBM) can also enter aquatic environments through anthropogenic activities as well as the natural processes [8]. Abiotic and biotic losses determine the persistence of TBM in aquatic systems. Microcosm studies were conducted to study the biodegradation potential of TBM in water. Undefined microbial cultures from different soils along with a pure culture were explored to study the biodegradation of TBM under aerobic conditions. Experimental details and results from this laboratory study are presented in this manuscript.

Material and Methods

Microbial cultures

Undefined mixed microbial cultures from six different soils (Table

Property	Value
Molecular Formula	$C_4H_{10}S$ Molecular Weight 90.2 $g.mol^{-1}$
Synonyms	tert-butanethiol; 2-methyl-2-propanethiol Color and odor
	Colorless liquid; malodorous (skunk-like odor) Odor threshold
threshold	0.4 $mg.m^{-3}$
Density	0.807 g/mL at 16°C
Boiling point	62 to 68°C Freezing point 1.1°C
Flash point	-26°C
Vapor pressure	300 mm Hg at 16°C
Henry's constant	0.00719 $atm.m^3.mol^{-1}$ at 24°C
Solubility in water	1470 $mg.L^{-1}$
Octanol-water partition coefficient	138

Table 1: Chemical and physical properties of tertiary butyl mercaptan (TBM) [7].

2) were extracted as follows: One gram soil sample was extracted with 9 mL 0.7% NaCl buffer in a 10 mL test tube. The test tube was vortexed for 2 min to disperse microorganisms from soil particles. The solution was serially diluted to result in 10-2 dilution. One milliliter solution was taken from this dilution using a sterile disposable pipette and added to corresponding treatment bottle. The rest of the solution was stored at -10°C. Pure culture bacterium identified as *Alcaligines faecalis* subsp. *Phenolicus* subsp. Nov [9] was obtained from the researchers and maintained in M9 mineral medium at pH 7.0. M9 medium is a defined medium containing per liter of water, 7 g Na_2HPO_4, 3 g KH_2PO_4, 0.5

*Corresponding author: R. Karthikeyan, Associate professor, Biological and Agricultural Engineering, Texas A&M University, College Station, TX 77845-2117, USA, E-mail: karthi@tamu.edu

g NaCl, 1 g NH_4Cl, 0.02 g $CaCl_2$, and 0.2 g $MgSO_4$. One milliliter of pure culture was taken using a sterile disposable pipette and added to corresponding treatment bottle. The rest of the solution was stored at -10°C.

Water microcosms

Serum bottles (160 mL) containing 20 mL (approximately 20 g) autoclaved DI water (see next section for autoclaving protocol) and 10 µL of TBM were used as treatments and controls. One milliliter of microbial culture from six different soil solutions (10^{-2} dilution) and from the pure culture flask was added to each serum bottle. Serum bottles that did not receive any microbial culture were considered as controls. All treatments and controls were carried out in triplicates. All serum bottles were crimped tightly with a Teflon-lined septum that could withstand multiple punctures with negligible leakage, covered with aluminum foil to avoid any photodegradation, agitated continuously using an orbital-reciprocal shaker (Cole-Palmer, Vernon Hills, IL), and incubated at 22 ± 2°C.

Autoclaving protocol

Autoclaving of water was carried out as follows: the required amount of water was taken in Pyrex bottles (Fisher Scientific, Pittsburgh, PA) and placed in an autoclave at 125°C for one hour [10,11]. The bottles were allowed to cool for 2 hr, incubated at 30°C for 2 hr, and autoclaved again as before. This process was repeated three times to ensure maximum sterilization. All serum bottles, glassware, syringes, and other utensils were also sterilized. If not sterilized, pre-packed sterilized supplies were used throughout the experiment.

Analytical methods

All chemicals and standards were purchased through Fisher Scientific (Pittsburgh, PA). Headspace of each microcosm was sampled for TBM analysis at regular intervals. TBM was analyzed by gas chromatography using a method developed the first author of this manuscript and applied by [5]. Hewlett-Packard (HP, Avondale, PA; now Agilent Technologies) 5890 Series II Gas Chromatograph (GC) with ChemStation integration software was used in the analysis. TBM injection was 10 µL splitless. The column used was 30 m HP-1 (J&W Scientific, Folsom, CA) mega-bore column with internal diameter 0.53 mm and 4 µm film thickness. The carrier gas was H_2 (99.999%), the make-up gas was N_2 (99.999%), and the support gas was dry compressed air (zero-grade). The column oven temperature program began at 60°C for 2 min, increased at 10°C/min to 160°C, and then held at 160°C for 2 min. The injection port and flame ionization detector were kept at 160°C and 300°C, respectively.

The chromatogram obtained showed a distinct TBM peak appearing at about 2 min. This peak was verified using GC/MS by running standards and samples using the same temperature program with a minor modification (MS detector was kept at 280°C) and a similar column. The peak areas were obtained through automatic integration. Standard curves (with $r^2 = 0.999$) were developed using a set of standards prepared. Standards were run before, after, and in between every analysis to check for any deterioration. Blanks (acetone or methanol) were run in between samples to avoid any carryover due to column bleeding and cross contamination of samples and standards.

Bacterial enumeration

Heterotrophic bacteria were enumerated from each sample at the beginning of the experiments before being added to the microcosms. Ten-fold serial dilutions were performed as required to obtain appropriate colony numbers, and samples were plated in triplicate by the spread plate method on Difco nutrient agar. Plates were incubated at 34 ± 2°C for 24 hours and then counted. Plating was not done during the experiment because of the odor nuisance.

Results and Discussion

Degradation of tert-butyl mercaptan (TBM) in water with environmental microbial consortia

Initial incubation studies were conducted for a 14 day period (336 hrs) to observe the degradation of TBM in water. The dissipation of TBM in water microcosms over time is presented in Figure 1. There were about 17% losses in gas phase TBM concentrations in control microcosms over the period of 14 days (Figure 1). These losses may be attributed due to leakage in the septum, any biotic losses due to incomplete sterilization, and other abiotic losses such as photodegradation or dissolution of TBM in water. However, these losses are lower compared to microcosms inoculated with microbial cultures from different soils (Figure 1).

In general, treatments with microbial cultures from soils with no previous contamination history had significantly lower (p = 0.05) TBM dissipation rates compared to treatments with microorganisms from contaminated soils (Figure 1). This may be due to the fact that microorganisms extracted from contaminated soils were acclimated to organic pollutants (Table 2). About 85% of TBM was lost in 14 days due to abiotic and biotic degradation in water with environmental microbial consortium E, isolated from a petroleum contaminated site; whereas only 23 to 27% losses were observed in environmental microbial consortium resulting from sandy soils (A and B) with no known contamination history. It should be noted that these two sandy soils had the lowest organic matter and significantly lower total heterotrophic counts compared to other soils from which initial microbial cultures were obtained (Table 2,3).

TBM dissipation rates in microcosms with microbial cultures from methyl-tert-butyl-ether (MTBE) contaminated soil (mixed culture F) were slightly less than the dissipation rates with environmental microbial consortium E (Figure 1). Microcosms with mixed culture C and D had the dissipation rates of 64 and 55%, respectively. It is evident from the results that increase in the organic matter content and initial heterotrophic bacterial counts resulted in increase in TBM dissipation rates in water. If the soils had previous contamination history, the TBM dissipation rates were even higher. This phenomenon is very commonly observed in several field-scale and lab-scale biodegradation studies [12-18].

Degradation rates followed first-order decay in all microcosms except the ones with environmental microbial consortia A and B. The first-order rate constant for TBM degradation in water ranged from 0.002 to 0.005 hr^{-1} (Figure 1). Kalainesan et al. [5] reported similar degradation rates in soil, ranging from 0.0018 to 0.0058 hr^{-1}. However, it should be noted that even with dilute concentrations of heterotrophs inoculated in water, the organisms were able to degrade TBM comparable to the rates found in soils. Environmental microbial consortium E with the highest TBM degradation rate was utilized in further TBM degradation studies. We obtained reproducible TBM degradation rates in water (first-order decay constant = 0.005 hr^{-1}) during our second batch-incubation studies conducted over the period of 21 days using this consortium (Supporting Information, Figure S1). However, isolating an organism from this consortium that would degrade TBM in water was not successful. Addition of external carbon

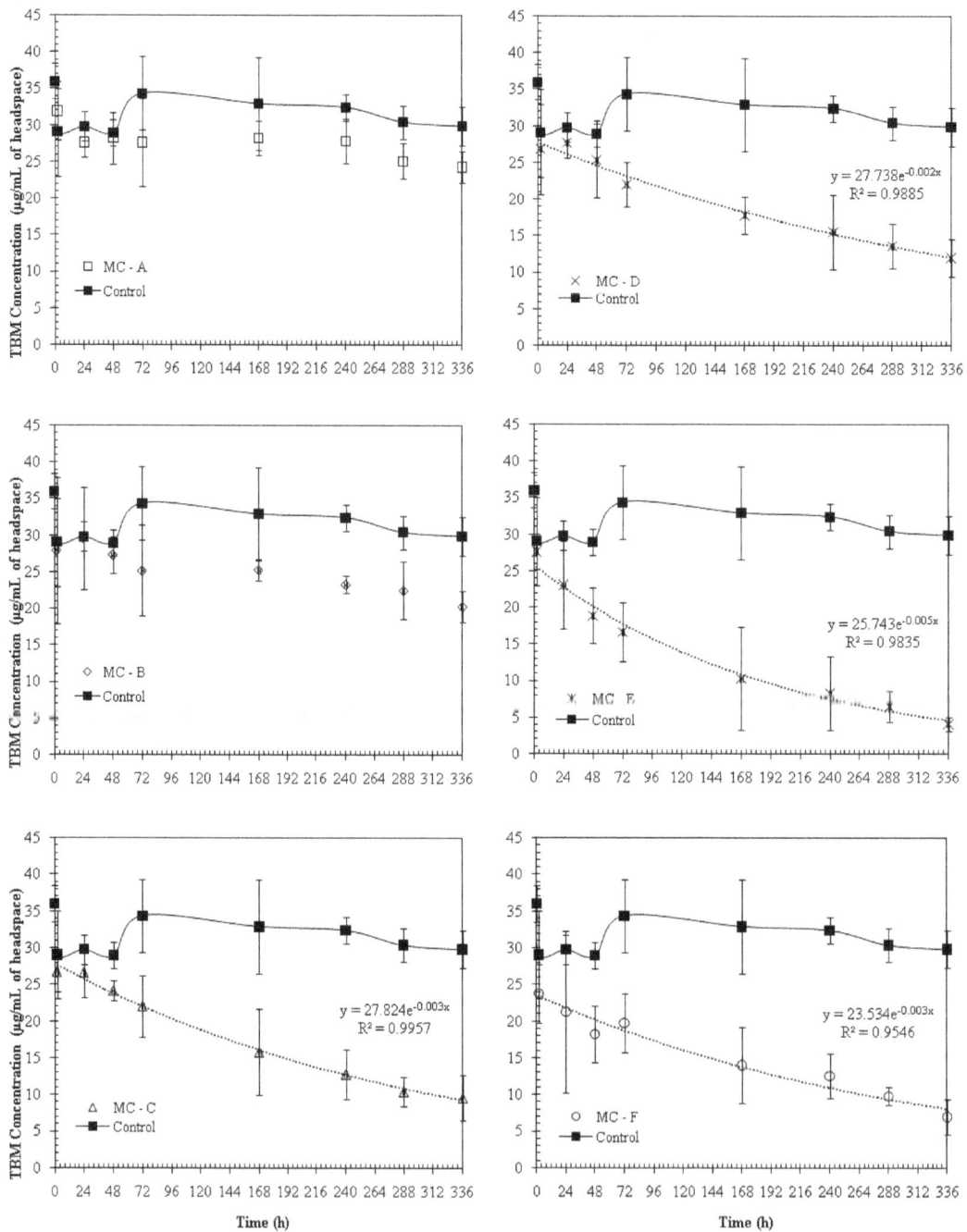

Figure 1: Gas phase TBM concentration in water microcosms with six different mixed microbial cultures (MC-A tough MC-F) and without microorganisms (control).

sources such as glucose and beef extract did not significantly increase the TBM degradation rates (*data not shown*).

Degradation of tert-butyl mercaptan (TBM) in water with *Alcaligines faecalis* subsp. *phenolicus* subsp. *nov*

The pure culture of *Alcaligines faecalis* subsp. *phenolicus* subsp. *nov* resulted in about 93% TBM degradation in water over 14 days. TBM degradation rates followed first-order kinetics with the rate constant of 0.007 hr^{-1} (Figure 2). This is a Gram-ve, coccobacillary bacterium

isolated from a grey-water bioprocessor in a previous study [9]. This organism was found to grow on either 0.1% phenol or 0.5% TBM as the sole carbon and energy source in solid media plates [5,9]. The doubling time of this bacterium in water with 0.5% TBM was reported to be 98 h [5]. However, that study was conducted with no replication and tert-butyl alcohol (TBA) was supplemented initially along with TBM. We were able to reproduce the results (doubling time approximately 99 h) with replications and no added TBA in our present study.

When TBA was added to the microcosms along with TBM at the

Soil Description	Organic Matter (%)	Sand (%)	Silt (%)	Clay (%)	Heterotrophs (CFU/g of soil) (n=6)
Sandy soil, no contamination history (MC-A)	0.3	96	3	1	14946 ± 508
Sandy loam soil, no contamination history (MC-B)	1.1	92	7	1	10648 ± 225
Farm soil, with some pesticide application history (MC-C)	2.4	26	57	17	219030 ± 1668
Greenhouse soil, with no known contamination (MC-D)	2	20	44	36	80285 ± 4210
Fort Riley Soil, with petroleum contamination (MC-E)	4.4	24	45	31	631830 ± 8652
Soil from MTBE lab study, with prior MTBE contamination (MC-F)	1.8	90	10	0	501386 ± 9232

Table 2: Characteristics of soils used in extracting microbial cultures.

Soil Description (n = 3)	Heterotrophs (CFU/mL
MC-A	215 ± 36a
MC-B	184 ± 40b
MC-C	4016 ± 435C
MC-D	1040 ± 320d
MC-E	6185 ± 506e
MC-F	5402 ± 294f

Table 3: Heterotrophic plate counts in 10-2 dilution at the beginning of the experiments. (Note: different letters mean the values are significantly different at p = 0.05).

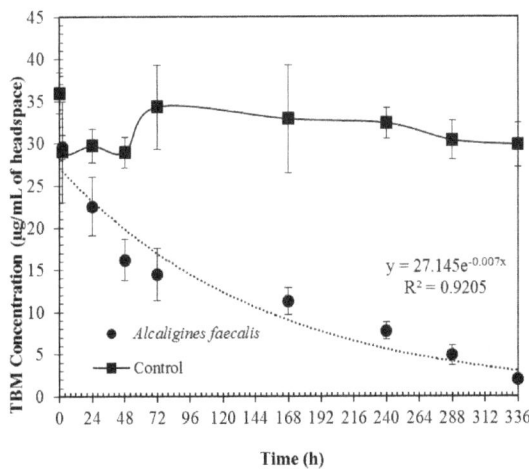

Figure 2: Gas phase TBM concentration in water microcosms with pure culture, *Alcaligines faecalis* subsp. *Phenolicus* subsp. *nov*.

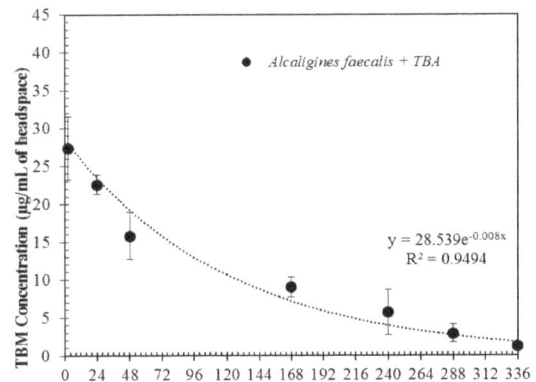

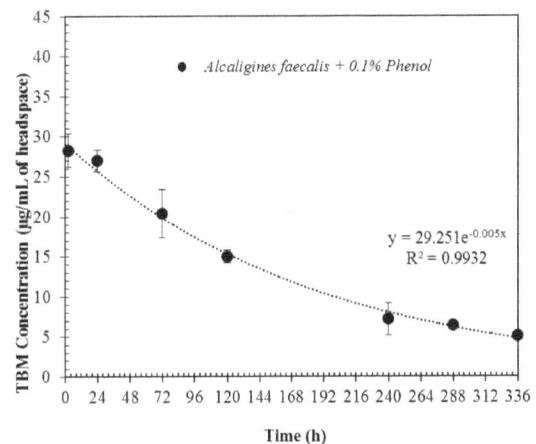

Figure 3: Gas phase TBM concentration in water microcosms with pure culture *Alcaligines faecalis* subsp. *Phenolicus* subsp. *nov* supplemented with external carbon sources.

beginning of the experiments, TBM degradation rates were slightly increased. The first-order rate constant was 0.008 hr^{-1} (Figure 3). This clearly shows that TBM can be cometabolically degraded when TBA is present. However, when 0.1% phenol was added to the microcosms with TBM at the beginning of the experiments, TBM degradation was inhibited. The first order rate constant of 0.005 hr^{-1} was obtained for TBM degradation. This may be due to *Alcaligines faecalis* subsp. *phenolicus* subsp. *nov* preferentially degrading phenol. Since we did not monitor the degradation of phenol, this statement is rather speculative than conclusive. However, it should be noted that this organism was originally isolated as a phenol-degrading organism [9].

Significance of the biodegradation study results

Figure 4 shows the relative gas phase TBM concentration with first-order decay over a period of one month for different treatments: Environmental microbial consortia C, D, E, and F; *Alcaligines faecalis* subsp. *phenolicus* subsp. *nov*, *Alcaligines faecalis* subsp. *phenolicus* subsp. *nov* with TBA, and *Alcaligines faecalis* subsp. *phenolicus* subsp. *nov* with 1% phenol. In general, the half-life of TBM in water ranged between

less than four days and 15 days (Figure 4). Half-life of TBM in water with microbial consortium obtained from soil with no contamination history (MC-D) was about 15 days where as the half-life was only four days in water with *Alcaligines faecalis* (Figure 4). Addition of external carbon sources such as TBA slightly decreased the half-life of TBM in water with *Alcaligines faecalis* while addition of phenol slightly increased the half-life at the given conditions. This demonstrates that while addressing the fate of TBM in water, the presence or absence of other organic substrates should be considered. When groundwater is contaminated with TBM either due to micro anaerobic conditions in the aquifer material or due to leakage of underground sewer lines and natural gas lines, *in situ* microorganisms can potentially degrade TBM. In our study, the environmental microbial consortia obtained from a previously contaminated site degraded about 97% of TBM in less than a month. However, in sandy aquifers where the microbial population

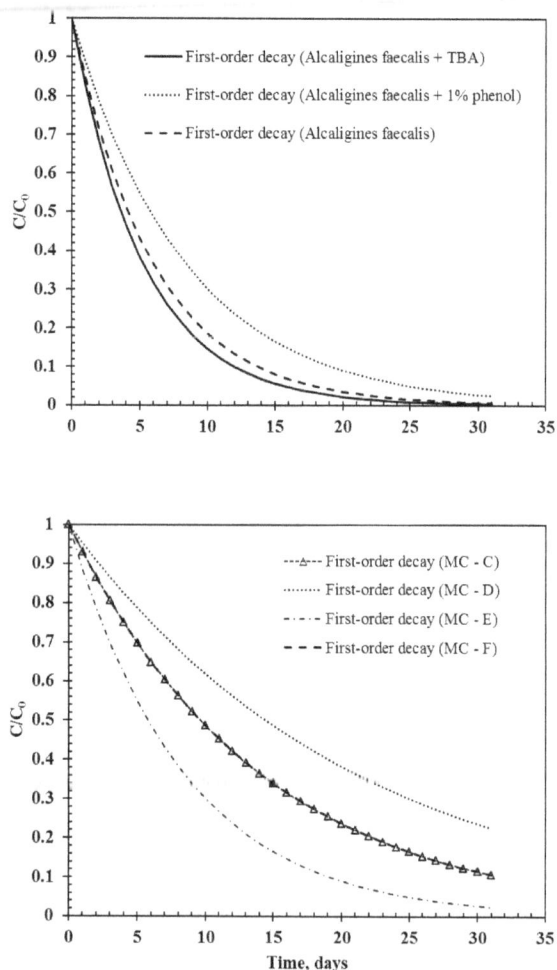

Figure 4: Relative gas phase TBM concentration with first-order decay with water microcosms over time.

is relatively low, TBM might persist for a longer time. These sandy aquifers can be enriched with pure cultures such as *Alcaligines faecalis* subsp. *phenolicus* subsp. *nov* used in this study along with trace amount of external organic carbon sources. Even though this culture is not yet tested at field-scale, bioaugmentation with pure cultures has resulted in promising end points in treating contaminated groundwater in the past [18-25].

Proposed biodegradation pathway

Headspace samples were analyzed using GC/MS during the second batch study with microbial consortia E. In all headspace sampling, two other significant peaks appeared in addition to TBM peak throughout the experiment (note: these two peaks appeared during the first experiment as well; however, these peaks were not confirmed using GC/MS). The first peak (retention time = 12.4 min) was identified as di-tertiary-butyl-disulfide (DTBD; molecular formula: $C_8H_{18}S_2$) with 95% similarity. The magnitude of this peak increased with time up to 10 days and then significantly decreased. Another peak appeared at 4.5 min, may be another intermediate product during the degradation of TBM, was identified as bis (1,1-dimethylethyl) disulfoxide (molecular formula: $C_8H_{18}O_2S_2$) with 88% similarity. The magnitude of this peak

decreased with time as well. Since we did not have standards for DTBD and sulfoxide, we could not report actual concentration of these two compounds in the gas phase.

The proposed TBM degradation pathway in water involves initial oxidation of TBM (specifically S-H bonds) to DTBD (S-S bond). This oxidation can be catalyzed by chemical catalysts such as MnO_2 or vanadium [26-29] as well as oxygenase enzymes present in microorganisms [30-37]. Kalainesan et al. [5] have reported that MnO_2 as well as soil microorganisms oxidize TBM to DTBD before complete mineralization. DTBD can be further oxidized to sulfoxides or sulfones. Our assumption is part of TBM is completely transformed to CO_2 and part is oxidized to disulfide and then further oxidized to sulfoxides before being oxidized to CO_2. It should be noted that even though 50% of TBM disappeared in six days during this treatment, only after 10 days was a significant CO_2 peak noticed in headspace. This clearly suggests that TBM undergoes intermediate oxidation steps; and complete mineralization in water could take longer than half life for mere disappearance of TBM [38].

In pure culture experiments with no external carbon sources, we did not notice any other significant peaks other than CO2. This may be due to complete mineralization of TBM to CO_2 by *Alcaligines faecalis* subsp. *phenolicus* subsp. *nov* in water. Since there are no toxic byproducts, bioaugmentation of TBM contaminated aquifers with this organism is advantageous to native organisms.

Conclusions

Laboratory studies were conducted to study the biodegradation potential of tertiary butyl mercaptan (TBM) in water. There were about 17% losses in gas phase TBM concentrations attributed to abiotic losses over the period of 14 days. Undefined microbial consortia from different soils with varying organic matter content and initial heterotrophic microorganisms were used to study the biodegradation of TBM under aerobic conditions. Environmental microbial consortium from sandy soils with low organic matter content and significantly lower heterotrophs resulted in the lowest biodegradation, only slightly higher than abiotic losses. Microbial consortia from soils with previous contamination history resulted in higher degradation rates.

In general, biodegradation of TBM in water followed first-order kinetics. The first-order kinetic constant ranged from 0.002 to 0.005h^{-1}. TBM was partly degraded to two significant intermediate products and partly mineralized to CO_2 in water with mixed culture isolated from a petroleum contaminated soil. The half-life of TBM in water with this mixed culture was only 6 days.

A Gram-ve bacterium isolated from a grey-water bioprocessor, *Alcaligines faecalis* subsp. *phenolicus* subsp. *nov*, was able to mineralize 50% of TBM within four days under laboratory conditions. The first-order rate constant was 0.007 h^{-1}. The degradation rate was slightly increased with the addition of tertiary butyl alcohol while slightly inhibited with the addition of phenol.

Acknowledgment

The authors thank Dr. James Urban for providing the pure culture of *Alcaligines faecalis* subsp. *phenolicus* subsp. *nov*. The authors acknowledge the help rendered by Dr. Scott Smith and Prini Gadgil in analyzing samples using GC/MS. The authors also thank Dr. Charles Rice for allowing us to use the autoclaving facility.

References

1. Gostelow P, Parsons SA (2000) Sewage treatment works odour measurement. Water Sci Technol 41: 33-40.

2. Hvitved-Jacobsen T, Raunkjaer K, Nielsen PH (1995) Volatile fatty acids and sulfide in pressure mains. Water Sci Technol 31: 169-179.

3. Hvitved-Jacobsen T, Vollertsen J, Nielsen PH (1998) A process and model concept for microbial wastewater transformations in gravity sewers. Water Sci Technol 37: 233-241.

4. Hvitved-Jacobsen T, Vollertsen J, Tanaka N (1998) Wastewater quality changes during transport in sewers–An integrated aerobic and anaerobic model concept for carbon and sulfur microbial transformations. Water Sci Technol 38: 257-264.

5. Kalainesan S, Erickson LE, Hutchinson SLL, Urban JE, Karthikeyan R (2006) Transformation of tertiary butyl mercaptan in aerobic environments. Environmental Progress 25: 189-200.

6. Suravajala A, Erickson LE, Bhandari A (2008) Sorption of Tertiary Butyl Mercaptan to Indoor Materials in Contact with Air or Water. J Environ Eng 134: 161-168.

7. MSDS (2004) Material Safety Data Sheet No. 35759. Natural Gas Odorizing Inc., Baytown, Texas.

8. Eberhardt A, Lopez E, Bucal'a V, Damiani DE (2003) Tertiary Butyl Mercaptan Adsorption in Soils. Determination of Kinetic and Transport Parameters from Experimental Data. International Journal of Chemical and Reactor Engineering 1:1542-6580.

9. Rehfuss M, Urban J (2005) Alcaligenes faecalis subsp. phenolicus subsp. nov. a phenol-degrading, denitrifying bacterium isolated from a graywater bioprocessor. Syst Appl Microbiol 28: 421-429.

10. Madigan M, Martinko J, Parker J (2003) Brock biology of microorganisms. Prentice Hall Upper Saddle River, NJ.

11. Paul E (2007) Soil microbiology, ecology, and biochemistry. Academic Press.

12. Dapaah SY, Hill GA (1992) Biodegradation of chlorophenol mixtures by Pseudomonas putida. Biotechnol Bioeng 40: 1353-1358.

13. Gibson DT, Cruden DL, Haddock JD, Zylstra GJ, Brand JM (1993) Oxidation of polychlorinated biphenyls by Pseudomonas sp. strain LB400 and Pseudomonas pseudoalcaligenes KF707. J Bacteriol 175: 4561-4564.

14. Nelson MJ, Montgomery SO, Mahaffey WR, Pritchard PH (1987) Biodegradation of trichloroethylene and involvement of an aromatic biodegradative pathway. Appl Environ Microbiol 53: 949-954.

15. Hopkins GD, Munakata J, Semprini L, McCarty PL (1993) Trichloroethylene concentration effects on pilot field-scale in-situ groundwater bioremediation by phenol-oxidizing microorganisms. Environ Sci Technol 27: 2542-2547.

16. Semprini L (1995) In situ bioremediation of chlorinated solvents. Environ Health Perspect 103: 101-105.

17. McCarty PL (1993) In situ bioremediation of chlorinated solvents. Curr Opin Biotechnol 4: 323-330.

18. Criddle CS, DeWitt JT, Grbić-Galić D, McCarty PL (1990) Transformation of carbon tetrachloride by Pseudomonas sp. strain KC under denitrification conditions. Appl Environ Microbiol 56: 3240-3246.

19. Dybas MJ, Barcelona M, Bezborodnikov S, Davies S, Forney L, et al. (1998) Pilot-scale evaluation of bioaugmentation for in-situ remediation of a car- bon tetrachloride-contaminated aquifer. Environ Sci Technol 32: 3598–3611.

20. Chang WK, Criddle CS (1997) Experimental evaluation of a model for cometabolism: Prediction of simultaneous degradation of trichloroethylene and methane by a methanotrophic mixed culture. Biotechnol Bioeng 56: 492-501.

21. Salanitro JP, Johnson PC, Spinnler GE, Maner PM, Wisniewski, HL, et al. (2000) Field-scale demonstration of enhanced MTBE bioremediation through aquifer bioaugmentation and oxygenation. Environ Sci Technol 34: 4152-4162.

22. Steffan RJ, Sperry KL, Walsh MT, Vainberg S, Condee CW (1999) Field- scale evaluation of in situ bioaugmentation for remediation of chlorinated solvents in groundwater. Environ Sci Technol 33: 2771-2781.

23. Strong LC, McTavish H, Sadowsky MJ, Wackett LP (2000) Field-scale remediation of atrazine-contaminated soil using recombinant Escherichia coli expressing atrazine chlorohydrolase. Environ Microbiol 2: 91-98.

24. Major DW, McMaster ML, Cox EE, Edwards EA, Dworatzek SM, et al. (2002) Field demonstration of successful bioaugmentation to achieve dechlorination of tetrachloroethene to ethene. Environ Sci Technol 36: 5106-5116.

25. Duba AG, Jackson KJ, Jovanovich MC, Knapp RB, Taylor RT (1996) TCE remediation using in situ, resting-state bioaugmentation. Environ Sci Technol 30: 1982-1989.

26. Murata S, Murata K, Kidena K, Nomura M (2004) A novel oxidative desulfurization system for diesel fuels with molecular oxygen in the presence of cobalt catalysts and aldehydes. Energy Fuels 18: 116-121.

27. Spivey JJ (1987) Complete catalytic oxidation of volatile organics. Ind Eng Chem Res 26: 2165-2180.

28. Shin MY, Park DW, Chung JS (2001) Development of vanadium-based mixed oxide catalysts for selective oxidation of H2S to sulfur. Appl Catal B 30: 409-419.

29. Margolis LY (1963) Catalytic oxidation of hydrocarbons. Advances in catalysis 14: 429-501.

30. Berry DF, Francis AJ, Bollag JM (1987) Microbial metabolism of homocyclic and heterocyclic aromatic compounds under anaerobic conditions. Microbiol Rev 51: 43-59.

31. Ohshiro T, Izumi Y (1999) Microbial desulfurization of organic sulfur compounds in petroleum. Biosci Biotechnol Biochem 63: 1-9.

32. Evans WC, Fuchs G (1988) Anaerobic degradation of aromatic compounds. Annu Rev Microbiol 42: 289-317.

33. Juliette LY, Hyman MR, Arp DJ (1993) Inhibition of ammonia oxidation in Nitrosomonas europaea by sulfur compounds: thioethers are oxidized to sulfoxides by ammonia monooxygenase. Appl Environ Microbiol 59: 3718-3727.

34. Xu P, Yu B, Li FL, Cai XF, Ma CQ (2006) Microbial degradation of sulfur, nitrogen and oxygen heterocycles. Trends Microbiol 14: 398-405.

35. Spain JC (1995) Biodegradation of nitroaromatic compounds. Annu Rev Microbiol 49: 523-555.

36. Benedik MJ, Gibbs PR, Riddle RR, Willson RC (1998) Microbial denitrogenation of fossil fuels. Trends Biotechnol 16: 390-395.

37. Karthikeyan R, Bhandari A (2002) Anaerobic biotransformation of aromatic and polycyclic aromatic hydrocarbons in soil microcosms: A review. Journal of Hazardous Substance Research 3: 1-19.

38. Dague RR (1972) Fundamentals of odor control. J Water Pollut Control Fed 44: 583-594.

11

Bioremediation Rate of Total Petroleum Hydrocarbons from Contaminated Water by *Pseudomonas aeruginosa* Case Study: Lake Albert, Uganda

Kiraye M*, John W and Gabriel K

Department of Chemistry, Makerere University, Kampala, Uganda

Abstract

Uganda is currently exploiting petroleum products. This is mainly around the fresh water bodies. However, these water bodies are habitants for several aquatic organisms and also the main drinking water sources. Despite the fact that they are known for several uses both ecological and economical, they are likely to be seriously polluted by crude oil petroleum hydrocarbons (PHs). Therefore they will require to be treated by ecologically friendly methods. *Pseudomonas aeruginosa* naturally habits Ugandan water bodies and it's known for no health hazards to human (after boiling the water) and to aquatic organisms. Therefore multiplying its numbers in aquatic environments has no health implications yet it's known for degrading PHs. Thus the current study aimed at determining the rate at which *Pseudomonas aeruginosa* can remediate PHs from water of Lake Albert.

Method: Water from Lake Albert was collected to laboratory, contaminated with 10% m/v PHs (100g/L). This was then inoculated with *Pseudomonas aeruginosa* (turbidity of 0.04 absorbance at a wave length of 600 nm) in a 1cm cuvette containing about 3.0×10^7 colony-forming unit (CFU)/mL. The waters were left at room temperatures to replicate the temperature of the natural water body in Uganda.

Results: Results showed that the initial rate, R_{biol} was 32.3 g/liter per day for n-hexane soluble PHs. Also, the maximum amount removed when the rate reduced to zero was 89.3/liter. The bioremediation process followed second order kinetics with half-life of 3.9 days. This means the original amount will reduce to half the original amount after 3.9 days (about 93.6 hrs). *Pseudomonas aeruginosa* significantly ($p=0.03$) ($p<0.05$) remediates PHs from Lake Albert water with maximum removal rate between day 1 and day 3. However, physico-chemical factors for example temperature, pH were not investigated in this current study.

Keywords: Bioremediation rate; Lake Albert; Petroleum hydrocarbons; *Pseudomonas aeruginosa*

Introduction

Uganda is presently trying to explore and engage in the exploitation of petroleum products. This is largely around the fresh water bodies in the Albertine Graben. On the other hand, the water bodies in Albertine graben are habitants for numerous water organisms and also the focal drinking water point sources. In spite of the fact that Lake Albert waters are known for quite a lot of uses both ecological and economical, they are expected to be extremely polluted by crude oil petroleum hydrocarbons (PHs). Consequently they will need to be dealt with by ecologically friendly methods. Bioremediation is one of such ways of which microorganisms of the species *pseudomonas aeruginosa* were used to biodegrade unsafe organic pollutants to ecologically safer toxic doses [1]. These microorganisms embrace other bacteria and fungi like yeast and moulds [2]. These microorganisms predominantly have been considered petroleum hydrocarbon biodegrading mediators living in the environment freely.

The PHs is progressively becoming water contaminants of great worry within the environment [3]. They have the potential to dissolve in lipids within vulnerable water organisms meaning they can bio accumulate in the food chain and can be delivered to other trophic levels of the food chain [4].

Furthermore, Spills of petroleum Hydrocarbons occurring on water usually are far more harmful than the spill on land [5]. The oil exploration industry in Uganda creates susceptibility of the country to petroleum related ecological encounters as well as spillage. Regrettably, there are no well-studied ecologically friendly means for bioremediation of petroleum hydrocarbons spillage in aquatic environment within the country. Oil exploration industry currently is in Buliisa District whose general population obtains water from point water sources that can be vulnerable to PHs contamination in case of crude oil spills [6].

The study aimed at exploring the rate at which *Pseudomonas aeruginosa* can reduce petroleum hydrocarbons spillage that may occur on Lake Albert water in future.

Materials and Methods

Study area and sample collection

Samples of water for this work were collected from Lake Albert at a point (01°32.032N, 03°57.958E) called Kaiso, selected because the oil exploration upstream facility is in the vicinity. Lake Albert is located in western Uganda in the Albertine region,

Two (2) liters of water from Lake Albert was collected in sterile bottles. These were then placed on Ice in cooler boxes and transported to Makerere University Chemistry pesticide laboratory. The bottles were then refrigerated for 10 hours.

*****Corresponding author:** Kiraye M, Department of Chemistry, Makerere University, Kampala, Uganda, E-mail: mickiraye@gmail.com

Contaminated sample preparation

The PHs was purchased from oil and gas market in Kampala (Uganda). The experimental bottles along with the control were contaminated with the PHs up to 10%m/v (100 g/L). The control sample of Lake Albert water was autoclaved before being contaminated with PHs. [Both the control and the natural (experimental) water were contaminated with 10% m/v PHs (100g/L)]. The experimental bottle was then inoculated with *Pseudomonas aeruginosa* (turbidity of 0.04 absorbance at 600 nm) (3.0×10^7 colony-forming units (CFU)/mL) [7]. Both control and the experimental bottles were left at room temperatures to replicate the temperature of the natural water body in Uganda.

Culturing

Growth of *Pseudomonas aeruginosa* was done in a sterile nutrient broth (100 ml) incubated at 37°C for a period of 1.5 hours that was expected for the log phase of these species of bacteria [7].

Bioremediation procedure

Aliquot volume of water contaminated with petroleum hydrocarbon (10 mL), was introduced in a 250 ml flask in which there was nutrient broth (100 mL). And aliquot of a starter culture (100 µl) containing *Pseudomonas aeruginosa* of turbidity absorbance of 0.04 at 600 nm was added. The unresolved complex mixture was shaken at a speed of 180 r/min for 24 hours at room temperature using a shaker model THZ-82. The activity of bacterial was momentary halted by decreasing the temperature of the resulting mixture to about 2°C to 8°C after every 24 hours.

Petroleum hydrocarbons extraction

The Petroleum hydrocarbons extraction from water was done using n-hexane following a method styled by A UNEP/IOC/IAEA 1992 method for PHs. Unresolved complex mixtures approach for Gas Chromatography-Mass Spectroscopy quantification was used for determination of amount removed.

Instrumentation analysis

An Agilent 6890N gas chromatograph (GC), combined with Mass spectroscopy detector (5975) was set. A sample of 1.0 µL aliquot of the extract was injected using the injector port held at150°C and run in split mode. Helium carrier was used to sense PHs at a split ratio of 1:20. The following temperature-programme was used: Preliminary temperature at 95°C for 1 min followed by 95-190°C at 20°C/min then190-250°C at 15°C/min and 250-300°C at 25°C/min for 3.0 min, resulting in a total run time of 18.5 min. The detector temperature was held at 150°C. Software, Agilent Chemstation was used for acquisition of the chromatogram data calculations using unresolved complex mixtures approach. The entire peaks were integrated to determine the total area counts for every sections of the chromatogram before and after bioremediation (Table 1). Dividing the difference in total area counts before and after bioremediation by the total area before bioremediation gave the biodegrading amount removed [8]. Biodegrading amount removed was used to represent the concentration of the total petroleum hydrocarbon used up during bioremediation activity in grams per 10ml of PHs spreading on Lake Albert water for every 24 hours for 7 days.

A software, Minitab17 [9] statistics package developed at the Pennsylvania State University by researchers was used to analyze results. From the analyzed results, various slopes of the tangents that signify bioremediation rate (R_{bio}) in grams per day at 0.60, 1.20, 2.10,

Time /days	Amount removed (m_p) (g/L)	Amount left (m_l) (g/L)	Amount removed (g/L)-control expt.
1	22	78	31
3	80	20	12
4	75	25	22
5	78	22	9
6	97	3	21
7	83	17	32

Table 1: Showing the relationship between amounts of PHs removed and amount left in grams per litre versus time in days.

Bioremediation Rate/gday^{-1}	Amount removed (m_p)(g/L)	Amount left (m_l) (g/L)	(Amount removed,m_p)^2/g^2/L^2
29	18	86.71	324
26	34	69.77	1156
20	55	48.36	3025
17	66	31.75	4356
13	75	27.28	5625
10	82	17.08	6724
6	87	12.94	7569
2	88	10.94	7744

Table 2: Showing the relationship between Bioremediation Rate per gram per day versus (Amount removed)^2/g^2/L^2.

3.00, 3.30, 4.20, 4.80 and 5.40 days of the remediation process along the curve were determine and the following results got as in Table 2.

Results and Discussions

Amount removed and left by *Pseudomonas aeruginosa* (m_l)(g/L)

This work is the first bioremediation study on Lake Albert fresh water. The concentration of the PHs removed from Lake Albert contaminated samples by *Pseudomonas aeruginosa*, are in Table 1.

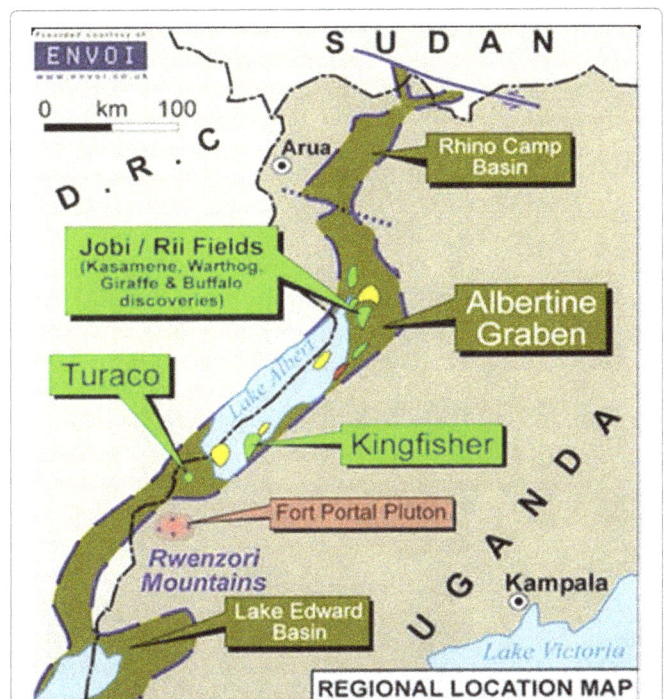

Figure 1: the map of Uganda, showing Albertine Graben, picture from regional location map.

The concentrations of what was left were also recorded. The highest amount removes was registered on day six, the point at which the rate of bioremediation was almost zero. A similar work [10] reported the maximum removed amount in this period of time. The observation from Table 1 showed that some PHs was also registered in the control experiment even though *Pseudomonas aeruginosa* were not present here. This suggested that, Lake Albert perhaps contained some PHs at the time when the water sample was picked. This agrees with the historic quotations that oil in Lake Albert (Uganda) was discovered following the seepage observations [11].

In Figure 1, the variation of amount of PHs removed and amount left in grams per litre with time in days were followed to generate Figure

2 that showed that bioremediation reaction followed second order though [12] suggested a first order kinetics, Figure 2 demonstrated this.

Bioremediation rate and its half-life

To make the relationship between bioremediation rate in gram per day versus amount removed squared stand out, a graph in Figure 3 below was drawn that depicted a linear state of connectedness typical of second order reactions [13] of the form Rate= $c+(-k)A^2$ with a second order rate constant,-k, since removal of PHs were followed indicated by a negative sign. The constant, c, put into consideration of other factors that could be involved and a concentration term-A, indicated the amount removed in grams per liter.

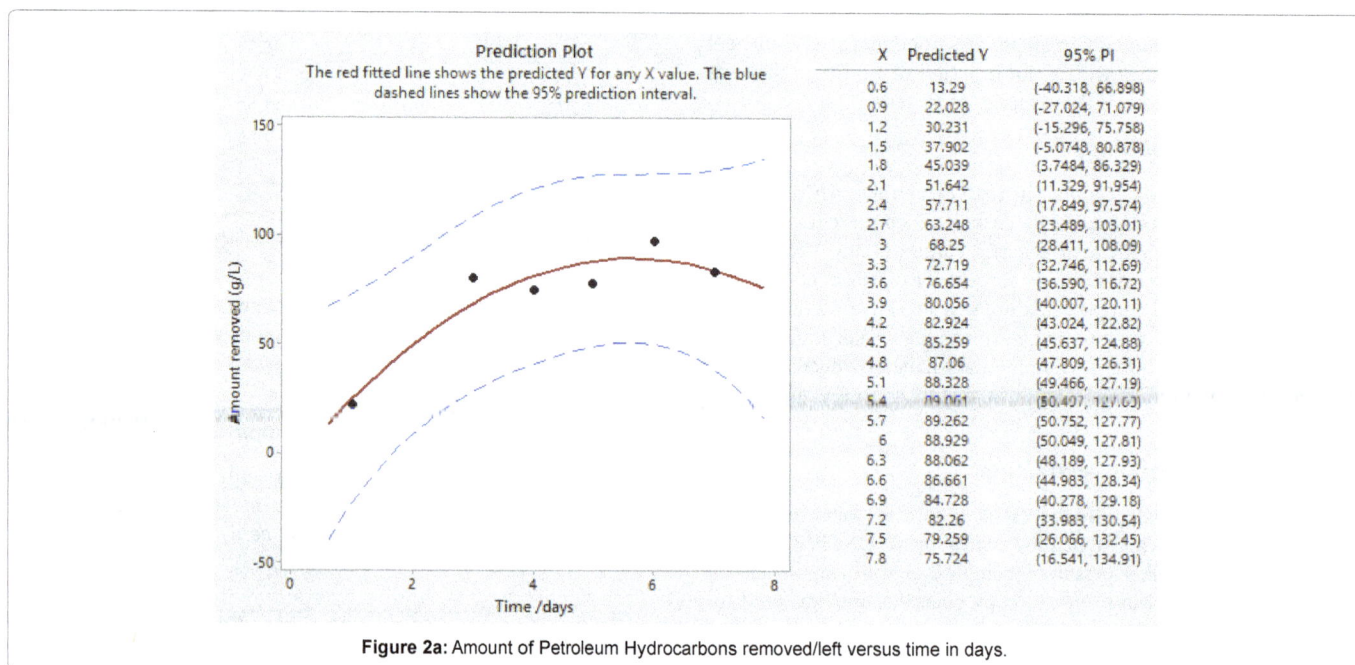

X	Predicted Y	95% PI
0.6	13.29	(-40.318, 66.898)
0.9	22.028	(-27.024, 71.079)
1.2	30.231	(-15.296, 75.758)
1.5	37.902	(-5.0748, 80.878)
1.8	45.039	(3.7484, 86.329)
2.1	51.642	(11.329, 91.954)
2.4	57.711	(17.849, 97.574)
2.7	63.248	(23.489, 103.01)
3	68.25	(28.411, 108.09)
3.3	72.719	(32.746, 112.69)
3.6	76.654	(36.590, 116.72)
3.9	80.056	(40.007, 120.11)
4.2	82.924	(43.024, 122.82)
4.5	85.259	(45.637, 124.88)
4.8	87.06	(47.809, 126.31)
5.1	88.328	(49.466, 127.19)
5.4	89.061	(50.407, 127.63)
5.7	89.262	(50.752, 127.77)
6	88.929	(50.049, 127.81)
6.3	88.062	(48.189, 127.93)
6.6	86.661	(44.983, 128.34)
6.9	84.728	(40.278, 129.18)
7.2	82.26	(33.983, 130.54)
7.5	79.259	(26.066, 132.45)
7.8	75.724	(16.541, 134.91)

Figure 2a: Amount of Petroleum Hydrocarbons removed/left versus time in days.

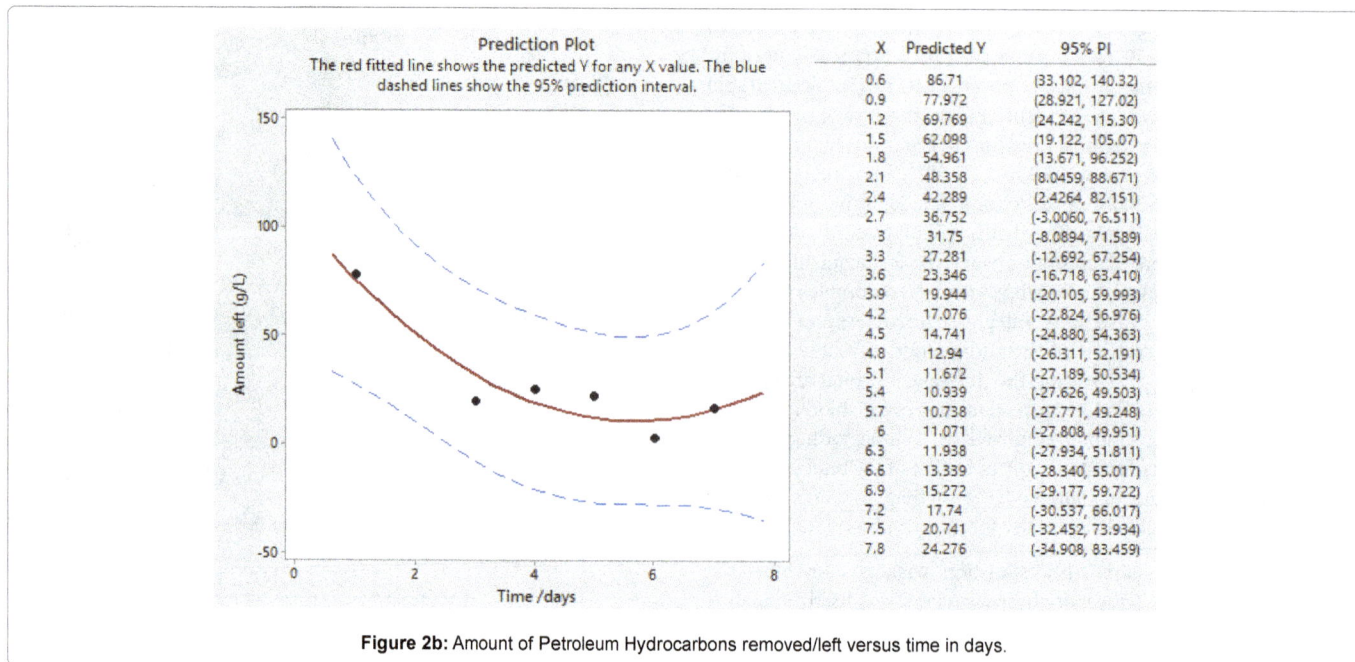

X	Predicted Y	95% PI
0.6	86.71	(33.102, 140.32)
0.9	77.972	(28.921, 127.02)
1.2	69.769	(24.242, 115.30)
1.5	62.098	(19.122, 105.07)
1.8	54.961	(13.671, 96.252)
2.1	48.358	(8.0459, 88.671)
2.4	42.289	(2.4264, 82.151)
2.7	36.752	(-3.0060, 76.511)
3	31.75	(-8.0894, 71.589)
3.3	27.281	(-12.692, 67.254)
3.6	23.346	(-16.718, 63.410)
3.9	19.944	(-20.105, 59.993)
4.2	17.076	(-22.824, 56.976)
4.5	14.741	(-24.880, 54.363)
4.8	12.94	(-26.311, 52.191)
5.1	11.672	(-27.189, 50.534)
5.4	10.939	(-27.626, 49.503)
5.7	10.738	(-27.771, 49.248)
6	11.071	(-27.808, 49.951)
6.3	11.938	(-27.934, 51.811)
6.6	13.339	(-28.340, 55.017)
6.9	15.272	(-29.177, 59.722)
7.2	17.74	(-30.537, 66.017)
7.5	20.741	(-32.452, 73.934)
7.8	24.276	(-34.908, 83.459)

Figure 2b: Amount of Petroleum Hydrocarbons removed/left versus time in days.

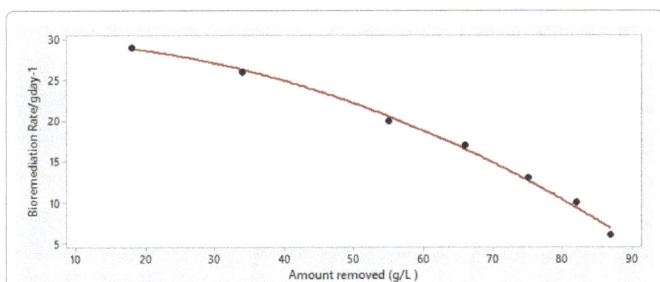

Figure 3: Regression for Bioremediation Rate/gday-1 vs amount removed (g/L).

In (Figure 2a) Regression for Bioremediation Rate/gday-1 vs. (Amount removed)^2/g^2/L^2

The relationship between bioremediation rate and the amount removed by microbes stood out as in Figure 2b indicating an upward curving line that is typical of the second order kinetics [13].

The rate equation for the reaction turned out to be: Rate = 30.31-(3.3×10^{-3}) $(R_{bio})^2$ in agreement with second order kinetics of the form Rate = $k(R_{bio})^2$ keeping other factor constant, giving a half-life (T½) period of $T_{\frac{1}{2}} = \frac{1}{k(R_{bio})_0}$ days and rate constant, k describing the degradation process as $k = \frac{1}{T_{\frac{1}{2}}(R_{bio})_0}$ Lg^{-1}day^{-1} the negative gradient indicated removal of PHs as the process was going on. The slope of the graph in Figure 2b gave the value of the rate constant, k, as 3.3×10^{-3} Lg^{-1}day^{-1} as and Figure 2b gave the initial amount of PHs remediation as 78.0 g/L at day one (start time of the experiment) resulting into a half-life period of 3.89 days or simply 3.9 day this is slightly higher than a half-life period of 2.19 days [10] reported in the bioremediation study.

Bioremediation rate versus reciprocal of amount of PHs removed

Figure 4 shows realistically the relationship between bioremediation rate as the reciprocal of time and dilution as the reciprocal of the

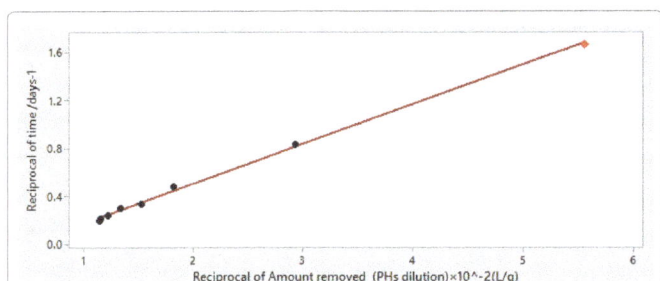

Figure 4: Bioremediation Rate/gday^{-1} versus reciprocal of Amount removed.

Time/days	Bioremediation Rate/gday^{-1}	Reciprocal of Amount removed (PHs dilution)×10^{-2}(L/g)	Reciprocal of time /days^{-1}
0.60	29	5.56	1.67
1.20	26	2.94	0.83
2.10	20	1.82	0.48
3.00	17	1.52	0.33
3.30	13	1.33	0.30
4.20	10	1.22	0.24
4.80	6	1.15	0.21
5.40	2	1.14	0.19

Table 3: showing the variation of bioremediation Rate per day (reciprocal of time) with reciprocal of Amount removed in litres per gram.

amount removed reflecting the fashion how *Pseudomonas aeruginosa* used up PHs, Table 3.

As depicted by the graph in Figure 4, it was observed that, bioremediation rate demonstrated direct proportionality with PHs reduction from the water surface (dilution). This is in agreement [14] who asserted that the biodegradation process follows kinetic models, and indeed it was found to follow 2nd order kinetics. To be precise, the rate of removal is proportional to the PHs diluted concentration; this was demonstrated in Figure 4. PHs reduction from the surface of Lake Albert water gets bulky as Bioremediation Rate gets greater and when PHs dilution grows lesser, the Bioremediation Rate grows lesser too. The meaning of this is that as the dilution increases, the removal rate also increases, basically applicable for heavy spills of Petroleum hydrocarbons on Lake Albert water surface that will elicit a heavily greater rate of removal by *Pseudomonas aeruginosa* of such a spill [15]. The explanation to this trend is done by means of the aspects that act as the source of Bioremediation Rate to surge when there is greater quantity of PHs. In this work, *Pseudomonas aeruginosa* behavior showed that the PHs were identified as their substrate, denoting that larger quantity of PHs would promote the growth of large population of these microorganism capable of removing the PHs until saturation point is re-reached as shown in Figure 1.

Conclusions

Pseudomonas aeruginosa remediates Petroleum Hydrocarbons from Lake Albert water significantly (p<0.05) with maximum removal rate between the first day and the third day. However, physico-chemical factors for example temperature, pH were not investigated in this current study.

References

1. Barathi S, asudevan N (2001) Utilization of petroleum hydrocarbons by pseudomonas fluoresces isolated from a petroleum-contaminated soil. Environment International 26: 413-416.

2. Prince RC (2002) Bioremediation: An overview of how microbiological processes can be applied to the cleanup of organic and inorganic environmental pollutants. Encyclopedia of Environmental Microbiology.

3. Lee JY, Cheon JY, Lee KK, Lee SY, Lee MH (2001) Statistical evaluation of geochemical parameter distribution in a ground water system contaminated with petroleum hydrocarbons. Journal of Environmental Quality 30: 1548-1563.

4. Barron MG, Kaaihue L (2001) Potential for photoenhanced toxicity of spilled oil in prince william sound and gulf of alaska waters. Marine pollution bulletin 43: 86-92.

5. Blumer M, Ehrhardt M, Jones J (1973) The environmental fate of stranded crude oil. Deep Sea Research and Oceanographic Abstracts 20: 239-259.

6. Vokes R (2012) The politics of oil in uganda. African Affairs ads017 111: 303-314.

7. Goldman E, Green LH (2015) Practical handbook of microbiology: CRC Press.

8. Reddy (1999) Gc-ms analysis of total petroleum hydrocarbons and polycyclic aromatic hydrocarbons after the north cape oil spill. Marine Pollution Bulletin 38: 126-135.

9. Ryan TA, Joiner BL, Ryan BF (2004) Minitab™: Wiley Online Library.

10. Van Gestel K, Mergaert J, Swings J, Coosemans J, Ryckeboer J (2003) Bioremediation of diesel oil-contaminated soil by composting with biowaste. Environmental Pollution 125: 361-368.

11. Wayland EJ (1925) Geological survey of uganda: Petroleum in uganda. Menoir 63: 431-433.

12. Buchanan R, Whiting R, Damert W (1997) When is simple good enough: A comparison of the gompertz, baranyi, and three-phase linear models for fitting bacterial growth curves. Food Microbiology 14: 313-326.

13. Robinson WR (2013) Rate laws from rate versus concentration data (differential rate laws).

14. Yan Z, Song N, Cai H, Tay JH, Jiang H (2012) Enhanced degradation of phenanthrene and pyrene in freshwater sediments by combined employment of sediment microbial fuel cell and amorphous ferric hydroxide. Journal of hazardous materials 199: 217-225.

15. Zhang Y (2008) Geochemical kinetics: Princeton University Press.

12

Patent Analysis on Bioremediation of Environmental Pollutants

Shweta Saraswat*

National Institute of Science, Communication and Information Resources (CSIR) 14, India

Abstract

The scientific literature published in peer reviewed journals is the limited source of information for most of the scientists and academicians. As neither all researches nor their technological strategies are published in papers, therefore, lots of information remains missing. The systematic patent analysis using authenticated patent database makes it possible to find out technological strategies adopted by potential researchers. In spite of that, the research gaps along with research directions in a particular research field can also be identified. The aim of present study was to analyze the patenting trends in bioremediation technologies for environmental pollutants. This analysis was based on various criteria i.e. patenting trend over time, country-wise and assignee-wise comparisons and types of technology used in various patents. The Delphion database was used to retrieve bioremediation patents filed during 1990 to 2013. Out of 443 patents, United States accounted for maximum patent publications, followed by China, Japan, Korea, India etc. Yearly analysis revealed that maximum (35) patents were published in 2005. Out of 31 patent assignees, the United States Secretary of Agriculture obtained maximum patents followed by Exxon Research and Engineering Co. Comparison among technological groups indicated that maximum patents (33%) were associated with technologies for reclamation of contaminated soil followed by wastewater and sewage treatment technologies (22%) and use of micro-organisms (20%) natural as well as genetically engineered for bioremediation of environmental pollutants. Further, identification of process/product based utility of screened patents revealed that 80% patents were directed towards processes of bioremediation while 20% were products for assisting bioremediation.

Keywords: Patent analysis; Bioremediation; Phytoremediation; Phytoextraction; Environmental pollutants

Introduction

To meet the world's growing needs; global industrial revolution has contributed in polluting the environment by producing extremely toxic chemical and biological wastes in bulk. To protect the environment from existing pollutants, there is a need to promote innovation in pollution abatement technologies. After various international environmental agreements, governments of different countries have focused on green product and process innovations to minimize environmental risk, but there has been little interest in developing eco-friendly technologies for removing the environmental pollutants. Considering this, bioremediation has now been emerged as a technological advancement to decontaminate the polluted sites either aquatic or terrestrial, in an effective, eco-friendly manner without generating secondary wastes. Bioremediation involves naturally occurring organisms to break down or neutralize inorganic and organic pollutants into less toxic or non-toxic forms. Some examples of bioremediation related technologies are phytoremediation, rhizofilteration, bioaugmentation, biostimulation, landfarming, bioreactor, and composting. As a part of bioremediation, Phytoremediation employs metal accumulating plants for removing toxic wastes via absorption, adsorption and extraction of contaminants from water and soil [1,2]. Thus, better environmental protection can be achieved by promoting eco-innovative bioremediation technologies.

Patent analysis of bioremediation could assist the scientists, stakeholders (technologists, business leaders, attorneys, etc.), policy planners and researchers to access technology updates, develop new process and product, plan future research strategies and take key decisions for developing R&D investment plans for more economic and environmental gain. Earlier workers also measured environmentally motivated innovations, such as pollution control technologies and green energy technologies, and for general purpose technologies with environmental benefits. Lanjouw and Mody [3] counted the number of patents in nine environmental fields (viz. Industrial air pollution, water pollution, vehicle air pollution, solid waste, incineration of waste,

alternative energy, oil spills, radioactive waste and recycling and reusing waste), Nameroff et al. [4] studied green chemistry patents, Johnstone [5] analysed renewable energy patents, and Sun et al. [6] studied the pattern of environmental patents by ownership in China. Since, patents are the best measure of innovative activities, therefore the objectives of present study were (i) to analyse overall patent activity in the field of bioremediation, (ii) to assess global distribution of bioremediation patents, their patenting trend over time and assignees, (iii) to identify potential areas for future research and development in bioremediation technologies.

Methodology

Patent databases searched

The relevant patents in the context of bioremediation were retrieved from different databases on the basis of priority of filing during 1990 to 2013. The databases searched included publicly available e.g. USPTO, EPO, SURFIP, WIPO and FREEPATENTSONLINE and subscription based database Delphion . Delphion is an authenticated databases which covered United States Patents - Applications (US), United States Patents - Granted (US), Derwent World Patents Index (DWPI), European Patents - Applications (EP-A), European Patents - Granted (EP-B), German Patents - Applications, German Patents - Granted, INPADOC Family and Legal Status, Patent Abstracts of Japan, Switzerland and WIPO PCT Publications. As Delphion do not cover

*Corresponding author: Shweta Saraswat, National Institute of Science, Communication and Information Resources (CSIR) 14, India
E-mail: shwetasara@gmail.com

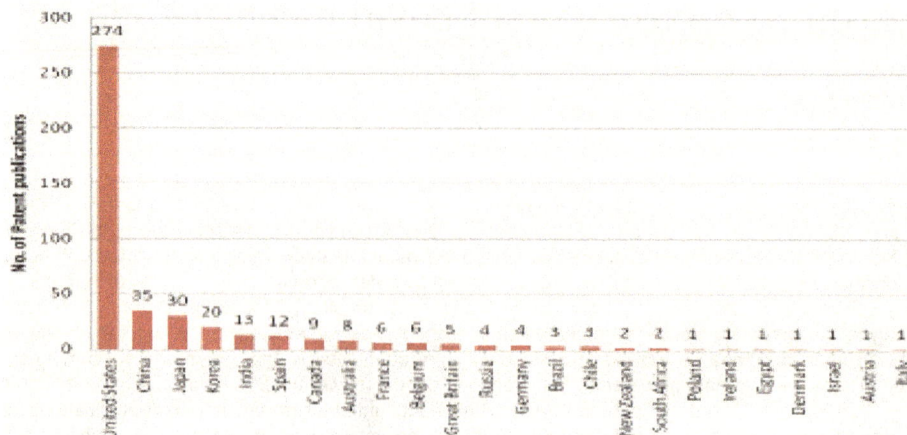

Figure 1: Geographical distribution of bioremediation patents.

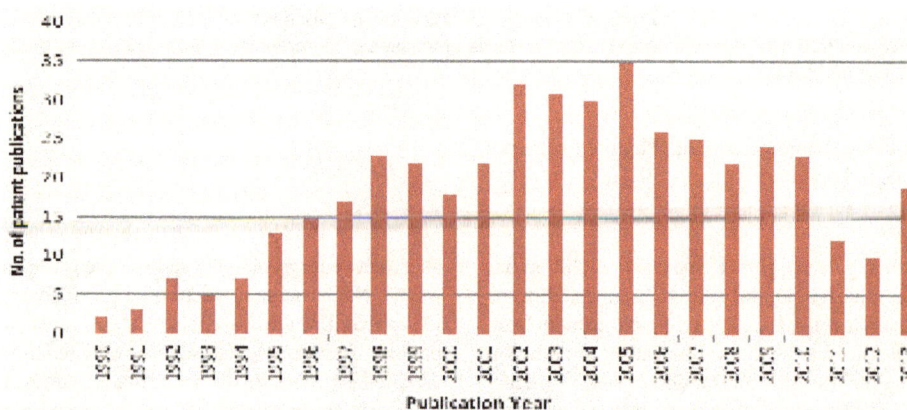

Figure 2: Yearly trend of bioremediation patents.

Indian Patent data of granted and filed patent application, therefore, the Indian patents on bioremediation were retrieved from Indian Patent Information Retrieval system separately.

Search strategy and analysis

To search patent publications in the field of bioremediation, the methodology was mainly based on appropriate keyword searches in title, abstract and claims fields of all patents available in different databases. The search algorithm for parent string was used as: ("Bioremediation" or "Phytoremediation" or "Phytoextraction") AND ("Plant" or "bacteria" or "fungal" or "microorganism"). A total of 627 patent documents were retrieved. The abstract of each single patent was scrutinized in order to determine relevancy of patent. The irrelevant patents, not related to bioremediation technologies, were eliminated. This step was undertaken to avoid the potential error of having unrelated patents, thereby improving the quality of searched patent data set. Finally, 443 patent documents were screened according to their relevance. The retrieved data was stored in a MS-Excel worksheet, analysed and interpreted in order to understand geographical distribution of patents, patenting trend over time and major assignees that holds proprietary rights. The patent data have been segregated primarily into five sectors comprising (i) Government (ii) Corporate (comprising industrial organizations and corporate bodies), (iii) Academic (comprising

Academic Institutes and Universities), (iv) Institutional (comprising Research Institutes), and (v) Individuals (comprising individual or group of individuals). The screened bioremediation patents were also classified as per International Patent Classification (IPC) based on their technology areas and utility.

Results and Discussion

Geographical distribution of patents

Patents are the indicator of country's economic growth [7] and patenting activities across the countries transfer and disseminate the technology contained in the patent [8]. Among top five countries, United States (US) was the forerunner sharing 61.85% patents in bioremediation technologies followed by China (7.9%), Japan (6.77%), Korea (4.51%) and India (2.93%) (Figure 1). The other countries viz. Spain, Canada, Australia, France, Belgium, Great Britain, Russia, Germany, Brazil, Chile were less active in bioremediation research. Out of 443 patents, 67% were filed by North American countries, 23% by Asian countries and 10% by European countries. There was a large gap in technological position of US and European countries. Even, Asian countries filed more patents (2 times) than European countries. The recent report suggested that the counties with the largest R&D expenditure are the US>UK>France>Germany>Japan>China>Korea [9]. The case was not the same in present study. The Asian countries filed more patents indicating more investment on R&D. The position of European countries was declined in patenting activity which may possibly due to less investment in R&D related of bioremediation technologies. China emerged as the world's second patent holder behind the US, which certainly be due to increasing its R&D spends even more than Russia, Britain, Germany, France and Italy combined. Although, other Asian countries particularly Japan and Korea are also rapidly increasing their R&D spends to become technologically strong by more patent filings. Although, highest innovative research in US might be due to having more scientific manpower, improved research facilities, infrastructure and financial resources [10]. As the patent protected technologies are too costly, therefore, government of developed nations should make provisions to disseminate such technologies to poor nations who are far behind in contributing scientific research due to inadequate infrastructure, less funding resources and high cost of importing scientific equipments.

Patenting trend over time

The results of patenting activity in the field of bioremediation revealed that maximum patents (35) were published in 2005 owing to more R&D activities (Figure 2). As such, the patenting activity was increased sharply since 1990 till 2005, which decreased gradually up to 2012. During 2002-2005, the patenting activities were high indicating that the activity has been very recent and this field has gained importance in recent years. Also, there might be more financial investment in R&D of bioremediation technologies during that period. However, before 1994, there was little patent activity in this field

Assignees of bioremediation patents

There were 31 assignees active in research related to bioremediation (Figure 3). The United States Secretary of Agriculture (USSA) got maximum number of patents followed by Exxon Research & Engineering Co. and University of Georgia Research Foundation, Inc, Regents of the University of California and Geovation Technologies, Inc. The USSA obtained patents on hyperaccumulating plant species capable of removing or recovering heavy metals from contaminated soils along with the processes for degrading and/or bioremediating waste wood containing organic and inorganic contaminants i.e. chromated copper arsenate (CCA), creosote and pentachlorophenol using a fungal inoculum. Exxon Research & Engineering Co. invented surfactant system for bioremediation of hydrocarbon contaminated soils and water that increases the interface between the hydrocarbon contaminant, microbial nutrients and microflora, and also stimulates the propagation of the microflora, thereby enhancing microbial degradation of the hydrocarbon contaminant. Georgia Research Foundation, Inc. developed process for producing metal resistant transgenic plants using recombinant DNA comprising a nucleic acid sequence encoding a metal ion resistance protein leading to enzymatic reduction of metal ions. The invention also involves a method of phytoremediation of a contaminated site by growing in the site a transgenic plant expressing a nucleic acid encoding a metal ion reductase and a nucleic acid encoding a phytochelatin biosynthetic enzyme. The Regents of University of California also developed method and compositions for enhancing heavy metal phytoremediation of metal contaminated soils

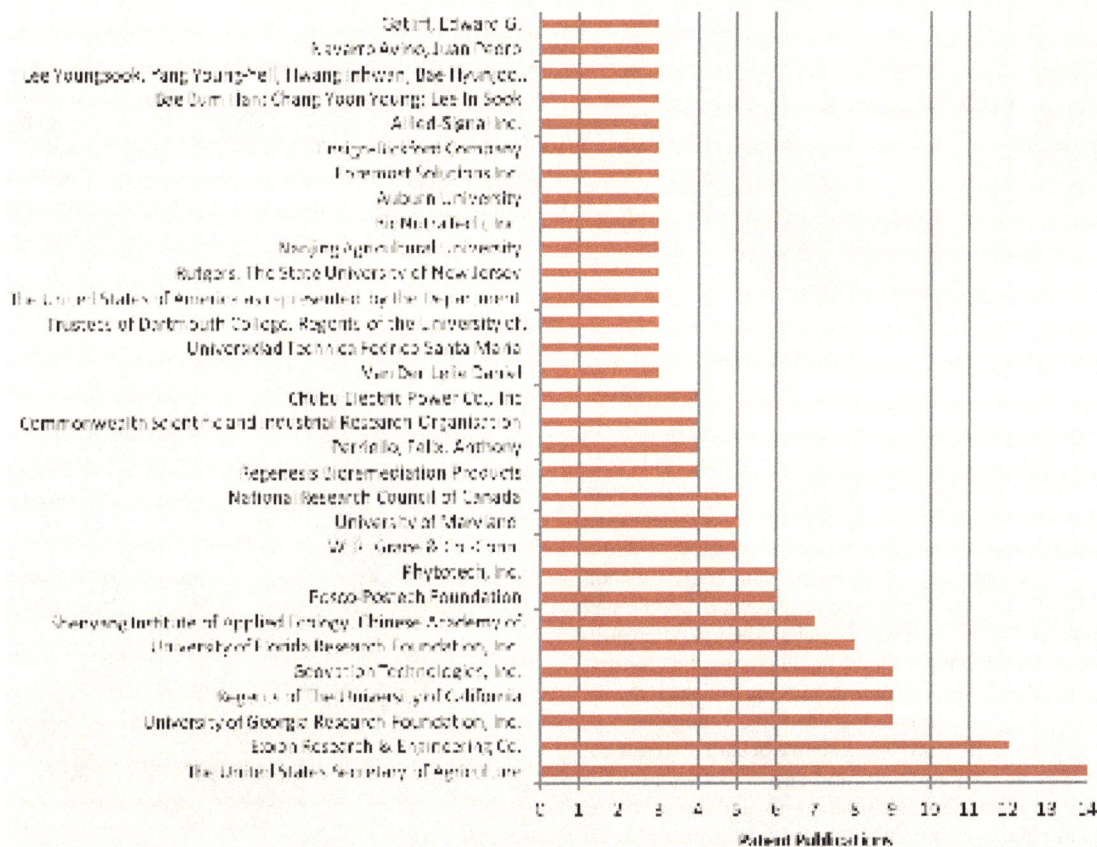

Figure 3: Major Assignees of bioremediation technologies.

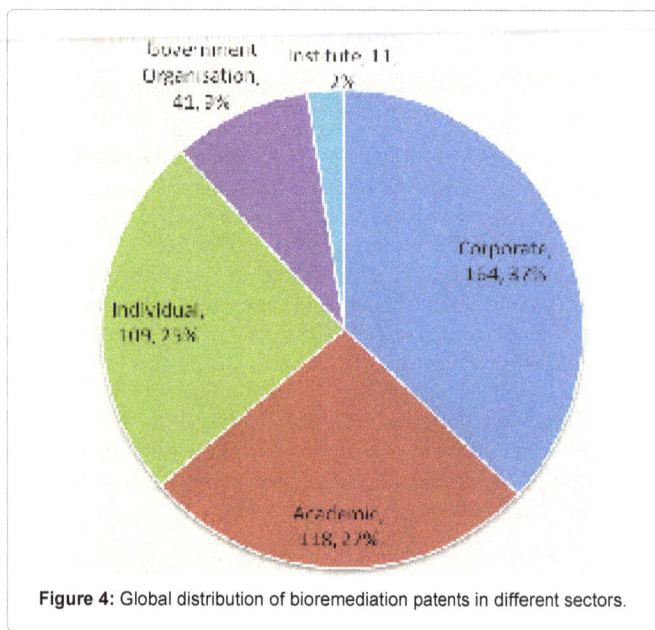

Figure 4: Global distribution of bioremediation patents in different sectors.

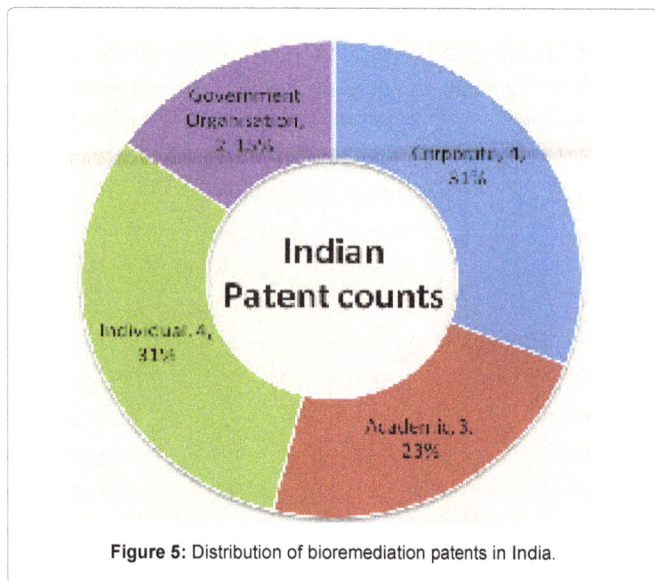

Figure 5: Distribution of bioremediation patents in India.

or waters employing genetically engineered plant (Brassicaceae family) comprising a gene encoding glutamylcysteine synthetase leading to enhanced heavy metal accumulation. They also invented a system for *in-situ* field water remediation using microbial filter. Geovation Technologies, Inc. got patents on solid and liquid chemical compositions for anaerobic biodegradation, detoxification, and transformation of toxic organic and inorganic compounds present in solid and liquid wastes, soils, sediments, and water bodies. The patent involves improved means of (i) promoting the solid-liquid phase extraction and absorption of recalcitrant contaminants from contaminated media, (ii) creating, enhancing, and maintaining anaerobic conditions (i.e., negative redox potential), (iii) providing a source of carbonaceous co-substrates, anaerobic electron acceptors, and nutrient to promote the growth of contaminant-degrading microorganisms, and (iv) providing sources of inoculum of naturally occurring microorganisms which act to promote the biodegradation of contaminants.

Further, the evaluation of technological competencies of different

assignees in the field of bioremediation revealed that maximum patents were filed by corporate sector, while government organizations and institutes showed least interests in patenting activity globally (Figure 4). Even in India, the government organizations were less active towards innovative R&D in bioremediation technologies (Figure 5). In India, government organizations viz. Council of Scientific and Industrial Research and Indian Council of Agricultural Research got 2 patents, whereas corporate sector viz. Bharat Petroleum Corporation Ltd. Indian Oil Corporation Ltd., and M/S Avestha Gengraine Technologies Pvt. Ltd. got 4 patents. Even individual assignees got more patents than government organizations within India. As the individuals are the catalyst and growth centre of every country, therefore government should promote such individuals and groups for developing eco-innovative pollution remediation technologies to leverage environmental protection. The environmental pollution is global issues and has no political boundaries, therefore, government of both developed and developing nations should make efforts for finding eco-friendly solution of pollution and international diffusion and transfer of bioremediation technologies in order to reduce their toxic wastes and contribute in global environmental protection.

Technology based classification of bioremediation patents

According to IPC, the bioremediation patents were categorized into various technological groups (Figure 6). The maximum (265) patents were covered under B09C followed by C02F (174) and C12N (160). Comparatively, there was less patent in other technology areas viz. A01H, A62D, B09B, C22B, C12P, A01N and A01G (Figure 6). The subclass B09C covers technologies related to reclamation of contaminated soil, however, C02F covers water treatment technologies, and C12N covers micro-organisms or enzymes and technologies for mutation or genetic engineering. The description of each subclass and associated groups along with patent counts are given in Table 1. Also, identification of process/product based utility of screened patents revealed that B09C, C02F, A01H, A62D, B09B, C22B, C12P, A01N and A01G represented process patents while C12N represented product patent, indicating that most of the patents were directed towards processes of bioremediation. Thus, there is a need to give more attention on product based bioremediation technologies employing microbes and enzymes degrading and/or detoxifying hazardous

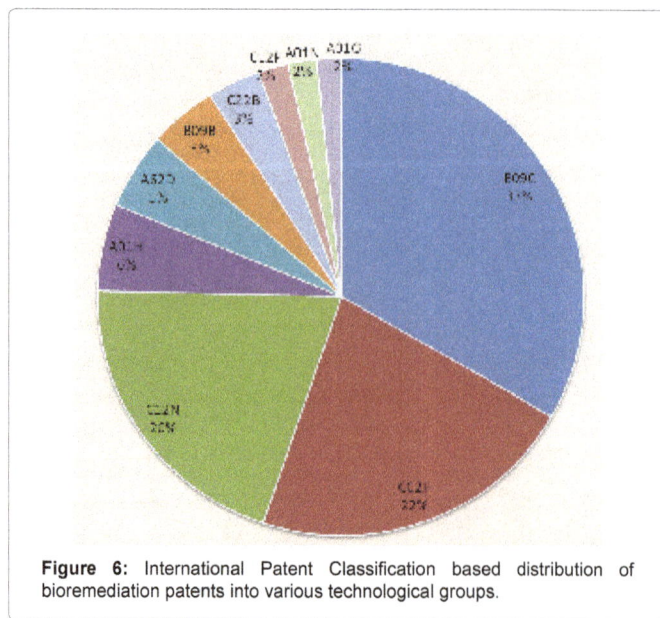

Figure 6: International Patent Classification based distribution of bioremediation patents into various technological groups.

Subclass	Title	Group	Patent records	Description
A01G	HORTICULTURE; CULTIVATION OF VEGETABLES, FLOWERS, RICE, FRUIT, VINES, HOPS, OR SEAWEED; FORESTRY; WATERING (plant reproduction by tissue culture techniques A01H 4/00; propagating unicellular algae C12N 1/12; plant cell culture C12N 5/00)	A01G 1/00	13	Horticulture; Cultivation of vegetables
A01H	NEW PLANTS OR PROCESSES FOR OBTAINING THEM; PLANT REPRODUCTION BY TISSUE CULTURE TECHNIQUES	A01H 1/00	11	Processes for modifying genotypes
A01H		A01H 5/00	36	Flowering plants, i.e. angiosperms
A01N	PRESERVATION OF BODIES OF HUMANS OR ANIMALS OR PLANTS OR PARTS THEREOF; BIOCIDES, e.g. AS DISINFECTANTS, AS PESTICIDES OR AS HERBICIDES; PLANT GROWTH REGULATORS (mixtures of pesticides with fertilisers)	A01N 63/00	15	Biocides, pest repellants or attractants, or plant growth regulators containing micro-organisms, viruses, microbial fungi, animals, e.g. nematodes, or substances produced by, or obtained from micro-organisms, viruses, microbial fungi or animals, e.g. enzymes or fermentates
A62D	CHEMICAL MEANS FOR EXTINGUISHING FIRES; PROCESSES FOR MAKING HARMFUL CHEMICAL SUBSTANCES HARMLESS, OR LESS HARMFUL, BY EFFECTING A CHEMICAL CHANGE; COMPOSITION OF MATERIALS FOR COVERINGS OR CLOTHING FOR PROTECTING AGAINST HARMFUL CHEMICAL AGENTS; COMPOSITION OF MATERIALS FOR TRANSPARENT PARTS OF GAS-MASKS, RESPIRATORS, BREATHING BAGS OR HELMETS; COMPOSITION OF CHEMICAL MATERIALS FOR USE IN BREATHING APPARATUS	A62D 3/00	41	Processes for making harmful chemical substances harmless, or less harmful, by effecting a chemical change in the substances (devices for rendering harmful chemical agents harmless
B09B	DISPOSAL OF SOLID WASTE	B09B 3/00	36	Destroying solid waste or transforming solid waste into something useful or harmless
B09C	RECLAMATION OF CONTAMINATED SOIL	B09C 1/00	64	Reclamation of contaminated soil (processes for making harmful chemical substances harmless or less harmful by affecting a chemical change in the substances
B09C		B09C 1/08	22	Chemically
B09C		B09C 1/10	179	microbiologically or by using enzymes
C02F	TREATMENT OF WATER, WASTE WATER, SEWAGE, OR SLUDGE (processes for making harmful chemical substances harmless, or less harmful, by effecting a chemical change in the substances A62D 3/00; separation, settling tanks or filter devices B01D; special arrangements on waterborne vessels of installations for treating water, waste water or sewage, e.g. for producing fresh water, B63J; adding materials to water to prevent corrosion C23F; treating radioactively-contaminated liquids	C02F 1/68	13	Treatment of water, waste water, or sewage by addition of specified substances, e.g. trace elements, for ameliorating potable water
C02F		C02F 3/00	49	Biological treatment of water, waste water, or sewage
C02F		C02F 3/10	13	Biological treatment of water, waste water, or sewage by using Packings; Fillings; Grids
C02F		C02F 3/32	32	characterised by the animals or plants used, e.g. algae
C02F		C02F 3/34	67	characterised by the micro-organisms used
C12N	MICRO-ORGANISMS OR ENZYMES; COMPOSITIONS THEREOF (biocides, pest repellants or attractants, or plant growth regulators containing micro-organisms, viruses, microbial fungi, enzymes, fermentates, or substances produced by, or extracted from, micro-organisms or animal material; PROPAGATING, PRESERVING, OR MAINTAINING MICRO-ORGANISMS; MUTATION OR GENETIC ENGINEERING; CULTURE MEDIA	C12N 1/14	11	Fungi; Culture media therefor
C12N		C12N 1/20	36	Bacteria; Culture media therefor
C12N		C12N 15/09	23	Recombinant DNA-technology
C12N		C12N 15/29	14	Genes encoding plant proteins, e.g. thaumatin
C12N		C12N 15/31	14	Genes encoding microbial proteins, e.g. enterotoxins
C12N		C12N 15/82	62	This group covers the use of eukaryotes as hosts for plant cells
C12P	FERMENTATION OR ENZYME-USING PROCESSES TO SYNTHESISE A DESIRED CHEMICAL COMPOUND OR COMPOSITION OR TO SEPARATE OPTICAL ISOMERS FROM A RACEMIC MIXTURE	C12P 1/00	16	Preparation of compounds or compositions by using micro-organisms or enzymes; General processes for the preparation of compounds or compositions by using micro-organisms or enzymes
C22B	PRODUCTION OR REFINING OF METALS; PRETREATMENT OF RAW MATERIALS	C22B 3/00	12	Extraction of metal compounds from ores or concentrates by wet processes
C22B		C22B 3/18	16	Extraction of metal compounds from ores or concentrates by wet processes with the aid of micro-organisms or enzymes, e.g. bacteria or algae

Table 1: International patent classification based description of Subclasses and Groups associated with bioremediation technology.

substances present in environment. The IPC results also demonstrated that bioremediation technologies for contaminated water are also available scarcely; therefore, attention should be given in this direction also. As such, the IPC mapping of bioremediation technologies could provide immense help to scientists, technologists, decision makers and researchers to plan future research strategies and R&D investment in thrust areas for more economic and environmental benefits.

Conclusions

The study concluded that number of patent applications in bioremediation domain is very less. The United States Secretary of Agriculture exhibited maximum patents. After US, China and Japan had strong patent activity. Most of the patents were directed towards processes for reclamation of contaminated soils using micro-organisms or enzymes. Comparatively, technologies for biological treatment of wastewater are less. The patents on genetically engineered micro-organisms and plants producing more metal chelating compounds e.g. phytosiderophores, phytochelatins and metallothioneins, and nucleic acid sequences encoding multimetal resistance protein are also reduced in number. Very few patents are available on production of metal hyperaccumulating plants by tissue culture techniques. Also, patents on processes for metal recovery using micro-organisms or enzymes are scarce. These identified research gaps in the field of bioremediation bring out a scope for generating intellectual property. Further, development of products (nucleic acid sequences, metal chelating agents, transgenic plants and micro-organisms) for removing and/or recovering harmful substances from the terrestrial and aquatic environment at a faster rate even in extreme environments could certainly ensure economic and environmental protection as well.

Acknowledgement

I gratefully acknowledge TIFAC (Technology Information, Forecasting and Assessment Council), Department of Science & Technology, Govt. of India, for providing research fellowship in the form of Women Scientist Scheme C (WOS-C). I would also thank NISCAIR (National Institute of Science, Communication and Information Resources) for providing authenticated patent database to perform patent searching and retrieving the data.

References

1. Salt DE, Blaylock M, Kumar NP, Dushenkov V, Ensley BD, et al. (1995) Phytoremediation: a novel strategy for the removal of toxic metals from the environment using plants. Biotechnology (N Y) 13: 468-474.

2. Salt DE, Smith RD, Raskin I (1998) PHYTOREMEDIATION. Annu Rev Plant Physiol Plant Mol Biol 49: 643-668.

3. Lanjouw JO, Mody A (1996) Innovation and the International Diffusion of Environmentally Responsive Technology. Research Policy 25: 549-571.

4. Nameroff TJ, Garant RJ, Albert MB (2004) Adoption of green chemistry: an analysis based on US patents. Research Policy 33: 959-974.

5. Johnstone N (2005) Environmental Policy, Technological Innovation and Patent Activity: Initial Empirical Results and Project Progress, OECD Report, Washington DC.

6. Sun Y, Lu Y, Wang T, Ma H, He G (2008) Pattern of patent-based environmental technology innovation in China. Technol Forecast Soc Change 75: 1032-1042.

7. Griliches Z (1990) Patent Statistics as Economic Indicators: A Survey. J Econo Lit 28: 1661-1707.

8. Eaton J, Kortum S (1996) Trade in ideas. Patenting and productivity in the OECD. J Int Econo 40:251-78.

9. Veugelers R (2013) The world innovation landscape: asia rising? Bruegel Policy Contribution. Issue 2013/02 February 2013.10.

10. Vose PB, Cervellini A (1983) Problems of scientific research in developing countries. IAEA Bull 25: 37-40.

13

Studies on Biosorption of Chromium Ions from Wastewater Using Biomass of *Aspergillus niger* Species

Korrapati Narasimhulu*[1] and Y. Pydi Setty[2]

[1]*Department of Biotechnology, National Institute of Technology, Warangal, India*
[2]*Department of Chemical Engineering, National Institute of Technology, Warangal, India*

Abstract

The biosorption capacity of immobilized biosorbents for Cr (VI) was found to depend on pH, contact time, biosorbent dose and initial concentration of Cr (VI). In this study, the maximum uptake of Cr (VI) was 93.4, 96.2 and 98.6 mg respectively at a pH 1.5 and with an increase in pH up to 11 the metal uptake decreased gradually upto 39.95, 52.35 and 66.48 mg respectively for acid treated, untreated and base treated fungal biosorbents. Increase in biosorbent dose up to 1g of biomass and contact time up to 60 min resulted in an increase in biosorption from 20.2, 16.8 and 28.3 mg at a biosorbent dose of 0.1g 100 ml^{-1} to 93.4, 96.2 and 98.6 mg at a biosorbent dose of 1.0 g 100 ml^{-1} and then further increase in adsorbent dose and contact time did not result in more Cr (VI) adsorption by per unit weight of biosorbent. The percentage metal uptake by the biosorbent was found to decrease upto 60.35, 50.64 and 82.5 percent respectively for acid treated, untreated and base treated fungal biosorbents at the 300 mgl^{-1} Cr (VI) ion concentration. The resulted data was found to fit well in Langmuir model of adsorption isotherm with a high value of correlation coefficient. The biosorbed metal was eluted from the biosorbent by using 0.1M H_2SO_4 as elutant.

Keywords: Biosorption; *Aspergillus niger*; Immobilization; Isotherm; Hexavalent chromium

Introduction

The main pollutant from industrial complexes is the effluent which contains heavy metals such as Cu, Ni, Zn, Pb, Cr, Hg, Cd etc. and various organic compounds such as phenols, formaldehyde etc. [1]. Wastewater usually contains about 5 ppm of chromium. To comply with the permissible limits, various techniques are used for the removal of heavy metals. The recovery of heavy metals using conventional techniques is neither economical nor eco-friendly [2]. So, there is a continuous search for an economic and ecofriendly approach of heavy metals removal. In this endeavor, biosorption, a biological method of environmental control, has emerged as an alternative to conventional waste treatment facilities as it has advantages of low operating cost, effective in dilute solutions, generates minimum secondary waste, completes with in a short time period and have no toxicity limits for heavy metals [3,4]. Biosorption is the process in which physico-chemical interaction between the charged surface groups of micro-organisms and ions in solution takes place by the process of complexation, ion exchange, microprecipitation etc [5]. Biosorption is attractive since naturally occurring biomass or spent biomass from various fermentation industries can be effectively utilized [6]. These advantages have served as potential incentives for promoting biosorption as viable cleanup technology for heavy metal pollution. Immobilization of biosorbent leads to its stability and it can be used repeatedly with ease for the process of adsorption/ desorption [7]. Biotechnological exploitation of biosorption technology depends on the efficiency of regeneration of biosorbent after metal desorption. Therefore, recovery of biosorbent by mild and cheap desorbing agents is desirable for regeneration of biomass for use in multiple biosorption cycles. In the light of these observations, studies on biosorption of Cr (VI) to immobilized fungal biomass were carried out in the present study. *Aspergillus niger* was used for the removal of Cr (VI) from electroplating industrial effluent and highlighted the prospects of its future uses as a biosorbent material for the Cr (VI) removal.

Materials and Methods

Characterization of industrial effluent and estimation of Cr (VI)

Stock metal solutions of Cr (VI) were prepared by dissolving appropriate quantities of pure analytical grade metal salts in double distilled water. The stock solutions were diluted further with deionized distilled water to obtain working solutions of different concentrations. The effluent was collected from Hyderabad electroplating works,Hyderabad, which is an elecroplating industry. The effluent collected from industry, was analyzed for different physico-chemical properties such as total dissolved solids, hydrogen ion concentration, colour, oil and grease, BOD, COD, sulphate, phosphate, chromium, copper, nickel and zinc (Table 1).

The concentration of each component was determined as per the procedure outlined in APHA [8]. Chromium analysis was carried out by spectrophotometric method by using 1,5-Di-phenyl carbazide accoding to APHA [8]. The hexavalent chromium was determined colorimetrically by reaction with diphenylcarbazide in acid solution. A red-violet color was produced. The reaction was very sensitive, the absorptivity based on chromium being about 40000 lg^{-1} cm^{-1} at 540 nm wavelength. *Aspergillus niger* was previously isolated from wastewater treatment plant of NIT Warangal and routinely maintained by streaking on a rose bengal agar medium and incubating at 25°C [9]. For mass culturing, liquid broth was used as a culture medium which was having the following composition (g l^{-1}): Bactodextrose, 20; Bactopeptone,

***Corresponding author:** Korrapati Narasimhulu, Assistant Professor, Department of Biotechnology, National Institute of Technology, Warangal, India
E-mail: simha_bt@nitw.ac.in

10; NaCl, 0.2; CaCl$_2$. 2H$_2$O, 0.1; KCl, 0.1; K$_2$HPO$_4$, 0.5; NaHCO$_3$, 0.05; MgSO$_4$, 0.25 and FeSO$_4$. 7H$_2$O, 0.005. The liquid phase pH was adjusted to 4.5 by using the 0.1M HCl and 0.1N NaOH. The liquid broth was inoculated with a loop of culture grown on rose bengal agar medium and incubated on an orbital shaker at 125 rpm and 25°C for five days in 250 ml conical flasks. The biomass produced was collected by vacuum filtration and washed twice with extra pure double distilled water. The washed biomass was dried at 60°C in an oven and dry material was ground and sieved through 100 mm screen. In order to generate active sites and enhance the metal biosorption capacity, the dry biomass was conditioned in batches by treatment with 0.1 M H$_2$SO$_4$ and 0.1N NaOH for 30 min at 60°C. The treated biomass was collected by centrifugation and washed twice with extra pure double distilled water. After washing, the biomass was dried at 60°C in an aluminum foil till the weight of biomass became constant. It was then ground and sieved through a screen with pore size of 100 mm and stored at 60°C.

Immobilization of biosorbent and batch studies

It was immobilized in calcium-alginate matrix to establish a simple and cost effective granulation of the biomass. The immobilization of fungal biosorbent via entrapment was carried out as follows: 3% (w/v) sodium-alginate was dissolved in distilled water and mixed with 5%(w/v) of fungal biosorbent. The mixture was stirred for 1 hour at 30°C and then, the slurry solution was dripped through a nozzle into 4% (w/v) CaCl$_2$ solution. Durable, spherical beads containing biomass were formed immediately by a phase inversion process as the alginate was crosslinked by Ca^{2+}. Bead size varied according to the diameter of injection nozzle. The gel beads with size 3.2 mm ± 0.1 mm, were moderately agitated in the double distilled water for 2 hr at 4°C. The curing procedure hardened the beads and resulted in the formation of a favourable micro-porous structure. Finally, the beads were stored at 4°C in ultrapure double distilled water until further use. The batch adsorption studies were carried out in triplicate in the 250 ml conical flasks at temperature 28°C ± 0.15°C [7]. The incubated shaker was used for batch studies at 150 rpm. A predetermined amount of biosorbent was mixed with 100 ml of heavy metal solutions in 250 ml conical flasks. After a desired contact period conical flasks were removed and the solutions were filtered through Whatman filter paper 42 and the filtrate was analyzed for residual Cr (VI) concentration. The Cr (VI) removal affinity of biosorbent was determined from batch experiments as a function of contact time, pH, biosorbent dosage and initial concentration of metal ions. The adsorption data was analyzed by employing Langmuir adsorption isotherm.

Metal desorption and biosorbent reutilization

The biosorbent was prior exposed to metal solutions under optimum conditions for metal desorption assays, as described above and then separated [7]. The metal loaded biosorbent was incubated for 1 hr at 28°C with 50 ml of 0.1 M H$_2$SO$_4$ in a rotary shaker for continuous stirring at 150 rpm. The solution was then filtered through whatmann filter paper 42 and the filtrate was used to determine the metal released. The total desorbed metal was established by comparing the metal released to the amount of metal previously adsorbed to the biosorbent. All experiments were run in triplicates. The metal stripped biosorbent was rinsed twice with 50 ml of ultrapure double distilled water for 15 min. The acid treated and base treated biomass was activated by treating them with 0.1 M H$_2$SO$_4$ and 0.1N NaOH respectively for 30 min at 150 rpm. The resulting biomass was then reloaded with metal solutions as described above and the desorption treatment was repeated.

Adsorption isotherm

Equilibrium occurring during adsorption at a definite concentration range could be represented by Langmuir isotherm for adsorption [10]. The equilibrium data was analyzed with the help of Langmuir adsorption isotherm. The Langmuir equation is given by,

$$Q_e = \frac{Q_{max}.b.C_e}{1+b.C_e} \quad (1)$$

Where, Q_e is the amount of metal adsorbed per unit mass of biosorbent (mgg^{-1}), Q_{max} and b are Langmuir constants and C_e is the equilibrium concentration of metal ion (mgl^{-1}). The value of r^2 (correlation coefficient) showed that the adsorption data for Cr (VI) was fitted well into Langmuir adsorption isotherm.

The Langmuir model assumes about constant adsorption energy. The agreement of the experimental data with the model showed that constant adsorption energy existed for the experimental conditions. The essential characteristics of Langmuir isotherm can be expressed in terms of dimensionless constant, R_L [10] separation factor for equilibrium parameter, which is defined by

$$R_L = 1 / (1 + b.C_i) \quad (2)$$

Where C_i is the initial concentration of metal ion (mgl^{-1}) and b is Langmuir constant (mgl^{-1}). The R_L values obtained in the present studies for different metal ions indicated favorable adsorption on the biosorbent surface. Thus, heavy metal removal by A. niger is considered to be a chemical, equilibrated and site specific mechanism on the biosorbent surface.

Free energy change (ΔG) was calculated using the following equation:

$$\Delta G = - RT \ln b \quad (3)$$

where, ΔG = Free energy change, R = Universal gas costant, T = Absolute temperature, b = Langmuir constant.

Results and Discussion

Characterization of industrial effluent

The turbid effluent with dark green colour was found to be odorless (Table 1). The amount of chromium was found to be higher than Central Pollution Control Board standards, which possess a great threat to the ecosystem. The main form of the metal present in the effluent was Cr (VI).

Effect of pH and biomass dose: Experiments were conducted with

S.No.	Property	Type/Value
1	Color	Dark green
2	Total dissolved solids	20,250 mgl^{-1}
3	Oil and grease	20 mgl^{-1}
4	BOD	168 mgl^{-1}
5	COD	447 mgl^{-1}
6	pH (1:2)	4.4
7	Sulphate	181 mgl^{-1}
8	Phosphate	0.82 mgl^{-1}
9	Cr (VI)	102 mgl^{-1}
10	Zn (II)	88 mgl^{-1}
11	Cu	0.52 mgl^{-1}
12	Fe	6.8 mgl^{-1}
13	Ni	13.1 mgl^{-1}

Table 1: Physico-chemical properties of electroplating industrial effluent.

1g biomass dose of biosorbent for a contact time of 60 min. The pH varied from 1.5 to 11 with the help of 0.1 M H_2SO_4 and 0.1 N NaOH. The results showed that the metal uptake was optimum at pH 1.5. Further incrase in pH beyond optimal value (1.5) resulted in decrease of metal uptake. This observation agrees with the earlier reports on Cr (VI) removal by different biosorbents. As the pH of the system increases, the number of negatively charged sites increases and the number of positively charged sites decreases. A negatively charged surface site on the adsorbent does not favour the adsorption of anions due to the electrostatic repulsion. The optimum pH from above experiments with contact time of 60 minutes was kept to find the optimum dose of biomass. Various doses (0.1 to 1.2 g) of biomass were tested for Cr (VI) removal from synthetic solutions of 100 mgl^{-1} concentration. 1g of biomass dose was sufficient for the optimum removal of Cr (VI) ions. Increase in biomass dose after optimum value did not show corresponding increase in the metal ion uptake from the solution. Interaction of biosorbent and metal ions is generally electrostatic in nature on the binding sites present on the surface of biosorbent. For a given constant biosorbent concentration, the initial metal ion adsorption increased up to the stage of saturation of all the binding sites and further increase in the dose of biosorbent did not change the metal adsorption to biomass ratio.

Rate of adsorption: Relationship between contact time and Cr (VI) removal was investigated by using immobilized biosorbent with 1g biomass dose of each. The pH of the 100 mgl^{-1} Cr (VI) solution was adjusted to 1.5. The maximum amount of Cr (VI) uptake was observed after 60 min. Increase in contact time from 60 to 180 min did not result in corresponding increase in uptake capacity of biosorbent. It was observed that biosorption is a rapid process as most of the adsorption (80-85%) was completed in the initial 45 min. These results indicated that the adsorption sites were bind up in the initial 60 min by the metal ions passively. After this, the increase in contact time might not help for more adsorption of metal ions with this biosorbent. As a result, 60 min was chosen as optimum contact time for further studies.

Effect of initial metal ion concentration on rate of adsorption of Cr (VI): The uptake of metal ions was observed for different initial ion concentration of Cr (VI) at optimum contact time and biosorbent dose. The results revealed that an increase in metal ion concentration resulted in gradual decrease in percent biosorption of Cr (VI) up to 100 mgl^{-1} and after that a sharp decrease in metal uptake percentage. An increase in initial concentration of metal ions resulted in the lowering of metal ion uptake due to reduction in ratio of sorptive surface to ion concentration.

Adsorption isotherm

It was found to be linear over a wide range of concentration. Langmuir constants 'Q_{max}' which is a measure of adsorption capacity expressed in mg g^{-1} and 'b' which is a measure of energy of adsorption expressed in mgl^{-1} were calculated from the slope and intercept of plots are shown in Table 2.

The values of ΔG (Table 2) observed to be negative, thus, indicating the spontaneous and exothermic nature of the adsorption proces. The higher negative values reflect a more energetically favorable adsorption process.

Reutilization of biosorbent: The total amount of metal biosorbed and desorbed in each subsequent cycle by immobilized biosorbents were calculated as shown in Figure 1, 2 and 3. The amount of Cr (VI) adsorbed in the 5th cycle was comparable to the first cycle. Also, the amount of metal desorbed after each loading cycle corresponded well

to the amount of metal biosorption, which showed that a complete elution took place. Performance of immobilized biosorbent in industrial effluent: The batch study was carried out with 100 ml industrial effluent at 28±0.15°C and 150 rpm in 250 ml Erlenmeyer flasks. Immobilized biosorbent with a biomass dose of 1 g was used for a contact period of 40 minutes and it was found that biosorption of Cr (VI) was less in industrial effluent as compared to single metal ion synthetic solutions as shown in Figure 4. This may be due to interference and binding of other metal ions to the binding sites which was not the case with single metal ion synthetic solutions as all the binding sites were available for just single metal species. Overall, the performance was good in present electroplating industrial effluent. The tendency of biosorbents for metal removal in the present effluent was found in order of base treated > untreated > acid treated.

Conclusion

The present study showed that base treated biosorbent performed better than acid treated and untreated biosorbent. The explanation

Biosorbent	Q_{max} (mg g^{-1})	B (mg l^{-1})	r^2	R_L	ΔG	K_{ad}
Base treated	24.246	0.3264	0.9882	0.0243	-5.2861	0.0328
Acid treated	18.265	0.1084	0.9964	0.0788	-9.2484	0.0245
Untreated	13.482	0.3012	0.9821	0.0296	-5.2873	0.0321

Table 2: Constants and correlation coefficient for immobilized *Aspergillus niger*.

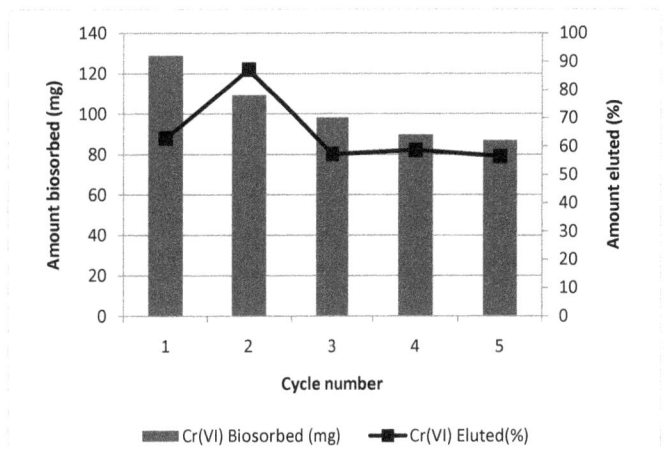

Figure 1: Biosorption and desorption cycles for chromium removal by immobilized acid treated *Aspergillus niger*.

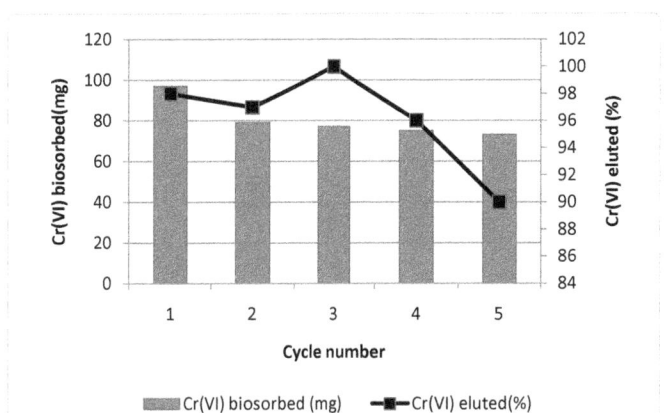

Figure 2: Biosorption and desorption cycles for chromium removal by immobilized untreated *Aspergillus niger*.

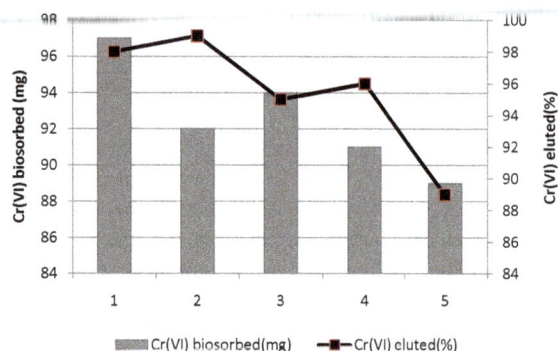

Figure 3: Biosorption and desorption cycles for chromium removal by immobilized base treated *Aspergillus niger*.

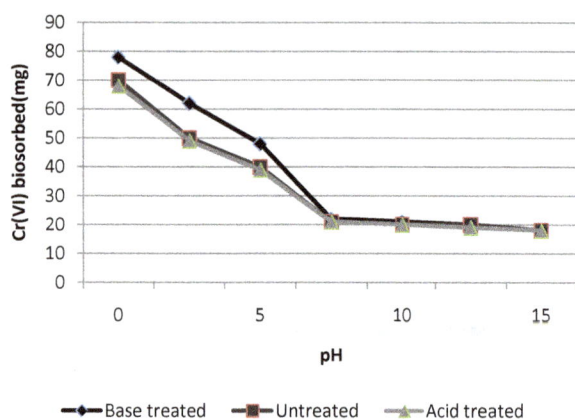

Figure 4: Removal of Cr (VI) from industrial effluent by immobilized base treated, untreated and acid treated *Aspergillus niger*.

offered is that the increase in the metal uptake may be due to unmasking of some of cellular groups, which cannot participate in the sorption process without the treatment of alkali [11] also found that alkali treatment of *Aspergillus niger* mycelia improved their capacity to chelate various metal ions [12], in trying to explain the increase in metal uptake by *A. niger* after NaOH treatment, thereby suggesting

that the removal of certain polysaccharides from the cell wall by alkali treatment generates more accessible spaces with in the b-glucan chitin Skelton, thus allowing more metal ions to be sequestered by this structure. Immobilization of biosorbent increases the efficiency and ease of handling the process. The effect of pH and the higher sorption capacity suggest the electrostatic attraction of metal ions onto the fungal surface. The favorable values of Langmuir constants, R_L factor and ΔG showed that *A. niger* is a good option for the removal of Cr (VI) from electroplating industrial effluent.

References

1. Rajendran P, Muthukrishnan J, Gunasekaran P (2003) Microbes in heavy metal remediation. Ind J Exp Biol 41: 935-944.

2. Sarkar S, Gupta A (2003) Treatment of chrome plating waste water (Cr⁶⁺) using activated alumina. Indian J Environ Health 45: 73-82.

3. Dadhich AS, Beebi SK, Kavitha GV (2004) Adsorption of Ni(II) using Agrowaste, Rice Husk. J Environ Sci Eng 46: 179-185.

4. Ahalya N, Kanamadi RD, Ramachandra TV (2007) Cr (VI) and Fe (III) removal using Cajanus cajan husk. J Environ Biol 28: 765-769.

5. Mise SR, Rajamanya VS (2003) Adsorption studies of chromium (VI) on activated carbon derived from Sorghum vulgare (Dried Stem of Jowar). Indian J Environ Health 45: 49-58.

6. Gupta R, Ahuja P, Khan S, Saxena RK, Mohapatra H (2000) Microbial biosorbents: Meeting challenges of heavy metal pollution in aqueous solutions. Curr Sci 78: 967-973.

7. Srinath T, Garg SK, Ramteke PW (2003) Biosorption and elution of chromium from immobilized Bacillus Coagulans biomass. Indian J Exp Biol 41: 986-990.

8. APHA (2005) Standard methods for the examination of water and wastewater. (21st edn), Washington, D.C.

9. Narasimhulu K, Rao PS, Vinod AV, Isolation and Identifi cation of Bacterial Strains and Study of their Resistance to Heavy Metals and Antibiotics. J Microbial Biochem Technol 2: 074-076.

10. Dhar NR, Khoda AK, Khan AH, Bala P, Karim MF (2005) A study of effects of acid activated saw dust on the removal of different dissolved tannery dyes (acid dyes) from aqueous solutions. J Environ Sci Eng 47: 103-108.

11. Muzzarelli RA, Tanfani F, Scarpini G (1980) Chelating,film forming and coagulat ing abi lity of the chitosan-glucan complex from Aspergillus niger industrial waste. Biotechnol Bioeng 22: 885-896.

12. Leuf E, Prey T, Kubiceck CP (1991) Biosorption of zinc by fungal mycelial waste. Appl Microbiol Biotechnol 34: 688-692.

14

Modelling Biogas Fermentation from Anaerobic Digestion: Potato Starch Processing Wastewater Treated Within an Up flow Anaerobic Sludge Blanket

Philip Antwi[1]*, Jianzheng Li[1], En Shi[1], Portia Opoku Boadi[2] and Frederick Ayivi[3]

[1]State Key Laboratory of Urban Water Resource and Environment, School of Municipal and Environmental Engineering, Harbin Institute of Technology, 73 Huanghe Road, Harbin 150090, PR China
[2]School of Management, Harbin Institute of Technology, 92 West Dazhi Street, Nan Gang District, Harbin 150001, PR China
[3]Department of Geography, University of North Carolina, 237 Graham building, 1009 Spring Garden St, Greensboro, NC27412, USA

Abstract

Herein, a modeling approach to predict biogas yield within a mesophilic (35 ± 1°C) upflow anaerobic sludge blanket (UASB) reactor treating potato starch processing wastewater (PSPW) for pollutant removal was conducted. HRTs and seven anaerobic process-related parameters viz; chemical oxygen demand (COD), ammonium (NH_4^+), alkalinity, total Kjeldahl Nitrogen, total phosphorus, volatile fatty acids (VFAs) and pH with average concentration of 4028.91, 110.09, 4944.67, 510.47, 45.20, 534.44 mg/L and 7.09, respectively, were used as input variables (x) to develop stochastic models for predicting biogas yield from the anaerobic digestion of PSPW. Based on the prediction accuracy of the models, it was established that, prediction of biogas yield from the UASB with the combination of COD, NH_4^+ and HRT, or COD, NH_4^+, HRT and VFAs as input variables proved more efficient as opposed to HRT, alkalinity, total Kjeldahl Nitrogen, total phosphorus and pH. Highest coefficient of determination (R^2) observed was 97.29%, suggesting the efficiency of the models in making predictions. The developed models efficiencies concluded that the models could be employed to control the dynamic anaerobic process within UASBs since prediction of biogas obtained in the UASB agreed with the experimental result.

Keywords: Up flow anaerobic sludge blanket (UASB); Potato starch processing wastewater; Biogas yield; Modeling; Multiple non-linear regression

Introduction

Potato is one of the most valuable food crop grown in many countries [1]. It has been reported that, a considerable proportion of the potato cultivated globally consumed through starch processing which subsequently generates tons of wastewater that goes to pollute water bodies [1-3]. Wastewater of raw potato processed into starch are classified as complex wastewater [4,5], and its concentration of chemical oxygen demand (COD), total suspended solid (TSS) and volatile suspended solid of (VSS) can yield concentrations of 50000, 9700 and 9500 mg/L, respectively [6]. Arhoun et al. argued that recovering valuable resource such as bioenergy (biogas) from such wastewater to supplement energy needs will be beneficial to humans and society at large [6,7]. Anaerobic digestion has severally been reported as a successful bioprocess treating various organic wastewaters and subsequently generating biogas [7-12]. However, the biological mechanism of anaerobic digestion is not well understood due to the complexity of the bacterial community structure and bioconversion [13]. Hu et al. asserted that process modeling is a good tool for predicting and describing the performance of biological processes [13]. Other reports also confirmed that process modeling based on previously acquired data is one technical route to enhancing the performance of anaerobic processes. These process models are often developed [14,15]. Nonetheless, modeling of anaerobic digestion is quite challenging and tough because performance of anaerobic systems is complex and varies considerably with influent characteristics and operational conditions [16].

Some predictive models have been developed in the past decades for biogas estimation during anaerobic treatment processes. For instance, a regression analysis model for estimating biogas generated in a landfill leachate treatment process was developed by Akaya et al.

using several leachate parameters [17]. In another study, Ozkaya et al. presented a neural network model for predicting the methane fraction in landfill gas originated from field-scale landfill bioreactors [18]. These models had provided more detail insight into the biological mechanism in anaerobic digestion [4].

Deterministic models also provide a good insight into the mechanism of biological relationships, but fluctuation of kinetic parameters and wastewater characteristics normally results in a laborious calibration, comprehensive computer analysis as well as laboratory work [14]. The International Water Association (IWA) Anaerobic Digestion Model No.1 (ADM1), a typical deterministic model, has been successfully used for modeling the whole anaerobic digestion process [19]. However, its mathematical complexity associated with extreme analytical difficulty of measuring kinetic parameters turns out to be laborious and time consuming [20,21]. On the other hand, stochastic based non-linear multiple regression model is preferably easy to handle as well as capable to estimate the relation between variables and numerical parameters [22,23]. The advantage of a regression based model compared to other models such as neural networks is its ability to write down relationships and to relate with underlying processes, whereas neural networks only produce an approximation that is opaque.

*Corresponding author: Philip Antwi, State Key Laboratory of Urban Water Resource and Environment, School of Municipal and Environmental Engineering, Harbin Institute of Technology, 73 Huanghe Road, Harbin 150090, PR China
E-mail: kobbyjean@yahoo.co.uk

In this study, the evaluation, feasibility and efficiency of biogas production from anaerobic digestion of PSPW within an upflow anaerobic sludge blanket (UASB) reactor was conducted. Thereafter, dynamic multiple non-linear regression models were developed for the timely prediction of biogas yield from the UASB. The proposed models could identify the most influential parameter(s) that could guide the control and operation of anaerobic systems. Based on residual analysis and diagnostic statistics, the best fit models were identified and their output performance compared with that of experimental data.

Materials and Methods

Feed and inoculum

Potato starch processing wastewater (PSPW) was collected from a local starch producing factory in Heilongjiang province, China and kept under 4°C in the State Key Laboratory of Urban Water Resources of Harbin Institute of Technology. The characteristics of the raw wastewater were as follows (averages): pH 5.0, total chemical oxygen demand (COD) 49179 mg/L, total organic carbon 9831 mg/L, ammonium (NH_4^+) 96 mg/L, total Kjehdahl nitrogen (TKN) 1023 mg/L, total nitrogen (TN) 1439 mg/L, total phosphorus (TP) 190 mg/L, TSS 25345 mg/L, VSS 24855 mg/L and alkalinity (ALK) 4945 mg/L. The raw wastewater was diluted with the pH adjusted to an average of 7.09 by sodium bicarbonate $NaHCO_3$, and then fed into the UASB. The feed was characterized by a COD, NH_4^+, TKN, TP, ALK and total volatile fatty acids (VFAs) of 4029, 111, 511, 45, 9177 and 534 mg/L, respectively.

The activated sludge for inoculation of the reactor was collected from a local municipal wastewater treatment plant operating in an anaerobic-anoxic-oxic (A^2/O) condition and were characterized as 11.5 g/L mixed liquor suspended solids (MLSS) and 5.6 g/L mixed liquor volatile suspended solids (MLVSS).

Experimental setup and reactor operation

The experiments were conducted in an UASB that was constructed with Plexiglas (Figure 1). The reactor was 120 cm high with an internal diameter of 10 cm. There were a conical bottom of about 0.4 L and a gas-solid-liquid separator at the upper part. The total volume and the effective working volume was 8.8 and 7 L, respectively. Five sampling ports at 25 cm interval from each other were allocated along the vertical height of the reactor. The first sampling port was 1 cm above the conical bottom whiles the topmost port was 3 cm below the reactor's head.

The UASB was operated under mesophillic condition (35 ± 1°C) and the heating source was obtained from a heat conducting wire wound around the stem. The heat conducting wire was connected to a temperature controller. The diluted raw PSPW was fed to the reactor by a peristaltic pump (BT100-2J, Langer Instruments, United Kingdom). The evolved biogas was collected by the gas-solid-liquid separator and was measured daily using the wet gas meters (Model LML-1, Changchun Filter Co., Ltd., China). The reactor was started up with an inoculum of 3.52 g/L MLVSS and a hydraulic retention time (HRT) of 48 h was kept during the first 49 days. The reactor was continuously operated with a decreased HRT of 24 h since the 50th day.

Analytical methods

The influent and the effluent of the reactor were sampled daily for the analysis of COD, ALK (in terms of $CaCO_3$), TKN, NH_4^+ and TP in accordance with the Standard Methods for the Examination of Water and Wastewater, APHA [24]. pH was determined using a DELTA 320 (Mettler Toledo, USA).

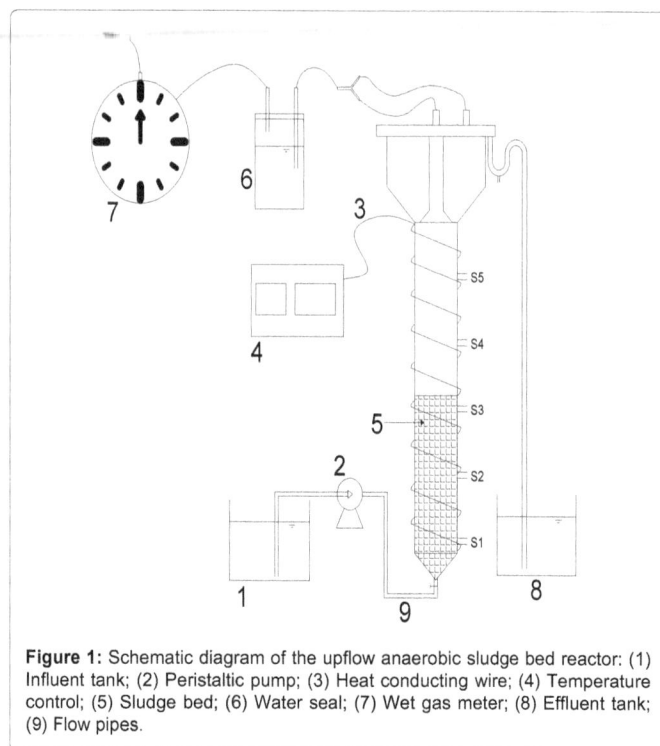

Figure 1: Schematic diagram of the upflow anaerobic sludge bed reactor: (1) Influent tank; (2) Peristaltic pump; (3) Heat conducting wire; (4) Temperature control; (5) Sludge bed; (6) Water seal; (7) Wet gas meter; (8) Effluent tank; (9) Flow pipes.

Volatile fatty acids (VFAs) were measured by a gas chromatograph (SP6890, Shandong Lunan Instrument Factory, China) equipped with a 30 m capillary column (Stabilwax-DA, i.d. 0.32 mm, 11054, Restek) and a flame ionization detector (FID). The operational temperatures of the injection port, oven and detector were 210°C, 180°C, and 210°C, respectively. Nitrogen gas was used as the carrier gas, with a 0.75 MPa column head pressure. The split ratio was 1:50. Liquid sample of 1 mL collected from the top most sampling port was centrifuged at 13000 rpm for 3 min, and 0.5 mL of the supernatant was pipetted and acidified with 25% H_3PO_4 and then 1 μL of the final solution was injected. The VFAs were measured in terms of CH_3COOH.

A 0.5 ml of biogas was sampled from the headspace of the reactor to determine CH_4 and CO_2 fractions. Fraction of CH_4 was analyzed by another gas chromatograph (SP-6800A, Shandong Lunan Instrument Factory, China) equipped with a thermal conductivity detector (TCD) and a 2 m stainless column packed with Porapak Q (60/80 mesh). Temperatures of the injector, column and the TCD were 80°C, 50°C and 80°C, respectively.

Data preparation and correlation analysis

The experimental data was used as an open database connectivity data source for the regression analysis. MINITAB (version 17) and Sigmaplot (version 13) statistical computing environment were used to carry out Pearson's correlation analysis. The few irregular biogas yield data were omitted prior to further analysis. Significances were corrected to avoid multiple comparisons [25]. A probability (p-value) less than 0.05 was used to determine the statistical significance of the regression coefficients during the correlation analysis and the prediction.

Model description

The general form of the models used in this study is expressed as Eq.1 [26-30]. The output variable y, written as a function of k, has input variables ($x_1, x_2, ..., x_k$) and a random error term $\hat{\varepsilon}$ that is added to make the model probabilistic rather than deterministic. In particular, the value of the coefficient β_i determines the contribution of the independent variable x_i given that the other $(k-1)$ independent variables are held constant. β_0 represents the y-intercept. The coefficients β_0, β_1, ..., β_k are usually unknown since they represents population parameters.

$$y = \beta_0 + \beta_1 x_1 + \beta_2 x_2 + \cdots + \beta_k x_k + \hat{\varepsilon} \tag{1}$$

The input variables can represent higher-order terms for quantitative predictors. Since most statistical tests are reliant on assumptions about the variables used [31], 5 assumptions were considered in this study which included linearity, independence among errors, non-multicollinearity, homoscedasticity and non-autocorrelation [32]. The first-order model (Eq.2), second-order model (Eq.3) and complete second-order model (Eq.4) were used in the model development. These models comprised two or more independent variables in different combinations and interactions among the input variables and the estimated unknown coefficients $(\beta_0, \beta_1, ..., \beta_k)$ [26,33].

$$y=\beta_0+\beta_1 x_1+\beta_2 x_2+\cdots+\beta_k x_k \tag{2}$$

$$y=\beta_0+\beta_1 x_1+\beta_2 x_2+\beta_3 x_1 x_2 \tag{3}$$

$$y=\beta_0+\beta_1 x_1+\beta_2 x_2+\beta_3 x_1 x_2+\beta_4 x_1^2+\beta_5 x_2^2 \tag{4}$$

where β_0 is the y-intercept of $(k+1)$; β_1 is change in y for one unit increase in x_1 when x_2, x_3, ..., x_k are fixed; β_2 is change in y for one unit increase in x_2 when x_1, x_3, ..., x_k are held; β_k is change in y for one unit increase in x_k when x_1, x_2, ..., x_{k-1} are held. $\beta_3 x_1 x_2$ is the interaction term. For Eq.3 and Eq.4, β_1 and β_2 cause surface shift along the x_1 and x_2 axes, β_3 controls the rate of twist in the ruled surface, β_4 and β_5 controls the type of surface and the rates of curvature, respectively.

Selection of input and output variables for model fitting

Seven process related parameters obtained from the feed, together with the HRT, were used as input variables (x_n) [34]. The seven parameters were influent COD, pH, NH_4^+, ALK, TKN, VFAs and TP. Variables which did not correlate with biogas yield significantly (p >0.05) were omitted [23]. A stepwise iterative variance inflation factor (VIF) analysis as described by Mac Nally was used to evaluate all of the input variables and the VIF values less than 5 were considered [26]. Thus, the developed models would contain fewer but consistent variables.

With the detected biogas yield (BgY) in the UASB as the output variable (Y), descriptive statistics of the model variables are given in Table 1. It was noticed that some input variables fluctuated remarkably. Akaya et al., however, argued that difference in values of input parameters is a preferable positive indicator in arriving at positive results for a general biogas prediction model [17].

Evaluation and selection of the models

All of the input variables were used to obtain the regression coefficients $(\hat{\beta}_0, \hat{\beta}_1, ...\hat{\beta}_k)$ and to estimate the value of regression residual $(\hat{\varepsilon})$ as given in Eq.5 [4].

$$\hat{\varepsilon} = y - \hat{y} = y - (\hat{\beta}_0 + \hat{\beta}_1 x_1 + \cdots + \hat{\beta}_k x_k) \tag{5}$$

Five unique model equations with multiple input variables in various combinations and interactions were developed and selected based on Goodness-of-Fit [35]. Nine regression coefficients, i.e., β_0, β_1, ..., β_8, were considered and its respective values were estimated and used in the models. The optimum models were selected based on the following statistical performance criterion: standard error of the

Variable	Term	Mean	Min	Max
COD (mg/L)	x_1	4029 ± 763	2469	5787
NH_4^+ (mg/L)	x_2	110 ± 18	68	169
pH (mg/L)	x_3	7 ± 1	5	8
ALK (mg/L)	x_4	4945 ± 1411	2297	7655
TKN (mg/L)	x_5	511 ± 96	281	841
TP (mg/L)	x_6	45 ± 6	34	58
VFAs (mg/L)	x_7	534 ± 171	27	895
HRT (h)	x_8	--	24	48
Biogas yield (L/d)	Y	10 ± 5	3	17

Table 1: Summary of descriptive statistics of input and output variables (confidence level 95.0%).

estimate (SEE) [36] (Eq.9), sum of squared residuals (SSR) (Eq.8), coefficient of multiple determination (R^2) (Eq.6), adjusted coefficient of multiple determination (Adj-R^2) [37] (Eq.7), VIF (Eq.10), Durbin-Watson statistics (DWS) (Eq.11) and p-value [15] (Eq.12).

$$R^2 = \frac{\sum_{i=1}^{n}(Y_p - \bar{Y})^2}{\sum_{i=1}^{n}(Y_o - \bar{Y})^2} \tag{6}$$

$$R_{adj}^2 = \left[\frac{(1-R^2)(n-1)}{n-k-1}\right] \tag{7}$$

$$SSR = \sum_{i=1}^{n}(Y_o - Y_p)^2 \tag{8}$$

$$SEE = \sqrt{\frac{\sum_{i=1}^{n}(Y_o - Y_p)^2}{n-m}} \tag{9}$$

$$VIF = \frac{1}{1-R^2} \tag{10}$$

$$d = \frac{\sum_{i=1}^{n}(e_i - e_{i-1})^2}{\sum_{i=1}^{n}e_i^2} \tag{11}$$

$$p = 2 \times P(TS >| ts || H_o\ is\ true) = 2 \times (1 - \text{cd } f(| ts |)) \tag{12}$$

where, Y_o, Y_p and $\bar{Y}$ denotes experimental data, predicted values and arithmetic mean of the observed data; n and m is the number of data points and parameters in the regression model, respectively; k is the number of independent regressors excluding the constant term; $e_i = yi - \hat{y}_i$, and y_i and $\hat{y}_l$ were, respectively, the observed and predicted values of the response variable for individual i; TS is random variable associated with the assumed distribution; ts is the test statistics calculated from sample, and cdf is the cumulative density function of the assumed distribution.

The suggested five models, named M1, M2, M3, M4 and M5 were expressed in Eq.13, Eq.14, Eq.15, Eq.16 and Eq.17, respectively.

$$\bar{Y} = \beta_0 + \beta_1 x_1 x_2 + \beta_2 x_1 x_5 + \beta_3 x_1 x_8 + \beta_4 x_2 x_5 + \beta_5 x_3 x_4 \tag{13}$$

$$\bar{Y} = \beta_0 + \beta_1 x_1 x_2 + \beta_2 x_1 x_5 + \beta_3 x_1 x_8 + \beta_4 x_2 x_5 + \beta_5 x_3 x_4 \tag{14}$$

$$\bar{Y} = \beta_0 + \beta_1 x_1 + \beta_2 x_2 + \beta_3 x_7 + \beta_4 x_8 \tag{15}$$

$$\bar{Y} = \beta_0 + \beta_1 x_1^2 + \beta_2 x_2^2 + \beta_3 x_8 \tag{16}$$

$$\overline{Y} = \beta_0 + \beta_1 x_1 + \beta_2 x_2 + \beta_3 x_3 + \beta_4 x_4 + \beta_5 x_5 + \beta_6 x_6 + \beta_7 x_7 + \beta_8 x_8 \qquad (17)$$

Results and Discussion

UASB performance

Performance of the UASB treating PSPW at 35 ± 1°C with HRTs of 48 h and 24 h by stages was presented in Figure 2 and Table 2. With an average influent COD of 3799 mg/L and an average organic loading rate (OLR) of 1.50 kgCOD/m³·d for HRT 48 h, the effluent COD averaged 267 mg/L with a removal ranged from 83.5% to 92.0% was obtained in the reactor (Figure 2a). As the influent COD was increased to about 4185 mg/L along with the shortened HRT of 24 h, the COD removal ranged from 90% and 94.5% with an effluent COD of about 280 mg/L, though the OLR had been increased to about 4.23 kg COD/m³·d. The higher COD removal at HRT 24 h resulted in an increase in biogas yield in the UASB. As shown in Figure 2b, the influent and effluent pH ranged from 5.35-8.05 (mean pH 7.00) and 7.35-8.86 (mean pH 8.00) for HRT 48 h and 24 h, respectively. The illustration in Figure 2c depicted biogas yield that ranged from 3.4 to 9.6 L/d obtained at HRT 48 h, while 11.3 to 17.4 L/d in HRT 24 h. The methane fraction throughout the performance of the reactor ranged from 56.2% and 84.5%.

Throughout the operation of the UASB, observed pH in both HRTs were almost similar in value even though a remarkable difference in ALK was observed in the reactor. Figure 2d indicated that no observable difference in NH_4^+ concentration was found when the reactor was operated at HRT 48 h or 24 h, with an influent and effluent concentration averaged 109 and 241 mg/L, respectively. The average influent and effluent ALK at HRT 48 h were 6010 and 10948 mg/L, while that of 3592 and 8638 mg/L for HRT 24 h, respectively (Figure 2e). The feasible pH and ALK enhanced the acetogenesis and methanogenesis in the reactor, resulting in the few VFAs observed in the effluent [38].

The average influent and effluent TKN at HRT 48 h were found to be 466 and 307, respectively (Figure 2f). With the shortened HRT 24 h, the influent and effluent TKN were increased to about 518 and 507, respectively. Within the 112 days' operation, the UASB showed no TP removal with the same concentration of about 45 mg/L in both influent and effluent (Figure 2f).

Correlations between output and input variables

Correlation analysis was performed during the data preparation to identify the potential input variables to build the model. The results as shown as Table 3 showed that influent COD, pH, NH_4^+, ALK, TKN, VFA, TP and HRT had remarkable influence on the biogas yield in the UASB. The eight variables correlated with biogas yield were therefore used as input and output variables in the models. Observably, NH_4^+ was the only variable included in all model types (Eq.13 to Eq.17), but it has seldom been used in predictive models before [23,39].

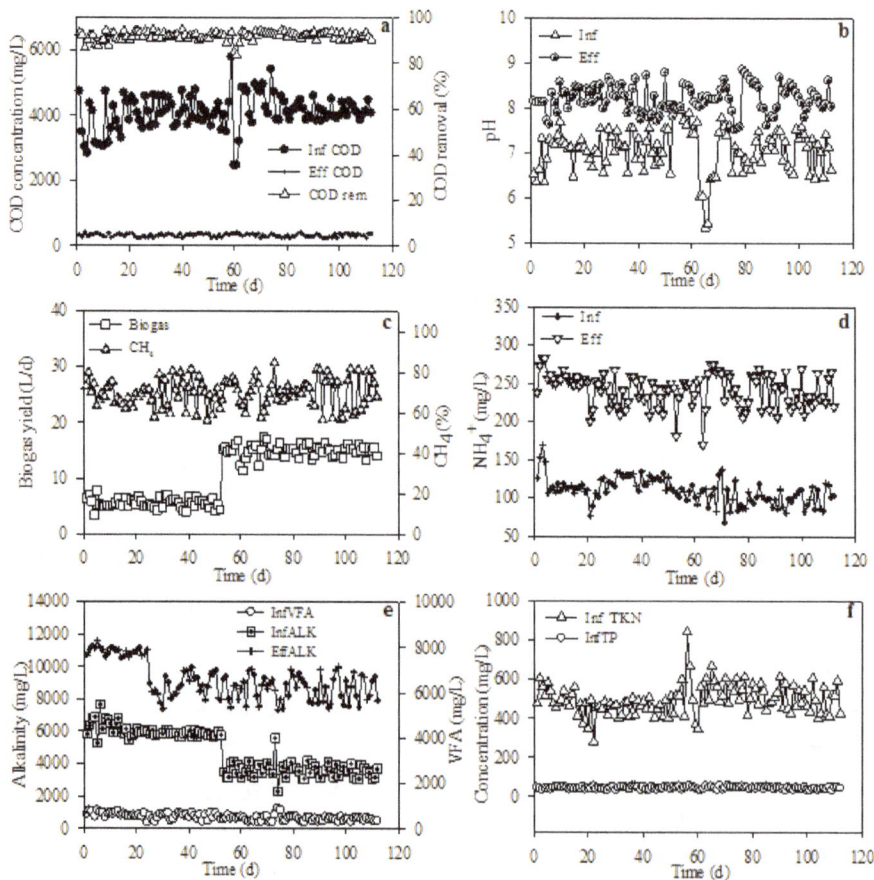

Figure 2: Fluctuation phenomenon of influent/effluent quality and reactor performance.

HRT	Water quality			OLR (kg COD/m³·d)	COD removal (%)	Biogas (L/d)	Methane (L/d)
		influent	effluent				
48 h (49 days)	pH	7.03 ± 1	8 ± 1	1.5 ± 1	91 ± 4	5.6 ± 1	3.8 ± 1
	COD(mg/L)	3799 ± 526	267 ± 109				
	VFAs(mg/L)	568 ± 126	--				
	NH₄⁺(mg/L)	119 ± 14	252 ± 17				
	TP(mg/L)	43 ± 5	45 ± 4				
	TKN(mg/L)	466 ± 58	307 ± 1				
	ALK(mg/L)	6010 ± 412	10948 ± 252				
24 h (63 days)	pH	7.08 ± 1	8.5 ± 1	4.23 ± 1	93 ± 4	14 ± 2	10.5 ± 2
	COD(mg/L)	4185 ± 546	280 ± 147				
	VFAs(mg/L)	433 ± 114	--				
	NH₄⁺(mg/L)	100 ± 13	243 ± 26				
	TP(mg/L)	45 ± 6	47 ± 4				
	TKN(mg/L)	518 ± 84	507 ± 1				
	ALK(mg/L)	3592 ± 456	8638 ± 751				

Table 2: The UASB performance throughout the 112-days operation (Confidence Level of 95%).

Input variables (influent)	Name	r	p value
COD	x_1	0.31	<0.001
NH₄⁺	x_2	-0.51	0.055
pH	x_3	-0.01	0.757
ALK	x_4	-0.93	0.747
TKN	x_5	0.33	0.801
TP	x_6	0.15	0.941
VFAs	x_7	-0.44	0.004
HRT	x_8	-0.98	<0.001
COD × NH₄⁺	x_1x_2	-0.18	<0.001
COD × TKN	x_1x_5	0.45	<0.001
COD × HRT	x_1x_8	-0.86	<0.001
NH₄⁺ × TKN	x_2x_5	-0.12	<0.001
NH₄⁺ × ALK	x_2x_4	-0.87	<0.001
NH₄⁺ × TP	x_2x_6	-0.31	<0.001
NH₄⁺ × VFAs	x_2x_7	-0.53	<0.001
TKN × TP	x_5x_6	0.34	0.042
TP × VFAs	x_6x_7	-0.35	<0.001
COD²	x_1^2	0.31	<0.001
(NH₄⁺)²	x_2^2	-0.50	0.018

Table 3: Pearson correlation coefficients (r) between biogas yield and input variables.

The correlation analysis demonstrated that some input variables correlated significantly with the biogas yield (r = 0.31 to 0.98) whiles others correlated poorly with r ranging from 0.01 to 0.18. Similarly, influent pH (x_3), ALK (x_4), TKN (x_5) and TP (x_6) were non-significant as their p values were relatively high ($p>0.05$). All other variables had a $p<0.001$ or <0.05 (Table 3). Among the input variables, COD (x_1), x_5, x_6, and the product terms of COD and TKN (x_1x_5), TKN and TP (x_5x_6) and COD (x_1^2) were positively correlated with the biogas yield. On the other hand, influent NH₄⁺ (x_2), pH (x_3), x_4, total VFAs (x_7), HRT (x_8) and the interaction terms x_1x_2, x_1x_8, x_2x_5, x_2x_4, x_2x_6, x_2x_7, x_6x_7, and x_2^2 were all negatively correlated with biogas yield.

Variable importance and model validation

There were 8 variable predictors (x_1, x_2, x_3, x_4, x_5, x_6, x_7, x_8) in the proposed 5 models (Eq.13 to Eq.17), and the entire sets of explanatory variables for the 5 models differed partially. Among the 8 variable predictors, NH4+ was noticed as a common predictor. Regression coefficients determined from the multiple regression analysis were

substituted into M3 (Eq.15) and M4 (Eq.16) to yield Eq.18 and Eq.19, respectively.

$$Y = 17.841 + 1.14 \times 10^{-3} x_1 + 1.11 \times 10^{-2} x_2 + 1.98 \times 10^{-3} x_7 - 0.4 x_8 \quad (18)$$

$$Y = 20.289 + 1 \times 10^{-7} x_1^2 + 8.8 \times 10^{-5} x_2^2 - 0.37 x_8 \quad (19)$$

The final structure of the model equations expressed in Eq.18 and Eq.19 were rewritten and given in Eq.20 and Eq.21, respectively.

$$BgY = 17.841 + 1.14 \times 10^{-3} COD + 1.11 \times 10^{-2} (NH_4^+)^2 + 1.98 \times 10^{-3} VFAs - 0.4 HRT \quad (21)$$

$$BgY = 20.289 + 1 \times 10^{-7} COD^2 + 8.8 \times 10^{-5} (NH_4^+)^2 - 0.37 HRT \quad (22)$$

Accordingly, independent variables x_1, x_2, x_7 and x_8 were used in M3 and M4 as shown in Eq.18 and Eq.19. Table 4 showed the results of the diagnostics statistics and performance criterion. Obviously, COD, NH₄⁺, VFAs and HRT were more useful than ALK, TKN, TP and pH in model M3 (Eq.15) and M4 (Eq.16). Conversely, NH₄⁺, ALK, TKN, TP and VFAs were more important in M1 (Eq.13). On the other hand, COD, NH₄⁺, TKN and HRT related better in M2 (Eq.14). Model M5 (Eq.17) engaged all eight variables as illustrated in Table 4.

Based on the diagnostics statistics, p-values associated with the variables in model M1, M2, M3 and M4 were statistically significant (Table 4). However, the entire variable set in M5 recorded high p-value. In particular, pH, TKN, ALK and TP used in M5 were 0.75, 0.74, 0.8 and 0.9 respectively, which were extremely > 0.05. This phenomenon was evident to conclude that M5 was not a good model to be considered by all standards although its R^2 yielded 97.30%. On the contrary, the multiple coefficient of determination (R^2) for M3 and M4 were 97.29% and 96.99%, respectively, with only 2.71% and 3.01% of the total variations not explained by both models in predicting biogas yield. For M1 and M2, about 13.46% and 3.85% of the variation existing among dependent variables were respectively not explained by these models, suggesting biogas yield predicted by M1 and M2 were unfit to the experimental data.

As shown in Table 4, the VIF for HRT and TP as regressors in M5 were 10.07 and 9.61 respectively. These huge values violated the assumptions specified in this study. In clarity, the VIF values confirmed that HRT and TP were highly correlated multicollinearity. Similarly, in M1, interaction variable of x_2x_7 (NH₄⁺ and VFAs) exhibited some multicollinearity as a value of 6.26 was estimated as VIF. In terms of

Model	Input variable	p-value	VIF	R^2 (%)	Adj-R^2 (%)	SEE	SSR	DWS
M1	β_0	0	--					
	x_2x_4	0	3.42					
	x_2x_6	0	2.42					
	x_2x_7	0	6.26	86.54	85.9	1.81	347.4	1.86
	x_5x_6	0.042	1.78					
	x_6x_7	0	4.01					
M2	β_0	0	--					
	x_1x_2	0	2.14					
	x_1x_5	0	1.56	96.15	96	0.96	99.4	1.9
	x_1x_8	0	1.69					
	x_2x_5	0	1.51					
M3	β_0	0	--					
	x_1	0	1.04					
	x_2	0.05	1.49	97.29	97.19	0.81	69.88	2.05
	x_7	0.003	1.37					
	x_8	0	1.69					
M4	β_0	0	--					
	x_1^2	0	1.04	96.99	96.9	0.85	77.79	2.02
	x_2^2	0.018	1.43					
	x_8	0	1.19					
M5	β_0	0	--					
	x_1	0	1.2					
	x_2	0.055	1.56					
	x_3	0.757	1.03					
	x_4	0.747	9.61	97.3	97.09	0.825	69.71	2.05
	x_5	0.801	1.2					
	x_6	0.941	1.11					
	x_7	0.004	1.39					
	x_8	0	10.07					

SEE, standard error of the estimate; SSR, sum of squared residuals; R^2, coefficient of multiple determination; Adj-R^2, adjusted coefficient of multiple determination; VIF, variance inflation factor; DWS, Durbin–Watson statistics; p-value, calculated probability.

Table 4: Diagnostics statistics and performance criterion of the models.

SEE, M3, M4 and M5 recorded the lowest values of 0.80, 0.84 and 0.82, respectively, suggesting a more precise evaluation of the variation in the estimated mean for set of the predictor values.

Furthermore, the SSR obtained for M3, M4 and M5 were the lowest among the 5 models. The values 69.88, 77.79 and 69.71, respectively, represented least variation or deviation of predictions from the mean as compared to M1 and M2. M3, M4 and M5 values in terms of DWS were 2.05, 2.02 and 2.05, respectively, suggesting no autocorrelation among input variables used in the models. On the other hand, DWS values for M1 (1.86) and M2 (1.90) indicated that input variables were approaching a positive autocorrelation.

Further analysis and applicability of selected optimum models

Base on the results shown in Table 4, M3 (Eq.15) and M4 (Eq.16) had been identified as the best models among the 5 proposed models in predicting biogas yield of the UASB. Residual analysis was carried out to determine the adequacy of the models and their compliance with the assumptions of regression. The normal probability plots for M3 (Figure 3a) and M4 (Figure 3e) showed some minimum deviations of data points from the straight line at the extremes. The less visible pattern observed on the plot of standardized residuals against the fitted (predicted) values (Figure 3b and 3f), verified the assumption that the residuals are randomly distributed and has constant variance.

In terms of residual distribution with histogram, M3 (Figure 3c)

and M4 (Figure 3g) were relatively well distributed with no trace of outliers. The standardized residuals for M3 and M4 are shown in Figure 3d and 3h, respectively. No obvious increasing or decreasing, cyclical or sudden shift of the data points was observed. The residuals versus order plot verified the assumption that the residuals were independent from one another. The linear regression analysis of the MnLRM output and the corresponding experimental data had a relatively good cohesion (Figure 4). The head-to-head comparisons of predicted data versus experimental data illustrated that model M3 (Figure 5a) and M4 (Figure 5b) were in perfect agreement, indicating their effectiveness in making predictions from the UASB treating PSPW.

Above all, the developed model M3 and M4 could serve as a valuable and practical management tool that could support the control of anaerobic wastewater treatment processes for biogas generation. All parameters used in the model development could be obtained from experimental observations or by rapid measurements. Nevertheless, to ensure reliable performances, the introduced model(s) could be fitted with large dataset to offer a significant improvement in prediction accuracy. Therefore, future research should target at collecting prolonged time-series data to improve the model performance and to minimize effects, errors or/and possible unrealistic predictions.

Conclusion

The UASB was feasible and efficient in treating potato starch processing wastewater. With an average organic loading rate (OLR)

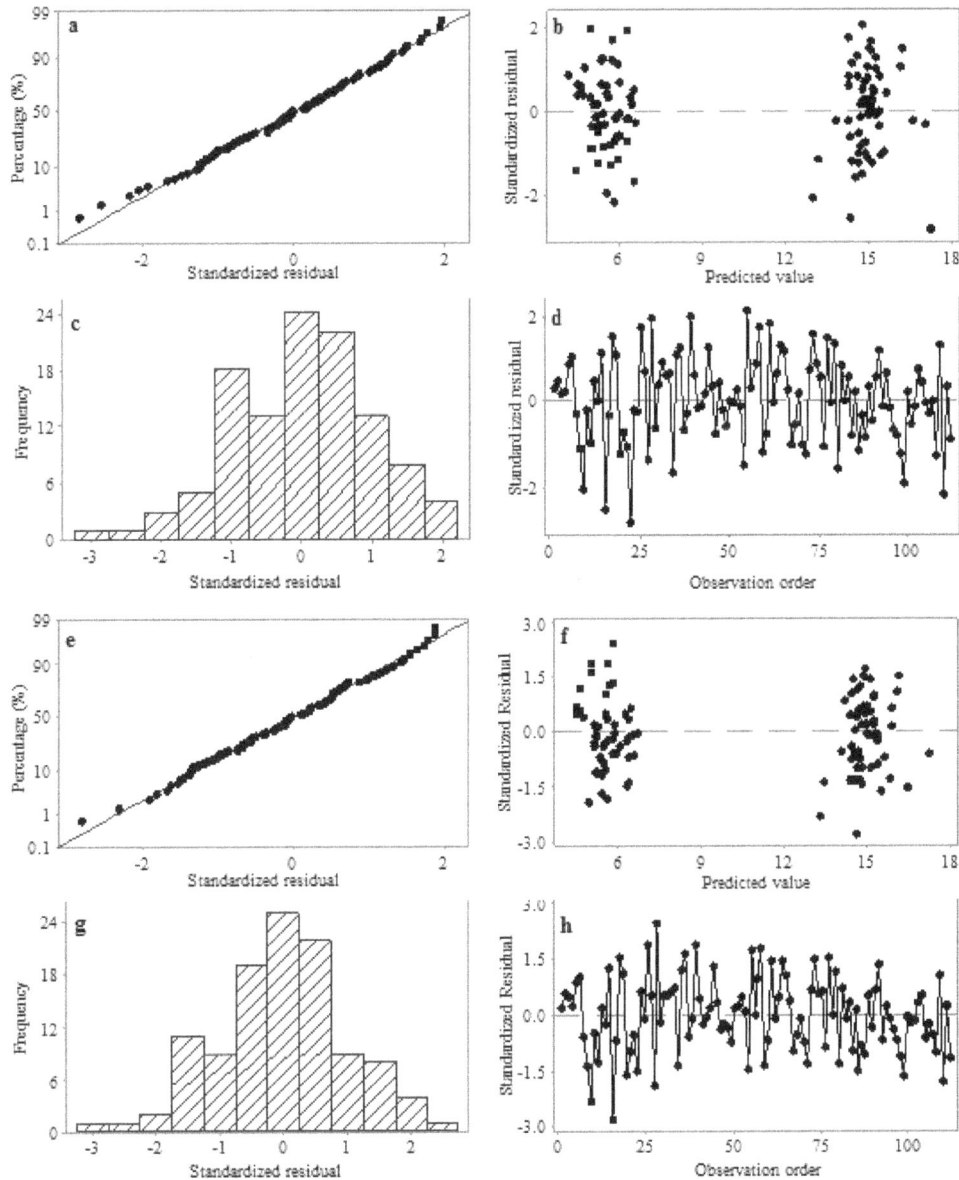

Figure 3: Residual plots of Normal Probability Plot (a, e), Versus Fits (b, f), Histogram (c, g) and Versus Order (d, h).

of 1.50 kg COD/m³·d, COD removal efficiency ranging from 83.5% to 92.0% was obtained when HRT was 48 h. As the influent COD was increased to about 4185 mg/L along with the shortened HRT of 24 h, the COD removal reached 94.5%, although organic loading rate (OLR) had been increased to about 4.23 kg COD/m³·d. The higher COD removal at HRT 24 h resulted in an increase in biogas yield in the UASB. Biogas yield at HRT 48 h ranged from 3.4 to 9.6 L/d, whiles 11.3 to 17.4 L/d were observed at HRT 24 h. The methane fraction throughout the performance of the reactor reached 84.5%. No signs of acidity were encountered in the UASB as effluent pH observed ranged from 7.35-8.86 (mean pH 8.00) for both HRT of 48 h and 24 h.

To predict the biogas yield in the UASB treating potato starch processing wastewater (PSPW), the dynamic relationship among PSPW parameters, reactor operational parameters and the biogas yield were modeled based on MnLR model and validated with residuals analysis. Among the 5 developed models, M3 and M4 were identified as the optimum ones due to their superior predictive performance on biogas yield. The R^2 emerged from M3 and M4 were 97.29% and 96.99%, respectively. COD, NH_4^+, VFAs and HRT were the most useful and favourable predictive parameters compared to ALK, TKN, TP and pH. Both model M3 and M4 turned out to be a good tool for predicting biogas yield in UASBs. These models can also contribute to the understanding of the factors that influence anaerobic processes, and subsequently be used as a guide to control the processes to enhance biogas yield.

Acknowledgements

The authors gratefully acknowledge the financial support from the Major Science and Technology Program for Water Pollution Control and Management (Grant No. 2013ZX07201007), and the State Key Laboratory of Urban Water Resource and Environment, Harbin Institute of Technology (Grant No. 2016DX06).

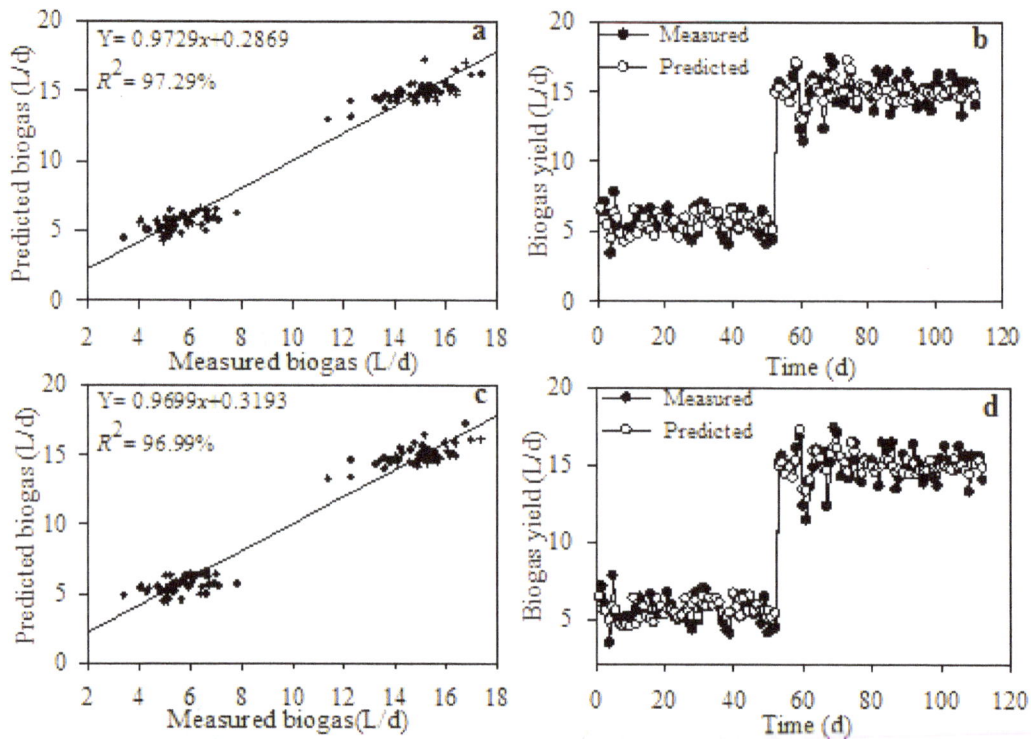

Figure 4: Correlation (a, c) and visual agreements (b, d) of the predicted and the experimental data in model M3 and M4, respectively.

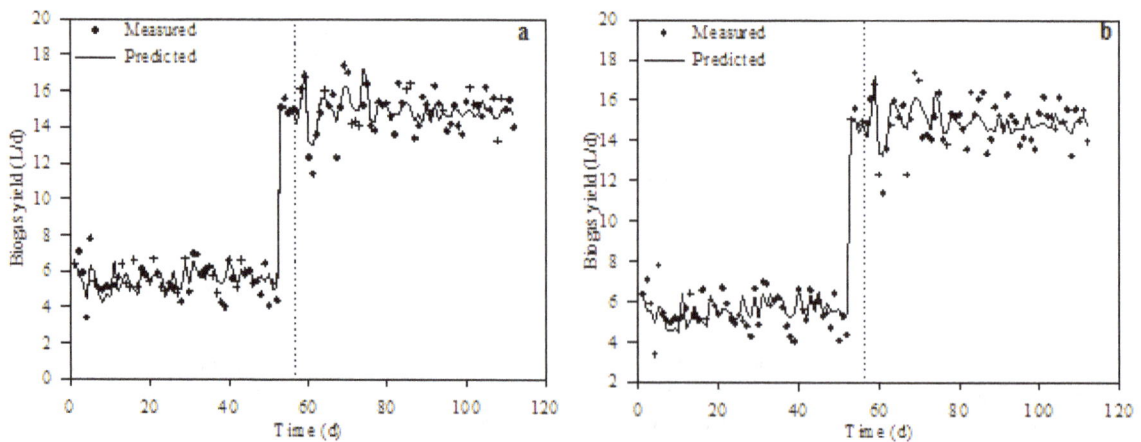

Figure 5: Head-to-head comparisons of predicted and experimental results.

References

1. Keijbets M (2008) Potato processing for the consumer: developments and future challenges. Potato Res 51: 271-281.

2. Wang RM, Li FY, Wang XJ, Li QF, He YF, et al. (2010) The application of feather keratin and its derivatives in treatment of potato starch wastewater. Functional Materials Letters 3: 213-216.

3. Li Y, Song J, Yang Q (2015) Utilization of potato starch processing wastes to produce animal feed with high lysine content. Journal of microbiology and biotechnology 25: 178-184.

4. Barampouti E, Mai S, Vlyssides A (2005) Dynamic modeling of biogas production in an UASB reactor for potato processing wastewater treatment. Chemical Engineering Journal 106: 53-58.

5. Wang RM, Wang Y, Ma GP, He YF, Zhao YQ (2009) Efficiency of porous burnt-coke carrier on treatment of potato starch wastewater with an anaerobic–aerobic bioreactor. Chemical Engineering Journal 148: 35-40.

6. Fang C, Boe K, Angelidaki I (2011) Biogas production from potato-juice, a by-product from potato-starch processing, in upflow anaerobic sludge blanket (UASB) and expanded granular sludge bed (EGSB) reactors. Bioresource technology 102: 5734-5741.

7. Arhoun B, Bakkali A, El Mail R, Rodriguez-Maroto J, Garcia-Herruz F (2013) Biogas production from pear residues using sludge from a wastewater treatment plant digester. Influence of the feed delivery procedure. Bioresource technology 127: 242-247.

8. Linville JL, Shen Y, Schoene RP, Nguyen M, Urgun-Demirtas M, et al. (2016) Impact of trace element additives on anaerobic digestion of sewage sludge with in-situ carbon dioxide sequestration. Process Biochemistry 51: 1283-1289.

9. Şentürk E, Ince M, Engin GO (2010) Kinetic evaluation and performance of a mesophilic anaerobic contact reactor treating medium-strength food-processing wastewater. Bioresource technology 101: 3970-3977.

10. Ratanatamskul C, Manpetch P (2016) Comparative assessment of prototype digester configuration for biogas recovery from anaerobic co-digestion of food waste and rain tree leaf as feedstock. International Biodeterioration & Biodegradation 113: 367-374.

11. Ghaniyari-Benis S, Martín A, Borja R (2010) Kinetic modelling and performance prediction of a hybrid anaerobic baffled reactor treating synthetic wastewater at mesophilic temperature. Process biochemistry, 45: 1616-1623.

12. Jiang J, Wu J, Poncin S, Li HZ (2016) Effect of hydrodynamic shear on biogas production and granule characteristics in a continuous stirred tank reactor. Process Biochemistry 51: 345-351.

13. Hu W, Thayanithy K, Forster C (2002) A kinetic study of the anaerobic digestion of ice-cream wastewater. Process Biochemistry 37: 965-971.

14. Yetilmezsoy K, Sapci-Zengin Z (2009) Stochastic modeling applications for the prediction of COD removal efficiency of UASB reactors treating diluted real cotton textile wastewater. Stochastic environmental research and risk assessment 23: 13-26.

15. Yetilmezsoy K, Turkdogan FI, Temizel I, Gunay A (2013) Development of ann-based models to predict biogas and methane productions in anaerobic treatment of molasses wastewater. International Journal of Green Energy 10: 885-907.

16. Cakmakci M (2007) Adaptive neuro-fuzzy modelling of anaerobic digestion of primary sedimentation sludge. Bioprocess Biosyst Eng 30: 349-357.

17. Akkaya E, Demir A, Varank G (2015) Estimation of biogas generation from a UASB reactor via multiple regression model. International Journal of Green Energy 12: 185-189.

18. Ozkaya B, Demir A, Bilgili MS (2007) Neural network prediction model for the methane fraction in biogas from field-scale landfill bioreactors. Environmental Modelling & Software 22: 815-822.

19. Batstone D, Keller J, Angelidaki I, Kalyuzhnyi S, Pavlostathis S, et al. (2002) The IWA Anaerobic Digestion Model No 1 (ADM1). Water Science and Technology 45: 65-73.

20. Nasr N, Hafez H, El Naggar MH, Nakhla G (2013) Application of artificial neural networks for modeling of biohydrogen production. International journal of hydrogen energy 38: 3189-3195.

21. Ahn JY, Chu KH, Yoo SS, Mang JS, Sung BW, et al. (2014) Determination of optimal operating factors via modeling for livestock wastewater treatment: Comparison of simulated and experimental data. International Biodeterioration & Biodegradation 95: 46-54.

22. Mas DML, Ahlfeld DP (2007) Comparing artificial neural networks and regression models for predicting faecal coliform concentrations. Hydrological Sciences Journal 52: 713-731.

23. Herrig IM, Böer SI, Brennholt N, Manz W (2015) Development of multiple linear regression models as predictive tools for fecal indicator concentrations in a stretch of the lower Lahn River, Germany. Water research 85: 148-157.

24. APHA, AWWA, WEF (2005) Standard Methods for the Examination of Water and Wastewater. 21th edn. American Public Health Association, Washington, USA.

25. Ramette A (2007) Multivariate analyses in microbial ecology. FEMS Microbiol Ecol 62: 142-160.

26. Mac Nally R (2000) Regression and model-building in conservation biology, biogeography and ecology: the distinction between - and reconciliation of - 'predictive' and 'explanatory' models. Biodiversity and Conservation 9: 655-671.

27. Huang L, Chen JC (2001) A multiple regression model to predict in-process surface roughness in turning operation via accelerometer. Journal of Industrial Technology 17: 1-8.

28. Yetilmezsoy K, Sakar S (2008) Development of empirical models for performance evaluation of UASB reactors treating poultry manure wastewater under different operational conditions. Journal of Hazardous materials 153: 532-543.

29. Singh KP, Basant N, Malik A, Jain G (2010) Modeling the performance of up-flow anaerobic sludge blanket reactor based wastewater treatment plant using linear and nonlinear approaches—a case study. Analytica chimica acta 658: 1-11.

30. Turkdogan-Aydınol FI, Yetilmezsoy K (2010) A fuzzy-logic-based model to predict biogas and methane production rates in a pilot-scale mesophilic UASB reactor treating molasses wastewater. Journal of hazardous materials 182: 460-471.

31. Rostami I, Juhasz AL (2013) Bioaccessibility-based predictions for estimating PAH biodegradation efficacy—comparison of model predictions and measured endpoints. International Biodeterioration & Biodegradation 85: 323-330.

32. Wold S, Sjöström M, Eriksson L (2001) PLS-regression: a basic tool of chemometrics. Chemometrics and intelligent laboratory systems 58: 109-130.

33. Preacher KJ, Hayes AF (2008) Asymptotic and resampling strategies for assessing and comparing indirect effects in multiple mediator models. Behavior research methods 40: 879-891.

34. Faul F, Erdfelder E, Buchner A, Lang AG (2009) Statistical power analyses using G*Power 3.1: Tests for correlation and regression analyses. Behavior Research Methods 41: 1149-1160.

35. Schermelleh-Engel K, Moosbrugger H, Müller H (2003) Evaluating the fit of structural equation models: Tests of significance and descriptive goodness-of-fit measures. Methods of psychological research online 8: 23-74.

36. Xu Y, Ma C, Liu Q, Xi B, Qian G, et al. (2015) Method to predict key factors affecting lake eutrophication–A new approach based on Support Vector Regression model. International Biodeterioration & Biodegradation 102: 308-315.

37. Abdul-Wahab SA, Bakheit CS, Al-Alawi SM (2005) Principal component and multiple regression analysis in modelling of ground-level ozone and factors affecting its concentrations. Environmental Modelling & Software 20: 1263-1271.

38. De-Sousa JT, Santos KD, Henrique IN, Brasil DP, Santos EC (2008) Anaerobic digestion and the denitrification in UASB reactor. Journal of Urban and Environmental Engineering 2: 63-67.

39. David MM, Haggard BE (2011) Development of regression-based models to predict fecal bacteria numbers at select sites within the Illinois River Watershed, Arkansas and Oklahoma, USA. Water, Air, & Soil Pollution 215: 525-547.

15

Development of Molecular Identification of Nitrifying Bacteria in Water Bodies of East Kolkata Wetland, West Bengal

Mousumi Saha[1]*, Agniswar Sarkar[2] and Bidyut Bandhophadhyay[1]

[1]Department of Biotechnology, Vidyasagar University, West Bengal, India
[2]Department of Biotechnology (Recognized by DBT, Govt. of India), The University of Burdwan, Burdwan, West Bengal, India

Abstract

Nitrifying bacteria plays a major role in converting the waste water to valuable renewable resources for the society. Ammonia is the toxic excretory product of most aquatic organisms and nitrite formed by the oxidation of ammonia is also toxic. Ammonia oxidizing bacteria converts ammonia to nitrite and Nitrite oxidizing bacteria convert nitrite to nitrate. Among them, *Nitrosomonas* sp. and *Nitrobacter* sp. has been the most widely studied organisms. An advanced molecular technique made it possible to explore the nitrifying bacteria in the environment and to enhance our knowledge of its functioning. In view of this it would be of prime importance for rapid detection of these strains/ isolates using molecular techniques. The present study was undertaken to develop and validate 16S rDNA sequence analysis of *Nitrosomonas* sp. and *Nitrobacter* sp. collected from different *bheries* in East Kolkata Wetland, West Bengal.

Keywords: Ammonia oxidizing bacteria; Nitrite oxidizing bacteria; 16S rDNA; East Kolkata Wetland

Introduction

East Kolkata Wetland (EKW) is the world's largest sewage-fed aquaculture and natural water recycling system and contains a series of wetland. It is renowned not simply as the sensitive ecosystem of biodiversity value, but more specifically as an example of best practice in the 'wise use' of wetlands (from 1879 onwards) for garbage farming (2500 metric ton of solid waste is dumped every day), sewage-treated fisheries (*bheris*) and sewage farming [1]. EKW located in the eastern part of the city (22024/- 22036/N to 88023/-88032/E). In addition, organic fertilizer produced by processing the solid waste from this area is found to be commercially important [2-4]. The wetland also supports the livelihoods of around 60,000 residents through the fisheries and other socio-economic activities.

It serves as a vast area for research purpose and environment protection through bioremediation or phytoremediation [3]. EKW declared as a Ramsar site on November 2002 as per "Ramsar Guidelines". Wastewater in EKW flows through various channels into the fish ponds covering about 4000ha, where a wide range of physical, biological and chemical processes take place which helps in improving the quality of the water [4,5]. Consequently this wetland system is popularly known as the "kidney of the city".

Various bacterial species distributed in different site or bheries in EKW and involved in bioremediation of heavy metals. These diverse microbial resources also produce some commercially important enzymes like protease, oxidase etc. *Streptococcus macedonicus* produces extracellular protease and food grade bacteriocine [6]. *Pseudomonas* sp. are involved in the production of extracellular protease [7]. *Klebsiella* sp. can help to overcome cadmium toxicity [8]. _Actinobacteria_ and *Fermicutes* participate in the degradation of nitrophenol, nitroaromatic compound, pesticide and herbicide [3]. The metal accumulating or removal property of these microbes can be exploited for treatment of contaminants rich in metals in various sites. Along with the above microbes, proteobacteria like *Nitrosomonas* sp. present in EKW and enable to oxidize ammonia into nitrite [3].

Ammonia is the principal excretory product of most aquatic organisms, but is toxic, acute and chronic to fish. Nitrite is formed either by the oxidation of ammonia (Nitrification) or the reduction of nitrate. Nitrite is also toxic to fish and another critical water quality factor. Ammonia affects on the central nervous system by causing "acute ammonia intoxication" which includes convulsions and death [9]. Acute concentrations of ammonia may cause loss of equilibrium, hyperexcitability, increased breathing, cardiac output, and oxygen uptake; in extreme cases convulsions, coma, and death. Lower concentrations of ammonia can cause a reduction in hatching success; reduction in growth rate and morphological development; pathologic changes in tissues of gills, livers, and kidneys [10]. Brown Blood Disease occurs in fish when water contains high nitrite concentrations. Nitrite enters the bloodstream through the gills and turns the blood to a chocolate-brown color. Hemoglobin, which transports oxygen in the blood, combines with nitrite to form Methemoglobin, which is incapable of oxygen transport [11]. Warm water fishes like Carps, Catfishes, and Tilapia are fairly sensitive to nitrite, and trout and other cool water fish are sensitive to extremely small amounts of nitrite. Nitrites also react directly with hemoglobin in human blood and other warm-blooded animals to produce methemoglobin.

Thus, it is very important to understand the dynamics of nitrification, where may be some microbe like ammonia oxidizing bacteria (AOB-*Nitrosomonas* sp.) and nitrite oxiding bacteria (NOB-*Nitrobacter* sp.) plays the major role in converting the waste water to valuable renewable resources for the society [12-14]. The genera *Nitrosomonas* and *Nitrobacter* both are chemolithoautotrophic and members of the *Proteobacteria* class [15]. Genus *Nitrosomonas* form a distinct group within the β subdivision where as nitrite-oxidizing bacteria like *Nitrobacter* occurring in α, δ, and γ subdivisions. Therefore, to maintain

*Corresponding author: Mousumi Saha, Department of Biotechnology, Vidyasagar University, West Bengal, India, E-mail: mou.cute@gmail.com

the water quality and fish health it's become very important to keep the ammonia and nitrite within the safe tolerance limit and these can be done some nitrifying bacteria. Various researchers described that only about 1 to 4% of the entire microbial community of EKW has been screened following the traditional and the conventional methods and about 96 to 99% are yet to be isolated and identified [16,17]. EKW, which can prove to be important from point of bioremediation and biotechnology and in this respect, it is necessary to develop the rapid and accurate technology for isolation and identification of these microbes using modern molecular techniques. Therefore, to gather more information about these diverse microbial communities, it is very important to use molecular technologies like 16S rRNA gene sequence, phylogenetic analysis etc [18,19]. Study of AOB and NOB by conventional cultivation techniques is also very difficult because of their long generation times and low growth rates [20]. Therefore, a rapid, culture-independent detection technique for nitrifiers would be useful to provide a record about the natural availability and role of these microbes in bioremediation of various bheries located in the East Kolkata Wetland. In view of this economic importance to aquatic animals, human and animal health hazard problems, it would be of prime importance for rapid detection of these strains/isolates using molecular techniques. With this in view, the present study was undertaken with following objective: To develop and validate different genotyping techniques viz 16S rDNA sequence analysis, Ribotyping and bioinformatics analysis for *Nitrosomonas* and *Nitrobacter* sp. collected from different bheries in West Bengal. To develop and standardize PCR technique, for accurate and prompt diagnosis of the species, to create a genomic fingerprint database for surveillance and monitoring genetic variability of isolates and also to establish a biosecurity protocol for the Bheries located in different areas of West Bengal.

Materials and Methods

Site selection

In West Bengal both sewage fed fisheries and floodplain wetlands plays an important fishery resources. The commonly cultivated fish includes Catla (*Catla catla*), Rohu (*Labeo rohita*), Mrigal (*Cirrhinus mrigala*), Silver carp (*Hypothalmichthys molitrix*), Grass carp (*Ctenopharyngodon idella*) and Common carp (*Cyprinus carpio*) are considered to be the best culturable species.

Flood plain wetland in West Bengal covers an area of about 42000 ha. Within the West Bengal, water bodies for fish culture are present in North and South 24 Praganas and Nadia Districts. Waste water fed fish ponds in East Kolkata Wetland, West Bengal, India (called Bheris) have a distinctly different architecture resulting in extensive purification of waste along with integrated resource recovery [3]. Bheri, the shallow (50-150 cm), flat bottom, waste water fed fish pond. Kolkata, Nadia, South 24 Parganas and Burdwan were selected for the present study because; they are important districts of West Bengal with respect to both aquaculture and agriculture. East Kolkata Wetland in Kolkata; Kulia, Bhomra in Nadia; Haruabhanga, Lakshmipur in Burdwan and Malancha, Jaynagar in South 24 parganas were identified and surveyed for our study. But for this present study EKW was chosen as it is a waste water fishery. Preliminary physico-chemical and biological survey indicates the presence of nitrifying bacteria in EKW as compared to other sites of West Bengal. Among which research work was initiated under East Kolkata Wetland at Nalban Fisheries (owned by Government of West Bengal and runed by State Fisheries Development Co-operation) and Jhagrasisa bhery (runed by private organization).

Collection of samples and process

Water samples (25 ml) were collected from Nalban Fisheries and Jhagrasisa bhery. A total 6 fish ponds or tanks were selected for the study and marked as N1, N2, N3 for Nalban Fishery and J1, J2, J3 for Jhagrasisa bhery in East Kolkata Wetland. Samples (two aliquots from each place) were also collected from Topsia (Sewage Pumping Station) and from different channels like Ambedkar Bridge, Bantala, Bamunghata, Ghosher Khal which is connected to the fish pond and Ghusighata (outlet). Samples from each site was collected at around 6.30-7.00 in the morning in sterile plastic bottle (Tarson) from a depth of 1-2.0cm below the surface of the water, kept in ice box (Milton, India) and then taken in laboratory for immediate culture. Water samples were collected in different seasons as winter, summer, post monsoon and monsoon. In each seasons two aliquots were collected from each tanks and a total of 96 aliquots were collected and studied.

Samples were collected and processed on Winogradskyi medium for the isolation of nitrifying bacteria. *Nitrosomonas* sp. was isolated using Winogradsky phase 1 medium, which contains $(NH_4)_2SO_4$, 2.0 g; K_2HPO_4, 1.0 g; $MgSO_4$, $7H_2O$, 0.5 g; NaCl, 2.0 g; $FeSO_4 7H_2O$, 0.4 g; $CaCO_3$, 0.01g, and miliQ water upto 1000 ml, pH 7.8). *Nitrobacter* sp. was isolated using Winogradsky phase 2 medium (KNO_2, 0.1 g; Na_2CO_3, 1.0 g; NaCl, 0.5 g; $FeSO_4$, $7H_2O$, 0.4 g, and miliQ water upto 1000 ml, pH 7.8) [21-23]. Tubes were incubated at 28°C for 90 hrs aerobically [21,24,25]. Cultures were initially identified by gram bacteria [26]. Suspected cultures were evaluated for presence of ammonia and nitrite and vice versa for production of nitrite and nitrate. These tests were performed and confirmed by a series of reaction. Selected cultures were preserved for further studies at -80°C with 30-40% glycerol stock.

Isolation of genomic DNA

Bacterial biomass from 10 ml of culture was collected by centrifugation for 5 min at 10000×g and resuspended in a 2 ml polypropylene tube (Eppendrof) containing glass beads with 500 ml buffer containing 20 mM sodium acetate, 1 mM EDTA (pH 5.5), 50 ml 25% SDS and 600 ml phenol: chloroform: isoamyl alcohol (25: 24: 1). Cells were lysed by the addition of 200 µl of 10 mg ml⁻¹ lysozyme and incubation at 37°C for 60 min. 20% Sodium dodecyl sulfate was added to a final concentration of 1% and incubated at 37°C for 60 min. Proteinase K solution was added (10 mg/ml) to a final concentration of 2 mg/ml and incubated at 50°C for 35 min and the mixture was then centrifuged at 10,000 g×10 min at 4°C. The aqueous phase was transferred carefully to a fresh tube, mixed with 600 ml chloroform: isoamyl alcohol (24: 1) and centrifuged 10000 g×10 min. The aqueous phase was transferred to a fresh tube and, after the addition of 0.1 vol. 3 M sodium acetate. DNA was precipitated by incubation with 0.6 volume of isopropanol and 5 ml glycogen (5 mg/ml) for 1 h and subsequently pelleted by centrifugation at 10 000 g×20 min at 4°C. Pellets were washed with 1 ml ice-cold 70% alcohol and dried it by Maxi Vacuum Dryer and resuspended in 500µl TE buffer (pH 8.0). The DNA concentration was estimated by visual comparison with the standard DNA size markers after electrophoresis through 0.8% agarose gel, stained with 0.5 mg ml⁻¹ ethidium bromide (Sigma Chemicals Co.).

16S rDNA-PCR amplification

The 16S rRNA genes were amplified by PCR using the universal primers: forward primer 27F 5'-AGA GTT TGA TCC TGG CTC AG-3' and reverse primer 1492R 5'- GGT TAC CTT GTT ACG ACT T-3' (Chromous Biotech Ltd., Bengaluru, India) [27]. PCR reactions has been carried out by following the series of standardizing experimental protocol as annealing temperature, concentration of MgCl₂, template DNA, Taq DNA polymerase, dNTP's and primers. The PCR reaction components consists of 200 mm dNTP, 20 pico moles of primer, 2 units of Taq DNA polymerase enzyme, assay buffer with working concentration of 1.5 mM MgCl2, 20-30 ng template DNA in an assay volume of 25 µl. These concentrations were determined by a series of preliminary standardizing experiments.

A cycling program was performed using a Thermal Cycler (Thermo Scientific) with an initial denaturation step at 95°C×3 min, 35 cycles of denaturation at 94°C×1 min, annealing at 55°C×1 min, and extension at 72°C×2 min, and final extension at 72°C×3 min. The presence of the PCR products were examined and visualized by electrophoresis in 1.5% agarose gel in TBE buffer. The gel was stained with EtBr (Sigma) and viewed in Gel Doc System (Biorad Gel Doc. 2000 system).

Sequencing of 16S rDNA amplicons

The PCR products were purified by the Exosap treatment in the Departmental Molecular Biology Laboratory. For the validation and identification, chosen strains were identified by 16S rDNA PCR pattern analysis on 1.5% agarose gel electrophoresis and those were selected for sequencing. In total, the PCR amplified products were sequenced by Ion torrent Sequencer from the Xcelris Lab Ltd, Ahmedabad, India.

Computer assisted pattern analysis

Chromatogram of sequence data confirmed the peak, reproducibility, quality and gene sequences of 16S rRNA amplicons. Raw sequence data are arranged as complementary and consensus through online Reverse-Complement tools (www.bioinformatics.org/sms/rev_comp.html) and Genefisher 2, the BiBiServ (Bielefeld BioInformatics Service) is a unit of the Institute for Bioinformatics (bibiserv.techfak.uni-bielefeld.de/genefisher2/submission.html). Identification of species and similarity calculations were performed, comparing sequences of approximately 1500 bases with sequences available in GenBank using BLAST network services (blast.ncbi.nlm.nih.gov/). Multiple Sequence Alignments (MSA) are carried out by Clustal W following all the sequence pairs were aligned separately for the calculation of distance matrix and constructed a guide tree and sequences are aligned according to the hiearchy.

Phylogenetic analysis between Ammonia Oxidizing and Nitrite Oxidizing Bacteria

Phylogenetic relationship, similarity index, distance between all isolates and maximum likelihood pattern analysis outlined and phylogenetic tree generated by Mega v5.05. The 16S rDNA sequences in this study were submitted to the GenBank database by using Sequin software (www.ncbi.nlm.nih.gov/projects/Sequin/) under nucleotide accession number. Analysis of the 16S rRNA gene proved as useful taxonomic tools with discriminatory properties for identification up to the species level.

Results and Discussion

Identification of strains

AOB and NOBs were initially identified by gram staining and produced gram negative characteristics. The nitrite accumulation test was performed to confirm AOBs and the nitrite consumption test was performed to confirm NOBs by following the standard protocol of APHA, (1998) [28]. A comparative 16S rRNA sequence analysis finally confirmed the genus and species (Table 1). Out of total seven strains were sequenced and tested, two were found as *Nitrosomonas* sp. and

another two *Nitrosomonas* includes *eutropha* and *euroraea* species; on the other hand, two strains were found as *Nitrobacter* sp. and another single strain was identified as *Nitrobacter vulgaris* (Figures 1 and 2).

Phylogenetic analysis

Grimont and Grimont (1986) described the importance of the 16S rRNA-based probes as taxonomic tools [29]. In the present study, we approached the use of the properties of the 16S rDNA for phylogenetic identification from a different perspective and results apparently covered the majority of NOBs and AOBs can be discriminated by 16s rDNA targeted primers therefore need to take into account the intragenomic diversity of the 16S rRNA gene. *Nitrosomonas europaea* cell are capable of doing nitrification in wastewater with high efficiency. A recent study

Strains code	Collection Site	Name of isolates	NCBI-GenBank Acc. No.
NSW1	Bantala	*Nitrosomonas eutropha*	KF618623
NSW2	Nalban fishery tank 2	*Nitrosomonas europaea*	KF618624
NSW3	Jhagrasisa bhery tank 1	*Nitrosomonas* sp.	KF618625
NSW4	Ambedkar bridge	*Nitrosomonas eutropha*	KF618626
NBW1	Bantala	*Nitrobacter* sp.	KF618620
NBW2	Nalban fishery tank 2	*Nitrobacter* sp.	KF618621
NBW3	Jhagrasisa bhery tank 1	*Nitrobacter vulgaris*	KF618622

Table 1: Details of samples and their source(s).

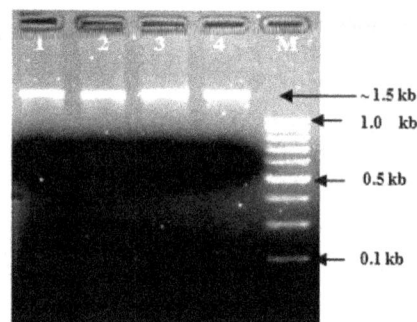

Figure 1: 1.5% Agarose gels showing atypical 16S rDNA-PCR amplicons of about 1.5 kb in size.
Lane 1, *Nitrosomonas eutropha* (KF618623); 2, *Nitrosomonas europaea* (KF618624); 3, *Nitrosomonas* sp. (KF618625); and 4, *Nitrosomonas eutropha* (KF618626). Samples isolated from Wastewater of East Kolkata Wetland; Lane M, 100bp DNA size marker (Promega).

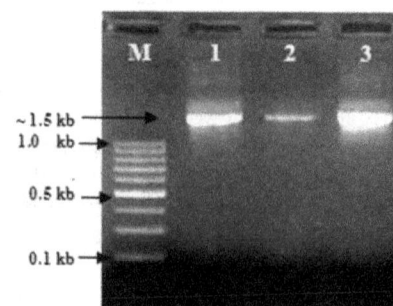

Figure 2: 1.5% Agarose gels showing atypical 16S rDNA-PCR amplicons of about 1.5 kb in size.
Lane 1, *Nitrobacter* sp. (KF618620); 2, *Nitrobacter* sp. (KF618621) and 3, *Nitrobacter vulgaris* (KF618622). Samples isolated from Wastewater of East Kolkata Wetland; Lane M, 100bp DNA size marker (Promega).

was developed by Niranjan Uemoto and Saiki and Kumar Shrestha et al. [12,30], where nitrogen removed using *Nitrosomonas europaea* and *Paracoccus denitrificans* from waste water. *Nitrobacter* population can be detected by using PCR- analysis [13].

As waste water is introduced into the bheries or the fish pond with high load of organics or heavy metals, the bacterial populations neutralizes or break down the organics or heavy metals to achieve maximum benefit and also enhance fish production. Thus, the application of probiotics in sewage fed bheries along with the details molecular characterization of microbes present in bheries can enhance the fish production and can reduce the level of ammonia, H2S stress. Analysis of different samples of *Nitrosomonas* spp. found in freshwater was done using 16S ribosomal DNA and PCR technique [18]. Classification of any bacterial group, taxonomically as well as molecular characterization is required, thus for studying ammonia- oxidizing bacteria polygenic and molecular analysis was done [19]. Along with this molecular characterization of the bacteria is also important for further investigation and to analyze these bacterial communities.

The isolation of samples collected from the distribution network in this study could be attributed to the effective strategy employed in the system, which may increase the survival of the organism as suspended or as part of the biofilm community. The similarity of isolated 16S ribosomal RNA gene, partial sequences were identical and ranging between 94 to 97% (Figures 3 and 4). To draw a proper biosecurity protocol, molecular diagnosis and phylogenetic relationships between the species in EKW, the present technique used in this study can be able to overcome threats and control the disease outbreaks. These data and methodology has been proved in this study as advance molecular tool. Our results emphasize the need to take into account the intragenomic diversity of the 16S rRNA gene. This method has shown discriminatory properties in the identification up to the species level. The study provided a quick identification tool.

Conclusion

Views were recognition of the heterogeneity detected in the sequences is important to avoid strain misidentification and faithful to the acknowledgement of diversity. Our study aimed towards prompt and accurate identification of members of the genus AOBs and NOBs as

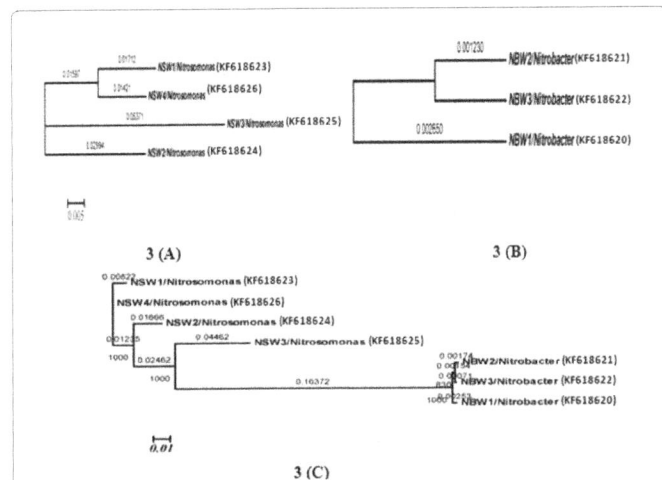

Figure 4: Phylogenetic distance tree displaying the evolutionary origin among different species of *Nitrosomonas* and *Nitrobacter* as reference, collected from NCBI- GenBank with query sequences within a lineage shared by the different species through Mega v5.05. Numbers indicate GenBank accession no. of reference sequences query sequences shared existence with them is indicated.

indicators in environment and as the most prevalent bacterium in the waterways of EKW in West Bengal, India. The method provided in this work, apart from being a reliable identifier at reasonable cost.

Acknowledgements

The author acknowledges to Dr. Bidyut Bandhophadhyay for his intellectual input. We also acknowledge Mr. S. D. Ghosh for his kind permission to work at EKW.

References

1. Raychaudhuri S, Thakur AR (2006) Microbial genetic resource mapping of East Calcutta Wetland. Cur. Sci 91: 212-217.

2. Chowdhury S, Mishra M, Adarsh VK, Mukherjee A, Thakur AR, et al. (2008) Novel Metal Accumulator and Protease Secretor Microbes from East Calcutta Wetland, American J Biochem Biotech. 4: 255-264.

3. Ray Chaudhuri S, Salodkar S, Sudarshan M, Mukherjee I, Thakur AR (2008) Role of water hyacinth-mediated phytoremediation in waste water purification at east Calcutta wetland. J Integ Environ Sci 5: 53-62.

4. Ghosh D, Sen S (1987) Ecological history for Calcutta's wetland conservation. Environ Conserv 14: 219–226.

5. Ghosh D (1998) Wastewater-fed aquaculture in the wetlands of Calcutta-an overview. International Seminar on Wastewater Reclamation and Reuse for Acquaculture, India, 6–9.

6. Georgalaki MD, Van Den Berghe E, Kritikos D, Devreese B, Van Beeumen J, et al. (2002) Macedocin, a food-grade lantibiotic produced by Streptococcus macedonicus ACA-DC 198. Appl Environ Microbiol 68: 5891-5903.

7. Guzzo J, Pages JM, Duong F, Lazdunski A, Murgier M (1991) Pseudomonas aeruginosa alkaline protease: Evidence for secretion genes and study of secretion mechanism. J Bacteriol 173: 5290-5297

8. Holmes JD, Richardson DJ, Saed S, Evans-Gowing R, Russell DA, et al. (1997) Cadmium-specific formation of metal sulfide 'Q-particles' by Klebsiella pneumoniae. Microbiology 143 : 2521-2530.

9. Randall DJ, Tsui TK (2002) Ammonia toxicity in fish. Mar Pollut Bull 45: 17-23.

10. McKenzie DJ, Shingles A, Claireaux G, Domenici P (2009) Sublethal concentrations of ammonia impair performance of the teleost fast-start escape response. Physiol Biochem Zool 82: 353-362.

11. Reutov VP, Sorokina EG, Kaiushin LP (1994)The nitric oxide cycle in mammals and nitrite reducing activity of heme-containing proteins. Vopr Med Khim 40: 31-35.

12. Uemoto H, Saiki H (1996) Nitrogen removal by tubular gel containing

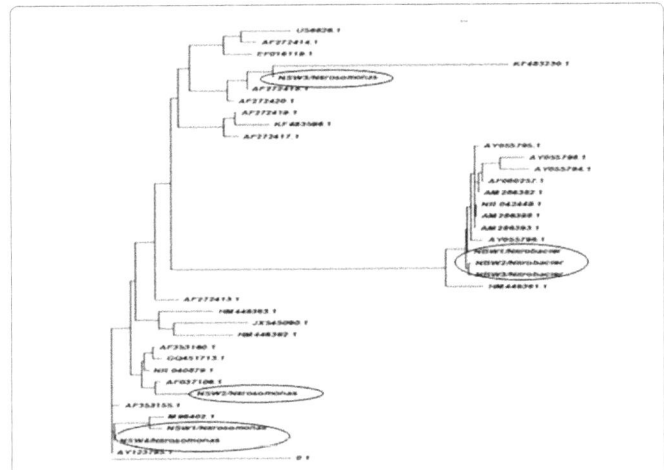

Figure 3: (A) Phylogenetic distance tree displaying the distance among *Nitrosomonas* sp.,
(B) shared the distance among *Nitrobacter* sp. and
(C) shared the linage of distance between both *Nitrosomonas* and *Nitrobacter* sp. The phylogenetic tree plotted through Mega v5.05.

Nitrosomonas europaea and Paracoccus denitrificans. Appl Environ Microbiol 62: 4224-4228.

13. Degrange V, Bardin R (1995) Detection and counting of *Nitrobacter* populations in soil by PCR. Appl Environ Microbiol 61: 2093-2098.

14. Brion N, Billen G (2000) Wastewater as a source of nitrifying bacteria in river systems: The case of the river seine downstream from paris. Wat Res. 34: 3213-3221.

15. Wheaton FW, Hochheimer JN, Kaiser GE, Kronos MJ, Libey GS, et al. (1994) Nitrification filter principles. Timmons MB and Losordo TM (ed.), Aquacult water reuse sys. Engineering design and management. Elsevier, Amsterdam, the Netherlands

16. Ward DM, Weller R, Bateson MM (1990) 16S rRNA sequences reveal numerous uncultured microorganisms in a natural community. Nature 345: 63-65.

17. Amann RI, Ludwig W, Schleifer KH (1995) Phylogenetic identification and in situ detection of individual microbial cells without cultivation. Microbiol Rev 59: 143-169.

18. Whitby CB, Saunders JR, Rodriguez J, Pickup RW, McCarthy A (1999) Phylogenetic differentiation of two closely related *Nitrosomonas* spp. That inhabit different sediment environments in an oligotrophic freshwater lake. Appl Environ Microbiol 65: 4855-4862.

19. Calvó L, Cortey M, García-Marin JL, Garcia-Gil LJ (2005) Polygenic analysis of ammonia-oxidizing bacteria using 16S rDNA, amoA, and amoB genes. Int Microbiol 8: 103-110.

20. Purkhold U, Pommerening-Roser A, Juretschko S, Schmid MC, Koops HP, et al. (2000) Phylogeny of all recognized species of ammonia oxidizers based on comparative 16S rRNA and amoA sequence analysis: implications for molecular diversity surveys. Appl Environ Microbiol 66: 5368–5382.

21. Odokuma LO, Akponah E (2008) Response of *Nitrosomonas*, *Nitrobacter* and *Escherichia* coli to drilling fluids. J Cell and Animal Biol 2: 43-54.

22. Odokuma LO, Akponah E (2004) Inhibition of nitrification and carbon dioxide evolution as rapid tools for ecotoxicogical assessment of drilling fluids. African J Appl Zool Environ Biol 6: 16-24.

23. Ogugbue CJ, Oranusi NA (2005) Inhibitory effect of Azo dyes on ammonia –N oxidation of *Nitrosomonas*. African J Appl Zool Environ Biol 7: 61-67.

24. Holt JG, Hendricks Bergey D (1993) Bergey's Manual of Determinative Bacteriology, R.S. Breed(Ed.,) USA.

25. Grundmann GL, Neyra M, Normand P (2000) High-resolution phylogenetic analysis of NO2--oxidizing *Nitrobacter* species using the rrs-rrl IGS sequence and rrl genes. Int J Syst Evol Microbiol 50 Pt 5: 1893-1898.

26. Beaumont HJ, Hommes NG, Sayavedra-Soto LA, Arp DJ, Arciero DM, et al. (2002) Nitrite reductase of *Nitrosomonas* europaea is not essential for production of gaseous nitrogen oxides and confers tolerance to nitrite. J Bacteriol 184: 2557-2560.

27. Stackebrandt E, Murray RGE, Truper HG (1988) *Proteobacteria* classis nov., a name for the phylogenetic taxon includes the 'purple bacteria and their relatives'. Int J Syst Bacterio 38: 321-325.

28. American Public Health Association (APHA) (1998). Standard Methods for the Examination of water and Waste-water Washington DC, USA.

29. Grimont F, Grimont PA (1986) Ribosomal ribonucleic acid gene restriction patterns as potential taxonomic tools. Ann Inst Pasteur Microbiol 137B: 165-175.

30. Shrestha NK, Hadano S, Kamachi T, Okura I (2001) Conversion of ammonia to dinitrogen in wastewater by *Nitrosomonas* europaea. Appl Biochem Biotechnol 90: 221-232.

16

Effect of Pesticide (Chlorpyrifos) on Soil Microbial Flora and Pesticide Degradation by Strains Isolated from Contaminated Soil

Hindumathy CK* and Gayathri V

Department of Biotechynology, Vinayaka Mission University, Salem, Tamil Nadu, India

Abstract

In the present study, the effect of pesticide (Chlorpyrifos) on rhizospheric soil and non–rhizospheric soil of two plants marigold and *Canna* has been investigated. Further, microorganisms have been isolated from Rhizospheric and Non–Rhizospheric soil, characterized and their pesticide degradation ability was investigated. Most of bio process materials have been taken and analyzed for microbial composition. The efficiency of microbial consortium obtained from each of this bio process material for chlorpyrifos degradation has been studied. The result indicates that presence of glucose supports more biomass, which in turn brings about higher degradation and dissipation of pesticide. Maximum 84.5% dissipation was observed through bacterial isolate in presence of glucose as compared to 73.3% dissipation in absence of glucose. In case of fungal isolate 76% dissipation occurred in presence of glucose and only 58% was dissipated in absence of glucose. Both the isolates showed resistance to chlorpyrifos at 10 ppm concentration and also brought about significant dissipation of this pesticide. Therefore, these isolated could be potential candidates for microbe mediated bioremediation of chlorpyrifos contaminated soils.

Keywords: Pesticide; Chlorpyrifos; Rhizospheric; Non–Rhizospheric; Degradation

Introduction

Bioremediation is an environmental clean–up technique that is currently being investigated for use on a wide variety of chemicals. It is the use of naturally occurring microorganisms to enhance biodegradation, or normal biological breakdown. It involves establishing the condition in contaminated environment so that appropriate microorganisms flourish and carry out the metabolic activities to detoxify the contaminants [1] and is also safe, viable remedy for the detoxification of environmentally hazardous chemicals [2]. There are three primary approaches to bioremediation; Biostimulation, Bioaugmentation {which may include genetically engineered microorganisms (Gem's)} and Phytoremediation. Biodegradation of organic pollutants is a natural process whereby, bacteria and other organisms alter and breakdown organic molecules into substances, eventually producing carbon dioxide and water or methane. Although the ultimate aim of the biodegradation is to degrade the organic contaminants completely into harmless constituents such as the carbon dioxide and water, many intermediate metabolites can also be formed. What makes bioremediation so desirable is that it is a permanent solution; it destroys the contaminant, focuses on detoxification rather than waste translocation [3].

In view of the above and the literature survey findings, present study is taken up with the following objective of biological dissipation of pesticide in the Chlorpyrifos contaminated soil and effect of pesticide on rhizosphere and non-rhizosphere soil micro flora in case of Marigold and *Canna* plants and pesticide degradation by isolate/consortium obtained from contaminated soil.

Chlorpyrifos (0,0-diethyl-3,5,6-trichloro-2-pyridylphosphorothioate) is a insecticide/acaricide for treatments of crops, lawns, ornamental plants. It is a widely used insecticide and effective against a broad spectrum of insect pests of economically important crops. It is also used for the control of mosquitoes (larvae and adults), flies, various soil pests, many foliar crop pests and household pests. Additionally it is used for ectoparasite control of cattle and sheep. It persists in the soil for 60-120 days, and has very low solubility in water (2 mgl⁻¹), but is readily soluble in most organic solvents. Chlorpyrifos undergo transformation in the soil by the abiotic hydrolysis and microbial degradation. A few studies have attempted to separate abiotic chemical hydrolysis from microbial processes and to determine their relative importance [4,5]. The half lives of in muck (48% Organic Matter [OM] and loam (2.7% OM) were determined in sterilized and natural soils at 3 temperatures (3, 15, and 28°C). The half lives of in muck (48% Organic Matter [OM] and loam (2.7% OM) were determined in sterilized and natural soils at 3 temperatures (3, 15, and 28°C). Degradation of the pesticide depends upon the type of the soil, soil property, moisture content of the soil and pH [6,7]. The bacterial systems were able to rapidly degrade fenamiphos and between 15 and 35°C. Singh et al. and Singh et al. [8,9] investigated the degradation behavior of three insecticides (bifenthrin, and imidacloprid) at termiticidal application rates under standard laboratory conditions [10,11].

Materials and Methods

In the present study, the effect of pesticide on rhizospheric and non–rhizospheric Marigold and *Canna* has been investigated. Further, microorganisms have been isolated from rhizospheric and non–rhizospheric soil, characterized and their pesticide degradation ability was investigated. Most of bio process materials have been taken and analyzed for microbial composition. The efficiency of microbial consortium obtained from each of this bio process material for degradation has been studied.

Samples used

- Rhizospheric soil without pesticide-"Marigold" control, and "*Canna*" control.

*Corresponding author: Hindumathy CK, Department of Biotechynology, Vinayaka Mission University, Salem-636308, Tamil Nadu, India
E-mail: hindumathyck@rediffmail.com

- Rhizospheric soil with pesticide-"Marigold" with pesticide and "*Canna*" with pesticide.

- Non-Rhizospheric soil with pesticide–"Soil" with pesticide.

- Non-Rhizospheric soil without pesticide-"Soil only".

- Bioprocess materials such as Heap manure, Biogas slurry, Vermicompost, and Mushroom spent were used as soil amendments.

Conventional method for the isolation of microflora in pesticide contaminated soil

Microorganism isolation and cultivation conditions: Microorganism which was employed here was isolated from the organophosphorus pesticides contaminated soils using an enrichment culture technique [12]. The enrichment medium (Czapek-Dox) containing (in gram per litre) 30 g of sucrose, 2 g of $NaNO_3$, 0.5 g of KCl, 0.5 g of $MgSO_4$, 1 g of K_2HPO_4, 0.01 g of $Fe_2(SO_4)_3$, 0.5 g peptone and the Mineral Salt Medium (MSM) containing (in gram per litre) 2.0 g of $(NH_4)_2SO_4$, 0.2 g of $MgSO_4.7H_2O$, 0.01 g of $CaCl.2H_2O$, 0.001 g of $FeSO_4.7H_2O$, 1.5 g of $Na_2HPO_4·12H_2O$, 1.5 g of KH_2PO_4 were used for the isolation of fungal strains. Enrichment and isolation of fungi were performed as described by Chen et al. [13]. In brief, two gram of soil sample was transferred into a 250 ml Erlenmeyer flask containing 50 ml MSM with the addition of 50 mg·L^{-1} chlorpyrifos as the sole carbon source and incubated at 28°C for 7 days in a rotary shaker at 150 rpm. 5 ml of the enrichment culture was transferred into 50 ml fresh enrichment medium and incubated for another 7 days. After five rounds of transfer, the final culture was serially diluted and spread on Czapek-Dox agar plates. The organism that could make use of chlorpyrifos as the sole carbon source was used for further study.

Characterization of bacteria and fungi: Study of motility by hanging drop method and gram staining were performed to characterize bacteria.

Pesticide degradation by isolate: The isolate was identified from the plates in which the soil sample treated with pesticide was plated. 100 ppm of the pesticide was added to the pre-sterilized 100 ml Erlenmeyer flasks in 1ml acetone. After evaporation of acetone in 24 hours, 45 ml of the mineral salt medium which contains $MgSO_4.7H_2O$ (0.2 g); K_2HPO_4 (0.1g); $CaSO_4$ (0.4g); $FeSO_4.7H_2O$ (0.001g); distilled water, 1L; with pH 6.5 were placed in 100 ml Erlenmeyer flasks and the flasks were shaken at 26 ± 1 in 200 rpm on a orbital shaker (Syngenics Model). Medium not inoculated served as control. One flask with additional glucose (1000 mg/l) was added. Both the inoculated and non- inoculated samples were incubated under intermittent shaking to aerobic conditions. At periodic intervals, the samples were withdrawn aseptically from flasks and analyzed for by Gas Liquid Chromatography (GLC) after the extraction in hexane [13,14]. The growth of isolates was monitored by protein estimation.

Pesticide Degradation by Different Bioprocess Materials: Degradation was studied in the mineral salt medium inoculated with suspension of various bioprocess materials such as heap manure, vermicompost, mushroom spent, biogas slurry, soil with pesticide, and the concentration of 100 ppm was added to 50 ml sterilized Erlenmeyer flasks in 1 ml acetone. After evaporation of acetone, 45 ml of sterile mineral salt medium was added [15] dispensed into each flask. However to one set of flasks glucose (1000 mg/l) was supplemented as additional carbon source. The mineral medium supplemented was equilibrated on a rotary shaker for 24 hours to allow the solubilization of pesticides. Both sets of mineral salt medium, with or without glucose were inoculated with bioprocess materials suspension (prepared by suspending 10 gm in 90 ml distilled water). Uninoculated mineral salt medium served as control. Flasks were incubated at 27 ± 1°C on an orbital shaker. At regular intervals of 0, 5 and 10 days, about 5 ml aliquots were aseptically removed from all the experimental flasks.

Microbial biomass and pesticide residue analysis

Analysis of protein content of Biomass: Microbial biomasses in the various bioprocess materials flasks and from the isolate flasks with and without glucose were quantified at regular intervals (0, 5, and 10 days) by Bradford's method for the estimation of total protein. 10 ml of the culture was taken and centrifuged at 12,000 rpm for 12 minutes and the biomass pellet was taken and 2-3ml of 0.1 N NaOH was added in it. Kept in boiling water bath for 15 minutes, cooled it, and centrifuged to remove cell debris and ml supernatant was taken and to this 4.5 ml Bradford's reagent was added and Incubated at room temperature for 2-4 minute. Absorbance was noted at 595 nm and protein content (μg/ml) was calculated with the help of standard curve.

Extraction Procedure for Pesticide Residue Analysis: The Mineral Salt Medium (inoculated and non-inoculated) were extracted by shaking portions (1-2 ml) of the culture in flasks with 1-5 ml of hexane and 3 g of sodium sulphate for 5 minutes and it was evaporated using rotary vacuum evaporator (model-Heidolph) for extraction of residues from samples, the extracted pesticide residue was analyzed by Gas Chromatography (GC).

Methodology: 1 g of the soil sample was dissolved in 10 ml of distilled water. The bottle was shaken vigorously to suspend all the microbes in water. It is allowed to for a while to enable settling of coarse particles at the bottom, while microbes appear in Brownian motion in the suspension. 1 ml of the sample was taken from the suspension with the help of 5 ml capacity syringe. The sample was added in a sealed vial, mixed gently. The vials were kept for observation, and recorded the colour faded time and matched it with the microbial population chart.

Results and Discussion

Effect of on non-rhizospheric soil

In the non-contaminated soil bacterial population was higher than fungal population. In contrast, in pesticide contaminated soil the bacterial populations were greatly suppressed and fungal population raised and became dominant. These results show that fungal strains were more resistant and possibly had the capacity to degrade into harmless metabolites [16,17]. On the other hand bacterial population, in general is not able to survive and multiply well in presence of pesticide. It has been reported that one of the primary metabolites of (3,5,6-trichloro-2-pyridinol) possesses antibacterial properties [18]. Significant decline in bacterial populations observed in the present study could be attributed to generation of such antibacterial metabolites [19,20].

Effect on bacterial population in rhizospheric soils of marigold and *canna*

Rhizospheric soil of Marigold and *Canna* were compared. The roots of both the plants were well grown and proliferated. The rhizospheric soils of both plants were aseptically withdrawn and the bacterial colonies were quantified. In the absence of pesticide, *Canna* rhizospheric soil (3rd week sample) had significantly higher bacterial population as compared to the marigold rhizosphere. However, in presence of, the bacterial population in *Canna* rhizosphere crashed while there was no substantial impact or lowering of bacterial population in marigold rhizospheric soil (Tables 1 and 2).

Change in bacterial load of -contaminated rhizospheric soils of marigold and *canna*

The observations were recorded after 3rd week of plantation in contaminated soil. Repeated observations were made after 14th week of the experiment. The bacterial population in *Canna* rhizosphere was greater than Marigold rhizosphere both in the 3rd week as well as 14th week. In general an increase in bacterial colonies was observed in 14th week sample as compared to the 3rd week sample. This was more significant in pesticide contaminated rhizospheric soil. In case of *Canna* rhizospheric soil, the viable bacterial population almost doubled in the 14th week sample as compared to the 3rd week sample. Similarly enhancement in rhizospheric bacterial population was observed in case of Marigold. These results (Table 3) indicate that in the 3rd week of plantation in pesticide contaminated soil, the proliferation of the bacterial population is inhibited either due to the presence of pesticide itself or its toxic metabolites [21]. However, with time the pesticide residues might get dissipated due to combined impact of abiotic process as well phytoremediation effect [22]. Therefore, significantly higher bacterial population has been observed in the 14th week sample [23,24].

Effect of fungal population in rhizospheric soils of marigold and *canna*

The fungal population in rhizospheric soil of both Marigold and *Canna* was higher than the non-rhizospheric soil. The degradation was declined (by almost 50%) in fungal population was observed in the rhizospheric soils of both *Canna* and Marigold (Table 3).

Characterisation of microflora

The colonies from *Canna* rhizospheric soil with pesticide were taken and characterized. The bacteria by hanging drop method, flagella staining, gram staining and fungi by staining with lactophenol cotton blue.

Bacteria

Most of the bacterial isolates appeared to be non-motile (few are motile), rod shaped, gram-negative bacteria (Figure 1).

Fungi

Most of the fungal groups displayed aseptate hyphae and resembled *Aspergillus sps.* (Figure 2).

Pesticide degradation by isolate/coinsortium

The inoculated samples from the mineral salt medium were withdrawn periodically (0, 5 and 10th day) in aseptic condition and analyzed for total protein and pesticide residue analysis.

Marigold			Canna	
Dilutions	Cfu/g (3rd week)	Cfu/g (14week)	Cfu/g (3rd week)	Cfu/g (14week)
10^{-1}	3.3×10^3 Cfu g^{-1}	5.0×10^3 Cfu g^{-1}	3.4×10^3 Cfu g^{-1}	6.8×10^3 Cfu g^{-1}
10^{-2}	2.3×10^4 Cfu g^{-1}	2.3×10^4 Cfu g^{-1}	Smeared colonies, fungal growth	4.3×10^4 Cfu g^{-1}
10^{-3}	No colonies	No colonies	Fungal colonies	3.8×10^5 Cfu g^{-1}
10^{-4}	No colonies	No colonies	No colonies	No colonies
10^{-1}	3.4×10^3 Cfu g^{-1}	4.0×10^3 Cfu g-1	5.2×10^3 Cfu g^{-1}	7.6×10^3 Cfu g
10^{-2}	Too small colonies	2.0×10^4 Cfu g^{-1}	1.4×10^4 Cfu g^{-1}	3.9×10^4 Cfu g
10^{-3}	Fungal growth	TLTC	No colonies	9×10^4 Cfu g
10^{-4}	No colonies	No colonies	No colonies	No colonies

Table 1: Rhizospheric soil with and without pesticide.

Marigold			Canna	
Dilutions	Cfu/g (3 week)	Cfu/g (14 week)	Cfu/g (3 week)	Cfu/g (14 week)
10^{-1}	13×10^4 Cfu g^{-1}	10×10 Cfu g^{-1}	More filamentous and bacterial colonies	28×10^4 Cfu g^{-1}
10^{-2}	Filamentous growth	8×10^5 Cfu g^{-1}	Black spores and Hyphae	20×10^5 Cfu g^{-1}
10^{-3}	Small Hyphae	Filamentous Hyphae	Filamentous Hyphae	5×10^6 Cfu g^{-1}
10^{-4}	No colonies	No colonies	No colonies	Hyphae growth
10^{-1}	20×10^4 Cfu g^{-1}	12×10 Cfu g^{-1}	More filamentous Black spores.	18×10^4 Cfu g^{-1}
10^{-2}	9×10^5 Cfu g^{-1}	Filamentous growth	Filamentous Hyphae	10×10^5 Cfu g^{-1}
10^{-3}	No colonies	Hyphae	3×10^6 Cfu g^{-1}	5×10^6 Cfu g^{-1}

Table 2: Rhizospheric soil with and without Pesticide.

	Bacteria	Fungi	
Dilutions	Cfu/gram	Dilutions	Cfu/gram
10^{-1}	2.3×10^3 Cfu g^{-1}, more fungal growth	10^{-1}	6.8×10^3 Cfu g^{-1}
10^{-2}	5×10^4 Cfu g^{-1}	10^{-2}	4.3×10^4 Cfu g^{-1}
10^{-3}	Fungal hyphae	10^{-3}	3.8×10^5 Cfu g^{-1}
10^{-4}	Fungal growth	10^{-4}	No colonies
10^{-1}	4×10^4 Cfu g^{-1}, more fungal growth	10^{-1}	1.8×10^3 Cfu g^{-1} Fungal Hyphae
10^{-2}	5×10^4 Cfu g^{-1}	10^{-2}	1.3×10^4 Cfu g^{-1}, filamentous growth
10^{-3}	Fungal hyphae	10^{-3}	Fungal Hyphae
10^{-4}	Fungal growth	10^{-4}	No colonies

Table 3: Non rhizospheric soil with and without pesticides.

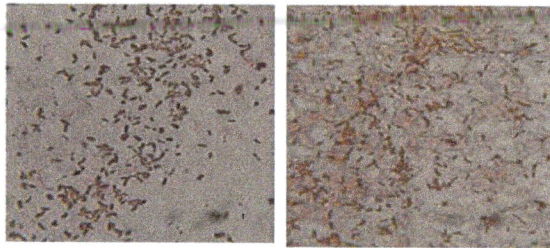

Figure 1: Gram staining of the bacterial isolates.

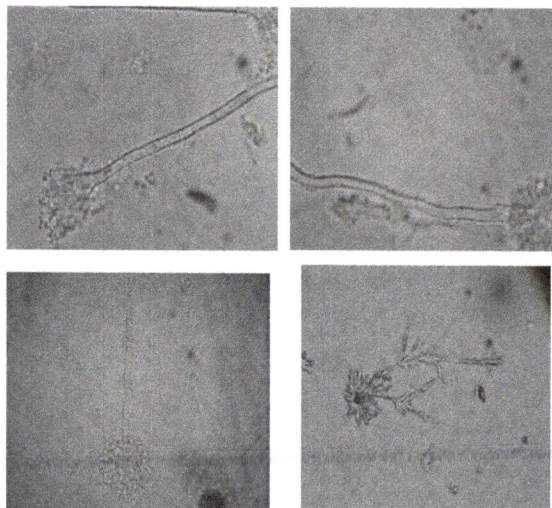

Figure 2: Lactophenol staining of the fungal isolates.

Protein estimation

The fungal isolate used in this study was isolated from *Canna* plate and the phase contrast micrograph of the same is shown in figure 1. In general the result implies that in all the case fungi growth increases slowly from 0 day and maximum growth is observed in 10th day, but the magnitude of growth is different in each cases. No significant inhibition of growth is observed in the presence of pesticide. In case when no glucose was added, the fungal isolate were able to survive even in the absence of any carbon source. These results indicate that the isolate was able to utilize the energy source and hence there is a continuous growth period during 0 to 10th days (Figure 3).

The bacterial isolate used in this study was isolated from *Canna* rhizospheric soil with pesticide and found to be rod shaped gram negative bacteria. It was reported that the bacterial isolate showed maximum growth in the presence of glucose. However, in the absence of carbon source also the bacterial isolate was able to survive and showed the continuous growth during 0 day-10th day. As compared to the fungal isolate, the bacterial isolate showed quite less growth in terms of the protein concentration (Figure 4). Nevertheless, results showed that the bacterial isolate was also able to utilize the energy source and carbon source [25]. Several investigations [26,27] carried out earlier also reported the requirement of electron donating co-substrates such as glucose and yeast extract for the reduction of azo bonds by bacteria [28], Moosvi et al. [25] have also reported the importance of carbohydrate and nitrogen source in the form of glucose and yeast extract respectively for the maximum degradation of Reactive Violet 5 by bacterial consortium RVM11.1.

The results of pesticide dissipation by the isolates are shown in table 1. After 5th day maximum pesticide dissipation occurred in when both glucose and pesticide were present in the case of bacterial as well as fungal isolate [21,22]. This result indicates that presence of glucose supports more biomass, which in turn brings about higher degradation and dissipation of pesticide. Maximum 84.5% dissipation was observed through bacterial isolate in presence of glucose as compared to 73.3% dissipation in absence of glucose. In case of fungal isolate 76% dissipation occurred in presence of glucose and only 58% was dissipated in absence of glucose. Both the isolates showed resistance to at 10 ppm concentration and also brought about significant dissipation of this pesticide. Therefore, these isolated could be potential candidates for microbe mediated bioremediation of contaminated soils.

Pesticide degradation by bioprocess materials

Microbial Enumeration of Bioprocess Materials: Bioprocess materials such as vermicomposting, heap manure, mushroom span, and biogas slurry act as a biostimulant as well as source of microflora [20]. For this enumeration biomanures diagnostic kit was used and this microbial kit helps to determine the quality of biomanures in terms of microbial count. Hence, high microbial population is desirable for quality manure.

Pesticide degradtion by consortium from bioprocess materials

- The total protein has been estimated in case of bioprocess

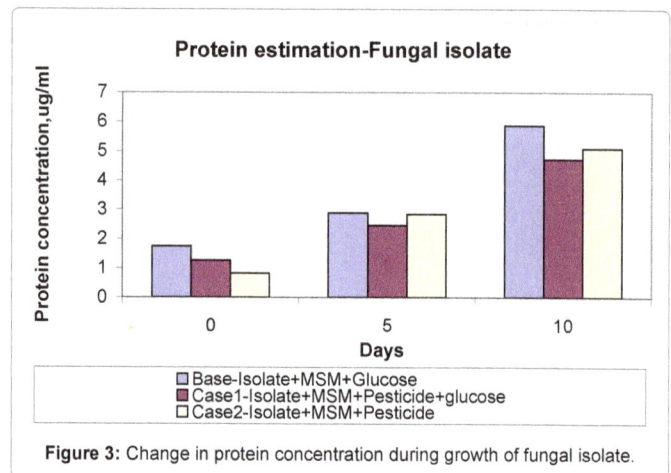

Figure 3: Change in protein concentration during growth of fungal isolate.

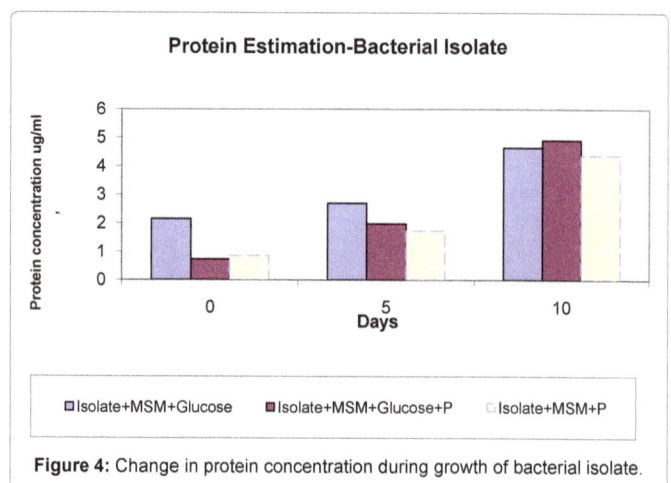

Figure 4: Change in protein concentration during growth of bacterial isolate.

materials by Bradford's method. Our results shows that in the different bioprocess materials vermicomposting biogas slurry and soil pesticide has shown an increased growth curve [29]. From the figure 5 it has been concluded that growth of biogas slurry is more than the other bioprocess materials from the period of 0 day to 10ᵗʰ day, in case of vermicomposting and soil pesticide the growth is slightly decreased while compared with the biogas slurry. So it has been concluded that: BS>VM>SP>HP>MS. The consortiums present in the materials are able to survive in the presence of pesticide as a energy source [20,21]. Further, many researchers have mentioned that a higher degree of biodegradation and mineralization can be expected when co-metabolic activities within a microbial community complement each other [26-29]

- From the figure 6 it has been concluded that in the presence of glucose the consortium present in the materials attains maximum growth on the 5ᵗʰ day which gets subsequently declined in 10ᵗʰ day. As compared to the case in absence of glucose, much higher protein concentrations were recorded in presence of glucose.

References

1. Singh BK, Walker A (2006) Microbial degradation of organophosphorus compounds. FEMS Microbiol Rev 30: 428-471.

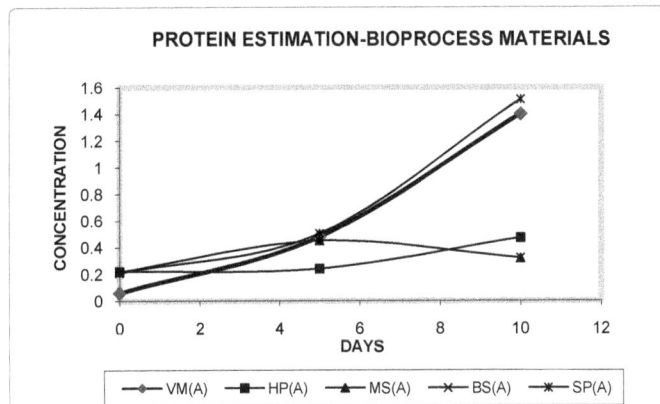

Figure 5: Growth of microbial consortium obtained from different bioprocess material in absence of glucose (Mineral salt medium+ Pesticide + consortium from Bioprocess materials).

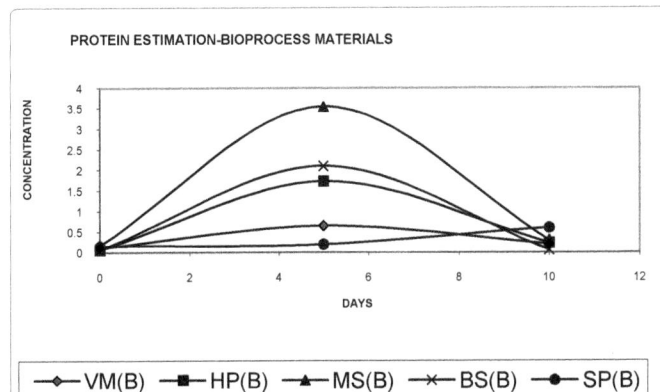

Figure 6: Growth of microbial consortium obtained from different bioprocess material in presence of glucose (Mineral salt medium+ Pesticide + glucose +consortium from Bioprocess materials).

2. al-Mihanna AA, Salama AK, Abdalla MY (1998) Biodegradation of chlorpyrifos by either single or combined cultures of some soilborne plant pathogenic fungi. J Environ Sci Health B 33: 693-704.

3. Singh BK (2009) Organophosphorus-degrading bacteria: ecology and industrial applications. Nat Rev Microbiol 7: 156-164.

4. Cycoń M, Wójcik M, Piotrowska-Seget Z (2009) Biodegradation of the organophosphorus insecticide diazinon by Serratia sp. and Pseudomonas sp. and their use in bioremediation of contaminated soil. Chemosphere 76: 494-501.

5. Thengodkar RR, Sivakami S (2010) Degradation of Chlorpyrifos by an alkaline phosphatase from the cyanobacterium Spirulina platensis. Biodegradation 21: 637-644

6. Li X, Jiang J, Gu L, Ali SW, He J, et al. (2008) Diversity of chlorpyrifos-degrading bacteria isolated from chlorpyrifos-contaminated samples. Int Biodeterior Biodegradation 62: 331-335.

7. Xu G, Zheng W, Li Y, Wang S, Zhang J, et al. (2008 Biodegradation of chlorpyrifos and 3,5,6-trichloro-2-pyridinol by a newly isolated Paracoccus sp. strain TRP. Int Biodeterior Biodegradation 62: 51–56.

8. Singh BK, Walker A, Morgan JA, Wright DJ (2004) Biodegradation of chlorpyrifos by enterobacter strain B-14 and its use in bioremediation of contaminated soils. Appl Environ Microbiol 70: 4855–4863.

9. Singh DP, Khattar JI, Nadda J, Singh Y, Garg A, et al. (2011) Chlorpyrifos degradation by the cyanobacterium Synechocystis sp. strain PUPCCC 64. Environ Sci Pollut Res Int 18: 1351-1359

10. Yu YL, Fang H, Wang X, Wu XM, Shan M, et al. (2006) Characterization of a fungal strain capable of degrading chlorpyrifos and its use in detoxification of the insecticide on vegetables. Biodegradation 17: 487-494.

11. Fang H, Xiang YQ, Hao YJ, Chu XQ, Pan XD, et al. (2008) Fungal degradation of chlorpyrifos by Verticillium sp. DSP in pure cultures and its use in bioremediation of contaminated soil and pakchoi. Int Biodeterior Biodegradation 61: 294-303.

12. Kulshrestha G, Kumari A (2011) Fungal degradation of chlorpyrifos by Acremonium sp. strain (GFRC-1) isolated from a laboratory-enriched red agricultural soil. Biology and fertility of soils 47: 219-225.

13. Chen S, Hu M, Liu J, Zhong G, Yang L, et al. (2011) Biodegradation of beta-cypermethrin and 3-phenoxybenzoic acid by a novel Ochrobactrum lupini DG-S-01. J Hazard Mater 187: 433-440.

14. Peng X, Huang J, Liu C, Xiang Z, Zhou J, et al. (2012 Biodegradation of bensulphuron-methyl by a novel Penicillium pinophilum strain, BP-H-02. J Hazard Mater 216-221.

15. Chen S, Hu Q, Hu M, Luo J, Weng Q, et al. (2011) Isolation and characterization of a fungus able to degrade pyrethroids and 3-phenoxybenzaldehyde. Bioresour Technol 102: 8110–8116.

16. Chen X, Tang L, Li S, Liao L, Zhang J, et al. (2011) Optimization of medium for enhancement of ε-poly-L-lysine production by Streptomyces sp. M-Z18 with glycerol as carbon source. Bioresour Technol 102: 1727-1732.

17. Moon C, Lee CH, Sang BI, Um Y (2011) Optimization of medium compositions favoring butanol and 1,3-propanediol production from glycerol by Clostridium pasteurianum. Bioresour Technol 102: 10561–10568.

18. Yong X, Raza W, Yu G, Ran W, Shen Q, et al. (2011) Optimization of the production of poly-γ-glutamic acid by Bacillus amyloliquefaciens C1 in solid-state fermentation using dairy manure compost and monosodium glutamate production residues as basic substrates. Bioresour Technol 102 : 7548-7554.

19. Singh BK, Walker A, Morgan JA, Wright DJ (2003) Effects of soil pH on the biodegradation of chlorpyrifos and isolation of a chlorpyrifos-degrading bacterium. Appl Environ Microbiol 69: 5198–5206.

20. Singh BK, Walker A, Wight DJ (2006) Bioremedial potential of fenamiphos and chlorpyrifos degrading isolates: Influence of different environmental conditions. Soil Biol Biochem 38: 2682–2693.

21. Mallick K, Bharati K, Banerji A, Shakil NA, Sethunathan N (1999) Bacterial degradation of chlorpyrifos in pure cultures and in soil. Bull Environ Contam Toxicol 62: 48-54.

22. Yang L, Zhao YH, Zhang BX, Yang CH, Zhang X (2005) Isolation and characterization of a chlorpyrifos and 3,5,6-trichloro-2-pyridinol degrading bacterium. FEMS Microbiol Lett 251: 67-73.

23. Liu X, Cole MA, Zhang L (1995) Remediation of pesticide contaminated soil by planting and compost addition. Compost science and utilization 3: 20-30.

24. Khadijah O, Lee KK, Faiz MF, Abdullah (2009) Isolation, screening and development of local bacterial consortia with azo dyes decoloursing capability. Malays J Microbiol 5: 25-32.

25. Moosvi S, Keharia H, Madamwar D (2005) Decolourization of textile dye Reactive Violet 5 by a newly isolated bacterial consortium RVM 11.1. World J Microbiol Biotechnol 21: 667-672

26. Singh N, Megharaj M, Kookana RS, Naidu R, Sethunathan N (2004) Atrazine and simazine degradation in Pennisetum rhizosphere. Chemosphere 56: 257-263.

27. Baskaran S, Kookana RS, Naidu R (1999) Degradation of bifenthrin, chlorpyrifos and imidacloprid in soil and bedding materials at termiticidal application rates. Pesticide science 55: 1222-1228

28. Mallick K, Bharati K, Banerji A, Shakil NA, Sethunathan N (1999) Bacterial degradation of chlorpyrifos in pure cultures and in soil. Bull Environ Contam Toxicol 62: 48-54.

29. Miles JRW, Harris CR, Tu CM (1984) Influence of moisture on the persistence of chlorpyrifos and chlorfenvinphos in sterile and natural mineral and organic soils. J Environ Sci Hlth B 19: 237-243.

17

Bioremediation of Polluted Soil obtained from Tarai Bhavan Region of Uttrakhand, India

Rajdeo Kumar[1], Ashish Chauhan[2]*, Nisha Yadav[1], Laxmi Rawat[1] and Manish Kumar Goyal[2]

[1]Forest Ecology and Environment Division, Forest Research Institute, Dehradun, India
[2]National Institute of Pharmaceutical Education and Research, Mohali, India

Abstract

The rapid industrialization in Tarai Bhavan region of Udham Singh Nagar, Uttarakhand has exposed the soil and water industrial effluent rich in pesticides like chlorophenols that are adversely affecting the ecosystem and disturbing the food chain. Soil is the basic requirement to sustain life on this earth for the living being including human beings, animals, planet or microorganisms (like bacteria and fungi). Bioremediation to remove pollutants is economic than the equivalent physico-chemical methods. It offers the potential to treat contaminated soil and ground water at the site without excavation. It requires lesser input and preserves the frame. The most attractive feature of bioremediation is the reduced impact on the ecosystems.

In this study, both the microorganisms of bacteria and fungi *P. fluorescence* and *P. chrysosporium* were inoculated in fresh minimal salt medium containing 0, 10, 50, 100, 200 and 500 ppm of pentachlorophenol (PCP) concentrations in separate flask for few hours (4, 8, 16 and 32) and their potentiality to degrade PCP was assessed and found to be fruitful.

Keywords: Industrialization; Pesticide; Bioremediation; Pentachlorophenol

Introduction

Industrialization plays a vital role in nation's socio-economic development as well as its political stature. Industries vary according to process technology, sizes, nature of products, characteristics and complexity of wastes discharged. Ideally citing of industries should strike a balance between socio- economic and environmental considerations. Although industrialization is inevitable, various devastating ecological and human disasters have continuously occurred over the last four decades that made industries responsible for various environmental pollutions. It has been widely reported that industrial effluent has a hazardous effect on the quality of flowing water. Industrial discharge contains toxic and hazardous substances, most of which affects human health. These include heavy metals such as lead, cadmium and mercury and toxic organic chemicals such as pesticides, PCBs, dioxins, poly aromatic hydrocarbons (PAHs), petrochemical and phenolic compound [1-5].

Bioremediation is the use of living organisms primarily microorganisms to degrade the environmental contaminants into nontoxic forms. The mechanism of microbial degradation is based on the general principles of physiology and ecology. Biological removal of chemo-pollutants becomes the method of choice since microorganisms can use a variety of xenobiotic compounds including pesticides for their growth, mineralize and detoxify Common soil bacteria and fungi can degrade the majority of the compounds. The most important soil factors that influence bio-degradation are temperature, moisture, presence or absence of oxygen, organic matter and clay content. Microorganisms possess the capable of degrading a large proportion of chemicals. Consequently many of the man-made pesticides introduced into the environment are microbial degraded, mostly by enzymes evolved in response to the presence of natural substrate [6-8].

In order to enhance the microbial degradation of organic pollutants for remediation of contaminated soils, it is essential to understand the enhancing mechanism, especially the relation between the degradation of chemicals in soil and the behavior of degrading microorganisms.

There are two types of microbial degradation of pesticides in soil. In the first, repeated application of a pesticide to soil enhance the degradation by enrichment of the pesticide-degrading microorganisms. The enriched microorganisms often metabolize the pesticide as carbon and energy source, which is designated as catabolism. In the second type of degradation, the population of degrading microorganisms in soil does not change even when a pesticide is repeatedly applied and no enhancement of degradation occurs. The microorganisms require other carbon sources to degrade the pesticide that are called as incidental metabolism or co-metabolism. Usually, as microorganisms carry out degradation, they obtain carbon and energy required for growth. An increase in the size of the pesticides degrading population leads to the faster rates of degradation. Microbial degradation of pesticides occurs ecologically through various dynamic and complex forces to support aerobic reactions simultaneously leading to more extensive degradation of many complex pesticides. The enhanced degradation occurs due to the repeated application of pesticides. This is favorable for environmental decontamination of toxic residues. The microbial degradation is the best means of detoxification of pesticides [9-20].

Pentachlorophenol is a polyhalogenated aromatic hydrocarbon of chlorophenol family. Chlorophenols are phenols carrying one chlorine atom attached to the benzene ring. The chemical formula for PCP is C_6HCl_5O and it has no isomer. Pentachlorophenol has been used extensively as a wood preservative, fungicide, bactericide, herbicide, moluscicide, algaecide, and insecticide. Although numerous reports

*Corresponding author: Ashish Chauhan, National Institute of Pharmaceutical Education and Research, Mohali, India
E-mail: aashishchauhan26@gmail.com

have shown that PCP undergoes biodegradation but its biodegradation in the environment is often slow. This coupled with its extensive use, has led to the contamination of many terrestrial and aquatic ecosystems world-wide.

Material and Methods

All the chemicals used during the course of this investigation were of A.R. grade and were supplied by E. Merck (India), Himedia (India), S.D. Fine chemicals (India), Qualigens (India) or Sigma (U.S.A). All glassware used of corning and borosil made.

Microorganisms

Pseudomonas fluorescence: Pure culture of *Pseudomonas fluorescence* was obtained from gene pool (G. B. Pant University of Agriculture and Technology, Department of Microbiology, Pantnagar).

Phanerochete chrysosporium: Pure culture of *Phanerochete chrysosporium* was obtained from Institute of Microbial Technology (CSIR Laboratory), Chandigarh.

Methods

Nutrient agar medium: This medium was used to culture bacterial community of individual strain (Tables 1 and 2).

Bacterial enrichment: Continuous enrichment of bacterial strains was facilitated by minimal salt medium (Table 3).

Fungal enrichment: The fungal community was incubated in Erlenmeyer flask containing a basal minimal medium (Table 4).

Trace elements (1 ml) were added and final volume was 1 liter by adding distilled water and the pH was adjusted to 7.0 [21-25].

Morphological characterization of bacteria

The selective medium was removed from culture flask under aseptic condition and growth of bacterial community was determined by measuring optical density OD at 600 nm. The culture medium was diluted serially in ten folds and 0.1 ml of diluents was spread on nutrient agar plates. The plates were incubated at 30°C. Colony forming units (CFU) were determined after 24 hours. The bacterial cells that has maximum CFU were characterized based upon morphology of colonies, diameter, colour, opacity, form elevation, margin, smoothness, texture

Constituents	Amount (g/l)
Peptone	5.0
Beef extract	3.0
Agar	20.0
Temperature	30°C

The final volume was made 1 litre by adding distilled water and the pH was adjusted to 7.0.

Table 1: Composition of nutrient medium.

Constituents	Amounts (g/l)
Yeast extracts	5.0 g
Glucose	10.0 g
Agar	15.0 g
Distilled water	1.0 liter
Temp	25°C

The final volume was made 1 liter by adding distilled water and the pH was adjust 5.8.

Table 2: Composition of fungal medium.

Constituents	Amount (g/l)
Na$_2$HPO$_4$2H$_2$O	7.8
K$_2$HPO$_4$	6.8
MgSO$_4$	0.2
NaNO$_3$	0.085
Ca (NO$_3$)$_2$4H$_2$O	0.050
Ferrous ammonium citrate	0.01
Trace elements	1 ml

Trace element (1 ml) was added and the final volume was made 1 liter by adding distilled water and the pH was adjusted to 7.0.

Table 3: Composition of bacterial enrichment (minimal salt medium).

Constituents	Amount (g/l
D-glucose	10.00
NH$_4$NO$_3$	0.079
KH$_2$PO$_4$	2.0
MgSO$_4$	0.05
CaCl$_2$.2H$_2$O	0.1
Mineral solution	1.0 ml
Vitamin solution	0.5 ml
pH	5.6

Table 4: Composition of fungal enrichment (minimal salt medium).

and spreading nature. Isolates were stored on nutrient agar slant as well as in 25% glycerol at 20°C [26-28].

Microscopic study of bacterial culture

Indirect staining: A loopful of liquid culture of each isolate was spread on separate clean grease free microscopic slides. A drop of nigrosine was added with culture on slide and made it into thin smear with the help of another slide. Slides were air dried for 40 seconds and examined under microscope and cells were characterized depending upon shape, size and arrangement [20-28].

Gram's staining: The gram's staining was performed by modified method of Hucker. For this propose a loopful of culture was taken on slide, smeared and heat fixed. The slides were stained with crystal violet for 30 seconds, washed with water and then few drops of gram's iodine were applied for 30 seconds. The slides were rinsed and decolorized with 95% alcohol and again rinsed with water. The slides were counter stained with safranin for 20 to 30 seconds. Finally; the sides were rinsed with water blot, dried and then examined under compound microscope with oil immersion lens [15-28].

Identification of fungi

Fungal strain that has maximum growth in basal mineral medium containing PCP, was taken and identified by visual observation of colony growth on potato dextrose agar plate and then by microscopic examination. The fungal hyphae was picked up from newly growing fungal colony on agar plate, with the help of alcohol sterilized needle and stained with cotton blue. The slide of fungal hyphae was prepared and placed on the stage of microscope. The slide was first examined under low power (10X) and then switched to high power oil immersion objective. The characteristics examined were noted and then consulted with standard reference literature.

Procedure of pentachlorophenol (PCP) degradation

Degradation of PCP was studied in laboratory and field conditions in three experiments.

Batch Experiment: Different concentration of PCP inoculated with microorganisms was analyzed for chloride release and Ring Cleavage at different time periods.

Degradation of pentachlorophenol: After enrichment the bacterial and fungal were inoculated in fresh mineral salt medium with PCP (100 ppm) as a sole carbon source for a few days in a gyratory shaker. The samples were taken out at 4, 8, 16, 32, hours and growth of microbes was measured.

Estimation of chloride ions release (Argentometric method)

24 ml of sample was taken in a beaker and added 5 drops of potassium chromate, an indicator solution. With the help of burette, $AgNO_3$ (0.0141 N) solution was added to the beaker. A red color formed, which disappeared soon. At a point where all the chloride ions were precipitated, a stable red color appeared referring to an end point of the reaction. Calculation of amount of chloride ions in the sample was done.

$$C_1V_1 (AgNO3)=C_2V_2 (Sample)$$

With the help of this formula C_2 concentration was calculated. Amount of chloride ions present in the sample was calculated by multiplying with 35.5 (atomic weight of chloride) with C_2 [29-35].

Estimation of ring cleavage

4 ml of cell suspension was dissolved in 0.02 M Tris buffer (0.1) and EDTA (0.1 ml) for lyses of bacterial and fungal cells. The pH was adjusted to 7.8. The mixture was treated with little toluene and 0.1 M catechol (4.0 ml). The development of color was noticed. Yellow color did not appear that suggested absence of meta cleavage. Mixture was shaken for 1 hour at 170 rpm and tested for formation of ß-ketoadipic acid (Roth era Reaction) that indicates the presence of ortho fission. In this procedure, 10 ml culture fluid was acidified with 2 ml HCl followed by addition of 1 ml $NaNOP_3$ (1%). After 2 minute, concentrated ammonia (15 ml) and 10% ferrous sulfate solution (10%) were added. The development of reddish brown color indicated typical Roth era reaction and the presence of ortho cleavage [29-35].

Soil sample

Soil sample were collected from the College of Basic Science and Humanities ground. Soil was uniformly grinded and screened for any apparent impurities. The collected soil was processed for determining its physico-chemical characteristics such as texture, moisture, organic carbon employing the following methods:

Texture: Particle size analysis of soil used in the experiment was done following the international pipette method using H_2O_2 (30%) for the removal of organic matter and sodium hexa meta phosphate as dispersing agent. Soluble salt and calcium carbonate were removed by following Jacksons Method.

Moisture: Moisture content of soil was determined by oven dry and weight loss method.

Potency of Hydrogen (pH): Soil pH was determined in soil extract prepared in a clean 50 ml glass beaker by suspending 20 g soil in 20 ml distilled water and filtered the same through whatman (no.1) filter paper. The filtrate was subjected to measure pH using a pH meter. The standard pH buffer (pH7.0 and pH 4.0) were used to calibrate the pH meter.

Organic carbon: Organic carbon of the soil was determined by wet digestion method of Walkey and Black.

Estimation of organic carbon: 1.0 g of soil sample was mixed with 10 ml of 1 N $K_2Cr_2O_7$ in a 500 ml conical flask. 20 ml of concentrated H_2SO_4 was gently added to it. The solution was shaken well for a minute or two and allowed to stand for about 30 minutes for complete digestion. The content was further diluted by addition of 200 ml of 0.2 g sodium fluoride. They were then added to the conical flask and titrated against standard ferrous ammonium sulphate solution (0.5 N) after addition of 1 ml Diphenyl amine indicator solution. The end point was detected as the violet color changed to purple and finally to brilliant green.

Calculation

% carbon in soil=(x-y) × 0.003/0.76 × w

Where, w=gram weight of soil taken, X=volume in ml of 0.5 N ferrous ammonium sulphate required for reducing 10 ml $K_2Cr_2O_7$ (blank reading), Y=Volume in ml of 0.5 N ferrous ammonium sulphate required for reducing the excess of dichromate (experimental reading) and 0.003=meq weight of carbon [20-35].

Results and Discussion

In the present study an attempt has been made to assess the potential of these two microorganisms i.e. *Phanerochaete chrysosporium* (fungi) and *Pseudomonas fluorescence* (bacteria) for the degradation of PCP in soil microcosms in order to suggest their degradation efficiency under natural conditions.

Degradation of pentachlorophenol by bacteria and fungi in liquid batch culture

Both the micro-organisms were separately inoculated in fresh minimal salt medium containing 0, 10, 50, 100, 200 and 500 ppm PCP concentrations in separate flask for few hours (4, 8, 16, and 32) and their potentiality to degrade PCP was assessed. The growth of bacteria and fungi and degradation of PCP were measured.

Growth of bacteria

Bacterial growth was measured by taking absorbance at 560 nm of culture broth. The OD values showed an increasing growth pattern in bacteria with time that indicated that bacteria degraded PCP as a carbon source. Bacterial growth was maximum in 100 ppm concentration for all the treatments that may be due to significant degradation of PCP as sole carbon source. However, bacterial growth decreased at higher PCP concentrations.

Growth of fungi

The growth was measured by spore count method. Spores were counted using the equation:

Spores/ml=(n) x 10^4

Where, n=the average cell count per square of the four corner squares counted. Fungal growth was maximum in 200 ppm concentration that may be due to significant degradation of PCP. However, the growth decreased at higher PCP concentrations (Table 5 and Figure 1).

Chloride ion released

The chloride ions released (mg/l) at different time interval were measured by Argentomatric method. The data signified that chloride release by the bacteria was maximum at 100 ppm of after 32 hr (i.e.; 52 mg/l) whereas, for fungus it was maximum at 200 ppm of PCP after 16 hr (i.e.; 62.5 mg/l) and declined later. The chloride release directly

Concentration of PCP (ppm)	No. of spores (spores/ml)
0	15000
10	150000
50	260000
100	300000
200	380000
500	310000

Table 5: Growth of *P. chrysosporium* (in terms of spore number) in culture having varied concentrations of PCP.

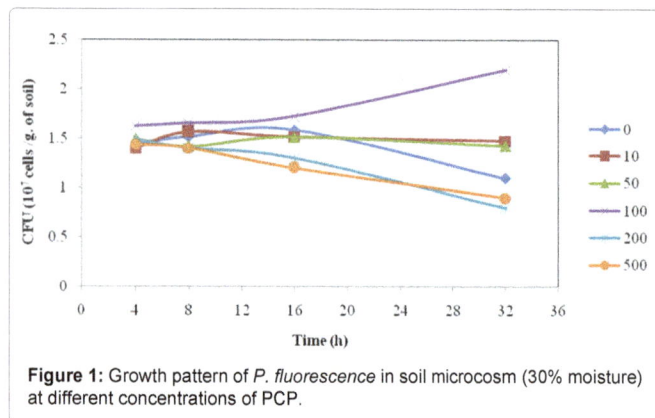

Figure 1: Growth pattern of *P. fluorescence* in soil microcosm (30% moisture) at different concentrations of PCP.

indicates dehalogenation of PCP. With due course of growth, results revealed that fungi had more potentiality to dehalogentae and degrade pentachlorophenol as compared to bacteria. The dehalogenation of PCP decreased its toxicity (Tables 6, 7, Figures 2 and 3).

Ring cleavage by both microorganisms (bacteria and fungi)

Mode of ring fission of pentachlorophenol by both the organisms was assessed by observing meta and ortho ring cleavage. No yellow color appeared after mixing catechol in cell suspension, that indicated the absence of meta cleavage in both the organisms. However, positive Rothera test (as violet color was appeared due to β ketoadipic acid formation) for both the organisms showed that ring cleavage occurred at ortho position and confirmed dechlorination of PCP. Flask inoculated with bacteria showed lower efficiency of ring cleavage than fungi (Table 8).

Microbial growth of soil microcosm

Soil microcosms were fabricated in plastic containers to evaluate potency of PCP degradation under natural soil conditions.

pH of soil

The pH of the soil was 7.8 which were suitable for the maximum degradation of PCP in soil. The pH gets decreased during experiment, which is primarily due to the release of chloride ion from PCP degradation.

Percent of organic carbon (%C)

The percentage of organic carbon in soil was 0.23%, which indicated the soil was poor in content of organic carbon.

Percent water holding capacity (%WHC)

The percentage of water holding capacity in soil was 63.7% which was quiet higher. The desirable moisture content was maintained by adding minimal salt media. The fungi and bacteria cultures were grown on nutrient broth for 24 hrs and were inoculated in both set of soil

microcosm containing 30% moisture and 60% moisture. Both the sets had different concentrations of PCP (0, 10, 50, 100, 200 and 500 ppm).

Growth of bacteria and fungi

Growth of bacteria and fungi was determined by total viable count. In case of 30% moisture, bacterial growth was highest at 100 ppm PCP. Fungal growth was highest at 200 ppm PCP. In case of bacteria at 500 ppm and 200 ppm concentration of PCP, growth decreased and in case of fungi at 500 ppm concentration of PCP, growth decreased indicating that high concentration of PCP had an inhibitory effect on bacteria and fungi. For control (0 ppm concentration) there was no significant increase in cell number due to lack of carbon content (Tables 9, 10, Figures 4 and 5).

In case of 60% moisture content growth of both the microorganisms was higher than 30% moisture content. PCP at 500 and 200 ppm concentration was also degraded by bacteria and fungi. Maximum bacterial growth was observed at 100 ppm PCP. Similarly maximum fungal growth was observed at 200 ppm of PCP (Figures 6 and 7). Both the organisms showed better growth at 60% moisture. Interphase between soil and water at soil surface was more prominent at 60% moisture as compared to 30% moisture resembling its native habitat (Tables 11 and 12).

The results of ring cleavage and chloride release in liquid batch culture concluded that both microorganisms were capable of utilizing PCP, which is a recalcitrant molecule capable of being degraded only by limited number of bacteria and fungi. The usual approach

Concentration of PCP (ppm)	4 h	8 h	16 h	32 h
10	23.2	25.1	26.2	27.9
50	30.3	32.3	35.6	37
100	43.1	47.2	49.2	52
200	35.7	37.2	39.1	40.2
500	30.3	34.5	37.5	39.8

Table 6: Chloride ion release (mg/l) by *P. fluorescence* at different PCP concentrations-Time Interval.

Concentration of PCP (ppm)	4 h	8 h	16 h	32 h
10	25.6	27.2	29.3	31.2
50	32.4	35.5	37.2	40.5
100	42.3	45.4	47.2	49.2
200	53.2	58.6	62.5	60.4
500	40.2	44.3	47.5	51.5

Table 7: Chloride ion release (mg/l) by *P. chrysosporium* at different PCP concentrations-Time Interval.

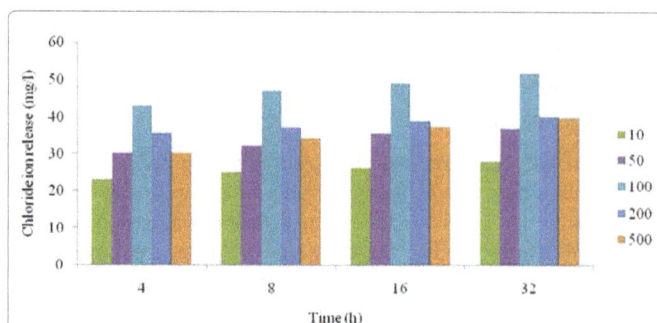

Figure 2: Chloride ion release (mg/l) by *P. fluorescence* at different PCP concentrations.

for isolation of xenobiotic compounds degrading bacteria and fungi is direct planting of active cultures on mineral salt medium with the toxicant as the only carbon source. The isolation of chlorophenols degrading bacteria has become a problem because of the toxic nature of the compounds and their recalcitrant nature [20-36].

Conclusions

Pentachlorophenol is a wide spectrum biocide with numerous

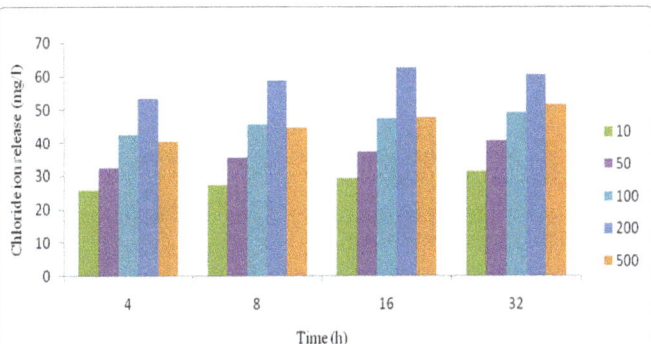

Figure 3: Chloride ion release (mg/l) by *P. chrysosporium* at different PCP concentrations.

	4 h	8 h	16 h	32 h
Fungi	-	++	+++	+++
Bacteria	-	-	+	++

(- refers to No cleavage while + indicates Color intensity for ortho-cleavage).

Table 8: Ortho ring cleavage by bacterial and fungal cultures.

Time (Hour)	CFU (10^7 cells/g. of soil)					
	0 ppm	10 ppm	50 ppm	100 ppm	200 ppm	500 ppm
4 h	1.45	1.4	1.5	1.63	1.49	1.44
8 h	1.52	1.57	1.42	1.66	1.4	1.4
16 h	1.59	1.52	1.52	1.73	1.3	1.2
32 h	1.1	1.48	1.43	2.2	0.8	0.9

Table 9: Growth pattern of *P. fluorescence* in soil microcosm (30% moisture) at different concentrations of PCP.

Time (Hour)	CFU (10^7 cells/g. of soil)					
	0 ppm	10 ppm	50 ppm	100 ppm	200 ppm	500 ppm
4	1.47	1.3	1.51	1.66	1.8	1.4
8	1.5	1.63	1.43	1.56	1.85	1.39
16	1.6	1.48	1.55	1.6	1.9	0.8
32	1.27	1.38	1.3	1.42	2.3	0.56

Table 10: Growth pattern of *P. chrysosporium* in soil microcosm (30% moisture) at different concentrations of PCP.

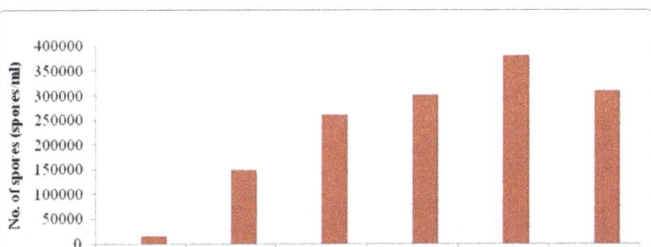

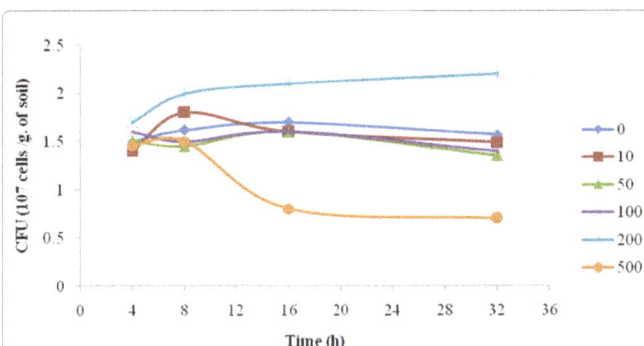

Figure 4: Growth of *P. chrysosporium* (interms of spore number) in culture having varied concentrations of PCP.

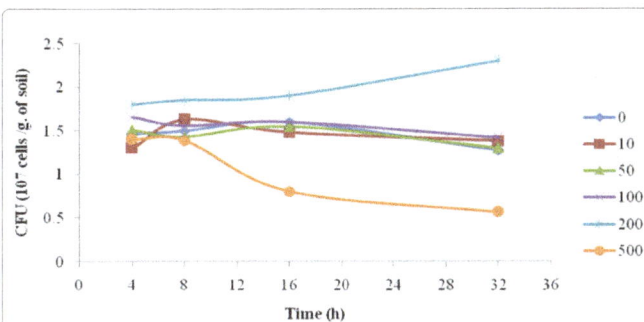

Figure 5: Growth pattern of *P. chrysosporium* in soil microcosm (30% moisture) at different concentrations of PCP.

Figure 6: Growth pattern of *P. fluorescence* in soil microcosm (60% moisture) at different concentrations of PCP.

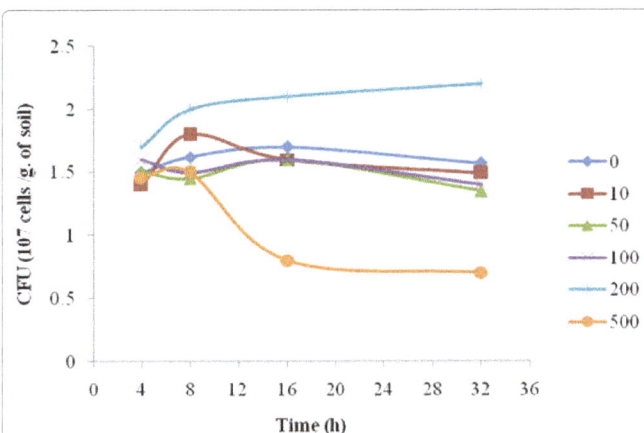

Figure 7: Growth pattern of *P. chrysosporium* in soil microcosm (60% moisture) at different concentrations of PCP.

Time (Hour)	CFU (10^7 cells/g. of soil)					
	0 ppm	10 ppm	50 ppm	100 ppm	200 ppm	500 ppm
4	1.7	1.75	1.49	2	1.51	1.53
8	1.76	1.8	1.52	2.1	1.46	1.49
16	1.5	1.6	1.85	2.3	1.48	1.5
32	1.47	1.5	1.7	2.45	1	1.3

Table 11: Growth pattern of *P. fluorescence* in soil microcosm (60% moisture) at different concentrations of PCP.

applications in agriculture, industries and public health. It is considered to be an environmental pollutant because of its board toxicity and persistence in soil for long. Biodegradation of PCP is challenging

Time (Hour)	CFU (10^7 cells/g. of soil)					
	0 ppm	10 ppm	50 ppm	100 ppm	200 ppm	500 ppm
4	1.5	1.4	1.51	1.6	1.7	1.46
8	1.62	1.8	1.45	1.5	2	1.5
16	1.7	1.6	1.6	1.6	2.1	0.8
32	1.57	1.49	1.35	1.4	2.2	0.7

Table 12: Growth pattern of *P. chrysosporium* in soil microcosm (60% moisture) at different concentrations of PCP.

because it unfolds oxidative phosphorylation and alters membrane fluidity for living beings. The result of the present study can be applied in industries that are the major PCP releaser. The contaminated soil and water in which PCP is a major pollutant could be reformed with the help of this proposed culture and optimized conditions. Biodegradation of pentachlorophenol by the proposed technique could be an effective means of bioremediation and eco-friendly transformation of the toxin accumulating in the environment.

Acknowledgments

Authors are thankful to the Director, FRI, Dehradun for his kind consent to publish this research articles. We also acknowledge the help of Mr. Rajendra Kumar and Manoj Kumar for providing all necessary chemicals, reagents etc.

References

1. Amandine C, Sylvie VM, Satish G, Jean-Paul S (2004) Lindane and Technical HCH Residues in Indian Soils and Sediments. JSS - J Soils & Sediments 4: 192-196.

2. Qureshi A, Mohan M, Kanade GS, Kapley A, Purohit HJ (2009) In situ bioremediation of organochlorine-pesticide-contaminated microcosm soil and evaluation by gene probe. Pest Manag Sci 65: 798-804.

3. Awasthi N, Manickam N, Kumar A (1997) Biodegradation of endosulfan by a bacterial coculture. Bull Environ Contam Toxicol 59: 928-934.

4. Barr DP, Aust SD (1994) Mechanisms white rot fungi use to degrade pollutants. Environ Sci Technol 28: 78A-87A.

5. Bajpai P, Bajpai PK (2004) Reduction of organochlorine compounds in bleach plant effluents, In Scheper,t.(ed) Advance in Biochemical Engineering Biotechnology. Springer-Verlag, Berlin 57: 213-260.

6. Balba M, Al-Awadhi N, Al-Daher R (1998) Bioremediation of oil contaminated soil: microbiological methods for feasibility assessment and field evaluation. Journal of Microbiological methods 32: 155-164.

7. Bampus J, Aust S (1987) Bioremediation of environmental pollutants by the white rot fungus Phanerochaete chrysosporium: involvement of the lignin degrading system. Bioassays 6: 166-170.

8. Büyüksönmez F, Rynk R, Hess TF, Bechinski E (2000) The Occurrence, Fate, and Degradation of Pesticides during Composting. Part II. Occurrence and Fate of Pesticides in Compost and Composting Systems. Compost Science and Utilization 8: 61-82.

9. Bhatanagar L, Li SP, Jain MJ, Zeikus JG (1989) Growth of methanogenic and acidogenic bacteria with PCP as co-substance, Biotechnology Applications in Hazardous Waste Treatment, pp. 383-393.

10. Canet R, Birnstingl JG, Malcolm DG, Lopez-Real JM, Beck AJ (2001) Biodegradation of polycyclic aromatic hydrocarbons (PAHs) by native microflora and combinations of white-rot fungi in a coal-tar contaminated soil. Bioresour Technol 76: 113-117.

11. Chen ST, Kang SY (1999) Pentachlorophenol and crystal violet degradation in water and soil-using heme and hydrogen peroxide. Water Research 33: 3657-3665.

12. Davis MW, Glaser JA, Evans JW, Lamar RT (1993) Field evaluation of the lignin degrading fungus Phanerochaete sordida to treat creosote contaminated soil. Environmental Sciences and technology 27: 2572-2576.

13. Edgehill RU, Finn RK (1983) Microbial treatment of soil to remove pentachlorophenol. Appl Environ Microbiol 45: 1122-1125.

14. Hestbjerg H, Willumsen P, Christensen M, Andersen O, Jacobsen C (2003) Bio-augmentation of tar-contaminated soils under field conditions using Pleurotus ostreatus refuse from commercial mushroom production. Environmental Toxicology and chemistry 22: 692-698.

15. Hendriksen, Ahring BK (2004) Anaerobic dechlorination of pentachlorophenol in fixed film and up flow anaerobic sludge blanket reactors using different inoculate. Biodegradation 3: 399-408.

16. Jin P, Bhattacharya SK (1996) Anaerobic removal of pentachlorophenol in presence of zinc. J Environ Engg 122: 590-598.

17. Kanda M, Takahama K, Waseda Y, Ishii Y, Miyazaki Y (1968) [Studies on the influences of organochloric pesticides, PCP, and endrin to mitochondrial respiration and oxydative phosphorylation of the rat brain]. Nihon Hoigaku Zasshi 22: 223-228.

18. Kodama T, Ding L, Yoshida M, Yajima M (2001) Biodegradation of an striazine herbicide, simazine, Journal of Molecular Catalysis B-Enzymatic 11: 1073-1078.

19. Kohring GW, Zhang XM, Wiegel J (1989) Anaerobic dechlorination of ,4-dichlorophenol in freshwater sediments in the presence of sulfate. Appl Environ Microbiol 55: 2735-2737.

20. Kudo A (2003) Degradation of pentachlorophenol by anaerobic digestion. Water Sci Technol 21; 1685-1689.

21. Lamar R, Evans JW, Glaser JA (1993) Solid phase treatment of a PCP-contaminated soil using lignin degrading fungi. Environmental Science Technology 27: 2566-2571.

22. Lamar RT, Dietrich DM (1990) In Situ Depletion of Pentachlorophenol from Contaminated Soil by Phanerochaete spp. Appl Environ Microbiol 56: 3093-3100.

23. Lee SG, Yoon BD, Park YH, Oh HM (1998) Isolation of a novel pentachlorophenol degrading bacterium pseudomonas sp. J Appl Environ Microbial 85: 1-8.

24. Madsen T, Aamand J (1991) Effects of sulfuroxy anions on degradation of pentachlorophenol by a methanogenic enrichment culture. Appl Environ Microbiol 57: 2453-2458.

25. Mester T, Tien M (2000) Oxidation mechanism of ligninolytic enzymes involved in the degradation of environmental pollutants. International Bio-deterioration & Biodegradation 46: 51-67.

26. Mourato S, Ozdemiroglu E, Foster V (2000) Evaluating health and environmental impact of pesticide use: implications for the design of eco-labels and pesticide taxes, 34: 1456-1461.

27. Nandish MS, Jagadeesh K S (2006) Pentachlorophenol degradation by entero bacter NV-5: optimization of process parameter. Int J Microbial6: 25-29.

28. Pointing SB (2001) Feasibility of bioremediation by white-rot fungi. Appl Microbiol Biotechnol 57: 20-33.

29. Premalatha A, Rajakumar GS (1994) Pentachlorophenol degradation by Pseudomonas aeruginosa. World J Microbiol Biotechnol 10: 334-337.

30. Pletsch M, de Araujo BS, Charlwood BV (1999) Novel biotechnological approaches in environmental remediation research. Biotechnol Adv 17: 679-687.

31. Resnick SM, Chapman PJ (1994) Physiological properties and substrate specificity of a pentachlorophenol-degrading Pseudomonas species. Biodegradation 5: 47-54.

32. Reddy C, Mathew Z (2001) Bioremediation potential of white rot fungi. In. Gadd G. (Eds.) Fungi in bioremediation Cambridge University Press. Cambridge, U.K.45: 321-334.

33. Singleton I (2001) Fungal remediation of soils contaminated with persistent organic pollutants, In. Gadd G. (Eds.) Fungi in bioremediation Cambridge University Press. Cambridge, UK, 34: 32-45.

34. Tekere M, Mswaka AY, Zvauya R, Read JS (2001) Growth, dye degradation and ligninolytic activity studies on Zimbabwean white rot fungi. Enzyme Microb Technol 28: 420-426.

35. Thakur IS, Saxena P (1998) Degradation of 4-chlorobenzoic acid and 4-dichlorophenoxy acetic acid by Pseudomonas sp. from a chemostat, In: Pandey, A. (ed.) Advances in Biotechnology. Educational Publishers and Distribution, 44: 423-431.

36. Zhang J, Chiao C (2002) Novel Approaches for remediation of pesticide pollutants. International Journal Environment and Pollution 18: 423-433.

18

Influence of Different Organic-Based Fertilizers on the Phytoremediating Potential of *Calopogonium mucunoides* Desv. from Crude Oil Polluted Soils

M. B. Adewole* and Y. I. Bulu

Institute of Ecology and Environmental Studies, Obafemi Awolowo University, Ile-Ife, Nigeria

Abstract

A greenhouse experiment was conducted to investigate the growth of *Calopogonium mucunoides* in soils contaminated by various concentrations of crude oil with a view of assessing its phytoremediating potential when different organic-based fertilizers were applied. The crude oil prepared at different concentrations of 0.0, 2.5, 5.0, 10.0 and 20.0% (v/v) acted as contaminants on 3 kg each of the air-dried soil collected from exhaustively cropped farm. Each treatment was replicated thrice in complete randomized design with four different fertility management levels, namely: 8t ha^{-1} compost organic fertilizer (CM); 8t ha^{-1} neem fortified organic fertilizer (NM); control$_1$, without fertilizer application (C1) and Control$_2$, where no fertilizer and no crop but crude oil was applied (C2). Significantly ($p<0.05$) highest total petroleum hydrocarbon (THC) uptake (10^{-2} mg kg^{-1}) of 1.08, 0.52 and 0.21; 1.01, 0.51 and 0.11 in the roots and shoots for CM, NM and C1 treatments respectively were obtained at 2.5% contamination. Also, significantly ($p<0.05$) higher values of (10^{-2} mg kg^{-1}) 2.57, 1.49 and 0.37; 3.02, 0.98 and 0.58 for Pb in the roots and shoots with CM, NM and C1 treatments respectively were phytoremediated at 5.0% contamination. Lower values of Cd were removed at different contamination levels and fertilizer treatments. With increased contamination, there was a reduction in the uptake of THC and Cd, while higher Pb bioaccumulated. The study concluded that *C. mucunoides* plant could be effectively used in the phytoremediation of crude oil contaminated soil when compost organic fertilizer is applied.

Keywords: *Calopogonium mucunoides*; Phytoremediation; Crude oil; Organic fertilizer; Soils

Introduction

There is an increasing petroleum hydrocarbon pollution of the soil ecosystem during oil exploration, exploitation, storage and transportation [1,2]. In Nigeria, tremendous increase in production and utilization of petroleum has led to a steady increase in the level of soil pollution by petroleum oil, especially in the Niger Delta region [3]. In addition to accidental discharge of petroleum oil [4], there exists the petroleum pollution of the soil ecosystem through the oil pipe vandalization and spillage [5]; and these have their attendant effects on the living soil organisms and vegetation [6].

Petroleum hydrocarbons are naturally occurring compounds that bind soil components thereby making the removal or degradation more difficult [7]. Also, their movement into the sub-soil leads to the expulsion of air thereby depleting oxygen reserves in the soil and impeding its diffusion to the deeper layers [8]. As the available soil oxygen diminishes, the soil microbial activities involved in the utilization of oxygen for biodegradation of the contaminants will reduce [9]. This will have adverse effect on the quality of soil and thereby reduce the growth performance of plants [10].

In the Niger Delta region of Nigeria, many of the petroleum hydrocarbon-polluted soils are found around the communities who practice subsistence fish and crop farming. Hence, the agri-business of many of the farmers in this region is in danger due to soil and freshwater pollution. Many of the soils that are good for agricultural purposes are being neglected because of their inability to support crops as a result of petroleum contamination. Phytoremediation is an emerging technology being used to clean-up polluted sites. Many of the established phytoplants such as *Brassica juncea* and *Triticum aestivum* [11] are temperate crops. However, tropical crops for example; *Helianthus annuus* [12] have been used at different times to remediate soils polluted with heavy metals, but is rarely cultivated by Nigerian farmers [13].

Since, these established phytoplants are not readily available; there is therefore the need to look for phytoweeds that could be adaptable to tropical environment. During our preliminary survey of petroleum hydrocarbon-polluted soils, among many of the weeds with short duration life-span that survived on petroleum hydrocarbon-polluted soils is *Calopogonium mucunoides*. The specific objectives of the study therefore, were to (i) investigate the growth of *C. mucunoides* as a phytoremediating plant at different levels of crude oil soil contamination, (ii) assess the uptake of the total hydrocarbon, Pb and Cd by *C. mucunoides* in crude oil soil contamination, and (iii) evaluate the effectiveness of different organic-based fertilizers application on the remediating potential of *C. mucunoides* plant.

Materials and Methods

The study was carried out in the greenhouse of the Faculty of Agriculture, Obafemi Awolowo University, Ile-Ife, Nigeria. Bulk surface soil samples (0-15 cm) from an exhaustively cropped farm land within the University was collected, the soil was air-dried for 7 days, sieved using a 2 mm mesh and then analyzed to determine the physico-

*Corresponding author: MB Adewole, Institute of Ecology and Environmental Studies, Obafemi Awolowo University, Ile-Ife, Nigeria
E-mail: adewoledele2005@yahoo.co.uk

chemical characteristics of the soil prior to sowing of the viable seeds of *C. mucunoides*. The soil properties obtained are presented in Table 1.

The air-dried and sieved soil was then filled into 60 buckets of radius 12.5 cm and height 17 cm leaving a space of 3 cm at the top end of the buckets to make allowance for watering. Each of the buckets contained 3 kg of the soil and each bucket was perforated at the base to avoid water logging and to increase the soil aeration. Crude oil obtained from Nigeria National Petroleum Corporation, Eleme, Rivers State, prepared to different concentration levels of 0.0%, 2.5%, 5.0%, 10.0% and 20.0% (v/v) were used as contaminants and the moistened buckets were left for one week to equilibrate. The crude oil was analyzed and has chemical composition [Total hydrocarbon (THC) 173.20 mg l^{-1}, Pb 4.87 mg l^{-1}, and Cd 0.85 mg l^{-1}].

The seeds of *C. mucunoides* collected from the Teaching and Research Farm, the Faculty of Agriculture, Obafemi Awolowo University, Ile-Ife were scarified and their germination percentages were determined. Cotton wool was spread in a petri dish and moistened with clean water. Fifty seeds of *C. mucunoides*, randomly counted were put on the moistened cotton wool and another petri dish was used to cover it. This was replicated 3 times. The number of seeds that germinated was counted on the fifth day and a mean germination percentage of 75 were obtained. This assisted to decide the planting rate of four seeds of *C. mucunoides* per pot.

Each of the pollution concentration was replicated thrice in completely randomised design with four fertilizer application levels viz: 8 t ha^{-1} compost organic fertilizer (CM); 8 t ha^{-1} neem fortified organic fertilizer (NM); control 1, soil with no fertilizer application (C1) and control 2, soil with no fertilizer and no crop but with crude oil only (C2). The constituents of the organic-based fertilizers which were purchased at a local market are presented in Table 2. The four seeds of *C. mucunoides* planted in each of the pots were thinned to two seedlings per pots at 2 weeks after planting (WAP) and thereafter fertilized. The thinned seedlings were dropped into their respective pots. Throughout the duration of the experiment, distilled water was supplied to plants as often as soil dryness was observed to field moisture capacity and the pots were maintained weed-free. At 12 weeks of planting, the experiment was terminated to prevent the *C. mucunoides* weed seeds from spreading.

Property	Value
pH (1:1 soil-water)	6.00
Organic carbon (g kg^{-1})	29.70
Total N (g kg^{-1})	2.56
Available P (mg kg^{-1})	9.40
Exchangeable cations (cmol kg^{-1})	2.75
Exchangeable acidity (cmol kg^{-1})	0.15
Total hydrocarbon content (mg kg^{-1})	0.17
Pb (mg kg^{-1})	0.67
Cd (mg kg^{-1})	0.11
Clay (g kg^{-1})	82.00
Silt (g kg^{-1})	60.00
Sand (g kg^{-1})	858.00
Textural class	Loamy sand

Table 1: Physico-chemical properties of the soil used for the study.

Type of fertilizer	N	P	K	Ca	Mg
CM	4.06	1.40	0.71	1.60	0.16
NM	6.40	0.64	0.46	12.42	0.59

Table 2: Chemical compositions (%) of organic-based fertilizers used.

Plant growth measurement

The plant length, number of leaves and stem girth were determined weekly for a period of 12 weeks. Plant length was measured with a meter rule from soil level to the terminal bud. The stem girth was derived after measuring the diameter of the plant with a vernier calliper using the formula πd, where π = 22/7 and d = diameter of the plant. At harvest, plant roots and shoots were oven-dried at a temperature of 80°C for 48 hours, allowed to cool and their dry weights were determined.

Laboratory analysis

Soil and organic-based fertilizers analyses: The soil pH was potentiometrically determined in 1:1 soil-water ratio [14]. The particle size analysis was determined using hydrometer method in 5 % sodium hexametaphosphate as outlined by Bouyoucos [15]. Soil organic carbon was determined using Walkey-Black method [16] and micro Kjeldahl procedure was used for the determination of total nitrogen [17]. The available phosphorus was determined by Bray P_1 method [18]. The exchangeable cations (Ca^{2+} + Mg^{2+} + K^+ + Na^+) were determined using 1M NH_4OAc (Ammonium acetate) buffered at pH 7.0 as extractant [19]. The Na^+ and K^+ concentrations in the soil extracts were read on Gallenkamp flame photometer while Ca^{2+} and Mg^{2+} were read using a Buck Scientific Model 210 VGP (Norwalk, Connecticut, USA) Atomic Absorption Spectrophotometer (AAS).

Exchangeable acidity (H^+ + Al^{3+}) in the soil samples was extracted with 1M KCl [19]. Solution of the extract was titrated with 0.05M NaOH to a permanent pink endpoint using phenolphthalein as indicator. The amount of NaOH used is equivalent to the total amount of exchangeable acidity in the aliquot taken [20]. Lead and Cadmium were determined using 5 ml of the mixture (concentrated HNO_3 and concentrated $HClO_4$ in the ratio 2:1) with 5 ml of concentrated H_2SO_4 to digest 0.5 g of each soil sample for 2 hours at 150°C [21]. The digests were allowed to cool and each was made up to 25 ml with distilled water. Concentrations of Pb and Cd in the extract were read on using AAS.

Root and shoot analyses: Lead and Cd concentrations in the root and shoot samples were determined using 5 ml of the mixture (concentrated HNO_3 and $HClO_4$ in the ratio 2:1) with 5 ml of concentrated H_2SO_4 to digest 0.5 g of each sample for 2 hours at 150°C [21]. The digests were allowed to cool and each was made up to 25 ml with distilled water. The concentrations of Pb and Cd in the extracts were read on using AAS.

Determination of total hydrocarbon content

Greenberg et al. [22] approach was used for total hydrocarbon determination. Ten grams of each of the air-dried and sieved soil samples and 0.5 g each of the oven-dried and ground plant samples was weighed into a 250 ml conical flask each. 20 ml of xylene was added to each of the sample and then placed on a reciprocating shaker for 30 minutes. Each sample was later filtered using Whattman No. 1 filter paper of 11 cm into filtering bottle. The crude oil was used to prepare a set of standard: 0.00, 5.00, 10.00, 15.00, 20.00 and 25.00 ppm, with xylene as the solvent and thereafter used to calibrate the spectrophotometer before the filtrates of soil, root and shoot samples' were read at 650 nm wavelength.

Translocation factor and bioaccumulation of the contaminants

The translocation factor (TF) was calculated as the concentration of Pb, Cd or THC in the shoots divided by the concentration in the roots. The bioaccumulation of the contaminants in the plant (shoots and roots) was calculated as the concentrations in the shoots and roots multiplied by their weights and thereafter, divided by the total weight of shoots and roots.

Statistical analysis

A statistical comparison of means was carried out with one-way analysis of variance (ANOVA) and treatment means were separated

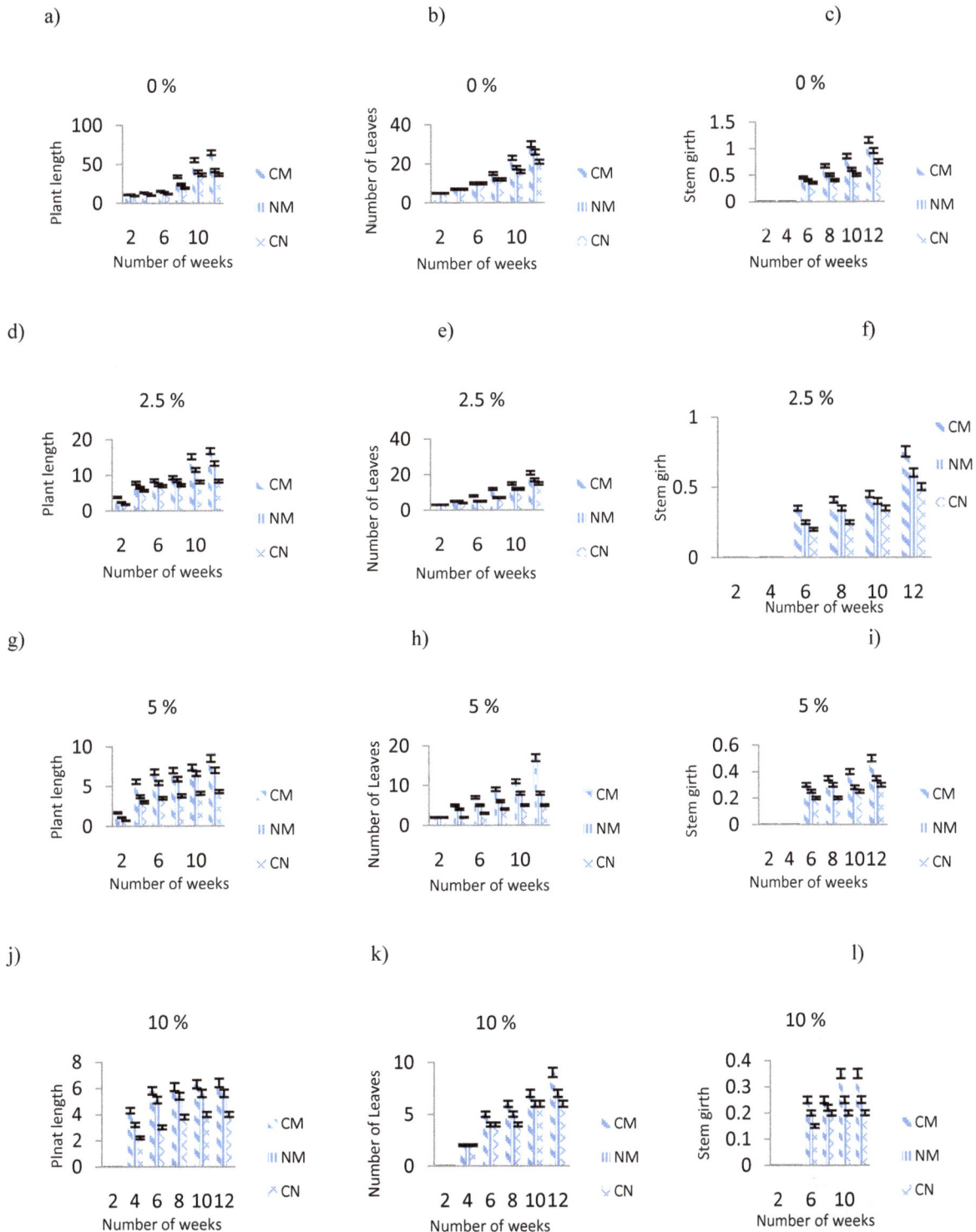

Legend: CM = Compost organic fertilizer, NM = Neem fortified organic fertilizer, CN = Control

Figure 1: Effects of fertilizer treatments on the growth parameters after planting; Bars indicate level of significance at p < 0.05.

using the Duncan range test available in SPSS 16 statistical package. Significance was set at p<0.05

Results and Discussion

Growth characteristics of *Calopogonium mucunoides*

The effects of fertilizer treatments and crude oil soil contaminations on the growth response of *C. mucunoides* are presented in Figure 1. The growth characteristics, namely, plant length, number of leaves and stem girth increased with fertilizer applications, but decreased with increase in crude oil contamination levels. From 8 WAP, the plant lengths were significantly ($p<0.05$) higher for 0% crude oil contamination among the fertilizer treatments; while significant difference was observed from 2 WAP for $\geq$ 2.5% soil contamination. Observations obtained in the plant lengths were similar to the number of leaves among the fertilizer treatments. However, significant ($p<0.05$) difference was observed in the stem girth of *C. mucunoides* from 6 WAP among different levels of crude oil contamination and fertilizer treatments. The order of increase in growth performance among the fertilizer treatments across the crude oil contamination levels was CM > NM > C1.

The higher the values of N and P in the fertilizers applied (Table 2), the higher was the growth performance of *C. mucunoides*. Makinde et al. [23] and Okokoh and Bisong [24] observed similar results of increased growth performance in a vegetable crop, *Amaranthus hybridus* with increase in organic N and P. Also, NM had higher organic N while CM had higher organic P in their available forms; and the possibility of good soil conditions created by organic manure in the two fertilizers could account for good vegetative yield compared to CN plots. Whereas, In contaminated pots but without fertilizers, the physical properties of the crude oil may have imposed some stressful conditions which may interfere with water and gaseous exchange [9]. This may disrupts the normal growth of plant roots within the soil thereby reducing the normal physiological growth of the plant [25]. The non-germinability of *C. mucunoides* at 20% contamination may have resulted from the effect of toxicity of the crude oil to the embryo of the test crop. On the whole, the dry weights of *C. mucunoides* decreased with increase in soil contamination (Table 3).

Bioaccumulation of THC, Pb and Cd by *Calopogonium mucunoides*

The bioaccumulation of THC, Pb and Cd in the roots and shoots of *C. mucunoides* under different organic-based fertilizer applications are given in Table 4. The *C. mucunoides* remediated optimally, 1.04×10^{-2}, 0.51×10^{-2}, and 0.16×10^{-2} mg kg^{-1} of THC at 2.5%; 2.77×10^{-2}, 1.18×10^{-2}, and 0.50×10^{-2} mg kg^{-1} of Pb at 5.0%; and 0.07×10^{-2}, 0.04×10^{-2}, and 0.02×10^{-2} mg kg^{-1} at 2.5% contamination with CM, NM, and C1 fertilizer applications, respectively. The increased organic manure in the soil brought about by the application of CM and NM fertilizers served as additional food for native micro-organisms to enhance their population and activities in the soil medium. Von Wedel et al. [26] earlier demonstrated that increased presence of mineral nutrients, either from organic or inorganic source enhanced the rate of microbial degradation of petroleum hydrocarbon, though in ground water. This also, agreed with the findings of Margesin et al. [27] that

CL (%)	Root			Shoot			Root + Shoot		
	CM	NM	C1	CM	NM	C1	CM	NM	C1
0	6.3 ± 0.7a	4.7 ± 0.5a	3.0 ± 0.3a	8.7 ± 0.7a	7.2 ± 0.7a	3.6 ± 0.1a	15.0 ± 0.7a	11.9 ± 0.6a	6.6 ± 0.2a
2.5	3.2 ± 0.2b	2.0 ± 0.1b	1.4 ± 0.1b	5.5 ± 0.5b	4.6 ± 0.5b	1.4 ± 0.1b	8.7 ± 0.4b	6.6 ± 0.3b	2.8 ± 0.1b
5.0	1.6 ± 0.2c	0.7 ± 0.1c	0.4 ± 0.1c	2.3 ± 0.2c	1.1 ± 0.2c	0.7 ± 0.1c	3.9 ± 0.2c	1.8 ± 0.1c	1.1 ± 0.1c
10.0	0.4 ± 0.1d	0.3 ± 0.0c	0.1 ± 0.0c	0.8 ± 0.2d	0.3 ± 0.1d	0.2 ± 0.0c	1.2 ± 0.2d	0.6 ± 0.1d	0.3 ± 0.0d
20.0	-	-	-	-	-	-	-	-	-

Means followed by the same letter within a column are not significantly different at p < 0.05 by Duncan's Multiple Range Test.
CM = Compost organic fertilizer, NM = Neem fortified organic fertilizer, C1 = Control, THC = Total hydrocarbon, CL = Crude oil contamination levels.

Table 3: Mean (± standard error) dry weight (10^{-3} kg /plant) of roots and shoots of *Calopogonium mucunoides*.

CL (%)	Root			Shoot			Root + Shoot		
	CM	NM	C1	CM	NM	C1	CM	NM	C1
				THC					
0	0.15 ± 0.04	0.10 ± 0.03	0.07 ± 0.03	0.11 ± 0.04	0.13 ± 0.02	0.04 ± 0.01	0.13 ± 0.04a	0.12 ± 0.03a	0.05 ± 0.02b
2.5	1.08 ± 0.08	0.52 ± 0.08	0.21 ± 0.04	1.01 ± 0.05	0.51 ± 0.04	0.11 ± 0.04	1.04 ± 0.06a	0.51 ± 0.06b	0.16 ± 0.04c
5.0	0.79 ± 0.10	0.40 ± 0.03	0.04 ± 0.01	0.65 ± 0.08	0.07 ± 0.02	0.02 ± 0.00	0.71 ± 0.09a	0.20 ± 0.02b	0.03 ± 0.00c
10.0	0.76 ± 0.08	0.29 ± 0.03	0.03 ± 0.00	0.41 ± 0.05	0.04 ± 0.01	0.00 ± 0.00	0.53 ± 0.07a	0.17 ± 0.02b	0.01 ± 0.00c
20.0	-	-	-	-	-	-	-	-	-
				Pb					
0	0.08 ± 0.02	0.06 ± 0.01	0.02 ± 0.00	0.05 ± 0.01	0.05 ± 0.01	0.02 ± 0.00	0.06 ± 0.01a	0.05 ± 0.01a	0.02 ± 0.00b
2.5	0.95 ± 0.06	0.61 ± 0.05	0.27 ± 0.02	0.87 ± 0.05	0.58 ± 0.02	0.11 ± 0.02	0.90 ± 0.06a	0.59 ± 0.04b	0.15 ± 0.02c
5.0	2.57 ± 0.13	1.49 ± 0.11	0.37 ± 0.03	3.02 ± 0.18	0.98 ± 0.08	0.58 ± 0.05	2.77 ± 0.15a	1.18 ± 0.10b	0.50 ± 0.04c
10.0	2.26 ± 0.12	1.41 ± 0.11	0.35 ± 0.03	1.78 ± 0.10	0.94 ± 0.08	0.47 ± 0.03	2.11 ± 0.11a	1.18 ± 0.10b	0.43 ± 0.03c
20.0	-	-	-	-	-	-	-	-	-
				Cd					
0	0.02 ± 0.00	0.01 ± 0.00	0.00 ± 0.00	0.01 ± 0.00	0.00 ± 0.00	0.00 ± 0.00	0.01 ± 0.00a	0.00 ± 0.00b	0.00 ± 0.00b
2.5	0.09 ± 0.02	0.07 ± 0.01	0.03 ± 0.00	0.06 ± 0.01	0.02 ± 0.00	0.01 ± 0.00	0.07 ± 0.02a	0.04 ± 0.00a	0.02 ± 0.00b
5.0	0.05 ± 0.01	0.01 ± 0.00	0.00 ± 0.00	0.01 ± 0.00	0.00 ± 0.00	0.00 ± 0.00	0.03 ± 0.00a	0.00 ± 0.00b	0.00 ± 0.00b
10.0	0.01 ± 0.00	0.00 ± 0.00	0.00 ± 0.00	0.00 ± 0.00	0.00 ± 0.00	0.00 ± 0.00	0.00 ± 0.00a	0.00 ± 0.00a	0.00 ± 0.00a
20.0	-	-	-	-	-	-	-	-	-

Table 4: Mean value (±standard error) of bioaccumulation of THC, Pb and Cd (10^{-2} mg kg^{-1}) in the roots and shoots of *Calopogonium mucunoides*.

Fertilizer CL (%)	THC	CM Pb	Cd	THC	NM Pb	Cd	THC	C1 Pb	Cd
0	0.690	0.256	2.000	0.271	0.123	0	0.571	0.294	0
2.5	0.990	0.916	0.667	0.981	0.951	0.286	0.524	0.407	0.333
5.0	0.855	1.444	0.333	0.175	0.658	0	0.500	1.568	0
10.0	0.519	1.336	0	0.138	0.667	0	0	1.343	0
20.0	-	-	-	-					

CM = Compost organic fertilizer, NM = Neem fortified organic fertilizer, C1 = Control, THC = Total hydrocarbon, CL = Crude oil contamination levels.

Table 5: Translocation factor of THC, Pb and Cd in Calopogonium mucunoides.

carbon addition through organic nutrient supplement increased the ability of soil microbes to degrade crude oil. As more hydrocarbons are degraded, there will be sufficient carbon and energy to support large numbers of soil microbes.

The binding force that brought the soil components together to make its biodegradation difficult may have been found easy to break in this experiment, by these active native soil micro-organisms. This broken binding force positively enhanced the removal rates of THC, Pb, and Cd when CM and NM were applied. This was because the increased microbial activity in oil-contaminated soil due to organic manures application, increased the availability of soil nutrients, including heavy metals such as Pb and Cd. Pots without organic fertilizer application had significantly ($p < 0.05$) least values of THC (0.16×10^{-2} mg kg^{-1}), Pb (0.50×10^{-2} mg kg^{-1}), and Cd (0.02×10^{-2} mg kg^{-1}) remediated. Parreira et al. [8] observed similar significant increase in the biodegradation and uptake of gasohol-contaminants due to increased microbial consortium brought about by organic manure application.

Translocation factor of THC, Pb and Cd in *Calopogonium mucunoides*

The translocation factors (TF) of THC, Pb and CD in the roots and shoots of *C. mucunoides* under different organic-based fertilizer applications are given in Table 5. The TF for THC, Pb and Cd increased when the soil contamination went up to 2.5% from zero and decreased thereafter from $\geq$ 5.0% in all the fertilizer applications. When the TF of THC, Pb and Cd were separately compared with the applied fertilizers, compost organic fertilizer was most effective, while control was the least at different levels of crude oil soil contamination.

Conclusion

The organic-based fertilizers enhanced the growth performance of *C. mucunoides* in crude oil polluted soils. The biodegradation and bioaccumulation of petroleum hydrocarbon are also enhanced with organic-based fertilizer applications. However, compost organic fertilizer performed better in the bioaccumulation of the total petroleum hydrocarbon, Pb and Cd than either neem fortified organic fertilizer or zero fertilization when *C. mucunoides* was the test phytoplant.

Refernces

1. Nicolotti G, Egli S (1998) Soil contamination by crude oil: impact on the mycorrhizosphere and on the revegetation potential of forest trees. Environ Pollut 99: 37-43.

2. Tyagi M, da Fonseca MM, de Carvalho CC (2011) Bioaugmentation and biostimulation strategies to improve the effectiveness of bioremediation processes. Biodegradation 22: 231-241.

3. Eriyamremu GE, Iyasels JU, Osubor CC, Anoliefo GO, Osagie VE, et al. (1999) Bonny light crude oil alters protease and respiratory activities in germinating bean Vigna unguiculata L. seeds. Journal of Science and Engineering Technology 6: 1589-1600.

4. Nwankwo JN, Ifeadi CN (1985) Case studies of environmental impacts of oil pollution and marketing in Nigeria. A paper presented at the seminar on policy, environmental issues and management in Nigeria, University of Benin, Benin-City, Nigeria.

5. Omodanisi EO (2008) Assessment of the effect of oil pipeline vandalization on vegetation in Ilado Area of Lagos State using remote sensing technology. Institute of Ecology and Environmental Studies, Obafemi Awolowo University, Ile-Ife, Nigeria.

6. Siddiqui S, Adams WA (2002) The fate of diesel hydrocarbons in soils and their effects on germination of perennial ryegrass. Environ Toxicol 17: 49-62.

7. Barathi S, Vasudevan N (2001) Utilization of petroleum hydrocarbons by Pseudomonas fluorescens isolated from a petroleum-contaminated soil. Environ Int 26: 413-416.

8. Parreira AG, Totola MR, Jham GN, Da Silva SL, Borges AC (2011) Microbial bioremediation of aromatic compounds in a soil contaminated with gasohol. British Biotechnology Journal 1: 18-28.

9. Ayotamuno MJ, Kogbara RB, Agunwamba JC (2006) Comparative analysis of some techniques in the biological reclamation of crude oil polluted agricultural soils in Nigeria. Niger J Technol 25: 15-26.

10. Pezeshki SR, Hester MW, Lin Q, Nyman JA (2000) The effects of oil spill and clean-up on dorminant US Gulf coast marsh macrophytes: A review. Environ Pollut 108: 129-139.

11. Cui Y, Wang Q, Dong Y, Li H, Christie P (2004) Enhanced uptake of soil Pb and Zn by Indian mustard and winter wheat following combined soil application of elemental sulphur and EDTA. Plant Soil 261: 181-188.

12. Adewole MB, Sridhar MKC, Adeoye GO (2010) Removal of heavy metals from soil polluted with effluents from a paint industry using Helianthus annuus L. and Tithonia diversifolia (Hemsl.) as influenced by fertilizer applications. Bioremediat J 14: 169-179.

13. Adewole MB, Adeoye GO (2008) Growth performance and yield response of sunflower to organomineral fertilizer application and season in derived savanna of southwestern Nigeria. Moor Journal of Agricultural Research 9: 17-25.

14. Mclean EO (1982) Soil pH and lime requirement. In: Methods of Soil Analysis. (2ndedn), Agronomy Monograph No. 9. Madison, WI: American Society of Agronomy.

15. Bouyoucos GJ (1951) A recalibration of the hydrometer method for making the mechanical analysis soils. Agronomy Journal 43: 434-438.

16. Nelson DW, Sommers LE (1982) Total carbon, organic carbon and organic matter. In: Methods of soil analysis. (2ndedn), Agronomy Monograph No. 9. Madison, WI: American Society of Agronomy.

17. Bremner JM, Mulvaney CS (1982) Nitrogen – Total. In: Methods of Soil Analysis. (2ndedn), Agronomy Monograph No. 9. Madison, WI: American Society of Agronomy.

18. Olsen SR, Sommers LS (1982) Phosphorus. In: Methods of Soil Analysis. (2ndedn), Agronomy Monograph No. 9. Madison, WI: American Society of Agronomy.

19. Thomas GW (1982) Exchangeable cations. In: Methods of Soil Analysis. (2ndedn), Agronomy Monograph No. 9. Madison, WI: American Society of Agronomy.

20. Odu CTI, Babalola O, Udo EJ, Ogunkunle AO, Bakare TA, Adeoye GO (1986) Laboratory manual for agronomic studies in soil, plant and microbiology. (1stedn), Department of Agronomy, University of Ibadan, Ibadan, Nigeria.

21. Juo ASR (1982) Automated and semi-automated methods for soil and plant analysis, International Institute of Tropical Agriculture (IITA), Ibadan, Nigeria.

22. Greenberg AE, Connors JJ, Jenkins D (1981) Determination of organic

constituents: standard methods for the examination of water and wastewater. The America public health association, America water works association and water pollution control federation.

23. Makinde EA, Ayeni LS, Ojeniyi SO (2010) Morphological characteristics of Amaranthus cruentus L. as influenced by kola pod husk, organomineral and NPK fertilizers in Southwestern Nigeria. New York Science Journal 3: 130-134.

24. Okokoh SJ, Bisong BW (2011) Effect of poultry manure and urea-N on flowering occurrence and leaf productivity of Amaranthus cruentus. Journal of Applied and Environmental Science Management 15: 13-15.

25. Odjegba VJ, Sadiq AO (2002) Effect of spent engine oil on the growth parameters, chlorophyll and protein levels of Amaranthus hybridus I Environmentalist 22: 23-28.

26. von Wedel RJ, Mosquera JF, Goldsmith CD, Hater GR, Wong A (1988) Bacterial Biodegradation of Petroleum Hydrocarbons in Groundwater: In Situ Augmented Bioreclamation with Enrichment Isolates in California. Water Sci Technol 20: 501-503.

27. Margesin R, Zimmerbauer A, Schinner F (2000) Monitoring of bioremediation by soil biological activities. Chemosphere 40: 339-346.

19

In-situ Biological Water Treatment Technologies for Environmental Remediation

Mohamed Ateia[1], Chihiro Yoshimura[1] and Mahmoud Nasr[2]*

[1]Department of Civil Engineering, Tokyo Institute of Technology, 2-12-1-M1-4 Ookayama, Tokyo 152-8552 Japan
[2]Department of Sanitary Engineering, Faculty of Engineering, Alexandria University, PO Box 21544, Alexandria, Egypt

Abstract

Water treatment technologies can be classified as in-situ or ex-situ technologies. In-situ biological techniques include the use of aquatic plants, aquatic animals, and microbial remediation. Approaches to alleviate surface water pollution should use bioremediation methods as a primary technique. These methods should be tested not only on rivers and lakes, but also on other polluted surface streams. Furthermore, bioremediation processes need to be optimized depending on flow condition and nutrient availability. This paper comprehensively reviews the latest surface water remediation techniques that are suitable for in-situ applications, focusing on bioremediation technologies as effective techniques to remedy polluted water.

Keywords: Biological treatment; Bioremediation; In-situ treatment; Removal mechanism; Surface water

Introduction

In developing countries, water pollution is a key problem, with high levels of contaminates being reported in many rivers [1]. Different pollution control and water treatment technologies methods can be applied to resolve this issue [2-4]. Water treatment technologies can be classified as physical, chemical, or biological treatment techniques. They can also be classified as in-situ or ex-situ technologies. In-situ remediation techniques involve treatment at the site, while ex-situ involves the removal of contaminants at a remote location. Understanding in-situ treatment systems is essential to maintaining and controlling hydraulic conditions in open streams [5,6].

Aeration, as a physical treatment approach, is used either as a stand-alone system [6,7] or as a support for other systems (e.g., wetlands) [8,9]. Other examples of physical treatment approaches are water diversion and sediment dredging. Water diversion, however, can be a large-scale, high-cost option while sediment dredging can cause the re-suspension of contaminated sediments [10,11]. Chemical water treatment methods are also an option, an example of which is flocculation. Flocculation is used for in-situ treatment of both surface water and groundwater [12]. However, caution should be taken when handling chemicals, as they are potentially hazardous and can be used in large quantities.

Biodegradation is the breakdown of organic compounds by living organisms resulting in the formation of carbon dioxide and water or methane [13]. These microorganisms are bacteria, fungi, and microfauna (e.g., protozoans, some worms, and some insects) [5]. In-situ bioremediation has many advantages when compared to other techniques, such as low costs, less adverse impacts on the environment, and no secondary production of pollutants [14].

Indeed, many in-situ remediation processes, such as ecological floating bed techniques and constructed wetlands, have been developed for the bioremediation of polluted surface water and have produced "satisfactory" results [15]. This paper provides a holistic review of the latest surface water remediation developments and technologies that can be applied in-situ.

Remediation Techniques

Aquatic plants

Plants with strong tolerance for pollutants can mitigate or fix water pollutants through adsorption, absorption, accumulation, and degradation [16,17]. Macrophytes such as water hyacinth (*Eichhornia crassipes*) and water lettuce (*Pistia stratiotes*) have been used for upgrading effluent quality [18]. Whorl-leaf watermilfoil (*Myriophyllum verticillatum*), pondweed (*Potamogeton spp.*), common reed (*Phragmites communis*), cattail (*Typha latifolia*), duckweed (*Lemna gibba*) and canna (*Canna indica*) are also used for wastewater treatment purposes [19]. Aquatic plants can be introduced for in-situ surface water remediation in different treatment systems, such as in constructed wetlands and floating bed systems (e.g., Ruan et al. [20] or submerged systems using algae [21]).

These types of remediation techniques work either by aquatic plants assimilating pollutants directly into their tissues, or by increasing biodiversity in the rhizosphere, thereby increasing the variety of chemical and biochemical reactions that can enhance purification [22]. The primary characteristics of aquatic plants include their extensive root systems and rapid growth rate, which make them attractive biological support channels for bacteria [18]. Motility and chemotaxis enable the bacteria to move towards plant roots where they can benefit from root exudates as sources of carbon and energy, and may therefore contribute to the survival and colonization of the rhizosphere [23]. In addition to being able to mitigate organic pollutants, aquatic plants, especially algae, can be also used for the removal of nonconventional pollutants such as uranium from wastewater [21,24].

Of course, there are certain disadvantages of using the planted floating-bed in lake restoration. First, it is difficult to control the hydraulic retention time (HRT) and the pollutants loading rate when this treatment system is applied at real field sites and secondly, these systems in tropical and sub-tropical areas are especially vulnerable to natural disasters such as hurricanes or typhoons [25]. Moreover,

*Corresponding author: Mahmoud Nasr, Department of Sanitary Engineering, Faculty of Engineering, Alexandria University, PO Box 21544, Alexandria, Egypt
E-mail: mahmmoudsaid@gmail.com

problem facing plant based systems is being sensitive to nutrient availability, pollutants load and seasonally changes, as a result of the change of natural metabolic activities [2,26]. Therefore, some treatment systems were invented to simulate the natural aquatic plants and to overcome the disadvantages of the living plants. Aqua Mats, for instance, are a type of artificial seaweed with a high surface area that is designed to encourage colonization and growth of anaerobic bacteria, aerobic bacteria, algae, zooplankton and other aquatic organisms [27]. Further, removal of pollutants by bacteria in the system can be enhanced by methods such as immobilized bacteria [28] and/or by utilizing biofilm carrier [25]. Increasing the plant coverage plays an important in enhancing the removal efficiency as well [29]. Plus, the choice of appropriate plant species has been shown to generally improve pollutant removal and this seems an important avenue to explore for optimizing treatment system efficiency [16].

Nevertheless, plant-based systems are regarded as a low-cost, solar-energy-based, eco-friendly technology for *in-situ* purification of surface water, and an important ecological remediation to control water eutrophication [30]. Thus, more studies should consider optimizing these systems depending on the flow conditions and nutrient availability.

Aquatic animals

Aquatic animals such as clams, snails and other filter-feeding shellfish have a prominent effect on nutrient removal from eutrophic water bodies [6]. The biological treatment of stocking filter-feeding silver carp (*Hypophthalmichthys molitrix*) in eutrophic water bodies has been widely applied to control excessive phytoplankton levels and improve the quality of water bodies [27,31].

Silver carp has a long lifespan in natural water bodies (6-10 years, even 20 years in some instances) [32]. Commonly stocked in water reservoirs in developing countries, silver carp is an omnivorous filter-feeder that can filter particles >10 μm, including zooplankton and phytoplankton [27].

Filter-feeding fish such as silver carp have been shown to select zooplankton on the basis of prey escape ability; for instance, cladocerans are more vulnerable than copepods to fish predation because they have lower escape ability [33]. An example of aquatic animal usage to improve water quality is the introduction of Asiatic clam (*Corbicula fluminea*) into the Potomac River, United States of America. This was done in the early 1980s, when chlorophyll-a concentration in the Potomac River appeared to be strongly depleted, at levels of less than 1 g/L [25]. The Asiatic clam can also promote nutrient regeneration. Therefore, the species imposes simultaneous top-down and bottom-up effects on the ecosystem [27].

Experiments have shown that filter-feeding fish are able to reduce phytoplankton biomass to a certain degree, although the final efficiency depends on the characteristics of the given ecosystem. However, the application of such biomanipulation may lead to different effects depending on the composition of the initial plankton community (i.e., zooplankton and phytoplankton), the species and stocking density of fish, and the water temperature [27].

Silver carp usage to control algal biomass remains controversial. For instance, several studies have shown that stocking silver carp fails to reduce phytoplankton biomass in the presence of large herbivorous cladocerans [27,33]. A key reason for this was the reduction of grazing pressure on phytoplankton by zooplankton as a result of fish predation [33]. Moreover, inorganic or organic pollutants present in untreated water and some bio-toxins released by *Microcystis* spp. are harmful

to silver carp, which therefore affects the efficiency of this biological treatment. Further studies in the toxicology and security of water quality should be conducted [32] with an increased focus on understanding the effects of bioremediation on local ecosystems and biodiversity.

Microorganisms

Microorganism-based technologies are used to decompose, transform, and absorb water pollutants. Results to date generally confirm the existence of the appropriate microbial functional groups responsible for removing specific pollutants from wastewater [34]. Practically, two microorganism-based methods are used for *in-situ* surface water remediation. The first method is microbial dosing and the second utilizes biofilms [35].

Microbial dosing: Microbial dosing uses specific and efficient microorganisms to remove pollutants present in the water. Commercial products, such as FLO-1200, could achieve remarkable results in the river pollution control under the conditions of river aeration. Bio-energizer, combined water mixing was added and strengthened the ability of microbial degradation artificially for water purification [6].

Sheng [36] utilized two kinds of microbial reagents to remediate a heavily polluted river in Fangcun District, China, which became a black and odorous river. The dominant microbes in these reagents were photosynthetic bacteria and *Bacillus subtilis*. HRT was around 20 h. The reagents were directly diluted with river water before inoculation. The results of the small-scale experiment indicated that the removal rate increased with the increase of photosynthetic bacteria (PSB) concentration. The chemical oxygen demand (COD) and NH_3-N removal (corresponding removal rate are all over 60%). Furthermore, Field-scale test was undertaken, Except for suspended solids (SS), the total removal rates for each pollutant all exceeded 70%. Eventually, they recommended applying this method to remediate other heavily polluted rivers.

Mingjun [14] carried out a field trial of bioremediation in 60 m^3 of eutrophic water body in a local park for four months. A little amount of natural humic acid was added to speed up flocculation and deposition of the superfluous algae. Thus, the multiple microbial preparations used were composed of nitrobacteria, mixed bacteria and humic acid. The following conclusions were drawn: Pollution indexes of total nitrogen (TN), total phosphorus (TP), NH_4-N, COD and turbidity were declined differently, and the rates were 77.8, 72.2, 94.2, 60.0 and 85.6%, respectively. After bioremediation, the color of lake turned light green from dark green and clearer. The turbidity declined and DO increased. The water environment improved. Thus the problem of Lake Eutrophication can be solved radically by bioremediation.

Bio-film: The bio-film technology utilizes bio-membrane attached to the natural river bed and micro-carrier to move the pollutants in the river through adsorption, degradation and filtration under the conditions of artificial aeration or dissolved oxygen. Gravel contact oxidation method, artificial packing contact oxidation method, thin layer flow method, underground stream purification method, etc. The strengthening purification technology of The bio-film technology for river purification in Japan and South Korea and other countries were river researched by Japanese were mainly indirect purification, which was to build the purification facilities on the side of the river [6].

Ref. [37] evaluated the role of biofilm attached on streambed in linear alkylbenzene sulfonates (LAS) degradation in the stream using Environmental observations and laboratory biodegradation experiments using biofilm collected from Nogawa river bed located in southern part of Tokyo, Japan. Three batch culture experiments and one continuous culture experiment were conducted. For most observations, greater than 80% of the LAS were removed within 2-3 h of the travelling time. The batch culture experiments clearly indicate that the existence of the biofilm accelerates the biodegradation of LAS [38].

For the same river (Nogawa river), gravel contact oxidation was utilized, the packing was gravel, and the removal rates of biological oxygen demand (BOD) and SS were 72.3% and 84.9% respectively. With new non-woven fabric as packing, the drainage ditch facilities in Chiba County was set on the side of the ditch, and the removal efficiency of SS reached 97%, the removal rates of BOD and COD were 88% and 70% respectively [6]. Moreover, Ruan et al. [20] used Plant-biofilm oxidation ditch for *in-situ* treatment of polluted water. The system was designed for *in-situ* treatment of municipal sewage or polluted lake water in combination with plant biofilms for performing N and P removal. And running experiments at pilot scale for about 1.5 years resulted in the following observations:

1) The system was quite satisfactory and stable for treatment of municipal sewage and polluted lake water in removing COD, NH_4-N and PO_4-P.

2) The direct uptake of nitrogen and phosphorus by plants was negligible in comparison with the total removal by the system, but indirect mechanisms via plant root exudates and biofilms merit further studies.

The proposed process could dramatically reduce the costs of sewage collection, the land-space requirement and the construction costs compared with conventional sewage treatment plants; might be suitable for treatment of both municipal sewage and polluted lake water; and could lead to the promotion of wastewater treatment in many developing countries.

Further, biofilm processes, such as aerated bio-filter biological fluidized bed, suspended carrier biofilm reactors (SCBR), etc., are commonly used in surface water remediation. Immobilization of biomass in the form of biofilms is an efficient method to retain slow growing microorganisms in continuous flow reactors. These systems operated as aerobic or anaerobic phases with freely moving buoyant plastic biofilm carriers [6]. More specifically, microorganisms grow attached on small carrier elements that are kept in constant motion throughout the entire volume of the reactor, resulting in uniform, highly effective treatment [39].

The moving bed reactors provide distinct advantages, including being simple in operation, at low risk of losing the biomass and less temperature dependent. In addition, they have better control of biofilm thickness, higher mass transfer characteristics, they are not subject to clogging and they have a lower pressure drop [39,40].

Given its specific advantages, moving bed reactors are the most common activated sludge modifications used for industrial wastewater treatment [40], secondary effluent from sewage treatment plant [41], and river water [35,42] investigated the removal of organic matter from agriculture drainage water using MBBR. It was concluded that COD removal could reach up to 95% when the biofilm was acclimated to the same salinity level.

The biological contact oxidation process (BCOP), also called submerged biological filter or contact aeration system, is a hybrid wastewater treatment system, taking the advantages of both activated sludge process and biofilm process, e.g., no bed clogging and sludge bulking. Previous studies reported two types of biological contact

oxidation processes (BCOP). Step-feed (SBCOP) unit and Inter-recycle (IBCOP) unit were designed to investigate the treatment of heavily polluted river water. When spring dry season arrived, considering the lower substrate concentration of the raw water and positive effect of temperature rise on biological treatment, the total influent of each unit was 71.3 L/h with an HRT of 2 h. During the summer rainy season, in order to enhance the nitrification in the two biological treatment units, the total influent of each unit was recovered to 26.4 L/h with an HRT of 5.4 h. Further, the recycling ratio was 200% for the IBCOP. The results showed that The SBCOP unit had higher adaptability and better performance in the reduction of pollutants, i.e., with the average removal efficiency for COD, TN, and TP of 58.0%, 9.7%, and 40.4% in the winter, 46.4%, 24.7%, and 45.1% in the spring, and 66.5%, 27.2%, and 47.3% in the summer, respectively. Therefore, SBCOP is more applicable for the treatment of river water.

Yu and Tsao [43] studied the treatment efficiency of a gravel contact oxidation treatment system located in Guandu, Taiwan. This system was constructed at the riverside. The river water was inducted into an influent well by piping, and then pumped to a storage tower by submersible pumps. Finally, the river water flew into the system by gravity. They reported that the removal rates of BOD, TSS and NH_4-N with an average of 46%, 71% and 24%, respectively. HRT for better removal of SS was 15-20 h, 13-17 h for BOD, and 10-15 h for NH_4-N.

Ref. [44] evaluated the treatment efficiency of a gravel contact oxidation treatment system which was newly constructed under the riverbed of Nan-men Stream located at the Shin Chu City of Taiwan. The design flow rate of this system was 10,000 CMD (m^3/day) and the HRT ranged between 1.5 ~ 3 h. River water flew through the whole system by gravity. During wet days, if the river flow rate is higher than the design flow rate, the superfluous flow will directly pass through the treatment system to the downstream of the river. The results showed that the average removal rates of five-day biological oxygen demand, total suspended solids and NH_4+-N were 33.6%, 56.3% and 10.7%, respectively. And they reported that since the river water flew through this system by gravity, no power was consumed in the whole treatment process and the operation and maintenance cost was apparently reduced. Plus, further studies might be required to confirm whether higher HRT will improve the treatment efficiency of this gravel contact oxidation system.

Bio-ceramics were used as the carrier to treat a polluted river in Shenzhen, and the average removal rates of NO_2-N, NO_3-N, COD, turbidity, color, Mn and alga were 90.8%, 84%, 21.4%, 62%, 47%, 89% and 68% respectively. Based on the use of sewage treatment technology by rubber packing inner loop fluidized bed bio-film, the average removal rates of COD and ammonia were 88.16% and 91.8%, and the highest removal rates were 94.64% and 94.08% respectively. Wang Shu mei installed aerators, bio-film and added special bacteria in the river, and the removal rates of COD, BOD, NH_4-N, TN, TP and SS were 67.4%, 87.7%, 34.3%, 30.3%, 53.3% and 39.7%, the dissolved oxygen and transparency in the river increased from 0.9 mg/L and 12.5 cm to 7.6 mg/L and 137.5 cm respectively. Yang Tao laid the biological filter media on the river surface, and the average removal rates of COD, ammonia nitrogen and total phosphorus were 40.00%, 36.43% and 43.02% respectively [6].

Biofilm carrier can be either artificial or biological media [15]. Cao and Zhang [15] used filamentous bamboo as a biofilm carrier (Biocarrier) for bioremediation of polluted river water. Besides, evaluating the system under continuous flow conditions, they assessed the COD bioremediation efficiency when glucose was added to the river water in a hybrid batch reactor. Raw water was taken from a polluted river and poured into a wastewater tank. The flow rate was regulated using a peristaltic pump, and the column was operated in up-flow mode. In addition, air was supplied into the reactor from the bottom. The microorganisms used in the experiments were cultivated in the reactor, which was a hybrid system composed of filamentous bamboo and suspended activated sludge. The continuous flow reactor kept the same packing of filamentous bamboo used in the batch experiment, and had HRT of 3.5 h. The bioremediation of polluted surface water by using biofilm on filamentous bamboo is feasible and effective. As a result, the mean COD removal rate reached 66.1% in a batch hybrid reactor, and glucose can be used to substantially increase the COD removal. Under continuous flow conditions, the removal rates of COD, NH_4-N, turbidity, and bacteria were 11.2-74.3%, 2.2-56.1%, 20-100%, and more than 88.6%, respectively. Therefore, Polluted surface water with refractory organic pollution, low transparency, and high nitrogen pollution can be remediated by using biofilms on filamentous bamboo. The filamentous bamboo is beneficial to forming a rich microbial community. It is recommended that filamentous bamboo be widely used for the bioremediation of polluted river water instead of conventional bio-carriers and phytoremediation techniques.

Biocord is a man-made bio-reactor substrate, developed and manufactured for water management using microbe activity to passively treat water in controlled flow or storage applications. Biocord can also be used to treat wastewater in oceans, rivers, lakes, marshes and manmade reed beds [45]. Research results illustrated that the bio-cord exhibited good filtration performance and effectively removed COD, NH_3-N and TN with 26%, 65%, and 50% respectively. The flow rate of 4 L/min for 120 min, resulted in the water being completely replaced once every 10 min. The bio-cord fibers also provided suitable conditions and support media for microbial growth.

Recirculating ration is an important to improve the treatment efficiency. Liehr and Rubin [46] compared peat filter and a recirculating sand filter (RSF) for onsite treatment. Both systems were able to meet secondary effluent standards for BOD and TSS. The RSF also was moderately effective at removing nitrogen (58%) while the non-recirculating peat filter was not (26%).

In addition, hydraulic loading rate, aspect ratio, granular medium size and water depth are determining factors in the performance of the biofilm-based systems [47]. However, these techniques have drawbacks, such as complex water and air distribution systems, backwash requirements, occasional biofilm sloughing and a high nitrite residue in the effluent [48,49].

Conclusions and Recommendations

In-situ bioremediation methods can overcome the shortcomings of chemical and physical methods. Bioremediation methods also have advantages such as low cost requirements, fewer environmental influences, and no secondary pollution. In addition, understanding *in-situ* treatment systems is essential for maintaining and controlling hydraulic conditions in open streams.

After comparing the latest surface water remediation technologies that can be applied *in-situ* (i.e., aquatic plants, aquatic animals, and microorganisms), the following conclusions are made:

• Aquatic plants are an efficient *in-situ* method to treat surface waters, removing both conventional (e.g., organics) and nonconventional (e.g., radioactive materials) pollutants,

• There is an increased focus on simulated (artificial) aquatic plants in order to overcome the disadvantages of resident plants,

• Aquatic animals can mitigate pollution in water bodies and promote nutrient regeneration,

• More studies in water quality security and toxicology should be conducted when using aquatic animals as *in-situ* treatment methods,

• Microorganism-based systems are promising *in-situ* treatment methods that are low-cost and efficient.

• These methods should be tested not only on rivers and lakes, but also on other polluted surface streams such as agricultural drains,

• Bioremediation processes should be optimized, taking flow conditions and nutrient availability into account,

• It is important to integrate primary bioremediation options into water quality models to ensure effective design and management, and

• Research efforts should focus on understanding the effects of bioremediation on local ecosystems and biodiversity.

Acknowledgements

The authors are grateful to JSPS Core-to-Core Program (B. Asia-Africa Science Platforms) and SATRESP (JST-JICA) for making this concise review possible.

References

1. Nasr M, Ismail S (2015) Performance evaluation of sedimentation followed by constructed wetlands for drainage water treatment. Sustainable Environment Research 25: 141-150.

2. Yudianto D, Yuebo X (2011) Numerical modeling and practical experience of Xuxi River's natural restoration using biological treatment. Water Environment Research 83: 2087-2098.

3. Cheepi P (2012) Musi River pollution its impact on health and economic conditions of down stream villages–a study. Journal of Environmental Science, Toxicology and Food Technology 1: 40-51.

4. Yu S, Yu G, Liu Y, Li G, Feng S, et al. (2012) Urbanization impairs surface water quality: eutrophication and metal stress in the Grand Canal of China. River Research and Applications 28: 1135-1148.

5. Hamby D (1996) Site remediation techniques supporting environmental restoration activities-a review. Science of The Total Environment 191: 203-224.

6. Wang J, Liu X, Lu J (2012) Urban river pollution control and remediation. Procedia Environmental Sciences 13, 1856-1862.

7. Mostefa G, Ahmed K (2012) Treatment of water supplies by the technique of dynamic aeration. Procedia Engineering 33: 209-214.

8. Dong H, Qiang Z, Li T, Jin H, Chen W (2012) Effect of artificial aeration on the performance of vertical-flow constructed wetland treating heavily polluted river water. Journal of Environmental Sciences 24: 596-601.

9. Fan J, Zhang B, Zhang J, Ngo HH, Guo W, et al. (2013) Intermittent aeration strategy to enhance organics and nitrogen removal in subsurface flow constructed wetlands. Bioresour Technol 141: 117-122.

10. Mackie JA, Natali SM, Levinton JS, Sañudo-Wilhelmy SA (2007) Declining metal levels at Foundry Cove (Hudson River, New York): response to localized dredging of contaminated sediments. Environ Pollut 149: 141-148.

11. Zhu Y, Zhang H, Chen L, Zhao J (2008) Influence of the South–North Water Diversion Project and the mitigation projects on the water quality of Han River. Science of The Total Environment 406: 57-68.

12. Della Rocca C, Belgiorno V, Meriç S (2007) Overview of in-situ applicable nitrate removal processes. Desalination 204: 46-62.

13. Nasr M, Elreedy A, Abdel-Kader A, Elbarki W, Moustafa M (2014) Environmental consideration of dairy wastewater treatment using hybrid sequencing batch reactor. Sustainable Environment Research 24: 449-456.

14. Mingjun S, Yanqiu W, Xue S (2009) Study on bioremediation of eutrophic lake. Journal of Environmental Sciences 21: S8-S16.

15. Cao W, Zhang H, Wang Y, Pan J (2012) Bioremediation of polluted surface water by using biofilms on filamentous bamboo. Ecological engineering 42: 146-149.

16. Gagnon V, Chazarenc F, Kõiv M, Brisson J (2012) Effect of plant species on water quality at the outlet of a sludge treatment wetland. Water Res 46: 5305-5315.

17. Fawzy M, Nasr M, Abdel-Gaber A, Fadly S (2015) Biosorption of Cr (VI) from aqueous solution using agricultural wastes, with artificial intelligence approach. Separation Science and Technology 51: 416-426.

18. Zimmels Y, Kirzhner F, Malkovskaja A (2008) Application and features of cascade aquatic plants system for sewage treatment. Ecological engineering 34: 147-161.

19. Allam A, Tawfik A, El-Saadi A, Negm A (2016) Potentials of using duckweed (Lemna gibba) for treatment of drainage water for reuse in irrigation purposes. Desalination and Water Treatment 57: 459-467

20. Ruan X, Xue Y, Wu J, Ni L, Sun M, et al. (2006) Treatment of polluted river water using pilot-scale constructed wetlands. Bulletin of environmental contamination and toxicology 76: 90-97.

21. Kalin M, Wheeler W, Meinrath G (2004) The removal of uranium from mining waste water using algal/microbial biomass. Journal of environmental radioactivity 78: 151-177.

22. Hadad HR, Maine MA, Bonetto CA (2006) Macrophyte growth in a pilot-scale constructed wetland for industrial wastewater treatment. Chemosphere 63: 1744-1753.

23. Steenhoudt O, Vanderleyden J (2000) Azospirillum, a free-living nitrogen-fixing bacterium closely associated with grasses: genetic, biochemical and ecological aspects. FEMS Microbiol Rev 24: 487-506.

24. Sawaittayothin V, Polprasert C (2007) Nitrogen mass balance and microbial analysis of constructed wetlands treating municipal landfill leachate. Bioresource technology 98: 565-570.

25. Li XN, Song HL, Li W, Lu XW, Nishimura O (2010) An integrated ecological floating-bed employing plant, freshwater clam and biofilm carrier for purification of eutrophic water. Ecological engineering 36: 382-390.

26. El-Shafai SA, El-Gohary FA, Nasr FA, van der Steen NP, Gijzen HJ (2007) Nutrient recovery from domestic wastewater using a UASB-duckweed ponds system. Bioresour Technol 98: 798-807.

27. Xiao L, Ouyang H, Li H, Chen M, Lin Q, et al. (2010) Enclosure study on phytoplankton response to stocking of silver carp (Hypophthalmichthys molitrix) in a eutrophic tropical reservoir in South China. International Review of Hydrobiology 95: 428-439.

28. Sun L, Liu Y, Jin H (2009) Nitrogen removal from polluted river by enhanced floating bed grown canna. Ecological engineering 35: 135-140.

29. Zhao Y, Yang Z, Xia X, Wang F (2012) A shallow lake remediation regime with Phragmites australis: Incorporating nutrient removal and water evapotranspiration. Water Res 46: 5635-5644.

30. Saeed T, Sun G (2012) A review on nitrogen and organics removal mechanisms in subsurface flow constructed wetlands: Dependency on environmental parameters, operating conditions and supporting media. Journal of environmental management 112: 429-448.

31. Ma H, Cui F, Liu Z, Fan Z, He W, et al. (2010) Effect of filter-feeding fish silver carp on phytoplankton species and size distribution in surface water: a field study in water works. J Environ Sci (China) 22: 161-167.

32. Ma H, Cui F, Liu Z, Zhao Z (2012) Pre-treating algae-laden raw water by silver carp during Microcystis-dominated and non-Microcystis-dominated periods. Water Science & Technology 65: 1448-1453.

33. Zhao SY, Sun YP, Lin QQ, Han BP (2013) Effects of silver carp (Hypophthalmichthys molitrix) and nutrients on the plankton community of a deep, tropical reservoir: an enclosure experiment. Freshwater Biology 58: 100-113.

34. Faulwetter JL, Gagnon V, Sundberg C, Chazarenc F, Burr MD, et al. (2009) Microbial processes influencing performance of treatment wetlands: a review. Ecological engineering 35: 987-1004.

35. Ateia M, Nasr M, Yoshimura C, Fujii M (2015) Organic Matter Removal from Saline Agricultural Drainage Wastewater Using a Moving Bed Biofilm Reactor. Water Science & Technology 72: 1327-1333.

36. Sheng Y, Chen F, Sheng G, Fu J (2012) Water quality remediation in a heavily polluted tidal river in Guangzhou, South China. Aquatic Ecosystem Health & Management 15: 219-226.

37. Takada H, Mutoh K, Tomita N, Miyadzu T, Ogura N (1994) Rapid removal of linear alkylbenzenesulfonates (LAS) by attached biofilm in an urban shallow stream. Water research 28: 1953-1960.

38. Chen JM, Hao OJ, Al-Ghusain IA, Lin CF (1995) Biological fixed-film systems. Water Environment Research 450-459.

39. Di Trapani D, Mannina G, Torregrossa M, Viviani G (2008) Hybrid moving bed biofilm reactors: a pilot plant experiment. Water Sci Technol 57: 1539-1545.

40. Moussavi G, Mahmoudi M, Barikbin B (2009) Biological removal of phenol from strong wastewaters using a novel MSBR. Water Res 43: 1295-1302.

41. Chiemchaisri C, Panchawaranon C, Rutchatanunti S, Kludpiban A, Ngo H, et al. (2003) Development of floating plastic media filtration system for water treatment and wastewater reuse. Journal of Environmental Science and Health 38: 2359-2368.

42. Chiemchaisri C, Passananon S, Ngo H, Vigneswaran S (2008) Enhanced natural organic matter removal in floating media filter coupled with microfiltration membrane for river water treatment. Desalination 234: 335-343.

43. Yu S, Tsao C, Lin C, Chen C (2006) Relationship between engineering design parameters and water quality for constructed cobble bed in Guandu. J Environ Protec 29: 73-90.

44. Juang D, Tsai W, Liu W, Lin J (2008) Treatment of polluted river water by a gravel contact oxidation system constructed under riverbed. International Journal of Environmental Science & Technology 5: 305-314.

45. Yuan X, Qian X, Zhang R, Ye R, Hu W (2012) Performance and microbial community analysis of a novel bio-cord carrier during treatment of a polluted river. Bioresource technology 117: 33-39.

46. Liehr SK, Rubin AR, Tonning B (2004) Natural treatment and onsite processes. JSTOR 76: 1191-1237.

47. Garci´a J, Aguirre P, Mujeriego R, Huang Y, Ortiz L, et al. (2004) Initial contaminant removal performance factors in horizontal flow reed beds used for treating urban wastewater. Water research 38: 1669-1678.

48. Li XY, Chu HP (2003) Membrane bioreactor for the drinking water treatment of polluted surface water supplies. Water Res 37: 4781-4791.

49. Li L, Xie S, Zhang H, Wen D (2009) Field experiment on biological contact oxidation process to treat polluted river water in the Dianchi Lake watershed. Frontiers of Environmental Science & Engineering in China 3: 38-47.

20

Rhizoremediation of Contaminated Soils by Comparing Six Roots Species in Al-Wafra, State of Kuwait

Danah Khazaal Al-Ameeri and Mohammad Al Sarawi*

Earth and Environmental Science, College of Science, Kuwait University, Kuwait

Abstract

Toxic heavy metals concentrations in soils are locally quite high in Kuwait due to the gulf war in 1991. There are many reasons that make heavy metals high in Kuwait soils such as massive oil spills and direct dumping of untreated sewage and hydrocarbons. Phytoremediation may offer a possible solution to this problem because it might treat the soils through the use of plants even without the demand of removing the contaminated material and disposing it elsewhere. This research project will discuss the best contribution among six types of plant (tree) species (*Conocarpus, Tamarix, Phoenix, Rhamnus, Vitex Agnus-Castus and Salix*) collected from Wafra 14 km south Kuwait and another six reference trees from the same plant species collected for correlation from Sabah Al -Salem residential area. The samples are divided into two sample categories (roots and soil). From the chemical analysis it was found that rhizoremediation is a unique process in decreasing the level of soil contamination of hydrocarbons and trace metals. The most effective part of the plant in accumulating hydrocarbons and heavy metals were the roots while the appropriate plant was *Tamarix*. As a recommendation there is a need for studying the fruits of Phoenix to detect accumulates of TPH and Trace Metals and to decrease the harmful effects of the fruits on the human.

Keywords: Bioremediation; Pollutant; Rhizoremediation; Hydrocarbons; Trace metals

Introduction

The study area is located in south of Kuwait along Wafra/Mina Abdullah road at 14 km (Figure 1). The area is about 1 km² and it receives over than 4500 m³ daily of industrial hazardous liquid waste and untreated sewage In one of the largest unlined evaporation ponds in the State of Kuwait. The study area now is completely rehabilitated and planted by several types of plants in four main sectors comprises 45000 trees. This research project will focus on the most efficient and beneficial plant out of six plants that were planted in Wafra/Mina Abdullah 14 km, to treat the contaminated soils in many areas of Kuwait. Thus, reducing and/or eliminating contaminants and restoring or partly restore the ecological service function of soil. The State of Kuwait is in Western Asia located at the north eastern corner of the Arabian Gulf on the Arabian Plate. It has a strategic location at the head of the Arabian Gulf and shares borders with Iraq and Saudi Arabia. Kuwait's area is about 17,818 km² and it lies between latitudes 28°30' and 30°05' N, and longitude 47°30' and 48°36' E.

The main objectives of this study:

- To understand the behavior of the six tree plants roots in absorbing pollutants.

- To study the rate of fate and flux of pollutants.

- To evaluate the roots and soils of the different six plants in cleaning the contaminated soil.

- To make a recommendation for the best out of six tree plants selected in the study area in absorbing hydrocarbons and trace metals from contaminated soil.

This will be achieved through:

- Determining the level of total petroleum hydrocarbon (TPH) in plant's (roots and soil).

- Measuring Trace Metals accumulation (TM) in plant's (roots and soil).

- Chemical comparisons between the plants and soil samples.

This research will focus on studying the plant roots specially and identifying the most efficient tree out of the six trees that were planted in the study area in absorbing contaminants (Figure 2).

Literature Review

Refractory organic harmful and toxic chemicals have accumulated for decades as a result of industrial activities and caused soil contamination. Soil contamination is the essential source that transmits toxins like heavy metals from environment to living beings. Plants adsorb heavy metals from soil. Heavy metals enter the animal's kingdom including people through the food chain, and cause a lot of health risks [1].

Many published research have been proved that plants play an important role in removing heavy metals from contaminated soils. Plants and Microorganisms are possessed by many mechanisms to avoid heavy metal poisoning. Metal efflux particularly which is represented in eubacteria is a famous example in addition to synthesis of metal-binding peptides like metallothionein (MTs). Also, it is representing in plants, fungi, blue-green algae and phytochelatins [2,3]. Additionally, Plant roots cause changes at the soil-root interface as they discharge inorganic and organic mixes (root exudates) in the rhizosphere. The number and activity of the microorganisms, the accumulation and stability of the soil particles around the root, and the availability of the contaminants are affected by the root exudates. The availability of the contaminants in the root zone (rhizosphere) of the plant can be increased or decreased directly or indirectly by root

***Corresponding author:** Mohammad Al Sarawi, Earth and Environmental Science, College of Science, Kuwait University, Kuwait
E-mail: sarawi500@gmail.com

Figure 1: Location map of Kuwait and its physiography (Source: mapofworld.com).

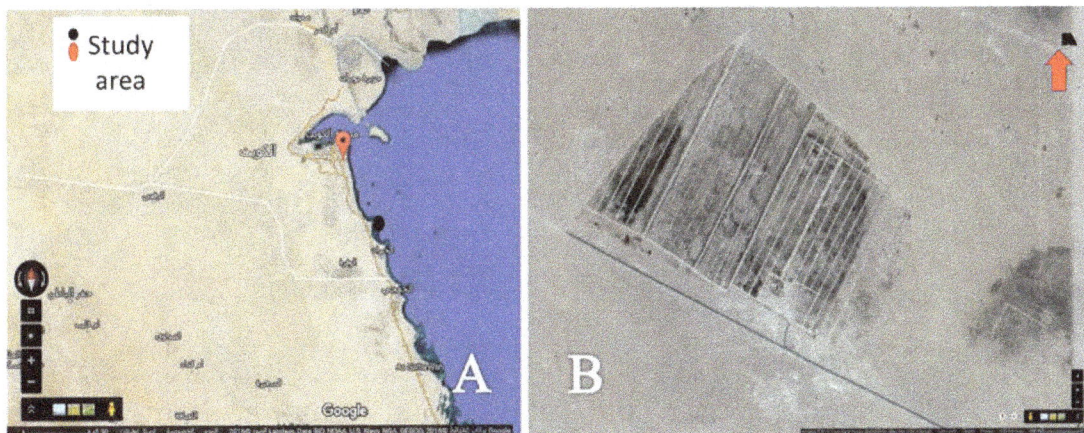

Figure 2: (A) Map showing the study area location across Wafra/Mina Abdullah road area (Source: Google maps, 2016). (B) Map showing the study area location across Wafra/Mina Abdullah road area in a close view (Source: Google maps, 2015).

exudates themselves through changes in soil characteristics, release of organic substances, changes in chemical composition, and/or increase in plant-assisted microbial activity (International Environmental Technology Centre (IETC).

Building up vegetation spread on generally contaminated perilous waste destinations may be a proficient way and low support strategy to waste remediation. Connection between plant roots and rhizosphere microflora altogether upgrades corruption of unsafe organic compounds in polluted soil [4]. Also, metals are cycling at low rates inside and between bio, geo, atmospheric and hydrosphere frameworks. However, still metals that are collected in soils are dependable to higher dangers since they are drained into ground and surface water. After that, uptake process assimilates and a store metal in plants and after that direct or indirect intake by human population occurs. According to Karenlampi et al. [5] metals are dangerous in the event that they are available at expanded levels of bioavailability for example essential (Mn, Zn, Cu, Mo, Fe, Ni) and non-essential metals (Pb, Cd, Cr, Hg). Metal hyper aggregating discovery properties in certain plants expressed that it might be conceivable to utilise these plants for the cleanup of heavy metal pollution in soil and water. This way to deal with environmental restoration has been termed phytoremediation [6,7].

The use of plants in bioremediation of soils has been proposed as an attractive technique and because of expanded metal fixations on most living creatures, numerous strategies were produced to remediate or phytoremediate contaminated soils. However, plants do not have the uncommon biodegradative abilities that microorganisms have, so another innovation, called rhizoremediation has emerged. The available methods are costly, environmentally invasive, and labor intensive. However, remediation technique is low. The utilization of green plants to expel, i.e., phytoremediation contains or diminishes innocuous contaminants from environment.

Plants cover can be an effective technique to prevent contaminants spread by reducing erosion process done by wind and surface run off. Furthermore, it is reducing percolation to the ground water. Plants may also be used to remove contaminants from soil by phytoextraction and then harvested for processing [8-11]. There are some promising results suggesting that these techniques might become practical alternatives to mechanical and chemical approaches in remediation of metal contaminated soils [5]. Recently, rhizoremediation objective is to upgrade and develop innovative, economical, and environmentally compatible approaches to remove heavy metals from the environment. Thus, phytoremediation targets currently include contaminating metals, metalloids, petroleum hydrocarbons, pesticides, explosives, chlorinated solvents, and industrial byproducts as stated by Cunningham et al. [8]. The biological technologies that can be employed in the remediation of contaminated soils are Bioremediation, phytoremediation and rhizoremediation. This study will focus on rhizoremediation.

Rhizoremediation

Rhizoremediation is a process where microorganisms degrade soil contaminants in the rhizosphere (Figure 3). It is use plant roots and associated microbial consortium to degrade environmental pollutants from soil with an aim of restoring area sites to a condition useable for intended purpose. Rhizoremediation takes advantage of plant roots

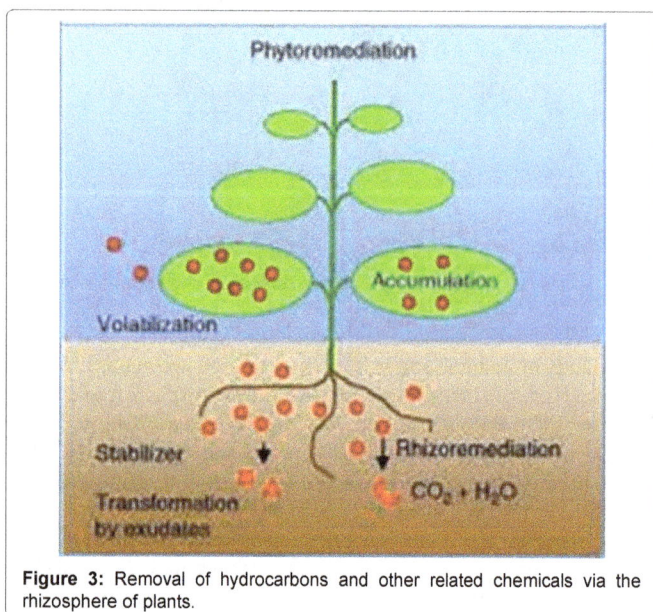

Figure 3: Removal of hydrocarbons and other related chemicals via the rhizosphere of plants.

natural symbiosis with a fungus which grows in association with the roots of a plant called mycorrhiza and root associated natural microbial flora for the enhanced degradation of pollutants in the rhizosphere [12]. This method can remediate the soil pollutants which are generally organic compounds that cannot enter the plant because of their high hydrophobicity. Generally, plants are not considered as the main mode of remediation in this procedure. Rather, the plant creates a niche for rhizosphere microorganisms to do the degradation. Rhizosphere microorganisms are served by the plant acting as a solar-powered pump that draws in water and the pollutant while producing substrates that benefit microbial survival and development. Root exudates and root turn over can serve as substrates for microorganisms that perform toxin degradation.

Determination of organisms has been attempted with great achievement to be useful in rhizoremediation. Utilizing pollutant soil as an initial media to select from, enrichment of bacteria that can survive on the contaminant. This enrichment can then be inoculated onto plants where selection for root colonization can be done. Utilizing this process will result in a host plant that backings a toxin degrader in its rhizosphere. Wild type organisms are selected this procedure and accordingly there is no constraint for usage as with genetically modified microorganisms [13] depicts the successful use of this technique to identify the rhizosphere PAHs degraders (Table 1). Rhizoremediation is usually less costly than competing alternatives such as soil excavation, pump-and-treat, soil washing, or enhanced extraction.

Role of plants in rhizoremediation: Certain plants are better at removing contaminants than others. There are many factors must be able in the plants used for phytoremediation. They must be able to tolerate the concentrations of contaminants present. Also, they must be able to grow and survive in the local climate. Another factor is

Plant	Pollutant	Microbes
Rice (cv.Supriya)	Parathion	Not identified
Mixture of grass, legume, herb and pine	TCE	Not identified
Prairie grasses	PAHs	Not identified
Prairie grasses	PAHs	Not identified
Grasses and alfalfa	Pyrene, anthracene, phenanthrene	Not identified
Sugar beet (cv.Rex)	PCBs	Pseudomonas fluorescens
Unidentified wild plant (compositae) and Senecus glaucus	oil	Arthobacter/Penicillium
Barley (Hordeum vulgare)	2,4-D	Bukholderia cepacia
Alfalfa and alpine bluegrass	Hexadecane and PAHs	Not identified
Wheat (Triticum aestivum)	2,4-D	Pseudomonas putida strains
Poplar (Populus deltoides nigra)	1,4-dioxane	Actinomycetes
Wheat	TCE	P. fluorescens
Oat, lupin, rape, dill, pepper, radish, pine	Pyrene	Not identified
Reed (Phragmitis australies)	Fixed nitrogen	Nitrospira sp. And Nitrosomas sp.
Poplar root extract	1,4-dioxane	Actinomycete Amycolata sp. CB 1190
Corn (Zea mays)	3-methylbenzoate	P. putida
Astragalus sinicus	Cd+	Mesorhizobium huakuii
Fern (Azolla pinnata)	Diesel fuel	Not identified

Table 1: Rhizoremediation of various environmental pollutants [15].

important such as the depth of contamination. If the contamination is shallow, small plants like ferns and grasses will be used. Trees such as poplars and willows are used to clean up deeper soil contamination and contaminated groundwater because tree roots grow deeper [14].

Plants may contribute to remediation in several ways, by reducing the leaching of contaminants, aerating soil, phytodegradation, phytovolatilization, evapotranspiration, and rhizoremediation. Nutrients in the rhizosphere are given by the mucigel emitted by root cells, lost root cap cells, the starvation of root cells, or the decay of complete roots [15,16]. In addition, plants discharge a variety of photosynthesis derived organic compounds [10,17]. Root exudates contain water soluble, insoluble, and volatile compounds including sugar, proteins, alcohols, amino acids, organic acids, nucleotides, flavonones, phenolic compounds and certain enzymes. Prevailing conditions that support the application of both microbes and plants depend on the selection of bioremediation or phytoremediation for cleanup of the contaminated site.

Factors affecting rhizoremediation: The different physical, chemical and biological properties of the root related to soil are soil condition, temperature, pH, organic matter, weather condition. Schroll et al. [18], Alberty [19], Guillot et al. [20], Mastronicolis et al. [21], Singh et al. [22] are responsible for changes in microbial diversity and for expanded numbers and metabolic activities of microorganisms in the rhizosphere microenvironment, this phenomenon called the rhizosphere effect [10,15,17,23]. Many studies have investigated the effects of soil moisture, temperature, aeration, pH, and organic matter content on the biodegradation of pesticides [11,24]. The age of a plant, the availability of mineral nutrients and the presence of contaminants are responsible of the rate of exudation changes [25]. The nature and the quantity of root exudates, and the timing of exudation are crucial for a rhizoremediation process. Plants might respond to chemical stress in the soil by changing the composition of root exudates controlling, in turn, the metabolic activities of rhizosphere microorganisms [25]. The root exudates mediate acquisition of minerals by plants and stimulate microbial growth and activities in the rhizosphere in addition to changing some physicochemical conditions. Some organic compounds in root exudates may serve as carbon and nitrogen sources for the growth and long-term survival of microorganisms that are capable of degrading organic pollutants [10,17,26].

Improvement in rhizoremediation: Rhizoremediation procedure can be intended to enhance in several perspectives like bioavailability of contaminant molecules, expression and support of genetically engineered plant microbial frameworks and root exudates for the effectiveness of the procedure. An interesting alternative to improve the removal efficiency is selection of bacteria. Bacteria are produce biosurfactants in the plants rhizosphere. Kuiper et al. [15] applied bioremediation and distinguished bacteria developing in contaminated area that produces biosurfactants that encourage the solubilisation of PAHs and thus biodegradation by microbes. This property is additionally

of interest because various biodegradative microorganisms show positive chemotaxis towards the contaminations. Thus, the consolidated activity of biosurfactant and chemotaxis can add to bacterial expansion and to microbial spread in contaminated soils, all together that more abundant zones can be cleaned. Microbial degradation of pollutants in the rhizosphere gives a constructive outcome to the plant; the toxin concentration is diminished in the zone close to the roots and the plant can develop better to those in contaminated zones. Because of this shared advantage it has been recommended that plants can choose particular genotypes to be available in their roots. Many experiments performed by Siciliano et al. [27] exhibited that the presence of the alkane monooxygenase quality was more pervasive in endophytic and rhizosphere microbial groups than in those present in bulk soil contaminated with hydrocarbons. This recommends if plants are impacting the rhizosphere, this impact is subject to the contaminant. Additionally, a few researchers concluded that the impact depended on the sort of the plant. This has led to the hypothesis that the viability of rhizoremediation methodologies is identified with the determination of the best plant bacterium pair in each case.

An environmental evaluation for the industrial and sewage wastes in Kuwait [28]

This study examined Al- Wafra 14 km since it is the main area that gets the industrial waste and the fluid from all other areas. The main point of this study was to evaluate and contribute groundwater, soil and air which are the probable contaminants that came from human activities to enhance the environment and people's health and helping the authorized people to plan for the future of country and environment actions by taking the fitting activities for ensuring these resources.

TPH analysis: TPH results showed high percentages in Wafra 14 km (Table 2). The acceptable TPH amount on earth should be 10 mg/l according to the International Bank while in Wafra the mean values of TPH showed 97.46 mg/l as shown in Table 3. Besides, natural Kuwaiti laws express the need to treat the consumed oils that contains high measure of hydrocarbons substances because of its hurtful impact on environment. Subsequently, they accepted a rate of 5 mg/l TPH.

Trace metals analysis (TM): Heavy metals such as Cd, Cr, Cu, Ni, Pb, and Zn in soils cause noteworthy dangers to human, animals as well as to environment. Human activities have expanded as an aftereffect of the Industrial Revolution and were the main causes of heavy metal deposition. Modern exercises, smelting, electroplating and Agriculture have brought about the deposition of undesirable concentrations of metals in soil. Addition of Trace metals might be hurtful to Humans, animals, plants and different life forms such as organisms reaching the soil or groundwater. In Table 3, the results of trace metals concentrations in Wafra 14 km showed that all of the values are above the acceptable local limits.

Phytoremediation for hazardous liquid industrial waste Al-

ID	V	Cr	Fe	Ni	Cu	Zn	Cd	Pb	TPH
Unit	(mg/L)	(mg/L)	(mg/L)	(mg/L)	(mg/L)	(mg/L)	(mg/L)	(mg/L)	(mg/L)
Mean	1.16	0.31	71.29	2.83	0.39	7.69	0.5	0.4	97.4
Minimum	0.07	0.15	19.9	0.49	0.11	3.98	0.0069	0.18	1.69
Maximum	6.55	0.81	282.9	9.95	1.10	17.72	3.1	1.07	523
Std.Dev.	1.76	0.19	78.2	2.7	0.26	4.41	1.04	0.26	164.5
1-KEPA standards	-	0.15	5	0.2	0.2	2.0	0.01	0.5	5
2-KEPA Standards	0.1	0.15	5	0.2	0.2	2.0	0.01	0.5	10

Table 2: Statistics analysis of trace metals (TM) and total petroleum hydrocarbons (TPH) samples conducted in Wafra 14 km [28].

Wafra [29]: In 2012 unpublished thesis study done by Alia Bado had done at the same study area Al Wafra. The chemical analyses were done systematically at the laboratory following the Vista-MPX CCD Simultaneous ICP-OES for five plant samples collected (*Salix Fragilis, Albizzia, OleaEuropaea, Ziziphus and Tamarix*). Table 4 showed the trace metals which were analyzed for the five plant species. The study concluded that the high concentration of trace metals absorbed by the Tamarix plant and is the most satisfactory plant that could be utilized in Kuwait for phytoremediation process [29].

Phytoremediation for hazardous liquid industrial waste using six famous trees species in Al Wafra [30]: Another unpublished thesis study done by Mohammed Al-Ibrahim in 2015 conducted in the same contaminated study area Wafra 14 km. He collected six types of plant tree species (*Conocarpus, Tamarix, Phoenix, Rhamnus, Vitex Agnus-Castus and Salix*) and studied different plant part (leaves and stem) from the contaminated area. The study showed that the best part absorbed and stored the contaminants and the Trace Metals is the leaves. *Vitex Agnus-Castus* species leaves got the highest storage of TPH amount. Stems samples TPH analysis revealed that the highest TPH storage is of *Rhamnus* and the least one is *Tamarix*. However, TPH analysis of soil samples collected from the study area showed that the least plant could absorb the contaminants from soil got the highest value of TPH, which is *Conocarpus* species.

The study concluded that Vitex Agnus-Castus species is the most adequate plant that could be used for phytoremediation process in Kuwait according to the high concentration of the TPH and trace metals absorbed by the plant leaves [30].

Regional case study of phytoremediation

Does phytoremediation work at every site? [31]: A study has been carried out examining the growth characteristics and performance of mangroves, halophytes and other plants in soil irrigated with saline water in the United Arab Emirates. Salinity is a serious threat for crop production in arid regions, where the high rate of evaporation, combined with the demand for scarce water reserves, causes salt build-up during the cultivation process. The study showed that some plants have the necessary physiological mechanisms and capacities to accumulate significant concentrations of iron, manganese and magnesium, calcium, sodium and chloride ions, thereby reducing the overall salinity of the soil system and potential related effects on crop physiological and growth performance. Based on this work, it has been suggested that by introducing highly salt-tolerant species (i.e., *Conocarpus erectus, Atriplex lentiformis*, etc.) that can be irrigated with saline water, higher plant and agricultural production levels in arid regions can be achieved.

International case study of phytoremediation

Petroleum hydrocarbons pollution in soil and its bioaccumulation in mangrove species, *Avicennia marina* from Alibaug mangrove ecosystem, Maharashtra, India [32]: A study done by Lotfinasabasl [32] had investigated the capability of mangrove species, *Avicennia marina* for bioaccumulation of petroleum hydrocarbons from contaminated soil. The results showed that the TPH concentration was vary in leaves, roots and seedlings at the same plant and it was showing higher uptake of petroleum hydrocarbons by roots took after by seedlings and leaves.

The study concluded that *Avicennia marina* is observed to be a potential species for protection of coastal ecosystem. Phytoremediation has been recognized as a practical, naturally benevolent, stylishly satisfying procedure for evacuation of ecological toxins [32].

Materials and Methods

From the study area Wafra 14 km as shown below (Figure 4), six types of plant tree species were collected (*Conocarpus, Tamarix, Phoenix, Salix, Rhamnus, and Vitex Agnus-Castus*). Another six reference trees were collected for correlation from Sabah Al -Salem residential area from the same plant species. The samples were divided

ID	V	Cr	Fe	Ni	Cu	Zn	Cd	Pb	TPH
Unit	(mg/L)	(mg/L)	(mg/L)	(mg/L)	(mg/L)	(mg/L)	(mg/L)	(mg/L)	(mg/L)
IA	0.5707	0.2476	37.9017	0.8922	0.4025	5.5232	0.0186	0.6762	54.3
1B	0.9354	0.4209	282.9234	6.2901	0.3624	15.7374	2.2618	0.6585	17.9
1C	6.5557	0.8118	176.6287	2.5396	1.1045	17.7218	0.0231	1.0749	9.73
2A	0.9241	0.5019	27.0217	0.4911	0.6518	4.1066	0.0156	0.2224	10.1
2B	0.0748	0.2803	45.081	2.7268	0.3044	7.7437	0.0747	0.3535	18.5
2C	0.5753	0.3197	34.2768	9.9575	0.1704	4.9234	0.0069	0.2792	14
3A	0.1764	0.1555	58.0832	1.8628	0.3318	6.374	3.1094	0.2613	206
3B	0.1015	0.1598	40.3729	1.8628	0.3318	7.9874	0.0242	0.3132	523
3C	0.7205	0.2089	58.9177	3.2001	0.1138	5.3476	0.4703	0.208	303
4A	1.0831	0.1916	29.3609	1.9591	0.338	6.4256	0.0114	0.2648	1.69
4B	0.4262	0.1587	19.9392	0.9415	0.3265	3.9842	0.0207	0.182	2.4
4C	1.8826	0.2657	45.05	1.8394	0.4364	6.5064	0.0079	0.3202	9

Table 3: Trace metals analysis results of Wafra 14 km [28].

Sample ID	Cr	Mn	Cd	Cu	Fe	Ni	V	Zn	Pb
Tamarix	6.607	52.77	0.204	41.201	181.146	4.432	BDL	94.05	BDL
Albizzia	3.945	49.486	BDL	24.234	201.538	0.957	BDL	28.027	BDL
Salix Fragilis	2.396	40.095	BDL	15.187	78.617	2.152	BDL	26.623	BDL
Ziziphus	1.682	67.103	BDL	9.192	69.121	0.213	BDL	33.776	BDL

All values are given in ppm (mg/g)

BDL: Below detection limit. In the method used for the analysis the minimum detection level of Cd is 0.205 ppm. V is 0.800 and Pb is 1.400 ppm.

Table 4: Trace metals analysis of plant samples from Al-Wafra 14 km [29].

into two sample categories (plant roots and soil) of each plant in the study. Therefore, the total numbers of samples collected are 24 samples including mixed treated soil (Figures 5 and 6). Total Petroleum Hydrocarbons (TPH) and Trace Metals analysis will be performed for duration of four months, starting from (November 2015 – March 2016). The samples collected and analyzed for determining the concentration of total petroleum hydrocarbon (TPH) and trace metals (TM).

Getting the plant roots samples from the study area was difficult so a bulldozer was used to extract the roots (Figures 5A and 5B) then the

Figure 4: Landscape view of plants in the contaminated study area Wafra 14 km.

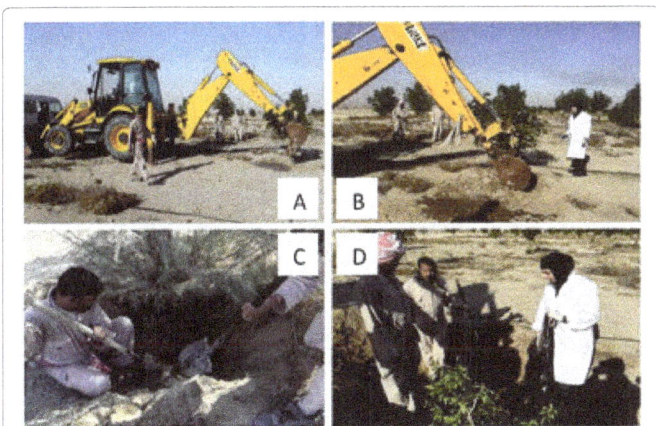

Figure 5: Study area Wafra 14 km A) Getting a root sample using the bulldozer. B) The bulldozer while taking roots samples of *Conocarpus* species. C) Labours digging using the shovel to get the roots. D) Collecting the plant roots.

Figure 6: Getting different plants roots from the study area A. *Tamarix* B. *Phoenix*.

plant was replanted again at the same time. Several labors were helping to continue the root extraction by using shovels (Figure 5C). Finally, the root samples were collected very carefully in a plastic bag (Figure 5D).

Field work

The fieldwork conducted on the contaminated soil in the Wafra/ Mina Abdulla road 14 km. The study area includes 12 samples which were divided as 6 samples of plant roots and 6 soil samples. In addition, 12 reference samples were collected from Sabah Al-Salem residential area, another six plants roots samples from the same species and 6 soil samples.

Results and Discussion

The utilization of plants for remediation might be particularly appropriate for soils polluted by organic chemicals to profundities of less than 2 m. Accordingly, the collaboration between plant roots and rhizosphere microflora fundamentally improves degradation of dangerous natural mixes in polluted soil. The root arrangement of plants can spread microbes through soil and infiltrate generally impermeable soil layers. The immunization of contamination degrading microscopic organisms on plant seed can be an essential added substance to enhance the proficiency of phytoremediation [15]. Through degradation or gathering plants can cooperate with dangerous natural mixes. Uptake of poisons in plants roots is an immediate capacity of the contaminant amount and concentration in the solution of soil and large contains compound isolating connected to the root surfaces took after by development over the cortex to vascular arrangement of plants [33]. The contamination may be bound or processed at whatever time among transference. Toxins in plants might be found as uninhibitedly extractable stores, extractable conjugate bound to plant material, and unextractable developments combined in plant tissue.

Plants may by implication add to the dispersal of contaminants in vegetated soil. The usage of plants to concentrate toxic mixes from soils (phytoextraction) is being made as a framework for remediation of metal sullied soils [11,34,35]. Phytoremediation is a suitable remediation technique for petroleum contaminated soil based on evidence from greenhouse and field studies. The utilization of vegetation for remediation of polluted sites is attractive because it is cheap and passive. Plants can gather trace metals, in or on their issues as a result of their capacity to adjust to variable chemical properties of the environment: in this way, plants supplies through which trace elements from soils, waters and air, move to man and animals [36]. According to Srinivas et al. [37] air and soil are the main sources of trace metals to plants from which the root foliage take up the trace metals. Generally, metals are concentrated in the root and absorbed by plant through root uptake.

Total petroleum hydrocarbons analysis

TPH is well -defined as the quantifiable amount of petroleum -based hydrocarbon in an ecological media. It is, thus, reliant on analysis of the medium in which it is found. This section explains the TPH analysis results that were conducted in the study area on different six plants species compared to similar ones from different areas as references samples. In this section we would explain the total petroleum hydrocarbons analysis of the six plants samples collected from the study area in Wafra 14 km (*Conocarpus, Vitex Agnus-Castus, Salix, Rhamnus, Tamarix and Phoenix*).

Conocarpus: *Conocarpus* or Button mangrove is a widespread species of terrestrial mangrove. On the coastal mainland, it grows as a small tree and has a variety of sizes on islands. *Conocarpus* has downy smooth leaves. It tolerates many things such as lowly drainage, compressed soil, air contamination and drought so that it is exceptionally adjustable as a decorative [38,39]. From total petroleum hydrocarbons analysis (TPH) of *Conocarpus* samples collected from the contaminated study area the soil had high rate of TPH value than the roots. On the other hand, the reference from same species collected from Sabah Al-Salem residential area depicted that the roots possessed the highest TPH value compared to soil which is about 9036 mg/kg as shown in Figure 7. The reference sample showed higher amount of TPH and that could be due to the age of plant and/or its nearness of a sewage area.

Vitex Agnus-Castus: *Vitex Agnus-Castus* is one of the common flowering plants and has many uses such as medical and herbal uses. It varies in size from small to tall tree. *Vitex Agnus-Castus* leaves are shaped liked a hand or palmate, and were once accepted to have sedative impacts. If *Vitex Agnus-Castus* soil kept too moist it would decline due to the root rot [38]. The results of Total Petroleum Hydrocarbons (TPH) of study and reference samples collected from Sabah Al-Salem residential area showed that *Vitex Agnus-Castus* roots stored the highest amount of TPH 4565-2091 mg/kg compared to soil 600-570 mg/kg (Figure 8).

Salix: Around 400 species of deciduous trees and shrubs form the genus *Salix* were found in many regions [40]. *Salix* leaves are typically elongated and sometime may be round to oval. The roots grow from aerial parts of the plant and they are remarkable for their toughness and size. *Salix* as a plant have many environmental uses. It is used for biofiltration, wastewater treatment systems, phytoremediation, windbreak and soil building. From the roots TPH results, the value ranges from 1820-1477 mg/kg whereas soil TPH ranges from 960-810 mg/kg (Figure 9). As mentioned before, the reference sample showed higher amount of TPH and that might be due to the age of plant and/or its nearness of a sewage area.

Rhamnus: *Rhamnus* species are varying in size from shrubs to medium sized trees. The *Rhamnus* leaves are undivided and innately veined. The collected sample of *Rhamnus* roots from study area and the reference area are highly stored the contaminants 2770-18822 mg/kg and the low amount of contaminants stored in soil 250-270 mg/kg (Figure 10).

Tamarix: *Tamarix* is one of flowering plants that composed of many species. It grows as tree in the arid area. *Tamarix* can tolerate a lot of environmental conditions. It is capable to exploit

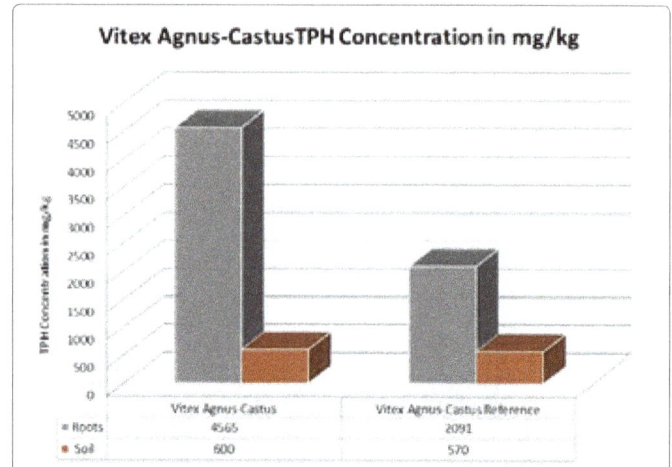

Figure 8: *Vitex Agnus- Castus* TPH concentrations in mg/kg results of the study area sample compared with the reference of same species collected from Sabah Al-Salem area.

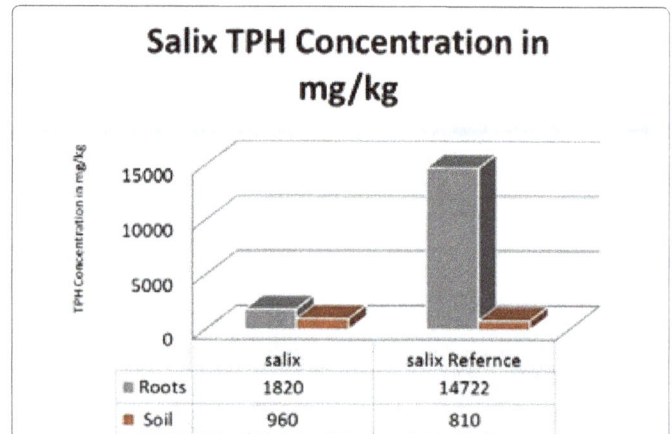

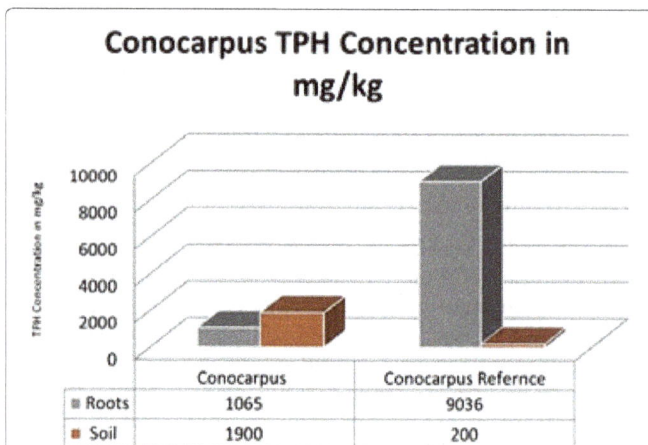

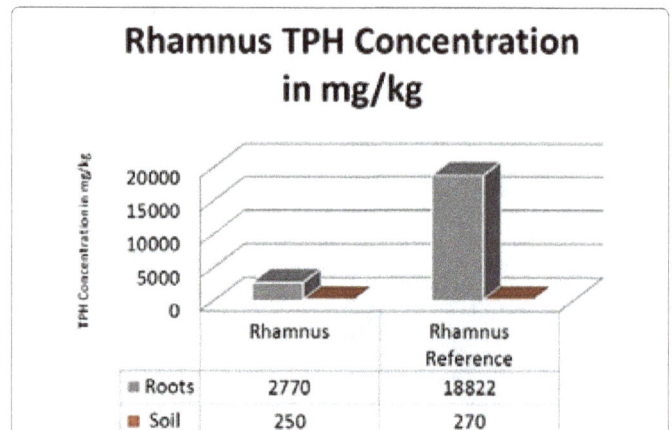

Figure 9: *Salix* TPH concentrations in mg/kg results of the study area sample compared with the reference of same species collected from Sabah Al-Salem area.

Figure 7: *Conocarpus* TPH concentrations in mg/kg results of the study area sample compared with the reference of same species collected from Sabah Al-Salem area.

Figure 10: *Rhamnus* TPH concentrations in mg/kg results of the study area sample compared with the reference of same species collected from Sabah Al-Salem area.

a wide spectrum of habitats because of the genotype in saltcedar [41]. Moreover, *Tamarix* TPH analysis reveals that the roots stored efficiently the soil contaminants in the study area and reference area at Sabah Al-Salem. Total Petroleum Hydrocarbons (TPH) values of *Tamarix* roots were the highest 7600-11263 mg/kg whereas the soil comprises 340-370 mg/kg. Soil sample collected from the contaminated study area is little lower than the TPH amount the one collected as a reference from the similar species in Sabah Al-Salem area (Figure 11).

Phoenix: *Phoenix* is a genus of palms family. *Phoenix* is widespread and commonly grown palm in the world. They are growing in many varied climates such as tropics, arid and even in cold climate. The TPH results of *Phoenix* samples collected from the study area showed that the plant's root and soil were storing the same amount of Total Petroleum Hydrocarbons (TPH) 350 mg/kg (Figure 12). However, the *Phoenix* reference sample collected from Sabah Al-Salem area were having higher amount of TPH 4573-210 mg/kg compared to the sample collected from Al-Wafra 14Km and that might be due to age of plant and/or its nearness of a sewage area.

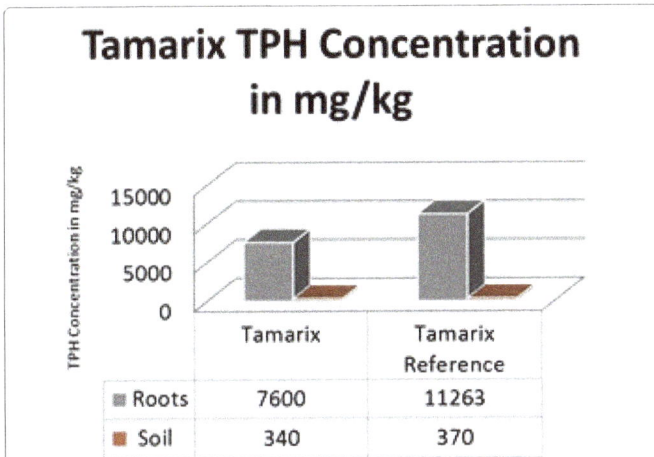

Figure 11: *Tamarix* TPH concentrations in mg/kg results of the study area sample compared with the reference of same species collected from Sabah Al-Salem area.

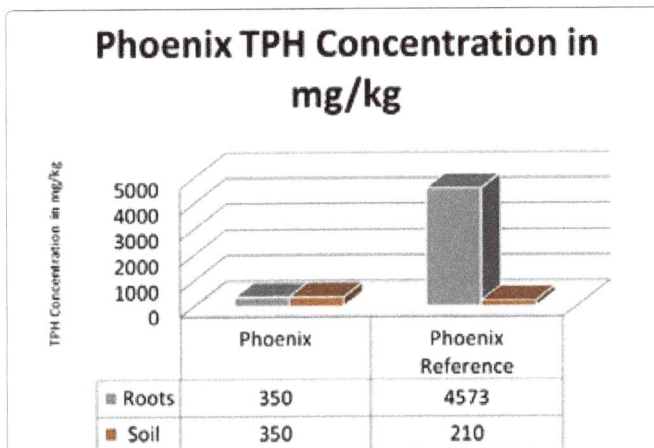

Figure 12: *Phoenix* TPH concentrations in mg/kg results of the study area sample compared with the reference of same species collected from Sabah Al-Salem area.

Total petroleum hydrocarbons (TPH) results comparison between plant's root and soil

Plant's root: TPH concentration results of the six plants species collected from the study area and their reference samples of the same species collected from Sabah Al-Salem area will be compared in this section. This comparison will show TPH concentration results for plant's root of all the plants species and the higher TPH value will be determined. TPH results of plant roots showed that the highest value goes for *Tamarix* 7600-11236 mg/kg in both reference and study area samples (Figures 13).

Soil: Soil TPH concentration results depicted that the introduced planted *Conocarpus* soil was having the highest amount of contaminants in the study area 1900 mg/kg (Figure 14).

Comparisons between TPH results in plant roots and soil from this study and TPH results in plant roots and soil from another study [32]

A study by Lotfinasabasl [32] for different plant species roots (mangrove) find out that the highest amount of TPH was 16300 and the TPH amount in the soil was 3760, while from this study the highest amount of TPH in plant root collected from the study area Al-Wafra 14 km was 7600 mg/kg and TPH amount in the soil was 1900 (Figure 15).

Trace metals analysis in plants and soil

In geochemistry, the term "trace metals" has been used for the chemical elements that occur in the Earth's crust. The amount of the trace metals in the Earth's crust is less than 0.1% which is about 1000 mg/kg [36]. Nowadays, the term "heavy metals" are used widely and that because of chemical hazards [42]. Plants are capable of absorbing a quantity of elements from soil, some are known to be lethal at low concentrations and some have no recognized biological purpose. Plants can accumulate high concentrations of heavy metals in their tissue. Serious danger to individual human health can result from the

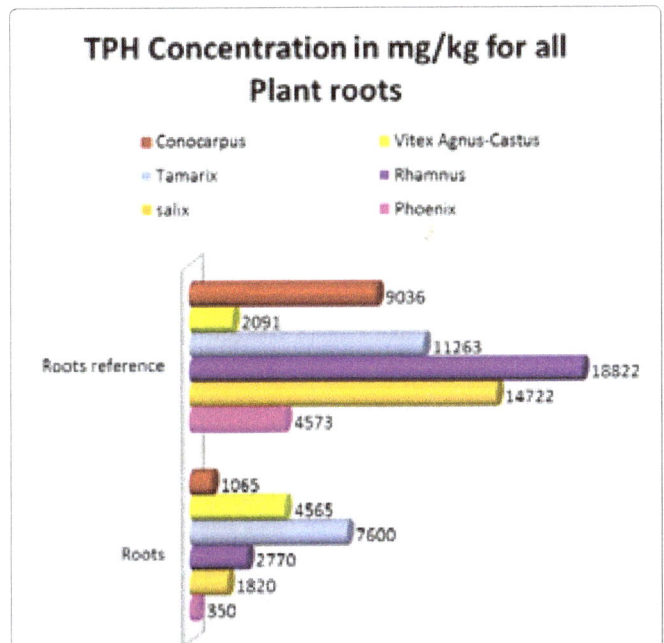

Figure 13: Plant roots TPH results comparison of six plant species collected from the study area and their reference from the same plant species.

TPH Concentration in mg/kg for Soil

■ Conocarpus ■ Vitex Agnus-Castus ■ Tamarix ■ Rhamnus ■ salix ■ Phoenix

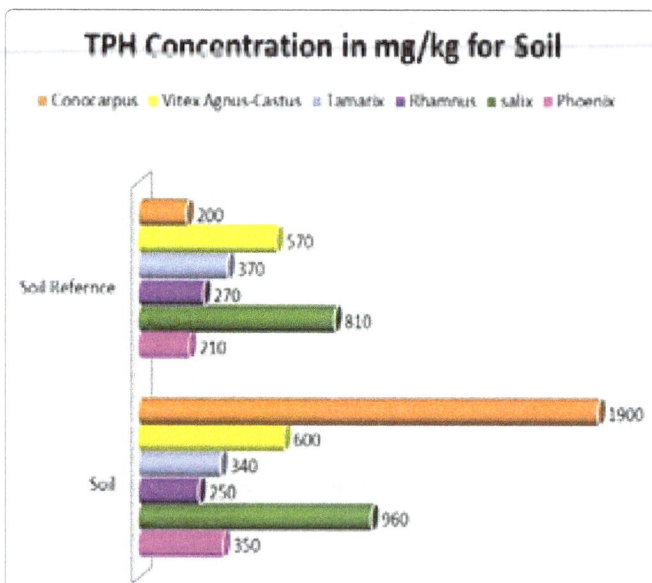

Figure 14: Soil TPH results comparison of six plant species collected from the study area and their reference from the same plant species.

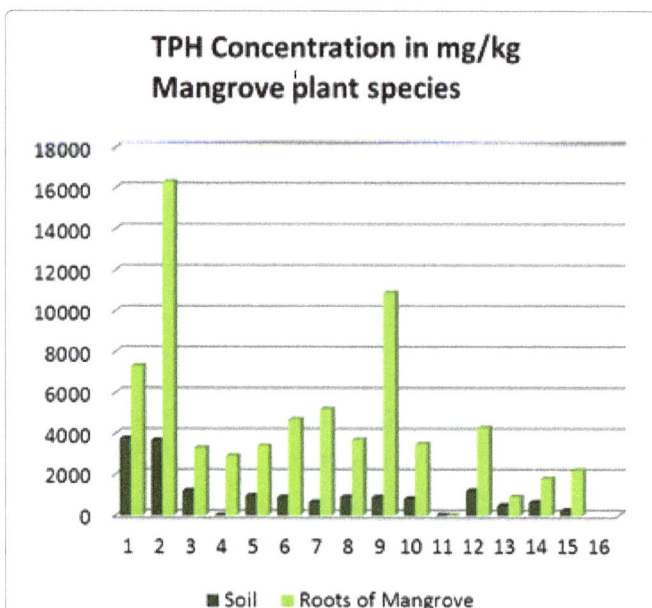

TPH Concentration in mg/kg Mangrove plant species

■ Soil ■ Roots of Mangrove

Figure 15: TPH concentration in plant roots and soil of mangrove plant species.

plants which are growing in a contaminate environment and collect trace elements at high concentrations. Roots depend on speciation of metal and soil types and kind of plant species to uptake of metal concentration root [37].

This section will explain the chemical analysis for the six plants root samples and soils were collected from the study area and the references area (*Conocarpus, Vitex Agnus-Castus, Salix, Rhamnus, Tamarix and Phoenix*).

Chromium (Cr): Chromium (Cr) is the 22nd most abundant element in Earth's crust. It has an average concentration of 100 ppm [6,7]. Chromium is a remarkable metal for being profoundly impervious to oxidation even at high temperatures. It has a high melting point. Chromium is discharged into the environment by industries. It

affects the soil fertility and the microbial activity of the soil. The high concentration of Chromium in the soil may cause the losses of plants. The permissible limit concentration for Cr in plants is 1.5 ppm (mg/g) [43]. Chromium (Cr) concentrations in study area plant's root samples are ranging from -0.7- 358.2 mg/kg and soil samples 13.6-31.8 mg/kg. The highest level of Cr recorded storage in roots goes for *Phoenix* 358.2 mg/kg whereas the lowest is *Rhamnus* at -0.7 mg/kg. The highest Cr recorded in the soil of the study area samples is *Conocarpus* 31.8 mg/kg and the lowest is *Tamarix* 13.6 mg/kg (Figure 16).

Manganese (Mn): Manganese is a naturally occurring metal that has silver color with no taste or smell. Manganese is usually found in the nature as a compound with elements such as iron, sulfur, oxygen, or chlorine [44]. It is hard and very brittle but easy to oxidize. Manganese has two forms in the environment. Inorganic manganese compounds are used in the manufacture of steel, batteries, ceramics, and nutritional supplements [44]. According to WHO [45] the Mn concentration in tend to range from 20-500 mg/kg in terrestrial plants (Figure 17).

As shown in Figure 18, Manganese (Mn) concentration in study area root samples ranges between 13-139.4 mg/kg while the reference root samples collected from Sabah Al-Salem ranges from 28.5-481.3 mg/kg. Manganese concentration of soil sample collected from study area ranges from 83.3-138.6 mg/kg whereas the reference soil sample is

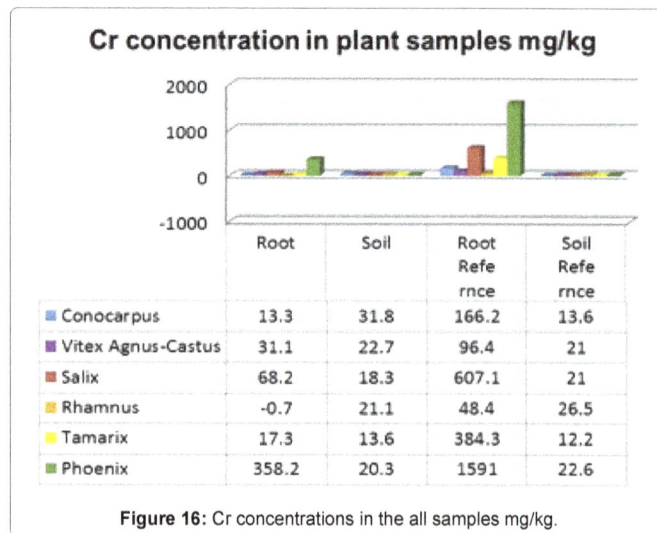

Cr concentration in plant samples mg/kg

	Root	Soil	Root Refernce	Soil Refernce
■ Conocarpus	13.3	31.8	166.2	13.6
■ Vitex Agnus-Castus	31.1	22.7	96.4	21
■ Salix	68.2	18.3	607.1	21
■ Rhamnus	-0.7	21.1	48.4	26.5
■ Tamarix	17.3	13.6	384.3	12.2
■ Phoenix	358.2	20.3	1591	22.6

Figure 16: Cr concentrations in the all samples mg/kg.

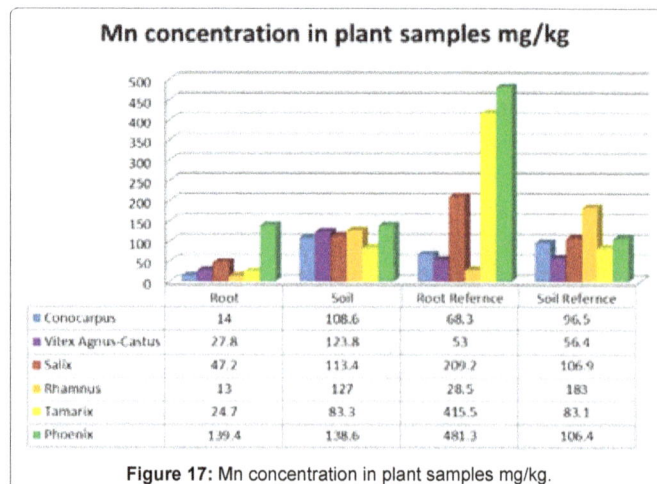

Mn concentration in plant samples mg/kg

	Root	Soil	Root Refernce	Soil Refernce
■ Conocarpus	14	108.6	68.3	96.5
■ Vitex Agnus-Castus	27.8	123.8	53	56.4
■ Salix	47.2	113.4	209.2	106.9
■ Rhamnus	13	127	28.5	183
■ Tamarix	24.7	83.3	415.5	83.1
■ Phoenix	139.4	138.6	481.3	106.4

Figure 17: Mn concentration in plant samples mg/kg.

ranging from 56.4-106.9 mg/kg. The lowest value of Mn concentration recorded in root samples of the study area is *Rhamnus* 13 mg/kg and the highest level recorded is *Phoenix* 139.4 mg/kg. Soil samples collected from study area that showed that the lowest Mn concentration recorded is *Tamarix* 83.3 whereas the highest value is from *Phoenix* 138.6 mg/kg.

Cadmium (Cd): Cadmium is one of the naturally occurring components in the earth's crust and waters. Many elements like: zinc, lead, and copper ores are associated with Cadmium. It is present everywhere in our environment. Cadmium is a poison and is known to cause birth defects and cancer. The pure one is a silvery metal with a bluish tinge to its surface. Cadmium chloride and cadmium sulfate are the soluble form of Cadmium in water. Cadmium is used for the following: rechargeable nickel-cadmium batteries, rods in nuclear reactors, pigments, protecting critical components of aero planes and oil platforms, stabilizers for plastics, nonferrous alloys, photovoltaic devices, and other uses.

From Figure 19, there was no concentration of Cadmium found in study area for both samples plant roots and soil whereas the highest Cd concentration found stored in reference roots samples is of *Tamarix* 5.8 mg/kg. However, the reference soil samples indicate that two plants species *Rhamnus* 0.5 mg/kg and *Tamarix* 0.1 mg/kg had higher Cd concentration. The World Health Organization (WHO) recommended that the maximum level of Cd in medicinal plants is 10 mg/kg.

Copper (Cu): Copper is a reddish metal that occurs naturally in rock, water, soil, sediment and at low levels in the air with concentration ranging between 8.85-13.44 mg/kg [46]. Copper is found as a pure metal in nature. Pure copper is usually too soft and ductile. Mixing copper with other metals make it strengthened. It is used as a conductor of heat and electricity, as a building material, and as a constituent of various metal alloys. Bronze and brass are the two most familiar alloys of copper. In the earth's crust the average concentration of Cu is about 50 parts copper per million parts soil (ppm) or 50 grams of copper per 1,000,000 grams of soil. Copper is found in all plants and animals. It is a crucial component for all perceived living creatures including people and different creatures at low ranks of consumption. At higher amounts, Cu can be harmful [47]. The concentration of Copper (Cu) in root plant samples of the study area was ranging from 17.5-145.5 mg/kg whereas soil samples of study area ranges from 4.4-34.2 mg/kg (Figure 20). According to Dutch Standards, the soil concentration of Cu that is below the permissible limit is 36 mg/kg. The highest level of Cu concentration in plant roots samples collected from the study area is of *Phoenix* 145.5 mg/kg whereas the lowest one goes to *Rhamnus* 17.5 mg/kg. For the soil sample taken from the study area, the highest level of Cu concentration is of *Conocarpus* 34.2 mg/kg whereas the lowest one goes to *Tamarix* 17.5 mg/kg. This present study indicate that the best plant stored Cu in Its root is *Phoenix*. We can notice the high percentage of Cu metal in reference area is more than the study area and that because the reference sample might be affected by its nearness from sewage pipe.

Iron (Fe): Iron is the most common element on Earth surface and is the fourth most common element in the Earth's crust. It is forming much of Earth's outer and inner core. About 4.6% of igneous rocks and 4.4% of sedimentary rocks contain iron as a regularly occurring metal [48]. Iron involves numerous biological processes. In plants, Iron plays a role in the creation of chlorophyll and is an essential part of hemoglobin in the blood. On the other hand, iron helps keep plants and animals alive. Iron classic concentrations in soils are range between 0.2%-55% [49] and that can differ in localized areas because of soil diversity and availability of other sources.

The present study finds that Iron (Fe) concentration in plant root samples collected from study area ranged between 3241-142.8 mg/kg. The lowest iron concentration is of plant root sample is of *Tamarix* 142.8 mg/kg and the highest one is of *Phoenix* 3241 mg/kg. Also, it indicates

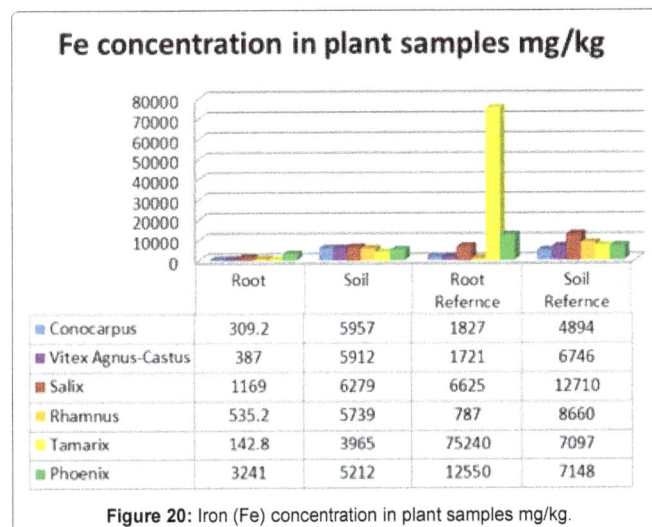

Cd concentration in plant samples mg/kg

	Root	Soil	Root Refernce	Soil Refernce
Conocarpus	0	0	0	0
Vitex Agnus-Castus	0	0	0	0
Salix	0	0	0	0.5
Rhamnus	0	0	0	0.1
Tamarix	0	0	5.8	0
Phoenix	0	0	0	0

Figure 18: Cadmium (Cd) concentration in plant samples mg/kg.

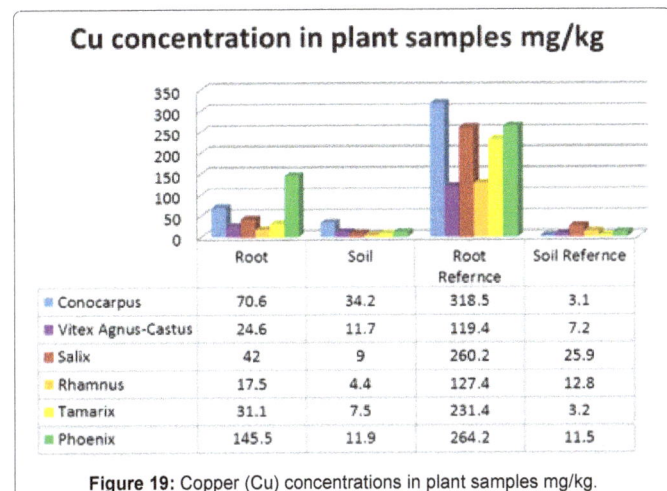

Cu concentration in plant samples mg/kg

	Root	Soil	Root Refernce	Soil Refernce
Conocarpus	70.6	34.2	318.5	3.1
Vitex Agnus-Castus	24.6	11.7	119.4	7.2
Salix	42	9	260.2	25.9
Rhamnus	17.5	4.4	127.4	12.8
Tamarix	31.1	7.5	231.4	3.2
Phoenix	145.5	11.9	264.2	11.5

Figure 19: Copper (Cu) concentrations in plant samples mg/kg.

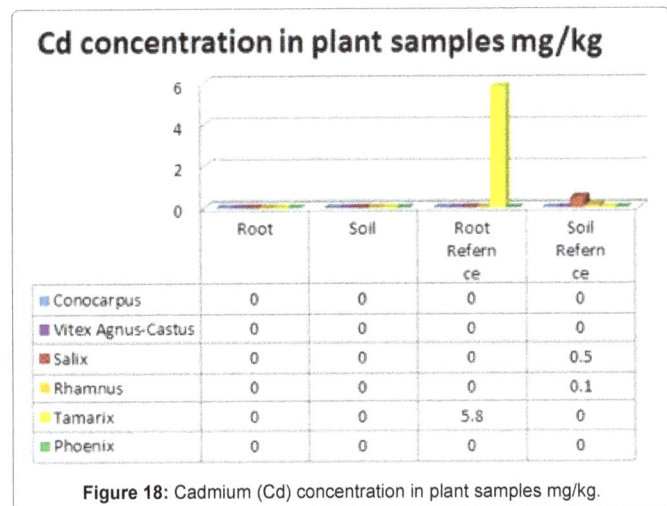

Fe concentration in plant samples mg/kg

	Root	Soil	Root Refernce	Soil Refernce
Conocarpus	309.2	5957	1827	4894
Vitex Agnus-Castus	387	5912	1721	6746
Salix	1169	6279	6625	12710
Rhamnus	535.2	5739	787	8660
Tamarix	142.8	3965	75240	7097
Phoenix	3241	5212	12550	7148

Figure 20: Iron (Fe) concentration in plant samples mg/kg.

that Iron (Fe) concentration in soil samples collected from study area ranged between 6279-3965 mg/kg. The lowest Fe concentration of soil goes for *Tamarix* 3965 mg/kg yet; the highest one is *Salix* 6279 mg/kg (Figure 21). The concentration of Fe is between 20000-250000 mg/kg in dry plant tissue.

Nickel (Ni): Nickel is a shiny white transition metal. Pure nickel is hard and ductile. According to McGrath [50] nickel appears naturally in soils due to parent rock weathering. Nickel has many properties that make it extremely attractive to be alloyed with different metals like: iron, copper, chromium and zinc. These combinations are used for making metal coins and in industry. Most nickel is utilized to make stainless steel. Nickel is considered as corrosion resistant because of its slow rate of oxidation. Nickel concentration in soil is ranging from 0.7 to 259 mg/kg [45].

In our study area the Ni in roots samples is ranging from 3.3-136.6 mg/kg whereas the soil sample is ranging from 12.5-19.3 mg/kg. The root samples of *Phoenix* contributed the best in storing Nickel in both reference and study area as it is shown in Figure 21.

Vanadium (V): Vanadium is a rare, soft, ductile gray-white element found combined in certain minerals and is widely distributed in the earth's crust. Vanadium occurs naturally and has many forms. Pure vanadium is a bright white, soft and ductile metal. Unintentionally discharged of coal bi-products may cause Vanadium Pollution of soil [36]. Vanadium is abundant in variable amounts in soils and it is taken up by plants at different levels. There are different ideas about the mean content of Vanadium in plants. Smith [51] mentioned that Vanadium concentration is lower than 0.5 mg/kg while others have referred concentrations ranges from 0.5-22 mg/kg. However, Cannon [52] mentioned that Vanadium concentration is from 0.8- 2.7 mg/kg.

As shown in Figure 22, the root samples analysis of the study area show that the range of Vanadium is between 0.6-12 mg/kg whereas reference samples range between 4.2-19.2 mg/kg. The least level recorded is of *Vitex Angus-Castus* 0.6 mg/kg in study area root samples yet; the highest one is of *Conocarpus* 12 mg/kg. Surprisingly, the soil samples for both study and reference area showed higher Vanadium amount [53-56]. The range of soil samples in the study area is 17.7-27 mg/kg whereas reference area samples showed a range of Vanadium concentration between 13.9-21.6 mg/kg. Moreover, the soil samples

in study area had the least level in *Tamarix* 17.7 mg/kg whereas the highest V recorded is of *Vitex Angus-Castus* 27 mg/kg.

Zinc (Zn): Zinc is a bluish-white, glittery, diamagnetic metal. It is one of the most common elements in the Earth's crust. It can be found in different places such as in air, water, and soil in addition to food. Metallic zinc has many uses in industry. It is most commonly used as an anti-corrosion agent. Metallic zinc is mixed with other metals such as bronze to form alloys. It is also used to produce dry cell batteries [57,58]. Zinc sulfide and zinc oxide is usually utilized in manufacturing such as white paints, ceramics, and other products. According to Emsley 2001, soil contains 5-770 mg/kg of zinc with an average of 64 mg/kg whereas seawater has only 30 mg/kg zinc. The permissible limit of Zinc in human consumption is 10-50 [37].

This present study shows that the least zinc concentration in root samples is in *Salix* at 16.2 mg/kg whereas the highest one is in *Conocarpus* plant species at 49.5 mg/kg. The concentration of Zinc in soil samples collected from study area ranged between 27.5-163.1 mg/kg. In another hand, the least level of Zn concentration stored in soil is of *Rhamnus* 27.5 mg/kg and the highest one is of *Conocarpus* 163.1 mg/kg (Figure 23).

Lead (Pb): Lead is a bluish to gray or bluish-gray metal that occurs naturally in the Earth's crust. It is a heavy metal with high density. Lead has the highest atomic number of any non-radioactive element. Yet, it is easily discovered naturally as a metal and it is easily mined and refined so it is not considered to be a rare element. It is corrosion

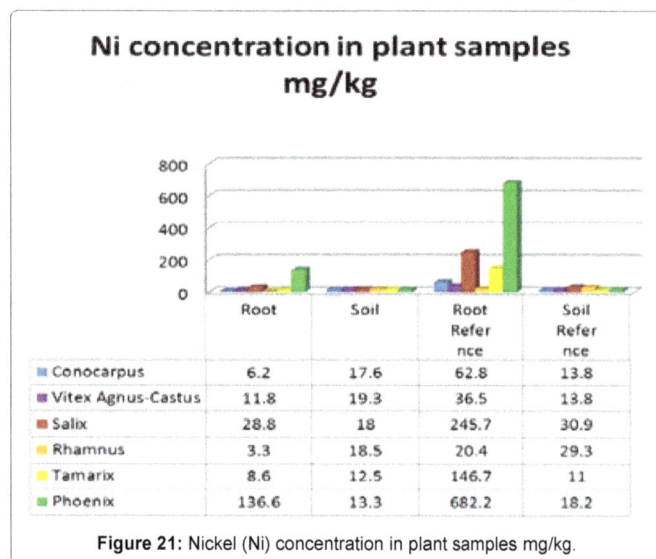

Zn concentration in plant samples mg/kg			
	Root	Soil	Root Refernce	Soil Refernce
Conocarpus	49.5	163.1	86.1	10.8
Vitex Agnus-Castus	20.3	53.9	43.4	31
Salix	16.2	39.2	50.3	47.9
Rhamnus	17.4	27.5	34.2	69.4
Tamarix	30	34	154.5	16.1
Phoenix	23.4	58.4	64.9	41.4

Figure 22: Vanadium (V) concentration in plant samples mg/kg.

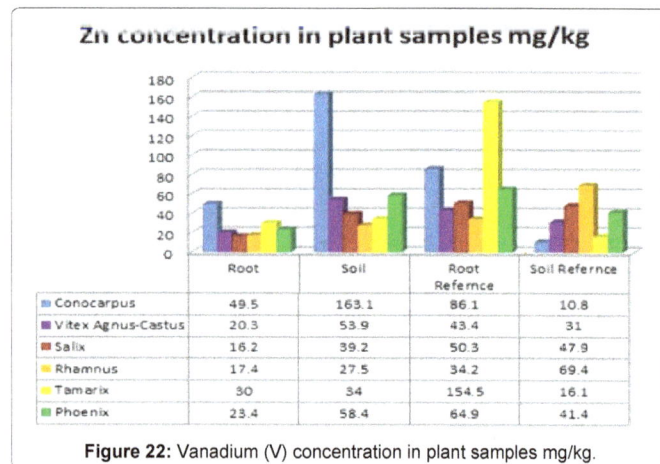

Zn concentration in plant samples mg/kg			
	Root	Soil	Root Refernce	Soil Refernce
Conocarpus	49.5	163.1	86.1	10.8
Vitex Agnus-Castus	20.3	53.9	43.4	31
Salix	16.2	39.2	50.3	47.9
Rhamnus	17.4	27.5	34.2	69.4
Tamarix	30	34	154.5	16.1
Phoenix	23.4	58.4	64.9	41.4

Figure 23: Zinc (Zn) concentration in plant samples mg/kg.

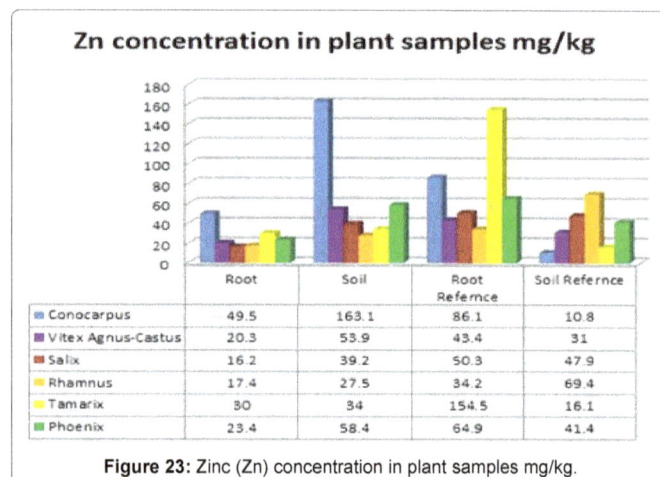

Ni concentration in plant samples mg/kg			
	Root	Soil	Root Refernce	Soil Refernce
Conocarpus	6.2	17.6	62.8	13.8
Vitex Agnus-Castus	11.8	19.3	36.5	13.8
Salix	28.8	18	245.7	30.9
Rhamnus	3.3	18.5	20.4	29.3
Tamarix	8.6	12.5	146.7	11
Phoenix	136.6	13.3	682.2	18.2

Figure 21: Nickel (Ni) concentration in plant samples mg/kg.

resistant material used in building construction, lead-acid batteries, bullets and shot, weights and as a radiation shield. In another hand, Lead is poisonous to animals and humans.

Pb concentration in plant samples is below detection limit except in root samples of with 1 mg/kg for *Phoenix* species. The study area sample has 7.3 mg/kg in the *Conocarpus* soil sample [59-61]. The increase of reference sample Pb concentration might be due to its nearness to a sewage pipe. For all samples, *Phoenix* species has the highest level of Pb concentration (Figure 24).

Percentage of total petroleum hydrocarbons (TPH) and trace metals absorbed in plants and soil

TPH: This section will show the percentage of TPH absorbed by the roots and soil of the six plants species collected from the study area and the reference samples collected from Sabah Al -Salem area. The study showed that the highest TPH for both reference and study area samples is storing in the roots 92% (Figure 25).

Trace metals: Figure 26, showed that the highest trace metal that was absorbed is iron (94%) for both reference and study area samples. Chromium is the second absorbed trace metal (2%) whereas zinc, manganese, nickel and copper all absorbed at the same percentage (1%).

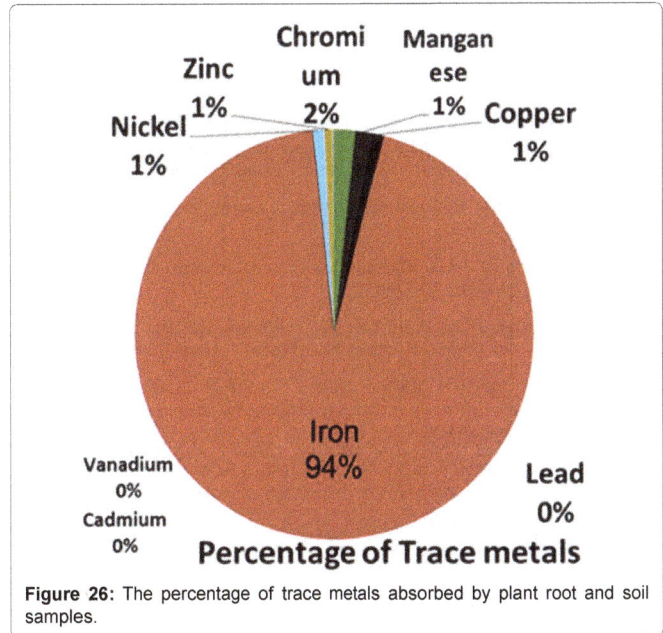

Figure 26: The percentage of trace metals absorbed by plant root and soil samples.

Conclusion and Recommendations

From the study area in Al-Wafra (14Km south of Kuwait), the Total Petroleum Hydrocarbons (TPH) and Trace Metals analysis for the six types of plant tree species collected (*Conocarpus, Tamarix, Phoenix, Rhamnus, Vitex Agnus-Castus and Salix*) shows that the concentration of Total Petroleum of Hydrocarbons is very high. Moreover, the concentration of conducted trace metals analysis in plant and soil depicted sometimes-high variation than the permissible limits of many trace metal elements [62-64].

Total Petroleum Hydrocarbons analysis (TPH) in this research study shows that the best part to absorbed and store the contaminants is the roots; the same conclusion was reached by Srinivas et al. [37] and Lotfinasabasl [32]. From the study area *Tamarix* species roots got the highest storage of TPH value of 7600 mg/kg while the lowest is *Phoenix* roots with TPH value of 350 mg/kg. However, TPH analysis of soil samples collected from the study area showed that *Conocarpus* species is the best plant for absorbing the contaminants from soil and has the highest value of TPH 1900 mg/kg.

The results and interpretation of trace metals (Cr, Mn, Cd, Cu, Fe, Ni, V, Zn and Pb) analysis in this research study show that the highest value of Trace metals in the study area was absorbed by *Phoenix* (4053.7 mg/kg) and the highest trace metal absorbed by all plants is iron.

The following recommendations should be followed after this research study:

Accordingly, roots show the most important part of plants in the process of phytoremediation.

1. Further study of the fruits of *Phoenix* should be conducted to detect accumulate of TPH and Trace Metals.

2. Research on rhizoremediation should be encouraged and understood to reduce soil contamination.

Acknowledgement

The authors like to thank Kuwait University, environmental laboratory and the central lab at college of science for the support and cooperation. Special thanks

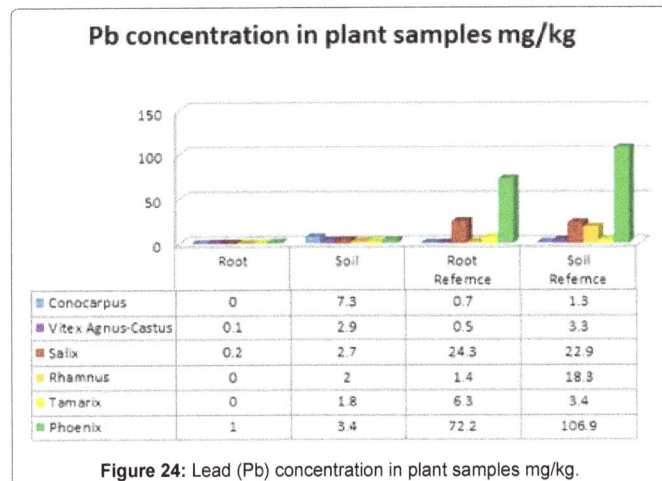

Figure 24: Lead (Pb) concentration in plant samples mg/kg.

	Root	Soil	Root Refernce	Soil Refernce
Conocarpus	0	7.3	0.7	1.3
Vitex Agnus-Castus	0.1	2.9	0.5	3.3
Salix	0.2	2.7	24.3	22.9
Rhamnus	0	2	1.4	18.3
Tamarix	0	1.8	6.3	3.4
Phoenix	1	3.4	72.2	106.9

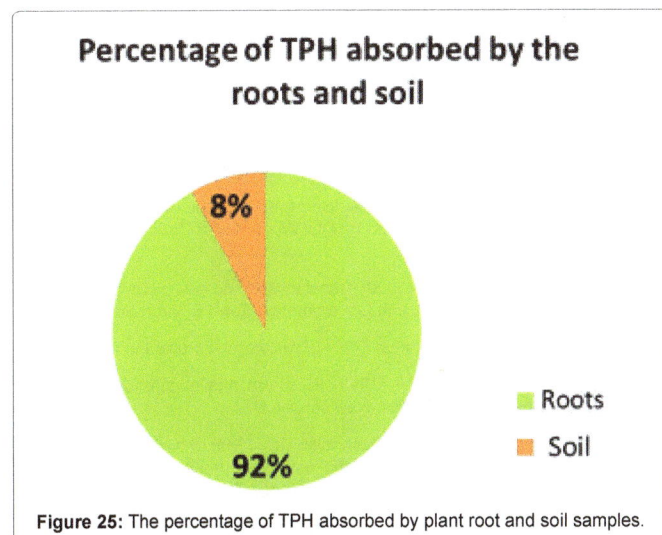

Figure 25: The percentage of TPH absorbed by plant root and soil samples.

to Florida International University, Department of Earth and Environment, United States of America for their technical guidance.

References

1. Ganesan V (2012) Toxicity of heavy metals to legumes and bioremediation. In: Rhizoremediation: A pragmatic approach for remediation of heavy metal-contaminated soil. Springer Vienna, Wien, Austria. pp: 147-161.

2. Macnair MR (1993) The genetics of metal tolerance in vascular plants. New Phytologist 124: 541-559.

3. Silver S, Phung LT (1996) Bacterial heavy metal resistances: new surprises. Annu Rev Microbiol 50: 753-789.

4. Fiorenza S, Oubre CL, Ward CH (2000) Phytoremediation of hydrocarbon contaminated soil. Lewis Publishers, Boca Raton, Florida, USA. p. 164.

5. Kärenlampi S, Schat H, Vangronsveld J, Verkleij J, Lelie D, et al. (1999) Genetic engineering in the improvement of plants for phytoremediation of metal polluted soils. Environ Pollut 107: 225-231.

6. Emsley J (2001) Chromium. In: Nature's Building Blocks: An A-Z Guide to the Elements. Oxford University Press, Oxford, England, UK. pp: 495-498.

7. Emsley J (2001) Zinc. In: Nature's Building Blocks: An A-Z Guide to the Elements. Oxford University Press, Oxford, England, UK. pp: 499-505.

8. Cunningham SD, Berti WR, Huang JWW (1995) Phytoremediation of contaminated soils. Trends Biotechnol 13: 393-397.

9. Salt DE, Blaylock M, Kumar NPBA, Dushenkov V, Ensley D, et al. (1995) Phytoremediation: A novel strategy for the removal of toxic metals from the environment using plants. Biotechnology 13: 468-474.

10. Salt DE, Smith RD, Raskin L (1998) Phytoremediation. Ann Rev Plant Phys Plant Mol Biol 49: 643-668.

11. Charnay M, Tuis S, Coquet Y, Barriuso E (2005) Spatial variability in 14 Cherbicide degradation in surface and subsurface soils. Pest Manago Sci 61: 845-855.

12. Kala S (2014) Rhizoremediation: A promising rhizosphere technology. IOSR Journal of Environmental Science, Toxicology and Food Technology 8: 23-27.

13. Kuiper I, Bloemberg GV, Lugtenberg BJJ (2001) Selection of a plantbacterium pair as a novel tool for rhizostimulation of polycyclic aromatic hydrocarbon-degrading bacteria. Molecular Plant Microbe Interaction 14: 1197-1205.

14. US Environmental Protection Agency (2012) Integrated risk information.

15. Kuiper I, Bloemberg GV, Lugtenberg BJJ (2004) Rhizoremediation: A beneficial plant-microbe interaction. Molecular Plant Microbe Interaction 17: 6-15.

16. Lynch JM, Whipps JM (1990) Substrate flow in the rhizosphere. Plant Soil 129: 1-10.

17. Pilon-Smits E (2005) Phytoremediation. Annu Rev Plant Biol 56: 15-39.

18. Schroll R, Becher HH, Dorfler U, Gayler S, Grundmann S, et al. (2006) Quantifying the effect of soil moisture on the aerobic microbial mineralization of selected pesticides in different soils. Environ Sci Technol 40: 3305-3312.

19. Alberty RA (2006) Biochemical reactions at specified temperature and various pHs. Biochemical Thermodynamics: Applications of Mathmatica. pp: 43-70.

20. Guillot A, Obis D, Mistou MY (2000) Fatty acid membrane composition and activation of glycine-betaine transport in Lactococcus lactis subjected to osmotic stress. Int J Food Microbiol 55: 47-51.

21. Mastronicolis SK, German JB, Megoulas N, Petrou E, Foka P, et al. (1998) Influence of cold shock on the fatty-acid composition of different lipid classes of the food-borne pathogen Listeria monocytogenes. Food Microbiol 15: 299-306.

22. Singh BK, Walker A, Wright DJ (2006) Bioremedial potential of fenamiphos and chlorpyrifos degrading isolates: Influence of different environmental conditions. Soil Biol Biochem 38: 2682-2693.

23. Barea JM, Pozo MJ, Azcon R, Azcon-Aquilar C (2005) Microbial co-operation in the rhizosphere. J Exp Bot 56: 1761-1778.

24. Rasmussen G, Olsen RA (2005) Sorption and biological removal of creosote-contaminants from groundwater in soil/sand vegetated with orchard grass (Dactylis glomerata). Adv Environ Res 8: 313-327.

25. Chaudhry Q, Blom-Zandstra M, Gupta S (2005) Utilising the synergy between plants and rhizosphere microorganisms to enhance breakdown of organic pollutants in the environment. Env Sci Pollut Res 12: 34-48.

26. Anderson TA, Guthrie EA, Walton BT (1993) Bioremediation in the rhizosphere. Env Sci Technol 27: 2630-2636.

27. Siciliano SD, Fortin N, Mihoc A, Wisse G, Labelle X, et al. (2002) Selection of specific endophytic bacterial genotypes by plants in response to soil contamination. Appl Environ Microbiol 67: 2469-2475.

28. Al-Sarawi MA, Al-Qassar R, Baby S (2010) Phytoremdiation technology to restore sewage and liquid industrial waste pond to develop National Forest in Kuwait. Proceedings of the 25th International Conference on Solid Waste Technology and Management, Philadelphia, USA.

29. Al-Sarawi MA, Al-Bado A (2011) Phytoremediation for hazardous liquid industrial waste site – Al-Wafra Project, Kuwait. Proceedings of the Twenty-Sixth International Conference On Solid Waste Technology And Management, Philadelphia, USA.

30. Al-Ibrahim M, Al Sarawi M (2017) Phytoremediation for hazardous liquid industrial waste using six famous trees species in Al-Wafra (14 km), Kuwait. J Bioremediat Biodegrad.

31. UNEP (2012) Does phytoremediation work at every site? Newsletter and Technical Publications Freshwater Management Series No. 2, Washington, USA.

32. Lotfinasabasl S, Gunale VR, Rajurkar NS (2013) Petroleum hydrocarbons pollution in soil and its bioaccumulation in mangrove species, avicennia marina from Alibaug mangrove ecosystem, Maharashtra, India. International Journal of Advancements in Research & Technology 2: 2278-7763

33. Crowdy SH, Jones DR (1956) Partition of sulphonamides in plant roots: A factor in their translocation. Nature 178: 1165-1167.

34. Baker AJM, McGrath SP, Sidoli CMD, Reeves RD (1994) The possibility of in situ heavy metal decontamination of polluted soils using crops of metal-accumulationg plants. Resour Conserv Recy 11: 41-49.

35. Raskin I, Ensley BD (2000) Phytoremediation of toxic metals: Using plants to clean up the environment. John Wiley and Sons, New Jersey, USA.

36. Kabata-Pendias A, Pendias H (2001) Trace elements in soils and plants (3rd edn). CRC Press LLC, Boca Raton, Florida, USA.

37. Srinivas N, Ramakrishna RS, Kumar SS (2009) Trace metal accumulation in vegetables grown in industrial and semi-urban areas – a case study. Appl Ecol and Environ Research 7: 131-139.

38. Gilman EF, Watson DG (1993) Conocarpus erectus, buttonwood. Fact Sheet ST-179, US Forest Service and Southern Group of State Foresters, Gainesville, Florida, USA. p. 3.

39. National Institute of Biodiversity (2002) List of specimens of Conocarpus erectus.

40. Mabberley DJ (1987) The plant-book: A portable dictionary of the vascular plants. Cambridge University Press, Cambridge, UK. pp: 1-706.

41. Brotherson JD, Winkel V (1986) Habitat relationships of saltcedar (Tamarix ramosissima). The Great Basin Naturalist 46: 535-541.

42. Duffus JH (2002) Heavy metals-a meaningless term? Pure Appl Chem 74: 793-807.

43. Saeed M, Khan H, Khan MA, Muhammad N, Khan SA (2010) Quantification of various metals accumulation and cytotoxic profile of aerial parts of *Polygonatum verticillatum*. Pak J Bot 42: 3995-4002

44. U.S. Centers for Disease Control (2000) Toxicological profile for manganese.

45. WHO (2004) Nickel in drinking-water: Background document for preparation of WHO guidelines for drinking-water quality. World Health Organization, Geneva, Switzerland.

46. Bu-Olayan AH, Thomas BV (2009) Translocation and bioaccumulation of trace metals in desert plants of Kuwait Governorates. Res J Environ Sci 3: 581-587.

47. U.S. Centers for Disease Control (2004) Toxicological profile for copper.

48. Morel F, Hering JG, Morel F (1993) Principles and applications of aquatic chemistry. John Wiley and Sons, New Jersey, USA.

49. Bodek I, Lyman WJ, Reehl WF, Rosenblatt DH (1988) Environmental inorganic chemistry: Properties, processes, and estimation methods. SETAC Special Publication Series, Pergamon Press, NewYork, USA.

50. Mcgrath SP (1995) Chromium and nickel. Heavy Metals in Soils 22: 152-178.

51. Smith JB (1983) Vanadium ions stimulate DNA synthesis in Swiss mouse 3T3 and 3T6 cells. Proc Natl Acad Sci (USA) 80: 6162-6166.

52. Cannon HL, Bowles JM (1962) Contamination of vegetation by tetraethyl lead. Science 137: 765-766.

53. Al-Sarawi M, Massoud MS, Wahba SA (1998) Physical properties as indicators of oil penetration in soils contaminated with oil lakes in the greater Burgan oil fields, Kuwait. Water, Air, & Soil Pollution 118: 281-297.

54. Amos PW, Younger PL (2003) Substrate characterization for a subsurface reactive barrier to treat colliery spoil leachate. Water Res 37: 108-120.

55. Bagnold RA (1941) The physics of blown sand and desert dunes. London: Methuen. 1953: The surface movement of blown sand in relation to meteorology, in Desert research. Research Council of Israel, Special Publication 2: 89-96.

56. Brown SL, Chaney RL, Angle JS, Baker AJM (1994) Phytoremediation potential of Thlaspi caerulescens and bladder campion for zinc and cadmium-contaminated soil. J Environ Qual 23: 1151-1157.

57. Chang SW, Lee SJ, Je CH (2005) Phytoremediation of atrazine by poplar trees: Toxicity, uptake, and transformation. J Environ Sci Health 40: 801-811.

58. Favas PJC, Pratas J, Varun M, D'Souza R, Paul MS (2014) Phytoremediation of soils contaminated with metals and metalloids at mining areas: potential of native flora. Environmental risk assessment of soil contamination. pp: 485-517.

59. Gunatilaka HA, Al-Zamel A, Shearman DJ, Reda AA (1987) A spherulitic fabric in selectively dolomitized siliciclastic crustacean burrows, northern Kuwait. J Sediment Petrol 57: 922-927.

60. Meagher RB (2000) Phytoremediation of toxic elemental and organic pollutants. Curr Opin Plant Biotecnol 3: 153-162.

61. Padmavathiamma PK, Li LY (2007) Phytoremediation technology: Hyperaccumulation metals in plants. Water Air Soil Poll 184: 105-126.

62. Pivetz BC (2001) United States Environmental Protection Agency. Phytoremediation of contaminated soil and ground water at hazardous waste site, Washington, USA.

63. Prasad MNV, Freitas H (2003) Metal hyperaccumulation in plants – Biodiversity prospecting for phytoremediation technology. Electron J Biotechnol 6: 75-321.

64. US EPA (2000) Benchmark dose technical guidance document (External Review Draft).

21

The Potential of Shea Nut Shells in Phytoremediation of Heavy Metals in Contaminated Soil Using Lettuce (*Lactuca sativa*) as a Test Crop

Quainoo AK[1]*, Konadu A[1] and Kumi M[2]

[1]*Department of Biotechnology, Faculty of Agriculture, University for Development Studies, Tamale, Ghana*
[2]*Water Research Institute, Tamale, Ghana*

Abstract

Contamination of soil and water by heavy metals cause serious risks to living organisms especially humans and the ecosystem in general through direct contact, inhalation and dermal contact. In this study, shea nut shells were used as adsorbent for heavy metals from contaminated soil. Shea tree (*Vitellaria paradoxa*) is one of the economic tree crops prevalent in Northern Ghana. Leafy vegetables depend on water for their growth and survival and have massive potential of accumulating heavy metals in their edible parts. Accumulation of heavy metals in leafy vegetables makes them dangerous to human health when consumed. It is within this perspective, which made it imperative for the application of shea nut shells to remove heavy metals such as manganese, iron, zinc and copper from contaminated soil with lettuce (*Lactuca sativa*) as a test crop. Plastic pots filled with soil from Nyankpala with drainage holes at the bottom and contaminated water from Zoomlion landfill site at Gbelahi in Northern Region of Ghana was used. Atomic Absorption Spectrophotometer was used to determine manganese, iron, zinc and copper in the test crop (lettuce). The mean concentration of heavy metals after 21 and 42 days of transplanting were; Fe (271.135 mg/kg and 457.791 mg/kg), Mn (45.245 mg/kg and 77.211 mg/kg) and Zn (20.049 mg/kg and 50.108 mg/kg). The concentration of copper was below the level of 0.001 mg/kg. According to the results obtained from this research, shea nut shells have the potentials of adsorption of heavy metals from contaminated soil and water hence it is recommended as suitable means of phytoremediation.

Keywords: Soil; Heavy metals; Waste water; Vegetables; Shea nut shells

Introduction

Soil pollution has currently been attracting substantial public attention since the magnitude of the problem in our soils call for instantaneous action [1]. Anthropogenic activities such as mining and smelting of metalliferous ores, gas exhaust, fuel production, fertilizer and pesticide application, and metal pollution, has become an important problem today. The unchangeable nature of metals has made them a group of pollutants of much concern [2]. Some of the heavy metals associated with soil pollution include copper, manganese, iron and zinc.

Copper is a reddish metal that is found naturally in rock, soil, water, sediment, and at low levels in air. Copper normally enters human body through drinking water or with food, soil or other substances that is known to have copper. Deliberately high intakes of copper can cause liver and kidney hurt and even death [3]. Manganese is a naturally occurring element and an important nutrient. Manganese toxicity can end up in a lasting neurological disorder known as manganism accompanied with signs such as tremors, complexity walking, and facial muscle spasms [4]. Iron is the second most copiously metal in the earth's crust, of which it accounts for about 5%. The normal lethal dose of iron is 200-250 mg/kg of body weight, but death has occurred after the ingestion of doses as low as 40 mg/kg of the body weight [5]. Zinc is one of the utmost frequent elements in the Earth's crust. It is present in the air, soil, water and is found in most food Ingesting high levels of zinc for several months may cause anemia, damage the pancreas, and decrease levels of high-density lipoprotein (HDL) cholesterol [6].

The implications derived from eating foods containing these heavy metals are very horrific and deadly and therefore necessitates characterization and remediation of the soil bionetworks. This research is undertaken to determine the ability of shea nut shells in bioremediation thus, removing heavy metals from contaminated water and polluted soil and also to evaluate concentrations of heavy metals that will be accumulating in the various parts of the test crop. It is expected that this research will determine the ability of shea nut shells as a suitable choice in bioremediation.

Materials and Methods

Experimental materials and site

The experiment was carried out at the Plant House of the University for Development Studies, Nyankpala Campus, Ghana. Plastic pots were used in carried out the experiment with drainage holes, created at the bottom of the pots for free drainage of excess water in order to prevent water logging. The media was in a ratio of two parts of sandy loam mixed with one part of grinded shea nut shells of which lettuce (*Lactuca sativa*) was used as the test crop for the experiment. The wastewater that was used in the experiment was collected from Zoomlion Landfill Site at Gbelahi near Tamale in the Northern Region of Ghana. Equipments used in digestion of the soil and the test crop included the Shimadzu Atomic Absorption Spectrophotometer model AA 6300, Satorious electronic weighing scale model TE612, Clamp, Pipette, Round bottom flask, Heat mantle and Crison pH meter Basic 20.

*Corresponding author: Quainoo AK, Department of Biotechnology, Faculty of Agriculture, University for Development Studies, Tamale, Ghana
E-mail: aquainoo@googlemail.com

Experimental procedure

Both the soil and water samples were initially digested using 2.5 ml of HNO_3 and 2.5 ml of H_2SO_4. The metals were then analyzed using Shimadzu Atomic Absorption Spectrophotometer model AA 6300. The soil was digested using 0.5 g of the soil sample. The pH of the soil and wastewater samples was 6.55 and 5.64 respectively.

Experimental set up

The set up comprised of four treatments each replicated three times. The containers were arranged on the platforms using the Completely Randomized Design. The treatments were as follows; T1=Shea nut shells+Unpolluted water+Soil, T2=Shea nut shells+Polluted water+Soil, T3=Unpolluted water+Soil, T4=Polluted water+Soil. The seeds of the test crop were nursed for three weeks and later transplanted into the plastic pots containing soil and grinded shea nut shells. At 21 days after transplanting, the first parts of the sample were analyzed and the second part 42 days after transplanting.

Analysis for Mn, Fe, Zn and Cu

Analysis of Mn, Fe, Zn and Cu was based on standard method of analysis of heavy metals adopted by the WHO/FAO. Heavy metals determinations in the samples were carried out in the laboratory of Water Research Institute of Council for Scientific and Industrial Research (WRI-CSIR), Tamale. Mean levels of each of the treatments were analyzed to determine accumulation levels using Microsoft Excel 2010.

Plates

Some pictures showed the materials used for the experiment, experimental activities and the test crop (lettuce) used for the experiment. These are arranged in Plates 1-6.

Results

Samples of the test crop were analyzed according to the leaf, stem and root at 21 and 42 days after transplanting. The Tables 1-4 indicated the mean and standard deviation levels of the heavy metals at 21 and 42 days after transplanting.

Comparison of heavy metals accumulated by lettuce

The comparison of mean level of Fe, Mn, Zn and Cu accumulated in the test crop (Lettuce) at 21 and 42 days after transplanting are presented in Table 1.

From Tables 1-4, it was observed that, there were an increased

Plate 2: Grinded shea nut shells.

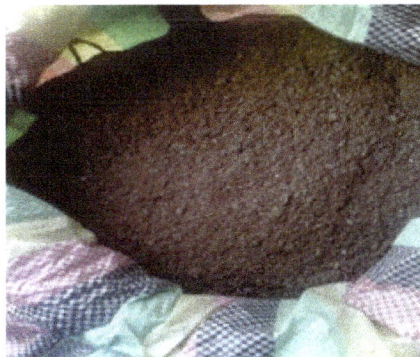

Plate 3: Zoomlion landfill site at Gbelahi.

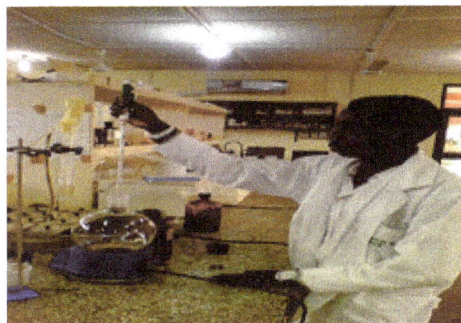

Plate 4: Soil digestion at WRI-CSIR.

accumulated mean in Fe, Mn, Zn and Cu levels by Lettuce after analysis at 42 days of transplanting as compared with the analysis at 21 days after transplanting in all the treatments.

The Table 5 shows the initial concentrations of Mn, Fe, Zn and Cu in the soil and water samples used in for the experiment and WHO/FAO standards for maximum permissible levels of heavy metals in vegetables.

Iron accumulation by lettuce

Iron recorded different levels of accumulation at 21 days after digestion of the test crop (lettuce), ranging from 175.686 mg/kg to 434.412 mg/kg (Figure 1a).

According to Figure 1a above, treatment one accumulated highest level of iron of 263.530 mg/kg in roots, followed by leaves with

Plate 1: Shea nut shells.

Plate 5: Lettuce at 21 days.

Plate 6: Lettuce at 42 days.

Heavy metals	21 Days		42 Days	
	Mean (mg/kg)	Standard deviation	Mean (mg/kg)	Standard deviation
Fe	230.929	48.100	398.101	116.518
Mn	44.091	7.778	63.457	20.165
Zn	14.485	1.154	42.397	5.529
Cu	0.001	0.000	0.001	0.000

Table 1: Mean level of heavy metals in treatment one at 21 and 42 days after transplanting.

Heavy metals	21 Days		42 Days	
	Mean (mg/kg)	Standard deviation	Mean (mg/kg)	Standard deviation
Fe	260.441	60.155	413.913	163.162
Mn	52.509	9.415	89.532	21.749
Zn	21.112	2.059	52.585	12.326
Cu	0.001	0.000	0.001	0.000

Table 2: Mean levels of heavy metals in treatment two at 21 and 42 days after transplanting.

Heavy metals	21 Days		42 Days	
	Mean (mg/kg)	Standard deviation	Mean (mg/kg)	Standard deviation
Fe	311.683	109.238	533.026	98.065
Mn	47.192	9.706	88.725	38.137
Zn	22.501	5.482	53.827	4.603
Cu	0.001	0.000	0.001	0.000

Table 3: Mean levels of heavy metals in treatment three at 21 and 42 days after transplanting.

accumulated value of 253.570 mg/kg and stem with accumulated value of 175.686 mg/kg. In treatment two, the same trend happened with roots accumulated highest value of 315.001 mg/kg, followed by

leaves with accumulated value of 270.392 mg/kg and stem with least accumulated value of 195.932 mg/kg. In treatment three, the leaves accumulated the highest value of 434.412 mg/kg, followed by roots with accumulated value of 275.540 mg/kg and stem with least accumulated value of 225.098 mg/kg. In treatment four, leaves accumulated highest value of 275.540 mg/kg, followed by roots with accumulated value of 245.344 mg/kg and stem with least accumulated value of 224.754 mg/kg.

Different accumulations were recorded in all the treatments as shown in Figure 1b above. In treatment one, the roots accumulated highest iron of 480.830 mg/kg, followed by the leaves with 448.624 mg/kg and the stem recorded least accumulated value of 264.848 mg/kg. This can be demonstrated as roots>leaves>stem. In treatment two, the same trend of highest root accumulation followed by leaf and stem occurred as it happened in treatment one but in treatment three and four, different accumulations were recorded. In treatment two, roots accumulated 574.668 mg/kg of iron, followed by leaves with accumulated value of 418.624 mg/kg and stem with least accumulated iron of 248.446 mg/kg. In treatment three, leaves accumulated more irons, followed by the roots and the stem with the same accumulations recorded in treatment four. Accumulated values in treatment three were as follows; leaves (640.470 mg/kg), roots (510.266 mg/kg) and stem (448.342). Accumulated values in treatment four were follows;

Heavy metals	21 Days		42 Days	
	Mean (mg/kg)	Standard deviation	Mean (mg/kg)	Standard deviation
Fe	201.407	81.009	400.120	90.520
Mn	37.188	6.753	67.129	16.899
Zn	22.099	5.703	51.623	19.755
Cu	0.001	0.000	0.001	0.000

Table 4: Mean levels of heavy metals in treatment four at 21 and 42 days after transplanting.

Heavy metals	Concentrations (mg/kg)			
	Polluted water	Unpolluted water	Soil	WHO/FAO standards
Fe	113.898	<0.001	10007.000	425.50[a]
Mn	4.200	<0.001	35.060	500.00[a]
Zn	<0.001	<0.001	0.580	99.40[a]
Cu	1.500	<0.001	8.748	73.30[a]

[a]Anonymous (2001).

Table 5: Initial concentrations of heavy metals in the soil and water samples and WHO/FAO standard.

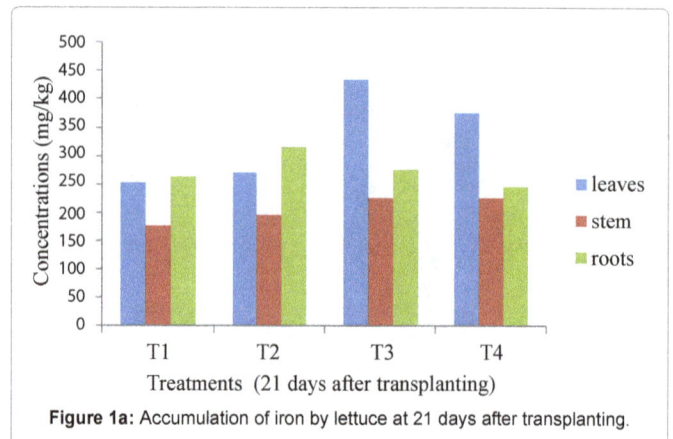

Figure 1a: Accumulation of iron by lettuce at 21 days after transplanting.

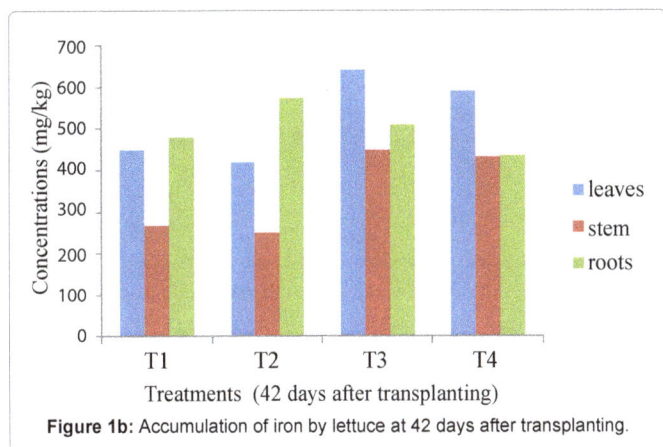

Figure 1b: Accumulation of iron by lettuce at 42 days after transplanting.

leaves (590.648 mg/kg), roots (435.062 mg/kg) and stem (432.664 mg/kg).

Manganese accumulation by lettuce

Manganese recorded varied levels of accumulation in all the treatments of the test crop (lettuce). The accumulation rate ranges from 31.886 mg/kg to 68.798 mg/kg (Figure 2a). In Manganese, the highest value of accumulation of 62.798 mg/kg was recorded in treatment three with the highest mean of accumulation of 52.509 mg/kg was recorded in treatment two (Figure 2a).

Manganese recorded different levels of accumulation in all the treatments of the test crop (Lettuce). In treatment one, manganese accumulated more in the leaves (50.944 mg/kg), followed by the roots (45.692 mg/kg) and the stem (35.638 mg/kg) with the same trend in treatment two and three. Accumulation levels in treatment two were as follows; leaves (60.396 mg/kg), roots (55.040 mg/kg) and stem (42.090 mg/kg). Accumulation levels in treatment three were as follows; leaves (62.798 mg/kg) roots (41.790 mg/kg) and stem (36.988 mg/kg). However, different accumulation values were recorded in treatment four. In treatment four, it was observed that, the manganese accumulated more in stem (44.798 mg/kg) followed by leaves (34.888 mg/kg) and roots (31.886 mg/kg).

Different accumulations of manganese were recorded in the leaves, stem and roots in all the treatments as shown in Figure 1b. Treatment one accumulated more manganese in the roots (80.224 mg/kg) followed by leaves (69.066 mg/kg) and stem (41.082 mg/kg) at 42 days after transplanting. In treatment two, manganese accumulated more in the leaves (112.668 mg/kg) followed by the roots(86.424 mg/kg) and stem (69.504 mg/kg).These same accumulations were recorded in treatment three, with leaves (128.62 8 mg/kg), roots (84.904 mg/kg) and stem (52.642). However, it was observed that, in treatment four, the stem recorded high manganese accumulation of 82.242 mg/kg, followed by roots (70.252 mg/kg) and leaves (48.886) as shown in Figure 2b above. It was observed that, highest accumulation value of 128.628 mg/kg was recorded in treatment three with least value of 41.082 mk/kg recorded in treatment one (Figure 2b). On the average, treatment two accumulated more manganese than the rest of the treatments.

Zinc accumulation by lettuce

Zinc, recorded different levels of accumulation rate in all the treatments. The highest mean of 22.500 mg/kg was recorded in treatment three with the least mean of 14.485 mg/kg recorded in treatment one.

Diverse accumulation levels of zinc were recorded at 21 days after transplanting (Figure 3a). Treatment one accumulated high level of zinc in roots (15.810 mg/kg) followed by stem (13.946) and leaves (13.698). In treatment two, leaves accumulated highest level of zinc of 22.476 mg/kg, followed by roots (22.118) and stem (18.744) with the same trend occurred at treatment three. In treatment three, leaves accumulated 31.414 mg/kg of zinc, followed by roots (18.772 mg/kg) and stem (17.318 mg/kg). Moreover, treatment four high accumulated more zinc in leaves (31.112 mg/kg), followed by stem (19.622 mg/kg) and roots (15.564 mg/kg).

At 42 days after transplanting, treatment one accumulated high amount of zinc in leaves (48.782 mg/kg), followed by the roots (39.244 mg/kg) and stem (39.166 mg/kg). In treatment two, zinc accumulated highest in the leaves (64.860 mg/kg), followed by the root (52.688 mg/kg) and stem (40.208 mg/kg). In treatment three, highest accumulation of zinc was recorded in the leaves (59.130 mg/kg) followed by stem (51.482 mg/kg) and roots (50.868). However, treatment four recorded highest accumulation of zinc in leaves (72.684 mg/kg), followed by stem (48.682 mg/kg) and roots (33.504 mg/kg) as shown in Figure 3b. Highest mean of 53.827 mg/kg was recorded in treatment three with the lowest mean of 42.399 mg/kg recorded in treatment one

Copper accumulation by lettuce

All the treatments of the test crop (lettuce) recorded accumulate values below detection level of 0.001 mg/kg.

All the treatments of the test crop (lettuce) recorded accumulation

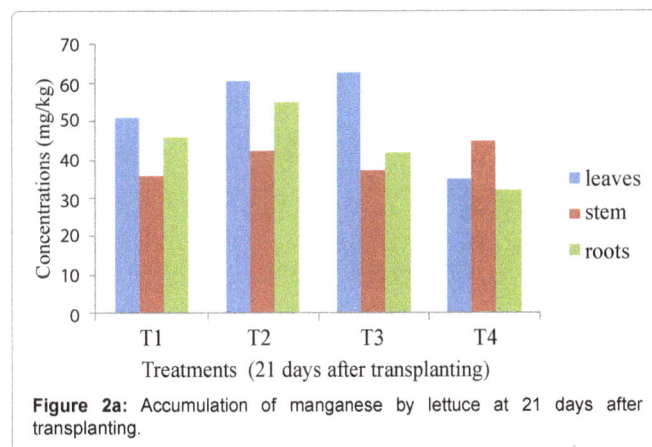

Figure 2a: Accumulation of manganese by lettuce at 21 days after transplanting.

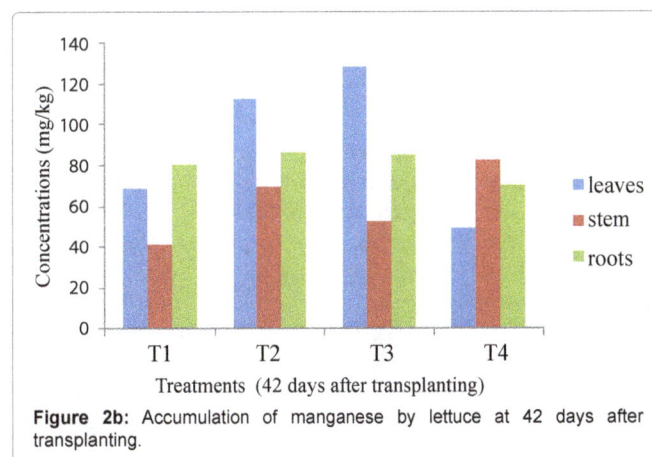

Figure 2b: Accumulation of manganese by lettuce at 42 days after transplanting.

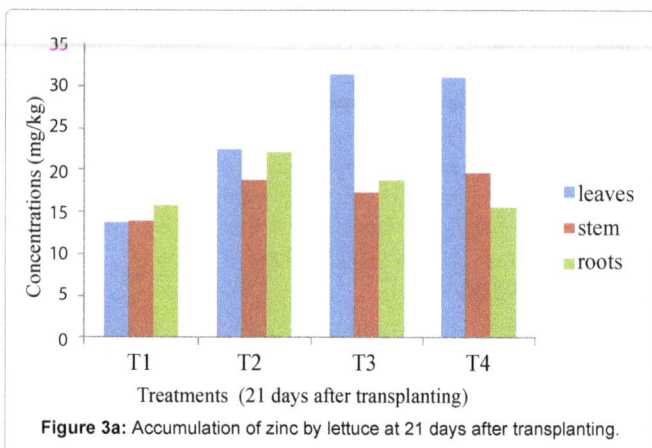

Figure 3a: Accumulation of zinc by lettuce at 21 days after transplanting.

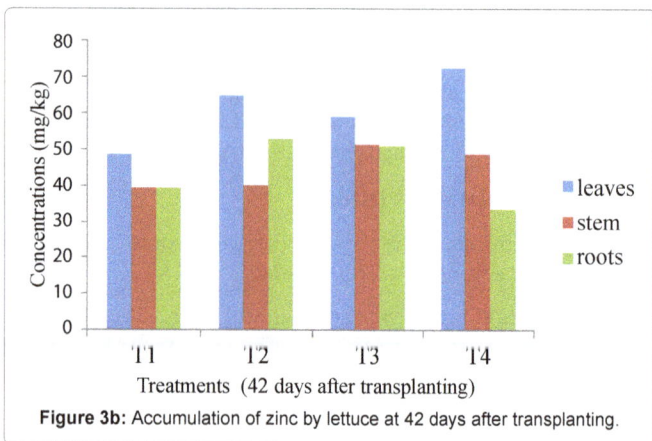

Figure 3b: Accumulation of zinc by lettuce at 42 days after transplanting.

values below detection level of 0.001 mg/kg at 21 days after transplanting as showed in Figure 4a.

All the treatments of the test crop (lettuce) recorded accumulation values below detection level of 0.001 mg/kg as happened at 21 days after transplanting as showed in Figure 4b.

Discussion

Iron (Fe) accumulation at 21 days and 42 days after transplanting

According to Figure 1a, iron recorded different levels of accumulation at 21 and 42 days after transplanting. However, treatment two accumulated value of 260.441 kg/mg which was higher than the initial value of 113.898 mg/kg of iron in the polluted water. Treatment four also accumulated value of 281.487 mg/kg iron which was higher than the initial value of 113.898 mg/kg of iron in the polluted water. It was observed that, all the treatments adsorbed values lesser than initial iron value of 10007.000 mg/kg in the soil. Initial concentration of iron in the unpolluted water was below detection level of 0.001 mg/ kg. Therefore, treatment one and three accumulations of irons were directly from the soil because unpolluted water was used in watering these treatments. This occurred because their accumulation values were above the detection level of 0.001 mg/kg of iron in the unpolluted water. High value accumulation of 281.487 mg/kg was recorded in treatment four for iron as compared to treatment two with 260.441 mg/kg which was totally different from accumulation of manganese at 21 days.

The same trend of low adsorption and removal of iron was recorded in treatment two as compared to treatment four at 42 days after transplanting. Thus, 413.913 mg/kg in treatment two compared to 486.125 mg/kg in treatment four. This trend differs from the adsorption rate of manganese accumulation at 42 days after transplanting.

However, the root of treatment two recorded the highest value of accumulation of iron of 574.668 mg/kg.

Iron accumulated more in the leaves and roots of (*Lactuca sativa*) in all the treatments both 21 and 42 days after transplanting as compare to Mn, Zn, and Cu accumulation at 21 and 42 days. This buttressed observation in (*Lactuca sativa* L.) by   Adu et al. [7] that heavy metals accumulated more in the roots and leaves as compared to other parts because both are the entry points of heavy metals from the air and soil.

Comparing the results obtained in this study with the Maximum Permissible Concentration of iron in vegetables by Anonymous [8], it was obvious that, the concentration of iron in the test crop on the average does not exceed [8] value of 425.50 mg/kg.

Manganese (Mn) accumulation at 21 days and 42 days after transplanting

Manganese recorded different levels of adsorption at both 21 and 42 days after transplanting. Treatments two adsorbed highest level of manganese than the initial concentrations of manganese in both the soil and polluted water of 35.060 mg/kg and 4.200 mg/kg respectively. Treatment two adsorbed 52.509 mg/kg and treatment four adsorbed

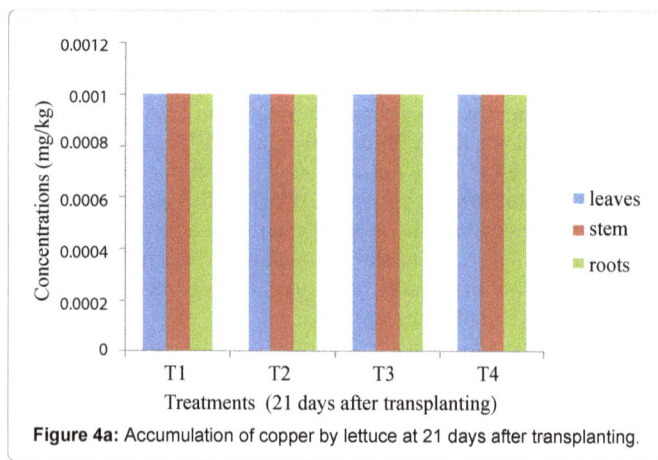

Figure 4a: Accumulation of copper by lettuce at 21 days after transplanting.

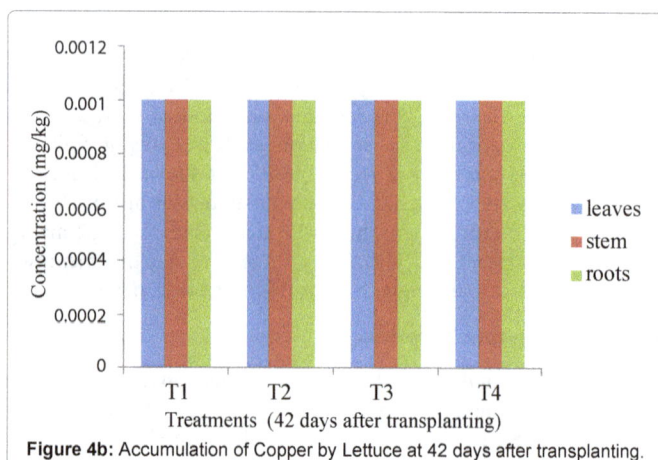

Figure 4b: Accumulation of Copper by Lettuce at 42 days after transplanting.

37.139 mg/kg. However, the total level of manganese that was present in both the soil and the polluted water was 39.260 mg/kg which was higher than treatment four accumulation value of 37.188 mg/kg at 21 days after transplanting. However, treatments one and three accumulated values of 44.091 mg/kg and 47.192 mg/kg were higher than the initial amounts of manganese in both the soil and the unpolluted water. This clearly indicated that it was treatment four that adsorbed level of manganese lower than the initial concentration of manganese in both the soil and the polluted water.

From Figure 2a, manganese generally recorded diverse levels of accumulation at 21 days after transplanting of the test crop ranging from 31.886 kg/mg to 62.798 mg/kg.

Treatment two with shea nut shells, soil and polluted water, on the average adsorbed high manganese compared to the rest of the treatments ranging from 42.090 mg/kg to 60.396 mg/kg. This demonstrated that, treatment two has strong potential for removal and adsorption of manganese indicating its strong affinity of available surface groups on shea nut shells for manganese.

Leaves of treatment two with shea nut shells, soil and polluted water at 21 days after transplanting, adsorbed high manganese, followed by the root and the stem. This might occurred because leaves of the test crop (lettuce) store high quantity of food in its leaves compared to the other parts such as root and stem and this statement is confirmed by Antwi [9] that, there was high adsorption of manganese in the leaves of cabbage than its roots and stem at 21 days after transplanting the test crop (cabbage). This was observed in maize tassel adsorption using cabbage as a test crop.

According to Figure 2b, manganese recorded different leaves of accumulation at 42 days after transplanting. Treatment two with shea nut shells, soil and polluted water on the average recorded the highest mean value of accumulation of manganese ranging from 69.504 mg/kg to 112.668 mg/kg. The root of treatment two with shea nut shells on the average recorded the highest mean value of accumulation of 86.424 mg/kg of manganese compared with other treatments. This implies that, shea nut shells in treatment two reduced the concentration of manganese in both water and soil.

Comparing the values obtained with Maximum Permissible Levels of manganese in vegetables, the concentration on average does not exceeded [8] recommended value of 500.00 mg/kg which makes the test crop suitable for consumption.

Zinc (Zn) accumulation at 21 days and 42 days after transplanting

Different levels of accumulation were recorded at both 21 and 42 days after transplanting. All the treatments adsorbed higher values of zinc than the initial value of 0.580 mg/kg in the soil and values below detection level of 0.001 mg/kg in the polluted water. Treatment one adsorbed 14.485 mg/kg with treatment two adsorbed 21.112 mg/kg, treatment three adsorbed 22.501 mg/kg and treatment four adsorbed value of 22.099 mg/kg of zinc. This clearly demonstrated that, accumulation of zinc in all the treatments were from the soil since both the polluted and unpolluted water samples iron values were below detection level of 0.001 mg/kg.

According to Figure 3a, zinc accumulation at 21 days after transplanting of the test crop ranged from 13.698 mg/kg to 31.414 mg/kg even though treatment two with shea nut shells, soil and polluted water recorded slightest reduction in average adsorption as compare to treatment three with unpolluted water and soil only.

Moreover, comparing treatment two to treatment four which is not having shea nut shells, the adsorption rate was higher in treatment two as compare to treatment four. Adsorption and removal of zinc by treatment two was not as strong as adsorption and removal of manganese by treatment two in Figure 2a.

On the average, zinc accumulation in the leaves of all the four treatments was 24.675 mg/kg which was greater than stems average accumulation rate of 17.407 mg/kg. This highest accumulation rate in leaves compared to stems was confirmed by Adu [7] that zinc content was highest in leaf and lowest in stem in (*Lactuca sativa* L).

According to Figure 3b, treatment two with shea nutshells, soil and polluted water recorded diverse levels of accumulation at 42 days after transplanting of the test crop. On the average, the accumulation rate of zinc of 52.585 mg/kg in treatment two was higher compared to treatment four with polluted water and soil accumulation value of 51.623 mg/kg.

However, it was treatment three with unpolluted water and soil that recorded the highest mean accumulation values of 22.501 mg/kg and 53.827 mg/kg of zinc at both 21 and 42 days respectively after transplanting of the test crop compared to the average accumulation values of 21.112 mg/kg and 52.585 mg/kg of treatment two at both 21 and 42 days respectively.

Comparing results obtained in zinc adsorption rate of the treatments with the Maximum Permissible levels of zinc in vegetables by Anonymous [8] it was clear that, the concentrations of zinc in the test crop on the average were below Anonymous [8] recommended value of 99.40 mg/kg.

Copper (Cu) accumulation at 21 days and 42 days after transplanting

At 21 and 42 days after transplanting, treatments two and four recorded low adsorption values compared to the initial values of 8.748 mg/kg and 1.500 mg/kg in the soil and polluted water respectively. The same low adsorption of copper was recorded in treatment one and three compared to the initial value of 8.748 mg/kg in the soil and below detection value of 0.001 mg/kg of copper in the unpolluted water. Accumulation values recorded in all the treatments were below detection level of 0.001 mg/kg. According to both Figures 4a and 4b, the accumulation of copper during 21 and 42 days after transplanting were all below detection level of 0.001 mg/kg. However, the value < 0.001 mg/kg demonstrated that various treatments adsorbed the copper content from the initial water and soil samples in smaller quantities which are not horrific to the health of human when consumed. The inability of the shea nut shells in treatment two with shea nut shells, polluted water and soil to adsorbed copper above detection level may be attributed to the fact that, copper contents have been retained in the soil sample used for the experiment.

The incapability of the shea nut shells in treatment two to remove copper above detection level of 0.001 mg/kg is contrary to Nanpuan [10] that shea nut shells were found to be a good adsorbent for the removal of heavy metals from contaminated groundwater and that adsorption capabilities of the shea nut shells could be used in heavy metal remediation from groundwater for farming, domestic and other industrial activities.

Comparing the results obtained in the copper accumulation rate of the treatments with Maximum Permissible levels of copper in vegetables by Anonymous [8], it was obvious that, the concentrations of copper in the test crop on the average were below Anonymous [8], recommended value of 73.30 mg/kg

Conclusion

Results obtained from this study demonstrated that, shea nut shells have adsorption capabilities although the highest adsorption capacity for heavy metals was recorded for manganese with zinc, iron and copper in that order. However, concentrations of manganese, iron, copper and zinc in the unpolluted water were below detection level of 0.001 mg/kg. The adsorption properties of shea nut shells may be used in remediation of contaminated water bodies that contain manganese and zinc to help curb negative health implications that will emanate from these hazards.

Recommendation

Use of shea nut shells in remediation of polluted environments is very economical and environmentally friendly. Shea nut shells are easily accessible because it is agricultural by-product that is prevalent in Northern Ghana. Further research on shea nut shells in phytoremediation should focus on determine the amounts of heavy metals retained in the soil sample after the experiment and varying the quantities of grounded shea nut shells for this study in different levels to help determine the adsorption capabilities of shea nut shells from contaminated water and soils. Also, more research should be conducted to determine the functional groups present on the surface of shea nut shells.

Acknowledgement

The authors acknowledged the support of the staff of the Department of Biotechnology and Water Research Institute of Council for Scientific and Industrial Research, Tamale.

References

1. Garbisu C, Alkorta I (2003) Basic concepts on heavy metal soil bioremediation. Eur J Min Proc & Environ Protect. 3:58-66.

2. Alkorta I, Hernandez-Allica J, Becerril JM, Amezaga I, Albizu I, et al. (2004) Recent findings on the phytoremediation of soils contaminated with environmentally toxic heavy metals and metalloids such as zinc, cadmium, lead, and arsenic. Reviews in Environmental Science and Bio/Technology 3: 71-90.

3. Agency for Toxic Substances and Disease Registry (ATSDR) (2004) Toxicological Profile for Copper.

4. Agency for Toxic Substances and Disease Registry (ATSDR) (2012) Toxicological Profile for Manganese.

5. WHO Guidelines for Drinking Water Quality (2003) Iron in Drinking-Water Quality, WHO/SDE/WSH/03.04/08.

6. Agency for Toxic Substance and Disease Registry (ATSDR) (2005) Toxicological Profile for Zinc.

7. Adu AA, Aderinola OJ, Kusemiju V (2012) Heavy Metals Concentration in Garden Lettuce (Lactuca Sativa L.) Grown along Badagry Expressway, Lagos, Nigeria. Transnational Journal of Science and Technology.

8. Anonymous (2001) Codex Alimentarius Commission (FAO/WHO) Food additives and contaminants- Joint FAO/WHO Food Standards program 2001, ALINORM 01/12A, Pp: 1-289.

9. Antwi KC (2012) Application of maize tassel to remove heavy metals from contaminated soil using cabbage as test crop. In BSc. Thesis, University for Development Studies.

10. Nanpuan MP (2013) The use of shea nut shells to phytoremediate heavy metals from contaminated ground water. In BSc. Thesis, University for Development Studies.

22

Statistical Methodology for Cadmium (Cd(II)) Removal from Wastewater by Different Plant Biomasses

Alaa El Din Mahmoud* and Manal Fawzy

Environmental Sciences Department, Faculty of Science, Alexandria University, Alexandria, Egypt

Abstract

The combined effects of metal ion concentration (X), hydrogen ion concentration (pH) and biomass dose (BD), on the biosorption of Cadmium Cd(II) were investigated. Two different plant biomasses; rice straw (Oryza sativa) and dragon tree leaves (Dracaena draca) were studied.

The optimum conditions were found at (X)=10 ppm, (pH)=7 and (BD)=0.5 g. Under these conditions, desirability values of 0.996 and 0.997 for rice straw and dragon tree leaves were obtained, showing that the calculated model may represent the experimental model and give the desired conditions. The samples before and after biosorption experiments were characterized by Energy Dispersive X-Ray Spectroscopy.

Keywords: Optimization; Cadmium; *Oryza sativa*; *Dracaena draca*; Response surface methodology

Introduction

The availability of water resources are becoming increasingly scarce; the consumption and exploitation of water resources, along with exponential increase in population have caused water pollution [1]. Toxic metals of particular concern in treatment of industrial wastewaters include: mercury, lead, cadmium, zinc, copper, nickel, and chromium [2]. So this study focuses on Cadmium (Cd(II)) that is attracting wide attention of environmentalists as one of the most toxic heavy metals. Currently methods that are being used to remove heavy metal ions include chemical precipitation, ion-exchange, adsorption, membrane filtration, electrochemical technologies. These methods are usually inadequate and expensive [3].

Biosorption is an emerging technology that is used to sequester toxic heavy metals and is particularly useful for the removal of contaminants from industrial effluents [4]. The biosorbent term refers to material derived from microbial biomass, seaweed or plants that exhibit adsorptive property [5]. Many biosorbents have been used in biosorption processes such as bacteria, fungi, algae [6] and agricultural wastes such as rice husk [7], Pequi Fruit Skin [8], *Psidium guajava* leaves powder [9], sugarcane bagasse, maize corncob, Jatropha oil cake [10] and cork waste [11].

The utilization of agricultural waste materials is increasingly becoming important concern because these wastes represent unused resources and, in many cases, present disposal problems [6]. So the use of natural biomaterials, especially crop wastes as biosorbents, is a promising alternative due to their relative abundance and their low commercial value [12]. Nearly 3 Million tons of rice straw is burned annually in the field of Egypt every year causing "Black cloud" [13]. However, no available literatures about using waste of ornamental plants as natural biosorbent.

In this work, the Central Composite Design (CCD), which is a type of Response Surface Methodology (RSM), was employed for Optimization the biosorption of Cd(II) using two different dried plant biomasses: rice straw (*Oryza sativa*) and dragon tree leaves (*Dracaena draca*); a common ornamental plant in Egyptian gardens. Samples before and after biosorption of Cd(II) were characterized using Energy Dispersive X-Ray Spectroscopy.

Materials and Methods

Biosorbent preparation

Plant biomasses were dried, then were washed with tap water to remove any dust or foreign particles attached to them and finally rinsed with deionized water. The washed biomasses were dried at 60°C for 48 hours and grounded to powder then sieved through a siever; mesh size ≤ 0.5 mm for biosorption experiments.

Reagents and equipments

Cadmium standard solution with initial concentration 1000 ppm was used to prepare experimental concentration of 10 and 100 ppm using deionized water. pH adjustment of the solutions was made by HNO_3 and NaOH utilizing a pH/mV hand-held meter (Crison pH meter, PH 25).

Biosorption experiments

Response surface methods are used to examine the relationship between response variable (RF%) and the studied factors (X, pH and BD). RSM is applied to optimize the studied factors that produce the best response and model a relationship between the factors and the response [14]. All data were analyzed using MINITAB°16 software. A2³ full factorial central composite design with two coded levels was performed. For statistical calculation, the variables were coded according to Eq. (1):

$$x_i = X_i - X_0 / \Delta X \tag{1}$$

Where x_i is the dimensionless coded value of the variable Xi, X0 the middle value of X_i, and ΔX the step change.

*Corresponding author: Alaa El Din Mahmoud, Environmental Sciences Department, Faculty of Science, Alexandria University, Alexandria, Egypt
E-mail: alaa-mahmoud@alexu.edu.eg

Batch experiments were conducted with the following conditions: 0.5 g of each biomass and 100 ml of Cd(II) solution with an agitation speed 300 rpm (round per minute) at room temperature. The influence of three factors i.e., initial metal ion concentration (X), hydrogen ion concentration (pH) of the solution, biomass dose (BD) have been investigated. The range and the levels of the variables investigated in this research are given in Table 1.

Then samples were collected after 2 hours to reach equilibrium in biosorption. Control samples were prior to batch biosorption experiment to determine initial metal concentration and all samples were conducted in triplicate. The metal ions contents in all the samples prior to and after batch biosorption experiments were analyzed by Varian Inductively Coupled Plasma (ICP-AES).

Removal efficiency (RF%) of biosorbent was calculated using the following equation

$$\text{Removal effeciency}\% = \frac{Ci - Cf}{Ci} \times 100 \qquad (2)$$

Where: C_i= Initial concentration of metal in solution, before the sorption analysis (mg/l), C_f = Final concentration of metal in solution, after the sorption analysis (mg/l).

Characterization of biosorbents

Energy Dispersive X-Ray Spectroscopy (EDAX): EDAX spectra can be collected from a specific point on the sample, giving an analysis of a few cubic microns of material. Each biosorbent was characterized by EDAX before and after Cd(II) biosorption.

Results and Discussion

Biosorption experiments

Batch experiments were conducted as tabulated in Table 2, '+1' for the higher level and '−1' for the lower level of the studied factors. Removal efficiency percentage (RF%) were calculated according to Eq.(1).

Regression coefficients (Coef) and the associated standard errors (SE Coef) of results are shown in Table 3. Results revealed that all the studied factors together with their interactions were significant at 95% confidence limits (P>0.05). The response variable (Cd(II) removal %) was fitted by the following equation:

$$Y = A + a_1x_1 + a_2x_2 + a_3x_3 + a_4x_1x_2 + \\ a_5x_1x_3 + a_6x_2x_3 + a_7x_1x_2x_3 \qquad (3)$$

Where: Y: Estimated of the response, A: represents the global mean (constant), a: Coefficients and x: Experimental Factors.

At X=10 ppm, pH=7 and BD=0.5 g, the highest percentage of Cd(II) removal by rice straw was 82.60% while that for dragon tree leaves was 79.60% (Table 2).

It worth noting that the effect of all the studied main factors (X, pH, BD) was identical for both biosorbents. As such, our results demonstrated that the factor (X) had the largest effect on biosorption process by rice straw and dragon tree leaves (Table 3). Results also

showed that Cd(II) biosorption was favored at low metal concentration values (X=10 ppm). This is in line with [15,16]. In the current work, the biosorption percentage was decreased as the metal ion concentration from 10 to 100 ppm. This is may be because the biomass surface area available for metal biosorption was higher the ratio of active adsorption sites to the initial Cd(II) ions is larger, resulting in higher removal efficiency [17]. This is in agreement with many researchers [6,11].

The second important main factor in the biosorption process was pH. Results indicated that as the pH value increases, Cd(II) biosorption increases by both biosorbents (Table 3). At lower pH values, the H_3O^+ ions compete with the metal ions for the active sites on the biosorbent [18]. In our work, the optimum higher pH value for Cd(II) biosorption was 7. The hydrolysis of Cd(II) ions occur beyond pH=7 as reported in [10].

In this account, the third main factor in the biosorption process was BD. Results indicated that as (BD) increases, Cd(II) biosorption increases by both biosorbents (Table 3). An increase in the biomass dosage generally increases the amount of solute biosorbed, due to the increased surface area of the biosorbent, which in turn increases the number of binding sites [19-21]. Data obtained from the response surface plots of both biosorbents are illustrated in Figures 1-3. These plots are used to visualize the relationship between response (%RF) and the level of each studied factors. Every one of them is mapped against two experimental factors while the third is fixed at two different levels [22].

Figure 1 illustrated the removal efficiency of Cd(II) by both biosorbents over (pH) and (BD). At constant metal ion concentration (100 ppm, 10 ppm), a remarkable increase in Cd(II) removal was attained as pH increases till reaching its maximum at pH=7 for both biosorbents. However, a slight increase in Cd(II) removal was observed as (BD) increases till reaching its maximum at BD=0.5 g.

Figure 2 illustrated the removal efficiency of Cd(II) by both biosorbents over (X) and (BD). When keeping pH constant (7, 2) for both biosorbents, a remarkable increase in Cd(II) removal was attained as (X) decreases till reaching its maximum at X=10 ppm for both biosorbents. However, a slight increase in Cd(II) removal was observed as BD increases till reaching its maximum at BD=0.5 g for both biosorbents.

Figure 3 illustrated the removal efficiency of Cd(II) by both biosorbents over (X) and (pH). At constant biomass dose (0.5 g, 0.1 g), a remarkable increase in Cd(II) removal was attained as (X) decreases till reaching its maximum at X=10 ppm for both biosorbents. However,

Factors			Rice straw	Dragon tree leaves
X	pH	BD	RF-R2 Average (%)	RF-D2 Average (%)
-1	-1	-1	31.00	48.38
-1	-1	1	61.99	67.63
-1	1	-1	72.78	76.06
-1	1	1	82.30	79.40
1	-1	-1	7.79	13.03
1	-1	1	28.09	35.32
1	1	-1	27.27	45.44
1	1	1	60.91	55.16

Table 2: Experimental factorial design results for Cd(II) biosorption(X: metal ion concentration, pH: hydrogen ion concentration, BD: biomass dose).

Factor	Low level	High level
X(mg/l)	10	100
pH	2	7
BD(g)	0.1	0.5

Table 1: High and low levels of the studied factors.

Biosorbents			Rice straw					Dragon tree leaves			
Term	Effect	Coef	SE Coef	T	p		Effect	Coef	SE Coef	T	p
Constant		46.51	1.652	28.16	0.000			52.55	0.5781	90.90	0.000*
Main Factors											
X	-31.00	-15.50	1.652	-9.39	0.000		-30.63	-15.32	0.5781	-26.49	0.000*
pH	28.60	14.30	1.652	8.66	0.000		22.93	11.46	0.5781	19.83	0.000*
BD	23.61	11.81	1.652	7.15	0.000		13.65	6.83	0.5781	11.81	0.000*
Two Factors Interaction											
X.pH	-2.45	-1.22	1.652	-0.74	0.480		3.20	1.60	0.5781	2.77	0.024*
X.BD	3.36	1.68	1.652	1.02	0.339		2.36	1.18	0.5781	2.04	0.076
pH.BD	-2.03	-1.02	1.652	-0.61	0.556		-7.12	-3.56	0.5781	-6.16	0.000*
Three Factors Interaction											
X.pH.BD	8.70	4.35	1.652	2.63	0.030		0.83	0.42	0.5781	0.72	0.491

* P>0.05

Table 3: Response surface regression for Cd(II) removal by rice straw and dragon tree leaves (T: t-test, p: probability value).

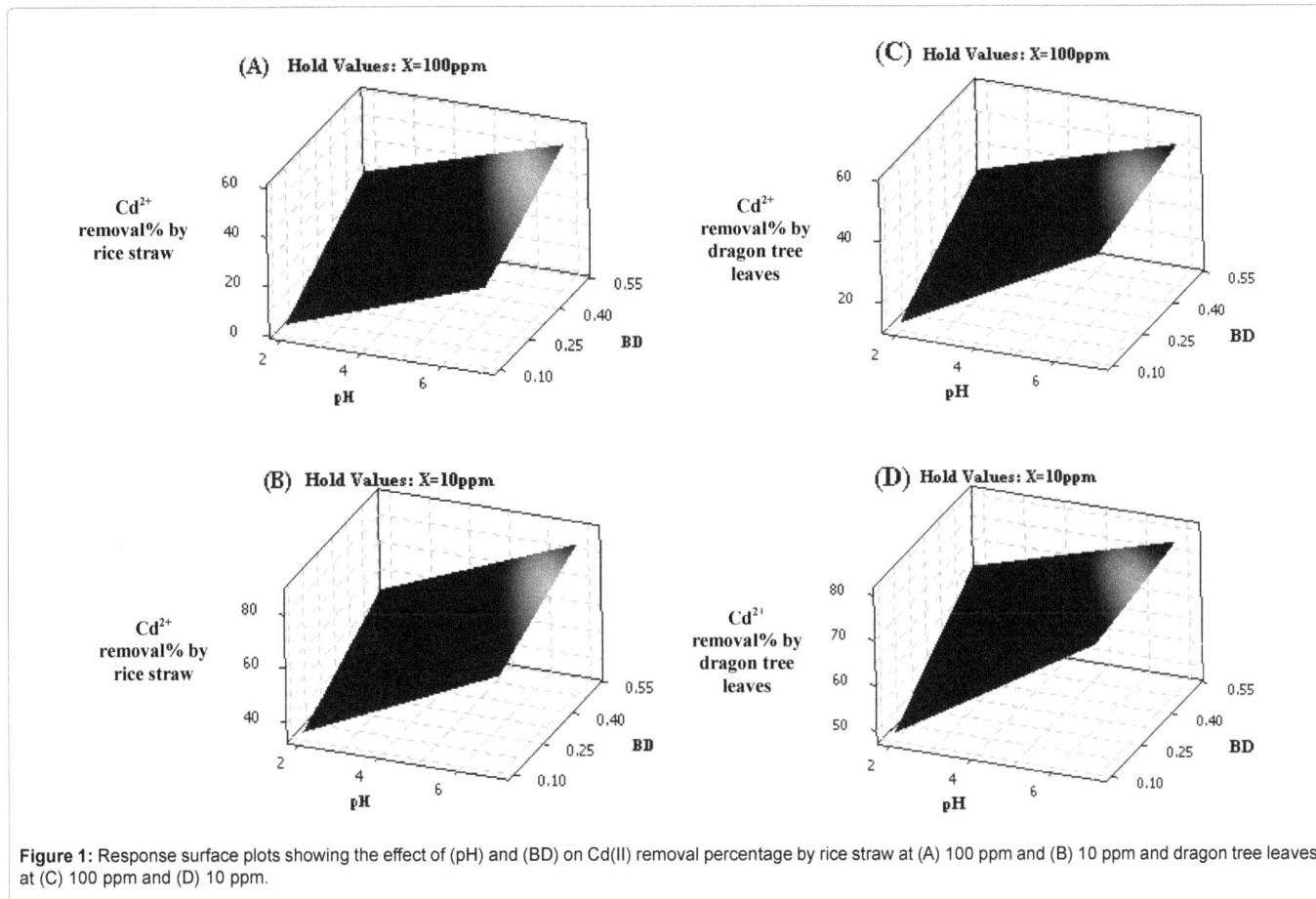

Figure 1: Response surface plots showing the effect of (pH) and (BD) on Cd(II) removal percentage by rice straw at (A) 100 ppm and (B) 10 ppm and dragon tree leaves at (C) 100 ppm and (D) 10 ppm.

a slight increase in Cd(II) removal was observed as pH increases till reaching its maximum at pH=7.

Analysis of variance (ANOVA - Table 4) showed the sum of squares used to estimate the factors' effect and the F-ratios defined as the ratio of the respective mean-square-effect and the mean-square-error. The significance of the present biosorption models as assessed by F-values and P-values indicated that the studied factors and their interactions (X.pH.BD) except (X.pH, X.BD, pH.BD) are statistically significant in the case of rice straw and the studied factors and their interactions except (X.BD and X.pH.BD) are statistically significant in the case of dragon tree leaves.

Characterization of biosorbents

The results of EDAX (Figure 4) showed that raw biosorbents did not contain any Cd(II) ions on their surfaces and these ions appeared only after batch biosorption experiments.

Response optimization

After Response Surface Methodology was carried out, Minitab's Response Optimizer was used to get the optimized factors and responses. The goal for the studied factors (X, pH, BD) was to maximize them as listed in Table 5.

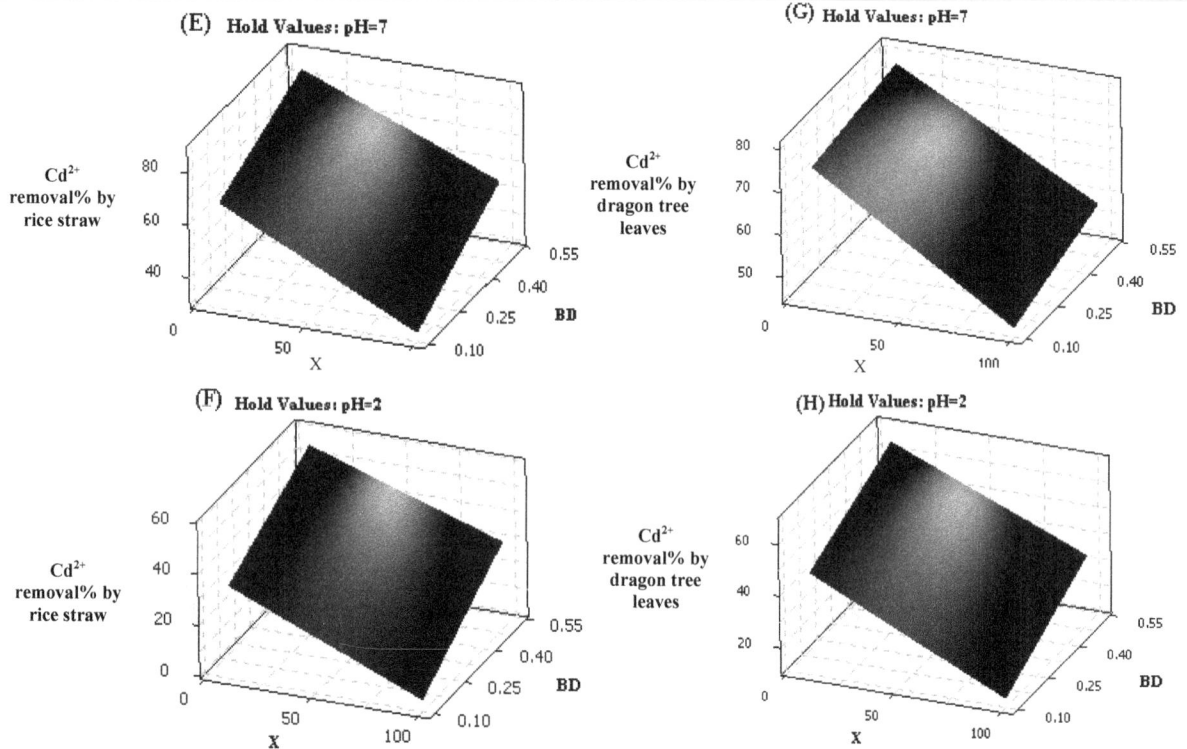

Figure 2: Response surface plots showing the effect of (X) and (BD) on Cd(II) removal percentage by rice straw at (E) pH=7 and (F) pH=2 and dragon tree leaves at (G) pH=7 and (H) pH=2.

		Rice straw					Dragon tree leaves			
Source	DF	Seq SS	Adj MS	F	P	DF	Seq SS	Adj MS	F	P
Main Effects	3	9346.4	3115.48	71.37	0.000	3	6601.06	2200.35	411.45	0.000
X	1	3844.9	3844.93	88.09	0.000	1	3753.16	3753.16	701.82	0.000
pH	1	3271.6	3271.55	74.95	0.000	1	2102.59	2102.59	393.17	0.000
BD	1	2230.0	2229.96	51.09	0.000	1	745.32	745.32	139.37	0.000
2-Way Interactions	3	85.5	28.50	0.65	0.603	3	266.08	88.69	16.58	0.001
X.pH	1	23.9	23.94	0.55	0.480	1	40.97	40.97	7.66	0.024
X.BD	1	45.1	45.06	1.03	0.339	1	22.23	22.23	4.16	0.076
pH.BD	1	16.5	16.50	0.38	0.556	1	202.88	202.88	37.94	0.000
3-Way Interactions	1	302.7	302.67	6.93	0.030	1	2.78	2.78	0.52	0.491
X.pH.BD	1	302.7	302.67	6.93	0.030	1	2.78	2.78	0.52	0.491
Residual Error	8	349.2	43.65			8	42.78	5.35		
Pure Error	8	349.2	43.65			8	42.78	5.35		
Total	15	10083.8				15	6912.70			

Table 4: Analysis of Variance.

	Goal	Lower	Target	Upper	Predicted Responses	Desirability
Rice straw (RF-R2)	Maximum	7.65	82.60	82.60	82.30	0.996
Dragon tree leaves (RF-D2)	Maximum	12.81	79.60	79.60	79. 39	0.997

Table 5: Parameters of Response Optimization.

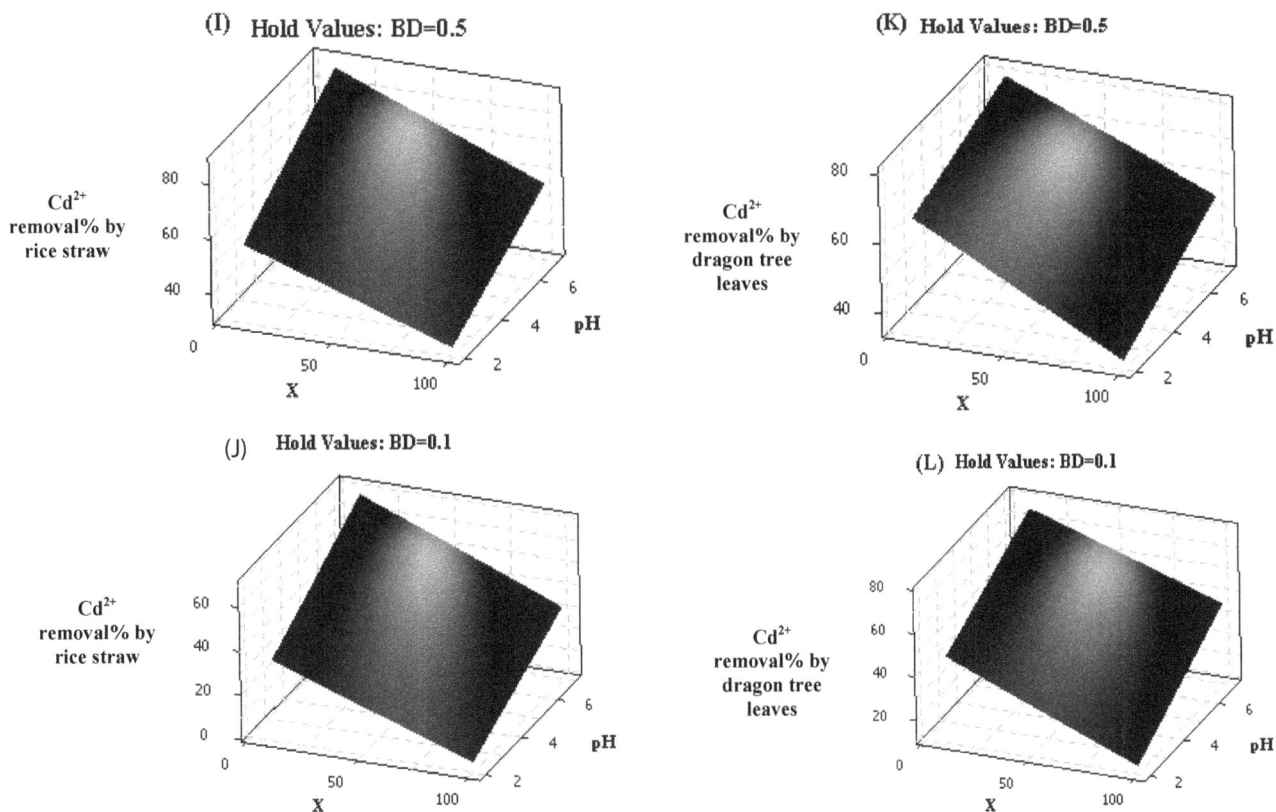

Figure 3: Response surface plots showing the effect of (X) and (pH) on Cd(II) removal percentage by rice straw at (I) BD=0.5 and (J) BD=0.1 and dragon tree leaves at (K) BD=0.5 and (L) BD=0.1.

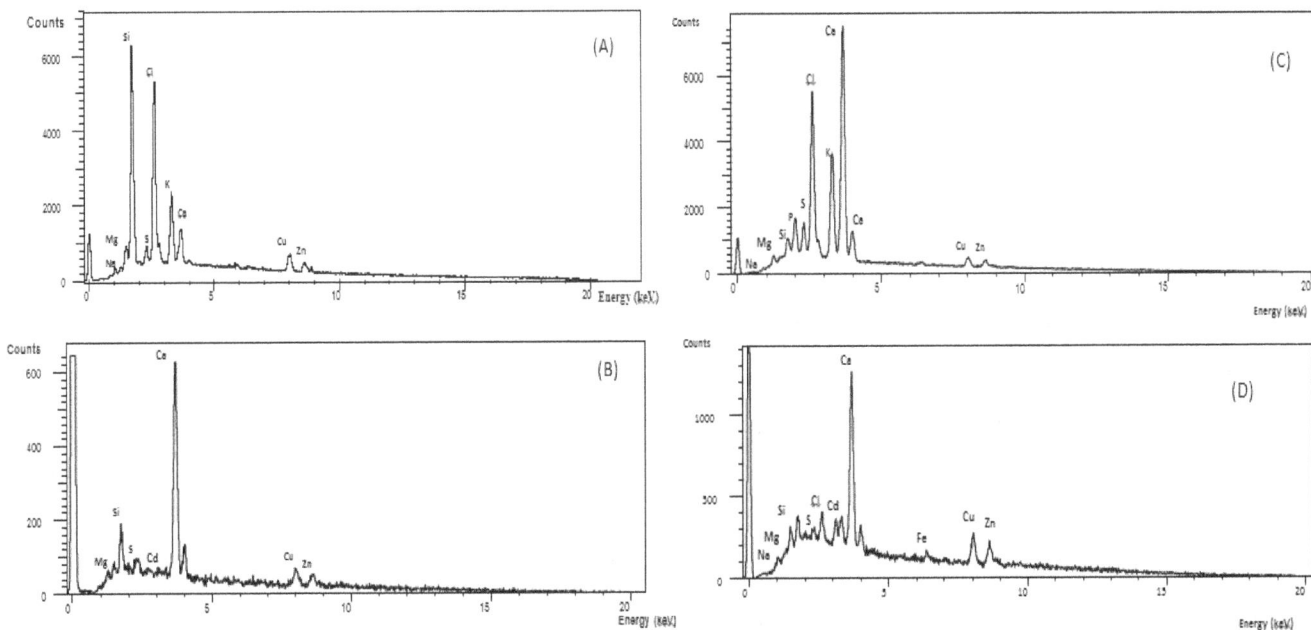

Figure 4: EDAX images of; (A) Raw rice straw, (B) Rice straw after Cd(II) biosorption, (C) Dragon tree leaves and (D) Dragon tree leaves after Cd(II) biosorption.

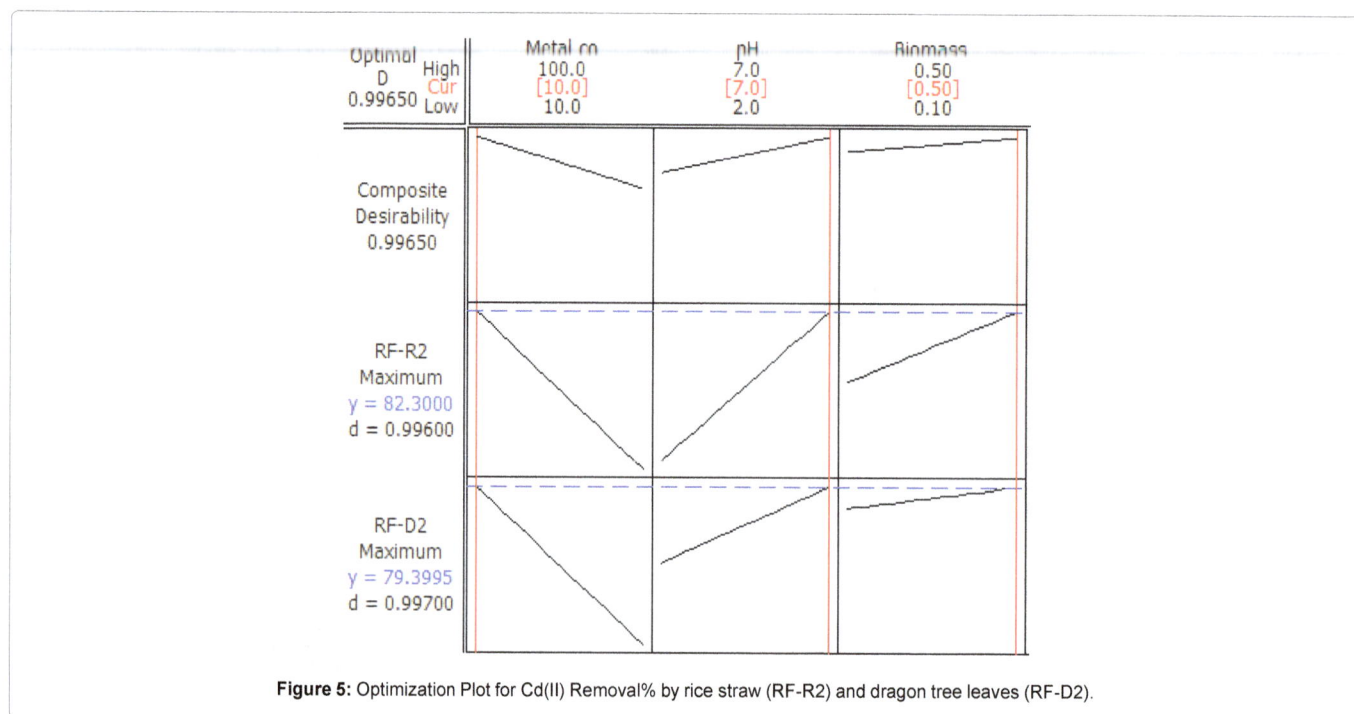

Figure 5: Optimization Plot for Cd(II) Removal% by rice straw (RF-R2) and dragon tree leaves (RF-D2).

All results had relatively high desirability scores of rice straw and dracaena draca were 0.961 and 0.970, respectively as listed in Table 5 because the predicted response of them were 82.30 and 79.39 which were quite close to the targets of each one of 82.60 and 79.60, respectively and optimization plot was shown in Figure 5. Desirability is an objective function that ranges from zero outside of the limits to one at the goal [23,24]. The composite desirability (D) of 0.99650 combined the individual desirabilities and it is high as it is closer to 1 and the best removal percentage of Cd(II) obtained at X=10 ppm, pH=7 and BD=0.5 g for each biosorbent where the vertical lines on the graph represent the current factor settings, the horizontal datch lines represent the current response values

Conclusion

It may be concluded that:

- The most significant effect for Cd(II) biosorption by rice straw and dragon tree leaves was ascribed to (X).

- Main factors exert more effect than interaction factors by both biosorbents.

- Ion exchange and complexation processes are the mechanisms of biosorption that occurred in rice straw and dragon tree leaves, respectively.

- EDAX confirmed biosorption process by the changes occurred on the surfaces of both biosorbents.

- Desirability values (0.996 and 0.997) indicated the calculated model can represent the experimental model and give the desired conditions for both biosorbents.

References

1. Singh R, Chadetrik R, Kumar R, Bishnoi K, Bhatia D, et al. (2010) Biosorption optimization of lead(II), cadmium(II) and copper(II) using response surface methodology and applicability in isotherms and thermodynamics modeling. Journal of Hazardous Materials 174: 623-634.

2. Mousavi HZ, Hosseynifar A, Jahed V, Dehghani SAM (2010) Removal of Lead from aqueous Solution using waste Tire Rubber Ash as an Adsorbent. Brazilian Journal of Chemical Engineering 27: 79-87.

3. Reddy DHK, Seshaiah K, Reddy AVR, Rao MM, Wang MC (2010) Biosorption of Pb2+ from aqueous solutions by Moringa oleifera bark: Equilibrium and kinetic studies. Journal of Hazardous Materials 174: 831-838.

4. Reddy KO, Maheswari CU, Reddy DJP, Guduri B, Rajulu AV (2010) Properties of ligno-cellulose ficus religiosa leaf fibers. International Journal of Polymers and Technologies 2: 29-35.

5. Nasr M, Mahmoud A, Fawzy M, Radwan A (2015) Artificial intelligence modeling of cadmium(II) biosorption using rice straw. Applied Water Science: 1-9.

6. Sulaymon AH, Mohammed AA, Al-Musawi TJ (2012) Competitive biosorption of lead, cadmium, copper, and arsenic ions using algae. Environmental Science and Pollution Research: 1-13.

7. Krishnani KK, Meng X, Christodoulatos C, Boddu VM (2008) Biosorption mechanism of nine different heavy metals onto biomatrix from rice husk. Journal of Hazardous Materials 153: 1222-1234.

8. Seolatto AA, Silva Filho CJ, Mota DLF (2012) Evaluation of the Efficiency of Biosorption of Lead, Cadmium, and Chromium by the Biomass of Pequi Fruit Skin (Caryocar brasiliense Camb.). Chemical Engineering Transactions 27: 1974-9791.

9. Varma DSNR, Srinivas C, Nagamani C, PremSagar T, Rajsekhar M (2010) Studies on biosorption of Cadmium on Psidium guajava leaves powder using statistical experimental design Journal of Chemical and Pharmaceutical Research 2: 29-44

10. Garg U, Kaur MP, Jawa GK, Sud D, Garg VK (2008) Removal of cadmium (II) from aqueous solutions by adsorption on agricultural waste biomass. Journal of Hazardous Materials 154: 1149-1157.

11. Lopez-Mesas M, Navarrete ER, Carrillo F, Palet C (2011) Bioseparation of Pb(II) and Cd(II) from aqueous solution using cork waste biomass. Modeling and optimization of the parameters of the biosorption step. Chemical Engineering Journal 174: 9-17.

12. Barka N, Abdennouri M, El Makhfouk M, Qourzal S (2013) Biosorption characteristics of cadmium and lead onto eco-friendly dried cactus (Opuntia ficus indica) cladodes. Journal of Environmental Chemical Engineering.

13. Bakker R, Elbersen W, Poppens R, Lesschen JP (2013) Rice Straw and Wheat Straw - Potential feedstocks for the Biobased Economy. Wageningen UR, Food & Biobased Research.

14. Pavan FA, Gushikem Y, Mazzocato AC, Dias SLP, Lima EC (2007) Statistical design of experiments as a tool for optimizing the batch conditions to methylene blue biosorption on yellow passion fruit and mandarin peels. Dyes and Pigments 72: 256-266.

15. Saadat S, Karimi-Jashni A (2011) Optimization of Pb(II) adsorption onto modified walnut shells using factorial design and simplex methodologies. Chemical Engineering Journal 173: 743-749.

16. Amarasinghe B, Williams R (2007) Tea waste as a low cost adsorbent for the removal of Cu and Pb from wastewater. Chemical Engineering Journal 132: 299-309.

17. Singh N, Gadi R (2012) Bioremediation of Ni(II) and Cu(II) from wastewater by the nonliving biomass of Brevundimonas vesicularis. Journal of Environmental Chemistry and Ecotoxicology 4: 137-142.

18. Arief VO, Trilestari K, Sunarso J, Indraswati N, Ismadji S (2008) Recent progress on biosorption of heavy metals from liquids using low cost biosorbents: characterization, biosorption parameters and mechanism studies. CLEAN-Soil Air Water 36: 937-962.

19. Lichtfouse E, Schwarzbauer J, Robert D, Mudhoo A, Garg V, et al. (2012) Heavy Metals: Toxicity and Removal by Biosorption. Environmental Chemistry for a Sustainable World, Springer Netherlands pp. 379-442

20. Garg UK, Kaur M, Garg V, Sud D (2007) Removal of hexavalent chromium from aqueous solution by agricultural waste biomass. Journal of Hazardous Materials 140: 60-68.

21. Badr N, Al-Qahtani KM (2013) Treatment of wastewater containing arsenic using Rhazya stricta as a new adsorbent. Environmental Monitoring and Assessment 185: 9669-9681.

22. Sarkar M, Majumdar P (2011) Application of response surface methodology for optimization of heavy metal biosorption using surfactant modified chitosan bead. Chemical Engineering Journal 175: 376-387.

23. Amini M, Younesi H, Bahramifar N (2009) Statistical modeling and optimization of the cadmium biosorption process in an aqueous solution using Aspergillus niger. Colloids and Surfaces A: Physicochemical and Engineering Aspects 337: 67-73.

24. Amini M, Younesi H, Bahramifar N, Lorestani AAZ, Ghorbani F, et al. (2008) Application of response surface methodology for optimization of lead biosorption in an aqueous solution by Aspergillus niger. Journal of Hazardous Materials 154: 694-702.

23

Monitoring of Pollution Using Density, Biomass and Diversity Indices of Macrobenthos from Mangrove Ecosystem of Uran, Navi Mumbai, West Coast of India

Prabhakar R. Pawar*

Veer Wajekar Arts, Science and Commerce College, Mahalan Vibhag, Phunde, MH, India

Abstract

In this study, density, biomass and diversity indices of selected macrobenthos were assessed from substations along Sheva creek and Dharamtar creek mangrove ecosystems of Uran [Raigad], Navi Mumbai, west coast of India from April 2009 to March 2011. A total of 86 species of macrobenthos representing 61 genera and 45 families were identified comprising of gastropods, pelecypods, cephalopods, polychaetes, sponges, crabs, prawns and shrimps. Higher values of density, biomass and diversity indices were recorded during pre-monsoon and post-monsoon than the monsoon. Diversity values in the study area ranged from 0.203 to 0.332 indicating heavy pollution and the macro benthic fauna is under stress due to discharge of domestic wastes and sewage, effluents from industries, oil tanking depots and also from maritime activities of Jawaharlal Nehru Port Trust [JNPT], hectic activities of Container Freight Stations [CFS], and other port wastes. This study reveals that macro benthic fauna from mangrove ecosystems of Uran is facing the threat due to anthropogenic stress.

Keywords: Biomass; Community structure; Diversity indices; Jawaharlal Nehru Port; Pollution; Population density; Species composition; Uran

Introduction

Plants ecosystems are a habitat for a wide variety of species, some occurring in high densities and provide food and shelter for a large number of commercially valuable finfish and shellfishes [1,2]. Mangroves are one of the biologically diverse ecosystems in the world, rich in organic matter and nutrients and support very large biomass of flora and fauna [3].

In India, 0.14% of the country's total geographic area is under mangroves and it account for about 5% of world's mangrove vegetation [4]. The Indian mangroves cover about 4827 Km², with about 57% of them along the east coast, 23% along the west coast, and 20% in Andaman and Nicobar Islands [5]. Anthropogenic activities involving development projects have resulted in depletion of coastal resources, destruction of mangrove habitats, disruption of ecosystem processes, and loss of biodiversity [6].

Mumbai, a major metropolis and generates 0.85 million m³/d of liquid effluent and 14,600 t/d of solid waste, which without any treatment are discharged in the coastal region in and around Mumbai [7]. Estimates of area of mangroves in Mumbai varied from 248.7 Km² [8] to 200 Km² [9] to 92.94 Km² [10] to 26.97 Km² [11,12] reported that Mumbai has lost 40% of all its mangroves in the past decade because of overexploitation and unsustainable demand for housing, slums, sewage treatment, and garbage dumps.

Mangroves are inhabited by a variety of macrobenthic invertebrates, which have a profound effect on sediment structure and their biochemical processes by enhancing the porosity and water flow through the sediments [13]. Macrobenthic fauna have been studied more widely than others because of their high commercial value [14]. They play an important role in the cycling of matter and energy in mangrove ecosystems [15,16]. Benthic communities are highly affected by all the environmental parameters governing the distribution and diversity variation of the macrofaunal community in Pondicherry mangroves [17].

Investigation of the population density and biomass of the organisms in creeks and estuaries is useful for the environmentalists to get enough information about the life span of important resource fauna [18]. It gives a lot of information about the inflow of the young ones, the fry, total biomass, maturity, spawning, breeding and fecundity of the organisms of that region [19]. Understanding the structure of the benthic faunal communities in relation to the impacts of pollution is an important part of monitoring changes in mangrove ecosystems in India [20]. Pearson and Rosenberg [21] have demonstrated that diversity indices provide important insights into faunal communities at different stages in succession. In biodiversity investigations, assessment of diversity indices is essential to determine the community structure of a particular ecosystem [22,23].

The coastal environment of Uran [Navi Mumbai] has been under considerable stress since the onset of industries like Oil and Natural Gas Commission [ONGC], Liquid Petroleum Gas Distillation Plant, Grindwell Norton Ltd., Gas Turbine Power Station, Bharat Petroleum Corporation Limited Gas Bottling Plant, Jawaharlal Nehru Port [JNP, an international port], Nhava-Seva International Container Terminal [NSICT], Container Freight Stations [CFS], etc. These activities affect the quality of mangrove ecosystems [24]. Although many studies have been undertaken to evaluate the macrobenthos of mangrove ecosystems in India, no scientific studies have been carried out on community structure of macrobenthos from mangroves of Uran, Navi Mumbai; hence, the present study is undertaken. Objective of the

---

*Corresponding author: Prabhakar Pawar R, Veer Wajekar Arts, Science and Commerce College, Mahalan vibhag, Phunde Tal. – Uran, Dist. Raigad, Navi Mumbai-400 702, Maharashtra, India
E-mail: prpawar1962@rediffmail.com, prabhakar_pawar1962@yahoo.co.in

present study is to evaluate the community structure of macrobenthos in relation to the impacts of pollution from mangrove ecosystems of Uran, Navi Mumbai with respect to population density, biomass and diversity indices.

Materials and Methods

Study area

Geographically, Uran [Lat. 180 50' 5" to 180 50' 20" N and Long. 720 57' 5" to 720 57' 15" E] with the population of 23,254 is located along the eastern shore of Mumbai harbor opposite to Coloba. Uran is bounded by Mumbai harbor to the northwest, Thane creek to the north, Dharamtar creek and Karanja creek to the south, and the Arabian Sea to the west. Uran is included in the planned metropolis of Navi Mumbai and its port, the Jawaharlal Nehru Port [JNPT] (Figure 1).

The mangrove ecosystem of Uran is a tide-dominated and the tides are semidiuranal. The average tide amplitude is 2.28 m. The flood period lasts for about 6-7 h and the ebb period lasts for about 5 hrs. The average annual precipitation is about 3884 mm of which about 80% is received during July to September. The temperature range is 12–36°C, whereas the relative humidity remains between 61% and 86% and is highest in the month of August. Four species of true mangroves representing three genera and three families were recorded during present study. The dominant species are Avicennia marina, Avicennia officinalis, Acanthus ilicifolius, and Ceriops tagal. The average tree height is 2.4 m and the canopy coverage is greater than 90%.

Sampling procedures

The present study was carried out for a period of two years, i.e., from April 2009 to March 2011. Two study sites, namely Sheva creek, site I [Lat. 18050'20" N and Long. 72057'50" E] and Dharamtar creek, site II [Lat.18050'50" N and Long. 72057'10" E] separated approximately by 10 km, were selected along the coast. At each site, three sampling stations separated approximately by 1 km were established for assessment of density, biomass and diversity indices of selected macrobenthos.

The selected sites were visited fortnightly at spring low tide from April 2009 to March 2011. The intertidal area was divided into 3 zones i.e. High water zone [HWZ], mid water zone [MWZ] and Low water zone [LWZ] following Bhatt [25] and Parulekar [26]. From selected sites, macrobenthos were collected and processed as per the recommendations of Holme and McIntyre [27]. Identification of macro benthos was done following the work of Hornell [28], Menon [29], Subrahmanyam [30-32], Chhapgar [33,34], Apte [35] and Khan and Murugesan [23].

Population study of selected macro fauna

The abundantly recorded macrobenthic fauna from mangrove ecosystem is considered for population studies and their density, biomass and diversity Indices were assessed following standard methods [27,36,37].

Population density: Number of selected macrobenthos present in one m^2 area was considered for assessment of population density. The macro benthos collected from fixed transects of one m^2 area, each from upper, middle and lower littoral zones were counted and average number of each species was recorded. Density of each species was expressed as average No/m^2 [36].

Biomass: Macrobenthos of different species collected from one m^2

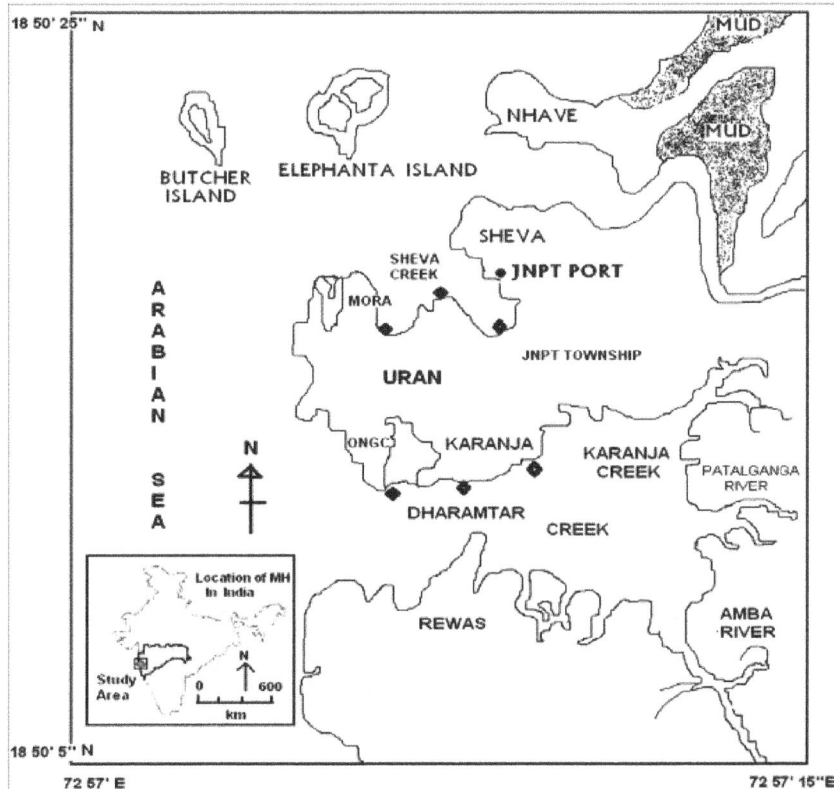

Figure 1: Location map of study area representing various sampling stations along Sheva creek and Dharamtar creek.

area were shelled and average wet weight was measured [38]. Biomass of each species was obtained by multiplication of average wet weight with average density and was expressed as g/m².

Diversity indices: Following indices were calculated for the quantification of biodiversity and comparison of species diversity.

Index of Frequency [F] OR Importance Probability [Pi] Smith and Smith [39]

$$Pi \ or \ F = \frac{ni}{N}$$

Where,

ni=Number of individuals of each species in the community.

N=Total number of all individuals of all species in the community.

Index of dominance [c] Simpson, [40]

$$C = \Sigma(ni \ / \ N^2$$

Where,

ni=Number of individuals of each species in the community

N=Total number of all individuals of all species in the community

Rarity Index [R] Ludwig and Reynolds, [41]

$$R = \frac{1}{F}$$

Where, F=Index of frequency

Shannon's Index of General Diversity [H'] Shannon and Weaver, [42]

H'=- Σ Pi. Log e Pi

Where, Pi=Importance Probability

log e=ln [log natural] OR

=log10 X 2.303

Results

Species composition of macrobenthos

A total of 86 species of macrobenthos representing 61 genera and 45 families were recorded from the mangroves of Uran coast. Varied diversity of macrobenthos belonging to gastropods, pelecypods, cephalopods, polychaetes, sponges, crabs, prawns and shrimps is recorded from both sites. Of the recorded species, 44.19% belonged to gastropods, 15.12% each to pelecypods, crabs and prawns and shrimps, 4.65% each to cephalopods and polychaetes and 1.16% to sponges (Figure 2).

Population studies of selected macrobenthos

Among the recorded macrobenthos, abundantly recorded species like Thais cariniferas, Perinereis cultrifera and Uca annulipes were selected for assessment of density, biomass and diversity Indices following standard methods [27,36,37].

Population density

Population density of macrobenthos in mangroves of Uran is high during pre-monsoon and post-monsoon than the monsoon (Figure 3 and Table 1). Maximum density of *P. cultrifera* in the range of 164 ± 8 to 222 ± 25 no/m² was recorded at site I and 188 ± 5 to 270 ± 17 no/m² at site II. Minimum density was noted for *T. carinifera* in the range of 24 ± 4 to 33 ± 2 at site I and 34 ± 4 to 50 ± 4 at site II. *U. annulipes* has moderate density in the range of 37 ± 4 to 55 ± 6 at site I and 46 ± 2 to 66 ± 3 at site II.

Biomass

The biomass follows the trend of population density with lowest value in monsoon and highest values in pre-monsoon and post-monsoon (Figure 4 and Table 2). Biomass of *T. carinifera* was highest in the range of 36.60 ± 5.81 to 48.47 ± 2.24 g/m² at site I and 54.90 ± 6.97 to 81.25 ± 8.08 g/m² at site II. *P. cultrifera* shows moderate biomass in the range of 30.80 ± 1.48 to 49.20 ± 5.19 g/m² at site I and 44.58 ± 3.74 to 70.39 ± 2.75 g/m² at site II. Lowest biomass in the range of 14.33 ± 1.58 to 26.96 ± 5.63 g/m² at site I and 22.13 ± 3.82 to 39.48 ± 1.66 g/m² at site II was recorded for *U. annulipes*. The degree of biomass can be put as *T. carinifera > P. cultrifera > U. annulipes*.

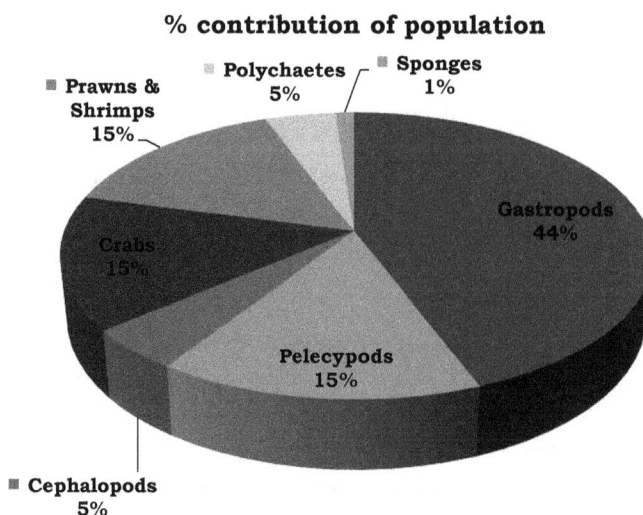

Figure 2: Percentage representation of species of macrobenthos in mangroves of Uran (Raigad).

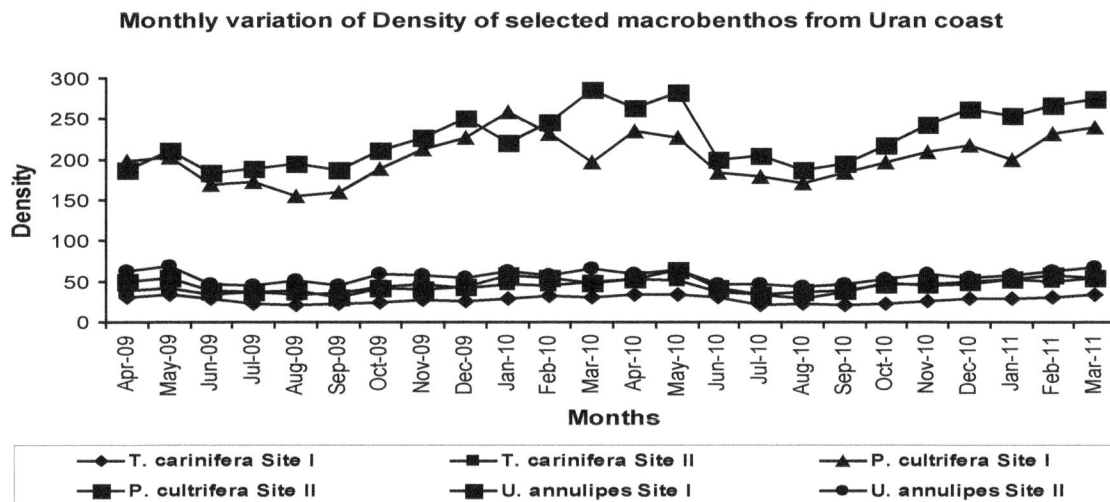

Figure 3: Monthly variation of density of selected macrobenthos from mangroves of Uran.

Macro-benthos	Site	Pre-monsoon 2009	Monsoon 2009	Post-monsoon 2010	Pre-monsoon 2010	Monsoon 2010	Post-monsoon 2011
Thais carinifera	I	32 ± 2	24 ± 4	27 ± 2	33 ± 2	24 ± 4	27 ± 3
	II	46 ± 7	35 ± 2	45 ± 3	50 ± 4	34 ± 4	48 ± 4
Perinereis cultrifera	I	217 ± 17	164 ± 8	222 ± 25	222 ± 16	179 ± 6	206 ± 9
	II	235 ± 36	188 ± 5	227 ± 15	270 ± 17	197 ± 7	244 ± 17
Uca annulipes	I	55 ± 4	37 ± 4	46 ± 7	55 ± 6	37 ± 3	49 ± 3
	II	66 ± 3	47 ± 3	59 ± 4	62 ± 4	46 ± 2	57 ± 3

Table 1: Seasonal variation of density of selected macrobenthos from mangroves of Uran.

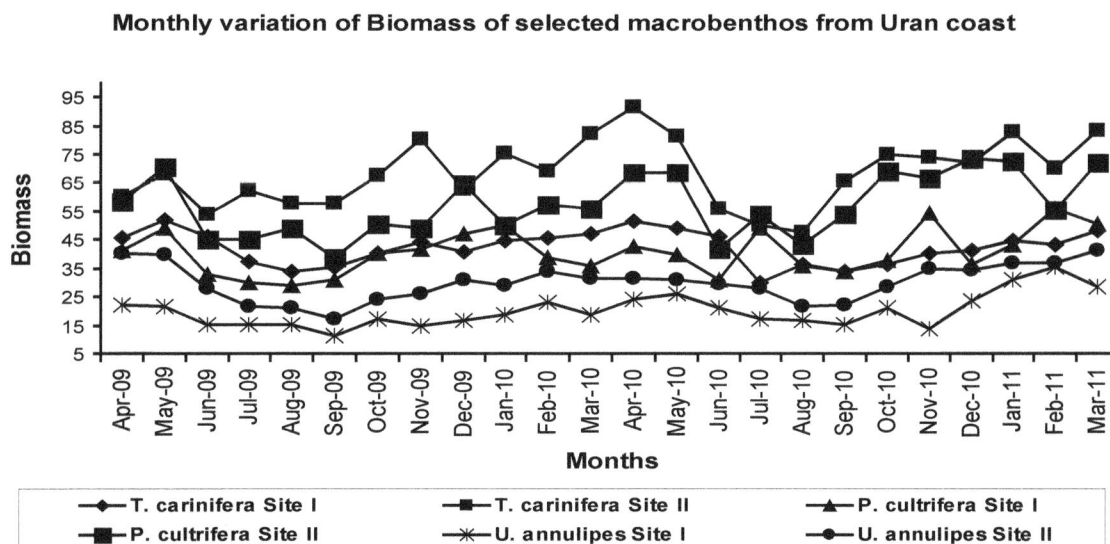

Figure 4: Monthly variation of biomass of selected macrobenthos from mangroves of Uran.

Diversity indices

The index of frequency [F] or Importance Probability [Pi] shows seasonal variation and the values were high in pre-monsoon and post-monsoon than the monsoon. Highest F was recorded for *P. cultrifera* during pre-monsoon [0.767 at Site I and 0.725 at Site II, Dec 2009] followed by *U. annulipes* during pre-monsoon [0.198 at Site I, May 2010 and 0.218 at Site II, Apr 2009]. The lowest values were recorded for *T. carinifera* [0.084 at Site I, Jan 2010 and 0.112 at Site II, Aug 2010]. The index of frequency can be placed in order of *P. cultrifera* > *U. annulipes* > *T. carinifera* (Figure 5).

The index of dominance was uniform at both sites where the dominance of the species can be placed in order as *P. cultrifera* > *U.*

Macro-benthos	Site	Pre-monsoon 2009	Monsoon 2009	Post-monsoon 2010	Pre-monsoon 2010	Monsoon 2010	Post-monsoon 2011
Thais carinifera	I	47.24 ± 3.20	37.89 ± 4.70	42.40 ± 2.00	48.47 ± 2.24	36.6 ± 5.81	40.59 ± 2.98
	II	70.52 ± 8.53	57.98 ± 2.85	71.57 ± 6.89	81.25 ± 8.08	54.90 ± 6.97	76.11 ± 4.05
Perinereis cultrifera	I	49.2 ± 5.19	30.8 ± 1.48	44.76 ± 4.07	39.31 ± 2.36	37.45 ± 6.99	43.05 ± 7.09
	II	64.18 ± 7.32	44.58 ± 3.74	53.58 ± 6.34	62.69 ± 6.02	48.22 ± 5.75	70.39 ± 2.75
Uca annulipes	I	26.96 ± 5.63	14.33 ± 1.58	16.89 ± 1.89	22.96 ± 2.75	17.58 ± 2.15	22.36 ± 6.10
	II	39.48 ± 1.66	22.13 ± 3.82	27.56 ± 2.63	31.97 ± 1.16	25.16 ± 3.47	33.65 ± 3.13

Table 2: Seasonal variation of biomass of selected macrobenthos from mangroves of Uran.

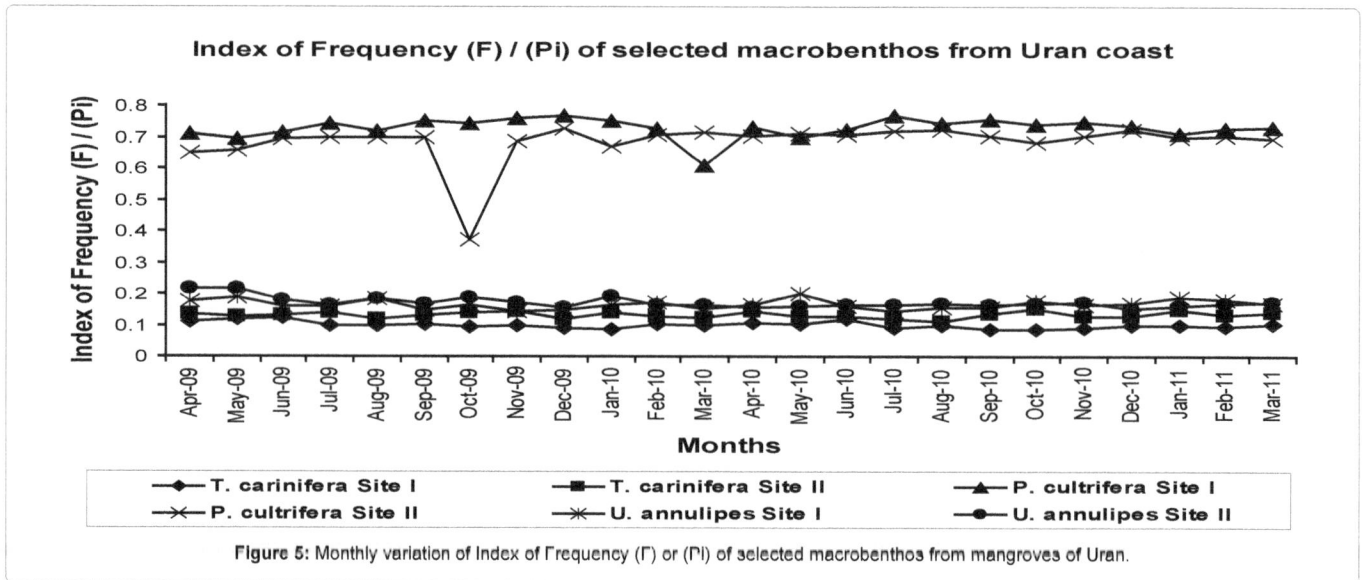

Figure 5: Monthly variation of Index of Frequency (F) or (Pi) of selected macrobenthos from mangroves of Uran.

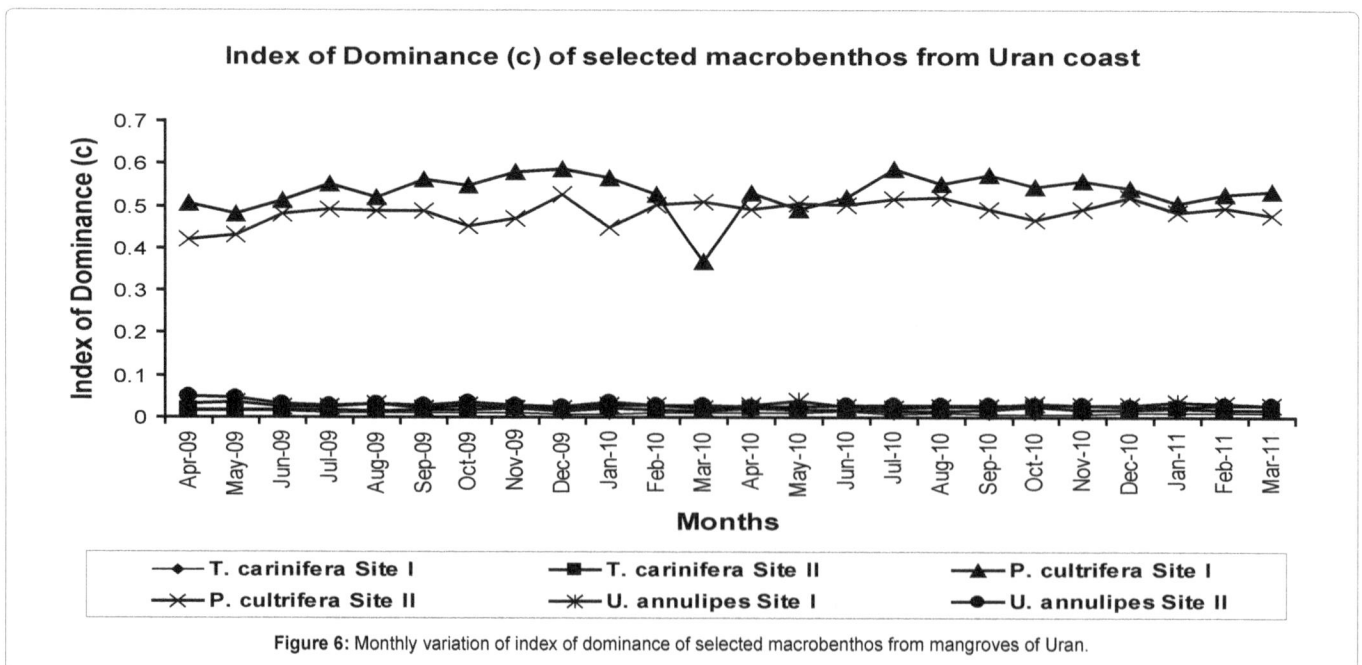

Figure 6: Monthly variation of index of dominance of selected macrobenthos from mangroves of Uran.

annulipes > T. carinifera. The data on index of dominance shows that the gastropod species T. carinifera is dominated over by other species (Figure 6).

The rarity index was according to the index of dominance. In this case, T. carinifera shows highest rarity index and found to be rare with respect to population density and biomass. Rarity index of P. cultrifera, found to be very common at both sites. Lower Rarity index of polychaetes marks the higher density of it. Among other species, U. annulipes found to be dominant over T. carinifera species assessed. The pattern of Rarity index was not significantly varied throughout the period of investigation, although, slight seasonal variation of Rarity index was recorded for all macro benthos (Figure 7).

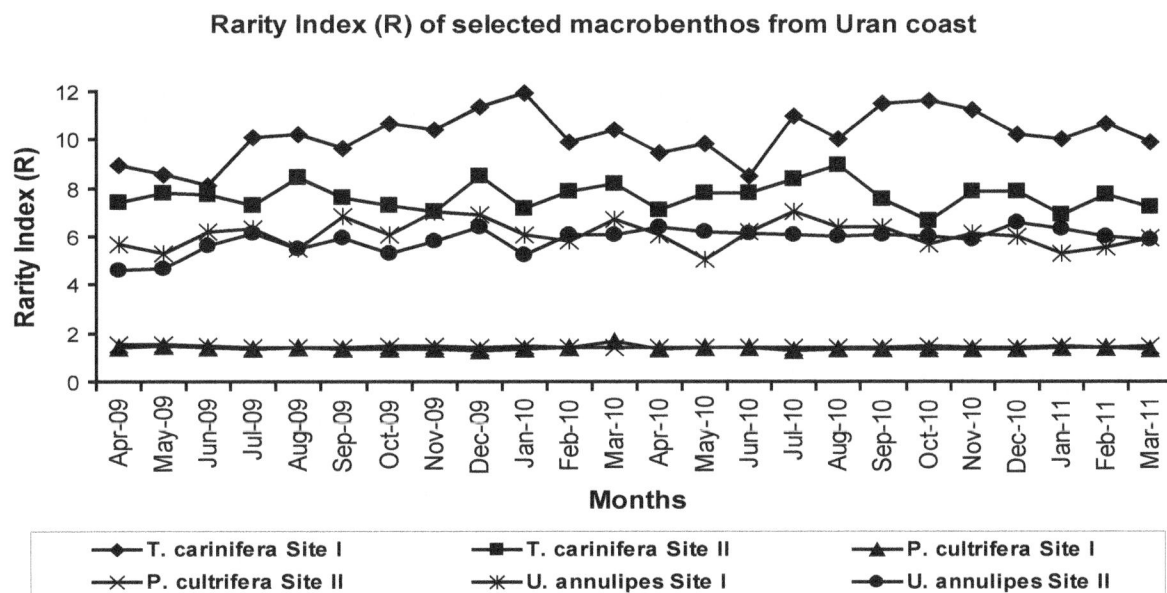

Figure 7: Monthly variation of Rarity index of selected macrobenthos from mangroves of Uran.

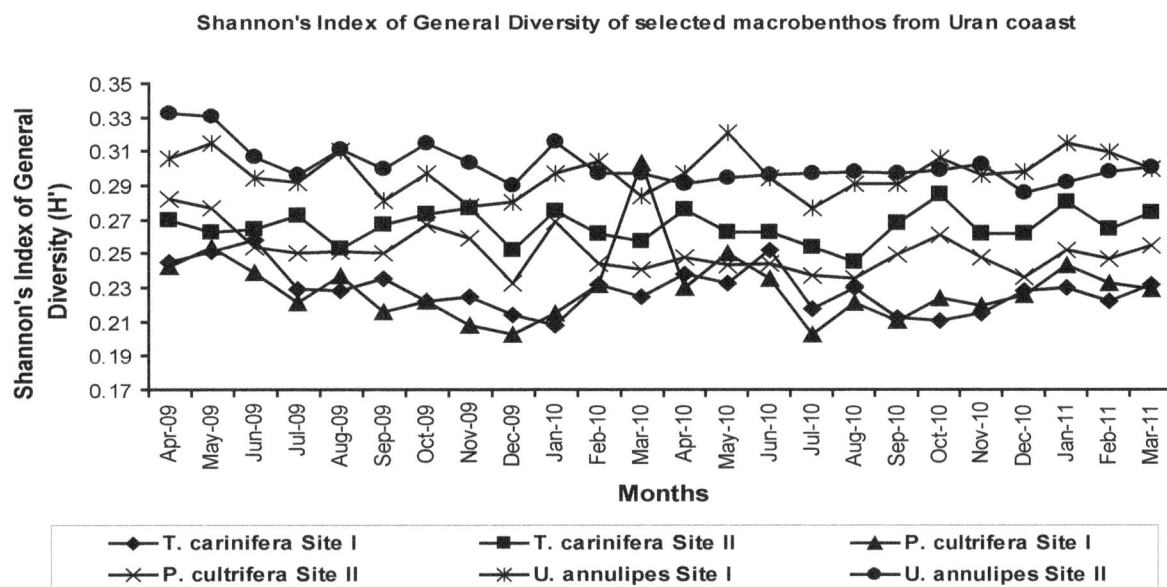

Figure 8: Monthly variation of Shannon's Index of General Diversity (H') of selected macrobenthos from mangroves of Uran.

The Shannon's Index of General Diversity [H'], was found to be uniform during the period of investigation. Maximum H' was observed for *U. annulipes*, which was followed by *T. carinifera* and *P. cultrifera*. The H' of all the species studied was uniform throughout the period of investigation. The data on H' shows that species richness of *P. cultrifera*, *U. annulipes* and *T. carinifera* was more or less same. These results of H' are in agreement with index of frequency, index of dominance and rarity index and did not varied significantly (Figure 8).

Discussion

Benthic macro fauna of the gastropods is dominant in Uran mangroves and is followed by pelecypods, crabs, prawns and shrimps,

cephalopods, polychaetes and sponges. Similar results on species composition of macrobenthos in mangrove ecosystems were reported by Ngo-Massou e [42], Thilagavathi [1], Pravinkumar [13] and Kumar and Khan [17].

Higher density of macrobenthos recorded during pre-monsoon and post-monsoon is attributed to higher total organic carbon coupled with a stable environment [1]. The results of the study are in agreement with Kurian [43], Saravanakumar [44] and Praveen kumar [13]. Results of dominance of gastropods, pelecypods, crustaceans and polychaetes in mangrove environment were also reported by Zhou [45] and Huang [46]. Low species diversity recorded during monsoon is attributed to the influx of freshwater, low temperature and lowered salinity. This is in conformity with Parulekar [26], Chandran [47], Devi [48] and Pravinkumar [13].

Density and biomass of *P. cultrifera* and *T. carinifera* shows an inverse relationship. In comparison of *P. cultrifera*, low density of *T. carinifera* was observed throughout the investigation period. However, the biomass of *T. carinifera* was higher than that of *P. cultrifera*. A size variation of these organisms can be correlated to their inverse relationship in between density and biomass [18]. As compared to *P. cultrifera*, well-developed body organs and thick coverings of *U. annulipes* can be correlated to their differential density and biomass [49,50].

Low biomass and high density of macrobenthos could be due to the recruitment process. Harkantra and Rodrigaes [51] have reported that salinity is the main significant factor influencing the species diversity, population density and biomass of macrobenthos in the estuarine system of Goa, west coast of India. Results of the study are in agreement with earlier reports of Kumar [52], Edgar and Barret [53], Liang [54], Tang and Yu [55] and Mahapatro [56].

The diversity indices calculated for two sites are given in Figures 5 and 8. The salinity acts as a limiting factor in the distribution of living organisms, and its variation was due to dilution and evaporation [13]. Maximum values of diversity indices recorded during pre-monsoon and post-monsoon are in agreement with the earlier reports [57,58].

The low species richness recorded during monsoon might be due to the low temperature, tidal fluctuation, freshwater runoff containing sewage and low salinity which in turn affected the distribution of benthos. Pawar and Kulkarni [59] have reported that the diversity indices can be used as a good measure for studying the effect of industrial pollution because industrial wastes and sewage almost always reduce the diversity of natural systems into which they are discharged.

Wilhm and Dorris [60] and Kumar and Khan [17] have stated that values less than 1.0 for diversity index [H] in estuarine waters indicate heavy pollution, values between 1.0 and 3.0 indicate moderate pollution, and values exceeding 3.0 indicate non-polluted water. Diversity values in the study area ranged from 0.203 to 0.321 at site I and 0.233 to 0.332 at site II. These values suggest that the mangrove ecosystem studied is heavily polluted and the macro benthic fauna is under stress due to anthropogenic factors.

Conclusion

Present study shows that the mangrove ecosystem of Uran, Navi Mumbai have heavy pollution from the sewage, industrial wastes, effluents, maritime activities of Jawaharlal Nehru Port [JNPT], Container Freight Stations, and other port wastes. This deteriorates the mangrove ecosystem affecting the community structure of macrobenthos with respect to population density, biomass and diversity indices. The macro benthic fauna from mangrove ecosystems of Uran is facing the threat due to anthropogenic stress. Present information on community structure of macrobenthos from mangrove ecosystem of Uran, Navi Mumbai would be helpful as a baseline data for further monitoring of anthropogenic inputs on mangrove ecosystem of Uran.

Acknowledgements

I thank Principal, Veer Wajekar Arts, Science and Commerce College, Mahalan Vibhag, Phunde [Uran], Navi Mumbai-400 702 for providing necessary facilities for the present study. This work was supported by University Grants Commission, Western Regional Office, Pune [File No: 47-599/08 [WRO] dated 2nd Feb 2009].

References

1. Thilagavathi B, Varadharajan D, Babu A, Manoharan J, Vijayalakshmi S, et al. (2013) Distribution and Diversity of Macrobenthos in Different Mangrove Ecosystems of Tamil Nadu Coast, India. Journal of Aquaculture Research Development 4: 199.

2. Manson FJ, Loneragan NR, Skilleter GA, Phinn SR (2005) An Evaluation of the Evidence for Linkages between Mangroves and Fisheries: A Synthesis of the Literature and Identification of Research Directions. Oceanography and Marine Biology - An Annual Review 43: 485-515.

3. Robin LE, Bazelevic NI (1966) The biological production of main vegetation types in the Northern Hemisphere of the Old World. Forestry Abstracts 27: 369-372.

4. Jagtap TG, Murthy P, Komarpant D (2002) Mangrove ecosystems of India: major biotic constituents, conservation and management. Wetland Conservation and Management. Pointer Publishers Jaipur India 34-64.

5. Venkataraman K, Wafar M (2005) Coastal and marine biodiversity of India. Indian Journal of Marine Sciences 34: 57-75.

6. Vijay V, Birader RS, Inamdar AB, Deshmukh G, Baji S, et al. (2005) Mangrove mapping and change detection around Mumbai (Bombay) using remotely sensed data. Indian Journal of Marine Sciences 34: 310-315.

7. Zingde MD (1999) Marine environmental status and coastal zone management issues in India. South Asia Regional Workshop on Estuarine modelling and Coastal Zone Management. A Joint START/LOICZ/IGBP-SL Workshop, 28th-30th April 1999 Colombo Sri Lanka 153-164.

8. Queshi IM (1957) Botanical silviculture features of mangrove forest of Bombay State. Proceedings of Mangrove Symposium. Government of India Press Calcutta 20-26.

9. Blasco F, Caratini C, Chanda S, Thanikaimoni G (1975) Main characteristics of Indian mangroves. Proceedings of the International Symposium on Biology and Management of Mangroves. G. E. Walsh, S. G. Snedkar and H. J. Reas 71-87.

10. Inamdar AB, Surendrakumar RK, Behera MC, Chauhan HB, Nayak S (2000) Land use mapping of Maharashtra Coastal Regulatory Zone. SAC/RESA/MWRD/CRZ/SN/02/00 Indian Space Research Organization, Ahmadabad, India 42.

11. Mukherji M (2002) Degradation of creeks and mangroves and its impact on urban environment - a case study of Mumbai. Proceedings of the National Seminar on Creeks, Estuaries and Mangroves - Pollution and Conservation 331-333.

12. Zingde MD (2002) Degradation of Marine habitats and Coastal management framework. Proceedings of the National Seminar on Creeks, Estuaries and Mangroves - Pollution and Conservation 3-7.

13. Pravinkumar M, Murugesan P, Krishna Prakash R, Elumalai V, Viswanathan C, et al. (2013) Benthic biodiversity in the Pichavaram mangroves, Southeast Coast of India. Journal of Oceanography and Marine Science 4: 1-11.

14. Hu ZY, Bao YX, Cheng HY, Zhang LL, Ge BM (2009) Research progress on ecology of natural wetlands zoobenthos in China. Chinese Journal of Ecology 28: 959-968.

15. Koch V, Wolff M (2002) Energy budget and ecological role of mangrove epibenthos in the Caet´e estuary, North Brazil. Marine Ecology Progress Series 228: 119-130.

16. Colpo KD, Chacur MM, Guimaraes FJ, Negreiros-Fransozo ML (2011) Subtropical Brazilian mangroves as a refuge of crab (Decapoda: Brachyura) diversity. Biodiversity and Conservation 20: 3239-3250.

17. Palanisamy Satheesh kumar, Basheer Anisa Khan (2013) The distribution and diversity of benthic macro-invertebrate fauna in Pondicherry mangroves, India. Aquatic Biosystems 9: 15.

18. Pawar PR, Kulkarni, Balasaheb G (2009) Population Density and Biomass of Selected Macrobenthos in Karanja Creek, (Dist. - Raigad), Maharashtra, West Coast of India. The Ecologia 9: 79-86.

19. Thakur Sandhya, Yeragi SG, Yeragi SS (2012) Population density and biomass of organisms in the mangrove region of Akshi creek, Alibaug taluka, Raigad district, Maharashtra. Marine biodiversity: Utter Pradesh State Biodiversity Board 135-140.

20. Pearson TH, Rosenberg R (1978) Macrobenthos secession in relation to organic enrichment and pollution of the marine environment. Oceanography and Marine Biology - An Annual Review 16: 229-234.

21. Dash MC (2002) Systems concept in Ecology. Fundamentals of Ecology. Tata McGraw-Hill Publishing Company Limited, New Delhi 35-144.

22. Khan SA, Murgesan P (2005) Polychaetes diversity in Indian estuaries. Indian Journal of Marine Sciences 34: 114-119.

23. Pawar PR (2013) Monitoring of impact of anthropogenic inputs on water quality of mangrove ecosystem of Uran, Navi Mumbai, west coast of India. Marine Pollution Bulletin 75: 291-300.

24. Bhatt YM (1959) A study of intertidal organisms of Bombay M. Sc. Thesis University of Mumbai.

25. Parulekar AH (1973) Studies on intertidal ecology of Anjidiv Island. Abstracts of the Indian National Science Academy 39: 611-631.

26. Holme NA, McIntyre AD (1984) Methods of the study of marine benthos. IBP Hand Book No. 16. Oxford and Edinburgh. Blackwell Scientific Publication 387.

27. Hornell J (1949) The study of Indian Mollusca. Journal of Bombay Natural History Society 4: 2-34.

28. Menon PK, Datta AK, Das Gupta D (1951) On the marine fauna of Gulf of Part II - Gastropoda. Journal of Bombay Natural History Society 8: 475-494.

29. Subrahmanyam TV, Karandikar KR, Murti NN (1949) The marine Pelecypoda of Bombay. Journal of University of Bombay 17: 50-81.

30. Subrahmanyam TV, Karandikar KR, Murti NN (1951) Marine Gastropods of Bombay, Part I. Journal of University of Bombay 20: 21-34.

31. Subrahmanyam TV, Karandikar KR, Murti NN (1952) Marine Gastropods of Bombay, Part II. Journal of University of Bombay 21: 26-73.

32. Chhapgar BF (1957) On the marine crabs (Decapoda - Brachyura) of Bombay State. Journal of Bombay Natural History Society 54: 399-439.

33. Chhapgar BF (1958) More additions to the crab fauna of Bombay State. Journal of Bombay Natural History Society 65: 608-617.

34. Apte DA (1988) The book of Indian shells, Bombay Natural History Society; Oxford University Press, India.

35. Parulekar AH, Nair SA, Harkantra SN, Ansari ZA (1976) Some quantitative studies on the benthos off Bombay. Mahasagar 9: 51-56.

36. Eleftheriou A, Holme NA (1984) Macrofauna techniques. Methods for the study of marine benthos. IBP Handbook No. 16. Blackwell Scientific Publications Oxford U. K 1-344.

37. Mehta P (1994) Bio-ecology of benthic organisms with reference to changing environment. Ph. D. Thesis, University of Bombay.

38. Smith RL, Smith TM (1988) Elements of Ecology. Wesley Longman Inc. C.A.

39. Simpson EH (1949) Measurement of diversity. Nature 163: 681- 688.

40. Ludwig JA, Reynolds JF (1988) Statistical Ecology: A Primer on Methods and Computing. John Wiley & Sons New York. 377.

41. Shannon CE, Weaver W (1949) The mathematical theory of communication. University of Illinois Press, Urbana 117.

42. Ngo-Massou VM, Essome-Koum GL, Ngollo-Dina E, Din N (2012) Composition of macrobenthos in the Wouri River estuary mangrove, Douala, Cameroon, African Journal of Marine Science.

43. Kurian CV (1984) Fauna of the mangrove swamps in Cochin estuary. Proceedings of the Asian Symposium on Mangrove Environment Research and Management 226-230.

44. Saravanakumar A, Sesh Serebiah J, Thivakaran GA, Rajkumar M (2007) Benthic macrofaunal assemblage in the arid zone mangroves of Gulf of Kachchh, Gujarat. Journal of Ocean University of China 6: 303-309.

45. Zhou H (2001) Effects of leaf litter addition on meiofaunal colonization of azoic sediments in a subtropical mangrove in Hong Kong. Journal of Experimental Marine Biology and Ecology 256: 99-121.

46. Huang B, Zhang B, Lu J, Ou Z, Xing Z (2002) Studies of macrobenthic ecology and beach aquaculture holding capacity in Dongzhai Bay mangrove areas. Marine Science, Haiyang-Kexue 26: 65-68.

47. Chandran R (1987) Hydrobiological Studies in the Gradient Zone of the Vellar estuary. IV. Benthic Fauna. Mahasagar Bulletin of National Institute of Oceanography 20: 1-13.

48. Devi Prabha L (1994) Ecology of Coleroon estuary, Studies on Benthic Fauna. Journal of Marine Biological Association of India 36: 260-266.

49. Ingole BN, Rodrigues NR, Ansari ZA (2002) Macrobenthic communities of the coastal waters of Dabhol, West coast of India. Indian Journal of Marine Sciences 31: 100-107.

50. Quadros G, Athalye RP (2002) Intertidal macrobenthos from Ulhas river estuary and Thane creek. Proceedings of the National Seminar on Creeks, Estuaries and Mangroves - Pollution and Conservation 216 - 222.

51. Harkantra Sadanand N, Rodrigaes Nimi R (2004) Environmental influences on the species diversity, biomass and population density of soft bottom macrofauna in the estuarine system of Goa, west coast of India. Indian Journal of Marine Sciences 33: 187-193.

52. Kumar RS (2001) Intertidal zonation and seasonality of benthos in a tropical mangrove. International Journal of Ecology and Environmental Sciences 27: 199-208.

53. Edgar Graham J, Barrett Neville S (2002) Benthic macrofauna in Tasmanian estuaries: scales of distribution and relationships with environmental variables. Journal of Experimental Marine Biology and Ecology 270: 1-24.

54. Liang CY, Zhang HH, Xie XY, Zou FS (2005) Study on biodiversity of mangrove benthos in Leizhou Peninsula. Marine Sciences 2: 18-25.

55. Tang YJ, Yu SX (2007) Spatial zonation of macrobenthic fauna in Zhanjiang Mangrove Nature Reserve, Guangdong, China. Acta Ecologica Sinica 27: 17030-1714.

56. Mahapatro D, Panigrahy RC, Sudarsan P, Mishra, Rajani K (2009) Influence of monsoon on macrobenthic assemblage in outer channel area of Chilika lagoon, east coast of India. Journal of Wetlands Ecology 3: 56-67.

57. Murugesan P (2002) Benthic biodiversity in the marine zone of Vellar estuary. Ph.D. Thesis, Annamalai University, India pp 330.

58. Ajmal KS, Raffi SM, Lyla PS (2005) Brachyuran crab diversity in natural (Pichavaram) and artificially developed mangroves (Vellar estuary). Current Science 88: 1316-1324.

59. Pawar PR, Kulkarni BG (2007) Diversity indices of selected macrobenthos in Karanja creek (District -Raigad), Maharashtra, West coast of India. Journal of Indian fisheries Association 34: 1-9.

60. Wilhm JL, Dorris TC (1966) Species diversity of benthic macro invertebrates in a stream receiving domestic and oil refinery effluents. The American Midland Naturalist Journal 76: 427-449.

24

A Novel Subspecies of *Staphylococcus aureus* from Sediments of Lanzhou Reach of the Yellow River Aerobically Reduces Hexavalent Chromium

Zhang X[1], Krumholz LR[2], Zhengsheng Yu[1], Yong Chen[1], Pu Liu[1] and Xiangkai Li[1]*

[1]*MOE Key Laboratory of Cell Activities and Stress Adaptations, School of Life Sciences, Lanzhou University, Lanzhou, Gansu, 730000, P. R. China*
[2]*Department of Botany and Microbiology and Institute for Energy and the Environment, University of Oklahoma, Norman, Oklahoma, 73019, U.S.A*

Abstract

Background: Lanzhou reach of the Yellow River is contaminated by heavy metals including chromium perennially. The microbial community within the sediment is very active. Yet the study on the bacteria in this distinctive microbial community is still scarce.

Results: LZ-01, a Gram-positive hexavalent chromium-reducing bacterium was isolated from the soil sample collected at a petrochemical corporation wastewater discharge site of Lanzhou reach. It was able to aerobically reduce 94.5% of 0.4 mM Cr (VI) to Cr(III) in 120 hours. Cd (II) and NaN_3 treatment both repressed Cr(VI) reduction in LZ-01 and Cr(III) precipitates were detected both on the cell membrane and in the cytoplasm by Transmission Electron Microscopy (TEM) imaging. LZ-01 also demonstrated resistance to 4 mM As (V) and 9 mM U (VI). LZ-01 was closely related to *Staphylococcus aureus* revealed by 16S rRNA sequence analysis. Comparison of cellular fatty acid components and Vitek phenotype identification provided further evidences that LZ-01 is a novel subspecies of *S. aureus*.

Conclusions: A chromate-reducing bacterium LZ-01 identified as a novel subspecies of *S. aureus* was isolated from Lanzhou reach of the Yellow River. Cd (II) and NaN_3 treatment and TEM images suggested that Cr(VI) was reduced not only intracellularly but also on the cell membrane. All the results indicate that the isolate has a great potential for bioremediation of Cr (VI)-contaminated environment.

Keywords: Cr (VI) reduction; Bacterial isolation and identification; *Staphylococcus aureus*; Bioremediation

Introduction

Lanzhou city, located at the geometric center of China, is the largest city with a massive industrial base on the Yellow River. The Yellow River provides water for industrial, agricultural and domestic consumption in Lanzhou. Lanzhou reach of the Yellow River is about 50 km long. The upstream region is largely industrial while the downstream region is residential with a population of about four million [1]. Effluents containing high levels of metals and hydrocarbons from manufacturing facilities are continuously discharged to the river [2]. A previous study has shown that the metal concentrations (in μg/g dry soil sample) in the river bank soil is 13.68-48.11 (As), 26.39-77.66 (Pb), 89.80-201.88 (Zn), 41.49-128.30 (Cr), 29.72-102.22(Cu), and 773.23-1459.69 (Mn) [3]. Metal contamination of Lanzhou reach has caused widespread concern in China. So far researches on the environmental governance of Lanzhou reach have been limited to physics and chemical analyses [2,3] and investigation of microbial activity in the river bank soil is less common.

Chromium, an important industrial metal is one of the most harmful contaminants in Lanzhou reach due to the presence of industries of textile dyeing and painting, electroplating, plastic processing, all of which have led to Cr (VI) being discharged into natural ecosystems. Like most heavy metals, chromium has toxic effects on both humans and environmental organisms [4]. Trace amount of chromium is required for a variety of enzymes [5-7], while accumulation in the human body can cause cancer, growth and developmental abnormalities, neuromuscular control defects, renal malfunction, mental retardation and other illnesses [8]. Though chromium exists in varied valance states ranging from Cr(II) to Cr(VI), Cr(III) and Cr(VI) are the most stable forms in nature [9]. Trivalent chromium is insoluble with low mobility, while the hexavalent chromium species is soluble and highly mobile [10]. Previous studies revealed that Cr(VI) is almost 1000-fold more mutagenic and cytotoxic than Cr(III) [10,11]. Recent work has shown that the chromium concentration within the reach is about 8-25 times higher than the China water safety standard with the majority in hexavalent form [3]. Hence, control of Cr(VI) becomes an essential step of environmental remediation of Lanzhou reach.

A number of methods have been applied to reduce metal concentrations in soil. Physicochemical method are practiced and biological remediations are also taken into consideration [12-15]. Compared with physicochemical method, microbial remediation is more cost-effective and rarely introduces secondary pollution [16-19], especially for metal reduction in bioremediation. A wide variety of microorganisms capable of reducing aqueous hexavalent chromium ions have been reported including *Staphylococcus epidermidis L-02* [11], *Acinetobacter* [20], *Arthrobacter* [21], *Bacillus* sp. [22], *Desulfovibrio vulgaris* [23], *Pseudomonas* sp. [24], *Serratia marcescens* [25], and *Ochrobactrum* sp. [26]. It is likely that microbes in Lanzhou reach river bank soil possess the ability to reduce Cr (VI) to Cr(III). Screening for these bacteria will be the first stage for microbial remediation of Cr(VI) in Lanzhou reach.

*Corresponding author: Xiangkai Li, MOE Key Laboratory of Cell Activities and Stress Adaptations, School of Life Sciences, Lanzhou University, Lanzhou, P. R. China, E-mail: xkli@lzu.edu.cn

Staphylococcus aureus is a facultative Gram-positive coccus-shaped bacterium [27] that is best known for its ability to cause a range of illnesses [27,28]. Several recent studies report that *S. aureus* is involved in environmental metal remediation [4,29] and *S. epidermidis* which is closely related to *S. aureus* is able to anaerobically reduce Cr(VI) to Cr (III) [11]. However, there is no direct evidence that *S. aureus* can reduce Cr(VI). In this study, we screened the Lanzhou reach river bank soil for Cr(VI) reducers and isolated a novel subspecies of *S. aureus* strain LZ-01, which reduces Cr(VI) to Cr(III) aerobically. 16S rRNA sequencing and fatty lipids profiling showed that strain LZ-01 is closely related to *S. aureus*. Comparison of Vitek phenotype identification showed that there were 7 different reactions out of 64 between LZ-01 and type strains of *Staphylococcus aureus*. LZ-01 reduces Cr(VI) in a highly efficient manner suggesting that it may be useful for Cr(VI) reduction studies in Lanzhou reach.

Materials and Methods

Soil sample collection

The sampling site was 60 kilometers west of central Lanzhou city and adjacent to the upper reach of the Yellow River. The actual location was close to a petrochemical corporation waste water discharge site. The latitude and longitude of the sampling site is 36°06'N 103°39'E, (Figure 1). The pH of soils at the sampling site ranged from 5.0 to 6.0 with temperature around 18°C at the time of collection. The soil samples were collected from 15 cm below surface horizon and stored in sterile aluminum boxes at -20°C.

Screening for Cr(VI) -resistant bacteria

M9 minimal salts medium [30] used for isolations contained: 17 mM Na_2HPO_4, 22 mM KH_2PO_4, 19 mM NH_4Cl and 0.1% yeast extract (w/v). The pH was adjusted to 7.4 prior to autoclaving and 2 mM sterilized $MgSO_4$ and 0.2% (w/v) glucose were added after autoclaving. Solid M9 medium was prepared as above with 1.5% (w/v) agar. Filter sterilized solution of 1 M Cr(VI) was prepared from $K_2Cr_2O_7$. Soil suspension was prepared by diluting 1 g fresh soil sample into 9 ml 0.85 mM NaCl solution and standing at room temperature for about 4 hours. For the isolation of Cr(VI)-resistant bacteria, 100 µl of soil

suspension was added onto M9 solid medium with 2 mM Cr(VI). The plates were incubated at 37°C for 12-16 hours and 13 colonies were selected and then transferred into M9 liquid medium containing 2 mM Cr(VI). The liquid cultures were incubated on a shaker (180 rpm) at 37°C for 24 hours and then streaked onto solid M9 medium with 2 mM Cr(VI) and single colonies were picked after 24 hours incubation at 37°C.

Cr(VI) reduction assay of isolated strains

The isolated strains were cultivated at 37°C for 60 hours in M9 medium with Cr(VI) (0.4 mM, 1 mM, 2.5 mM, 5 mM, 7.5 mM and 9 mM). The initial and the final Cr(VI) concentration were determined according to the method reported previously [31]. Colony-Forming Units (CFU) and Cr(VI) reduction rates were compared between 6 treatments.

The selected strains were grown in M9 minimal salts medium with 0.4 mM Cr(VI) at 37°C for 120 hours. Liquid cultures were serially diluted to 10^{-5} every three hours to determine CFU by plating on M9 medium. The strains grown in M9 medium without Cr(VI) was used as control. At each time point, 0.5 ml liquid culture was removed and centrifuged at 12000 rpm for 5 min and the Cr(VI) concentration in the supernatant was measured.

The effect of biosorption and abiotic reduction was analyzed with *Escherichia coli* strain DH5α as control. Cells were grown with 0.4 mM Cr(VI) and after 120 hours the culture was centrifuged and Cr(VI) concentration in the supernatant was measured. The cell pellets were resuspended and treated with lysozyme at 37°C for 30 min and at 95°C for 15 min , and subjected to Cr(VI) measurement to determine sorbed Cr(VI) levels. The presence of Cr(III) in the cells was determined by first diluting 5 ml cell resuspension with 75 ml H_2O, 10 ml H_2SO_4 solution ($V_{H2SO4}:V_{H2O}=1:1$), 10 ml 1% (w/v) $AgNO_3$ solution and 2 g $(NH_4)_2S_2O_8$. Then the mixture was boiled for 2 min and subjected to Cr(VI) measurement. The concentration of Cr(III) within cells was calculated by subtracting Cr(VI) concentration in the cell resuspension before thiosulfate treatment from that after thiosulfate treatment. Cultures that were not inoculated were set up as controls.

Washed cell experiments were also performed with 50 ml overnight

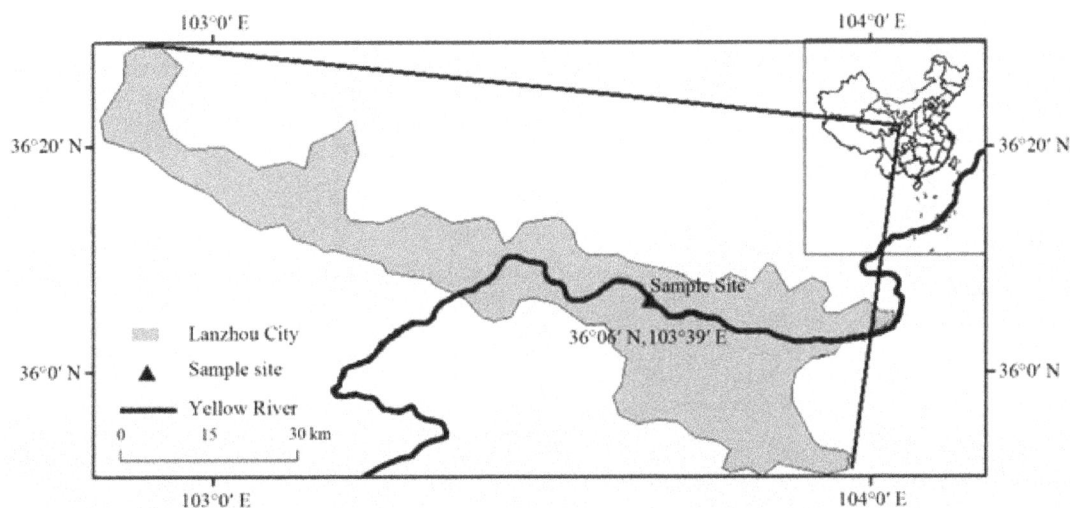

Figure 1: Map of the sampling site. The latitude and longitude of the sampling site is 36°06'N 103°39'E and the sampling site is 60 km west of central Lanzhou city and adjacent to the upper reach of the Yellow River.

culture. Then cells harvested from 50 ml culture were added to 5 ml M9 medium with 0.4 mM Cr(VI). After 3 hours at 37°C, the dissolved and sorbed Cr(VI) and Cr(III) concentrations were determined as described above.

To investigate the mechanism of Cr(VI) reduction, we investigated the effects of inhibitors NaN₃ and Cd(II). Since NaN₃ is an inhibitor of cytochrome c3 which is a transmembrane protein and an essential component of the respiratory electron transport chain, it can repress the extracellular pathway [32]. And Cd(II) inhibits reactions involving sulfhydryl groups. It can inhibit thioredoxin which is an electron donor for cytoplasmic metal reductase [33]. The effect of NaN₃ (3 mM) or Cd(II) (50 μM) on Cr(VI) reduction in washed cell experiments was determined by adding one inhibitor to M9 medium with 0.4 mM Cr(VI). After inoculation cells were incubated at 37°C for 12-16 hours and Cr(VI) concentrations were determined in culture supernatants. CFUs were measured after diluting the liquid culture to 10^{-5} in NaCl buffer (0.1% w/v). Cell suspensions were serially diluted to 10^{-5}, and 100 μl of each dilution was then spread onto a M9 plate. Plates were removed from incubator after 2 to 3 days of incubation, and those ones with 50 to 500 colonies were counted. Controls were incubated with Cr(VI) only and inhibitors only. The concentration of Cr(VI) before and after incubation were also measured. All the growth experiments and treatments were conducted in triplicates and error bars presented in the figures were Standard Error of Mean (SEM) of triplicates.

Determination of pH and temperature optimum for Cr(VI) reduction

Isolated strains were grown in M9 medium with 2 mM Cr(VI) at a range of pH values (5, 6, 7, 8, 9) at 180 rpm and 37°C. The temperature optimum was determined between 20°C and 42°C (20°C, 28°C, 30°C, 37°C, 42°C) at pH 7.4. All cultures were shaken at 180 rpm. Growth curves were measured by OD 600 and CFUs were counted after 12 hours incubation.

Determination of minimum inhibitory concentration (MIC) of As(V), Hg(II), U(VI) and Cr(VI)

Minimal Inhibitory Concentration (MIC) of each heavy metal was determined. The stock solution of Cr(VI), As(V), Hg(II), and U(VI) were prepared (0.5 M K₂Cr₂O₇, 0.5 M H₃AsO₄, 0.1 M HgCl₂ and 0.1 M UO₂(OAc)₂·2H₂O) in deionized water and autoclaved at 120°C for 20 min. To determine the MIC of Cr(VI), As(V) and Hg(II), M9 was used as base medium. To determine MIC of U(VI), a medium which contained 60 mM NaHCO₃, 20 mM NH₄Cl, 0.15 mM KH₂PO₄, 0.2 mM MgSO₄, 0.1% (w/v) yeast extract, 0.2% (w/v) glucose was considered as base medium. All strains were incubated at 37°C, 180 rpm for 12-16 hours to obtain log phase culture.

Determination of antibiotic resistance

Concentration gradients of different antibiotics were set up in M9 medium as follows: ampicillin (0-10 mg/ml), chloramphenicol (0-2 mg/ml), vancomycin (0-1 mg/ml), kanamycin (0-300 μg/ml), erythromycin (0-200 μg/ml), gentamycin (0-100 μg/ml), tetracycline (0-50 μg/ml). Cell density was measured at 600 nm after incubation at 37°C for 12-16 hours.

Identification and characterization of isolated strains

For 16S rRNA sequencing analysis, genomic DNA was isolated by CTAB method [34]. The 16S rRNA sequence was amplified using universal bacterial primers 27F (5'-AGAGTTTGATCCTGGCTCAG-3') and

1492R (5'-GGTTACCTTGTTACGAC TT-3'). The 50 μl PCR reaction system contained 1 μl 20 μM 27F and 1 μl 20 μM 1492R, 2.5 U/μl Taq DNA polymerase (0.3 μl) (TIANGEN Inc., China), 5 μl 10×PCR buffer (TIANGEN), 4 μl 2.5 mM deoxynucleoside triphosphate (TIANGEN) and H₂O. Temperature cycling comprised 30 cycles of 94°C for 1 min, 50°C for 1 min and 72°C for 2 min. Purified PCR products were cloned in pMD18-T vector and transferred into E. coli strain DH5α. The 16S rRNA gene of clones was sequenced at ShangHai Majorbio Bio-pharm Technology Co. Ltd and sequences were analyzed using NCBI database and EzTaxon database to identify the most closely related organisms. Phylogenetic trees were built by MEGA and Treeview program.

Transmission electron microscopy (TEM) analysis

LZ-01 was cultivated in M9 minimal salts medium with 1 mM Cr(VI) at 37°C, 180 rpm for 48 hours. Cells were harvested by centrifugation (10000×g) at 4°C for 3 min and resuspended in deionized water and washed twice. Cells were resuspended in 2.5% (v/v) glutaraldehyde in phosphate buffered saline (PBS) for at least 2 hours. Then cells were washed 3 times in PBS followed by fixation with 1% (w/v) osmium tetroxide for 2-3 hours and washing three times with PBS. Samples were gradually dehydrated in 50%, 70%, 90%, and 100% (v/v) ethanol for 10 min each. After incubation in the mixture of epoxy resin and acetone (v/v=1/1 for one hour, v/v=3/1 for 3 hours) and pure epoxy resin overnight, samples were embedded in solid resin blocks at 70°C for 24 hours. Blocks were sectioned on a microtome and ultra-thin sections were observed by using TEM (Tecnai™G²F30, FEI. USA). Isolated strains grown in M9 medium without Cr(VI) and E. coli strain DH5α grown in M9 medium with and without Cr(VI) were prepared as controls.

Phylogenetic analysis

LZ-01 was pre-cultured by growing in M9 medium overnight and then streaked on an M9 plate in four zones. Approximately 40 mg (wet weight) of cells was harvested from the third zone. The whole cell Fatty Acid Methyl Esters (FAMEs) were extracted by the Microbial Identification System (Microbial ID, Inc., Newark, Del.) and analyzed by gas chromatography. The Microbial Identification System Standard Software was used to integrate peaks and determine fatty acid identity and concentration. S. aureus strains ATCC 25923 and ATCC 29213 were used as controls.

Physiological tests were done using the Vitek 2 System (bioMérieux Industry, France). Strain LZ-01 was examined with the colorimetric identification GP card which tested 64 different carbon sources as well as enzyme activities and antibiotic resistance. The test was performed according to the manufacturer instructions for 8 hours. S. aureus strains ATCC 25923 and ATCC 29213 were tested in the same way as standard strains.

Results

Cr(VI) reduction of isolated strains

Thirteen Cr(VI)-resistant strains capable of growing with 2 mM Cr(VI) in liquid M9 media were isolated from the soil samples. The ability of four strains (LZ-01, LZ-02, LZ-03, LZ-04) to reduce Cr(VI) while growing with 0.4 mM to 9 mM Cr(VI) was investigated. Different concentrations of Cr(VI) had the similar effects on cell growth (Supplementary Figure 1A). Since LZ-01 showed the highest reducing ability, it was chosen for further study. Within 60 hours LZ-01 reduced 67.89%, 10.35%, 7.02%, 7.54%, 8.42% and 3.51% with 0.4 mM, 1 mM, 2.5 mM, 5 mM, 7.5 mM and 9 mM Cr(VI), respectively (Figure 2A and Supplementary Figure 1B). As reduction appeared to be most efficient at 0.4 mM Cr(VI), we chose this concentration for further reduction experiments. Similar results were obtained with strains LZ-01, LZ-02, LZ-03, LZ-04 with 94.5%, 89.1%, 88.6%, 85.2% of 0.4 mM Cr(VI) reduced respectively after 120 hours' incubation (Figure 2B). The Cr(VI) reduction rate of LZ-01 was 3.10 µM/hour and the bacterial doubling time with Cr(VI) was 5.61 hours. When the growth curve was compared with control, it showed that 0.4 mM Cr(VI) did not repress the growth of LZ-01 (Figure 2A).

Total hexavalent chromium level in the culture during the incubation with Cr(VI) was measured in growing cells and washed cells. *E. coli* strain DH5α and strain LZ-01 were both grown in 0.4 mM Cr(VI) M9 media. The soluble and sorbed Cr(VI) concentrations were measured before and after 120 h incubation (Figure 3A). Results showed that the *E. coli* strain DH5α removed 182 µM Cr(VI), of which 80 µM was sorbed Cr(VI). The recovery rate of Cr(VI) was about 75% of the initial Cr(VI) added. This data indicated that the decrease of Cr(VI) within the *E. coli* culture was mainly a result of biosorption.

On the other hand only 27.1 µM Cr(VI) was sorbed in LZ-01 cells, and total 369.5 µM Cr(VI) was removed from LZ-01 culture with about 5.5% of the initial Cr(VI) remaining, which suggested that the decrease of Cr(VI) in LZ-01 culture was mainly due to reduction. In addition, the presence of a much higher level of Cr(III) in the LZ-01 culture provided clear evidence that chromate reduction occurred (Figure 3A). The combined concentration of reduced Cr(VI), biosorbed Cr(VI) and remaining dissolved Cr(VI) almost equaled to the initial Cr(VI) concentration. Meanwhile, control medium without bacteria showed no change in Cr(VI) concentration. This confirmed that the Cr(VI) decrease was a biotic process. Similar results were obtained in the washed cells (Figure 3B).

Previous studies showed that microbial Cr(VI) reduction can take place on cell membrane [35] as well as in cytoplasm [36]. To determine whether the Cr(VI) reduction was associated with extracellular or surface associated proteins, NaN₃ and Cd(II) were added to reduction experiments. NaN₃ is an inhibitor of microbial respiratory chain [37]

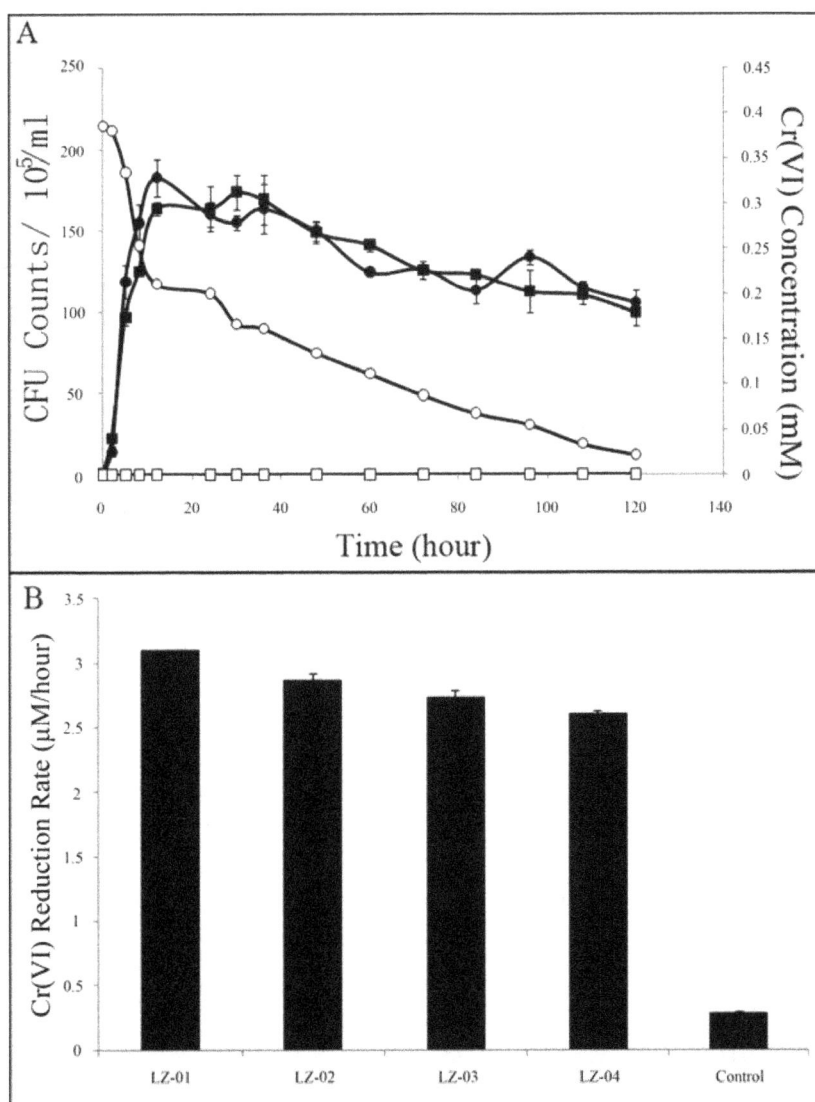

Figure 2: (A) Growth curve (•) and Cr(VI) reduction curve (○) for strain LZ-01 incubated with 0.4 mM Cr(VI) and growth curve (■) and Cr(VI) reduction curve (□) for strain LZ-01 incubated without Cr(VI). **(B)** Cr(VI) reduction rate calculated from the total Cr(VI) reduced over 120 hours for *S. aureus* strains LZ-01, LZ-02, LZ-03, LZ-04. Control was *E. coli* strain DH5α.

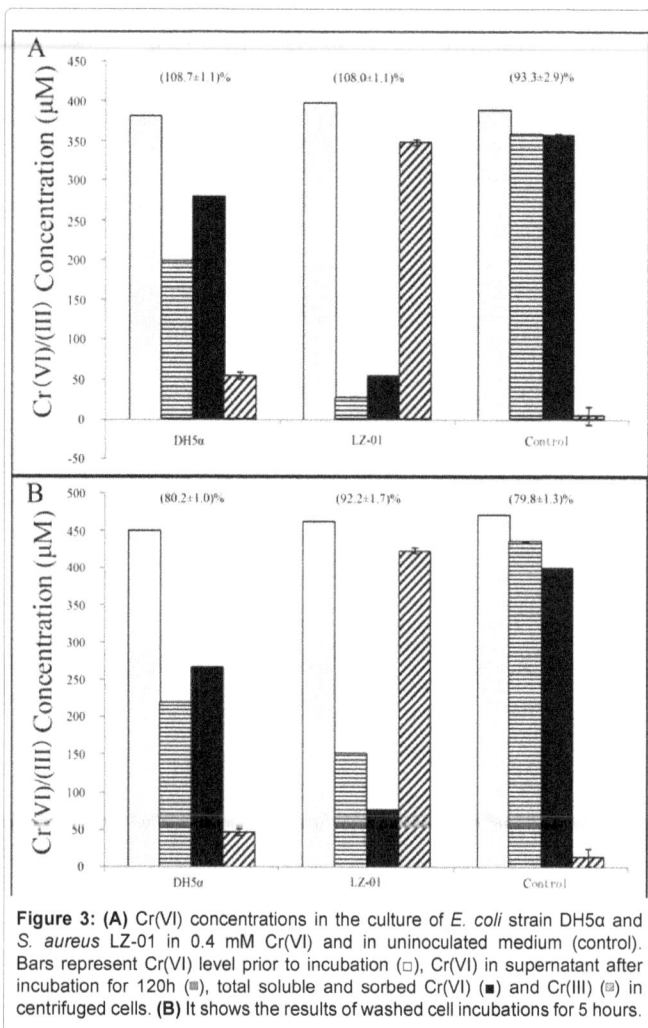

Figure 3: **(A)** Cr(VI) concentrations in the culture of *E. coli* strain DH5α and *S. aureus* LZ-01 in 0.4 mM Cr(VI) and in uninoculated medium (control). Bars represent Cr(VI) level prior to incubation (□), Cr(VI) in supernatant after incubation for 120h (▥), total soluble and sorbed Cr(VI) (■) and Cr(III) (▨) in centrifuged cells. **(B)** It shows the results of washed cell incubations for 5 hours.

and Cd(II) inhibits the proteins that make use of thioredoxin [36]. CFU counts were used in this test to evaluate the toxicity of Cd(II) and NaN$_3$ to living cells. Growth was not significantly influenced by the addition of either NaN$_3$ or Cd(II) (Figure 4A). The results showed that NaN$_3$ treatment inhibited Cr(VI) reduction in LZ-01 by 26.5% and Cd(II) treatment repressed the process by approximately 42.6% (Figure 4B) suggesting that Cr(VI) reduction in LZ-01 may require both components of the respiratory chain and cytoplasmic molecules that are sensitive to Cd. Control group in supplementary figure 2 showed that M9 medium had no effect on the measurement of Cr(VI) concentration.

Optimal growth conditions of LZ-01

Determination of growth curves and CFUs under a range of pHs showed that the growth was optimal at pH 7 (Figures 5A and 5B). The temperatures tested range from 20°C to 42°C and the growth of LZ-01 peaked at 37°C (Figure 5C and 5D). Therefore, pH 7 and 37°C was the optimum condition for LZ-01 growth and reduction with Cr(VI).

LZ-01 resistance to As(V), Hg(II), U(VI), Cr(VI) and to antibiotics

As Cr(VI) contamination is often associated with other metals including uranium (VI) and arsenic (V) [36,38,39], we tested the tolerance of strain LZ-01 to As(VI), Hg(II), U(VI) and Cr(VI). Strain

LZ-01 showed robust resistance to Cr(VI), As(V), and U(VI) (9 mM, 6 mM and 10 mM respectively), but not Hg(II) (MIC 0.025 mM) (Table 1). LZ-01 showed varied resistance to different antibiotics (Table 2).

Localization of reduced Cr(VI) in LZ-01 identified by TEM

To determine whether the reduction of Cr(VI) to Cr(III) occurred intra- or extracellularly, we used TEM to identify the presence of Cr(III) precipitate in LZ-01 culture (Figure 6). The TEM images showed that LZ-01 cells were all coccus-shaped and had smooth membrane which was consistent with typical *S. aureus* cells. The precipitate of Cr(III) was detected by EDX analysis after incubation while no Cr- containing material were observed in the negative control (Figure 6). Cr(III) precipitates were detected both free in the cells and adhering to the cell membranes suggesting that the reduction of Cr(VI) occurred both intracellular and extracellular. On the other hand, the images of *E. coli* strain DH5α showed no Cr(III) precipitates in cells with or without Cr(VI) treatment. This result suggest that reagents used during TEM sample preparation are not to able to reduce Cr(VI).

Phylogenetic analysis of isolated strains

Morphological characterization and 16S rRNA gene sequencing were carried out to identify four isolates (LZ-01, LZ-02, LZ-03, LZ-04). All four isolates were coccus-shaped, approximately 1 μm in diameter,

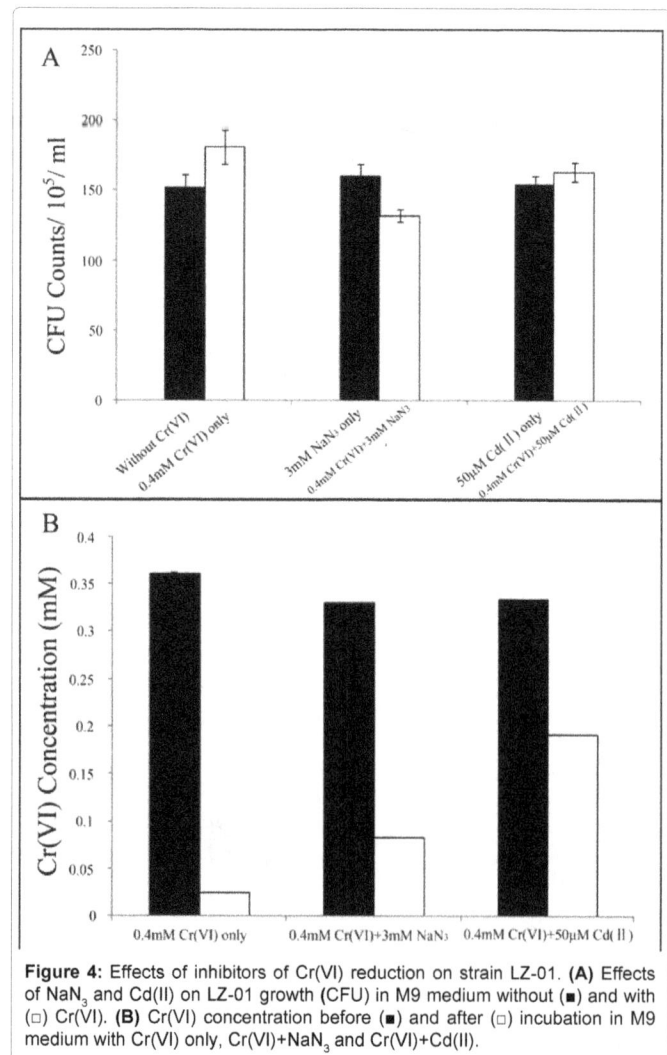

Figure 4: Effects of inhibitors of Cr(VI) reduction on strain LZ-01. **(A)** Effects of NaN$_3$ and Cd(II) on LZ-01 growth (CFU) in M9 medium without (■) and with (□) Cr(VI). **(B)** Cr(VI) concentration before (■) and after (□) incubation in M9 medium with Cr(VI) only, Cr(VI)+NaN$_3$ and Cr(VI)+Cd(II).

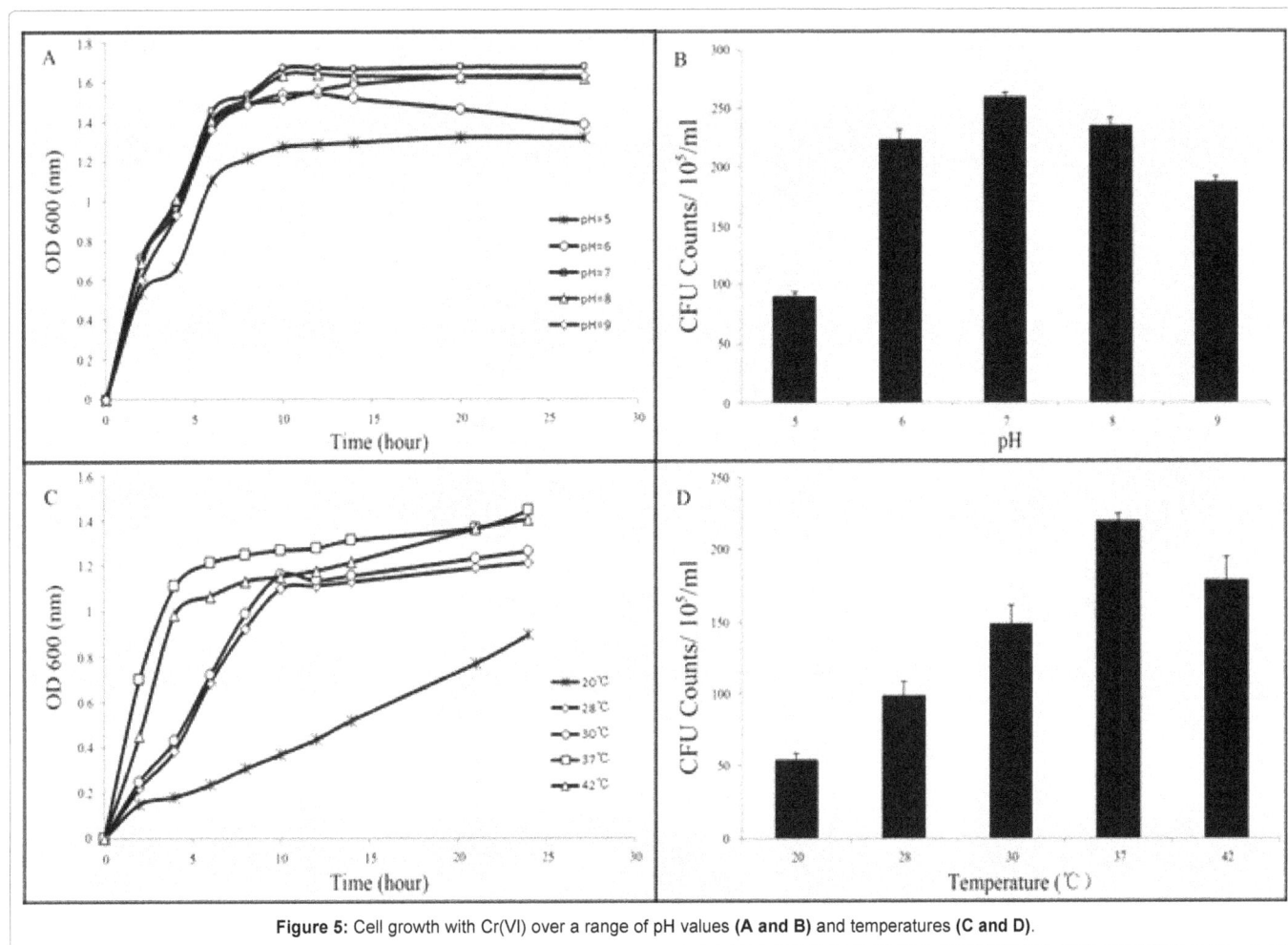

Figure 5: Cell growth with Cr(VI) over a range of pH values **(A and B)** and temperatures **(C and D)**.

Metals	Concentration (mM)											
	0.025	0.5	1	2	3	4	5	6	7	8	9	10
Cr(VI)	+	+	+	+	+	+	+	+	+	+	+	-
As(V)	+	+	+	+	+	+	+	-	-	-	-	-
Hg(II)	+	-	-	-	-	-	-	-	-	-	-	-
U(VI)	+	+	+	+	+	+	+	+	+	+	+	-

Table 1: Tolerance of heavy metals Cr(VI), As(V), Hg(II), and U(VI) in the isolated strain LZ-01 with concentration ranging from 0~10 mM.

	Antibiotics (Concentration mg/ml)						
	Amp	Van	Chlo	Kan	Eryth	Gent	Tetra
MIC	10	1	1.4	0.29	0.11	0.07	0.015

Table 2: Antibiotics ampicillin, vancomycin, chloramphenicol, kanamycin, erythromycin, gentamycin and tetracycline MIC of isolated strain LZ-01.

and Gram-positive. The 16S sequencing data indicated that all four strains were closely related to S. aureus. LZ-01 was 98.5% similar to S. aureus subsp. anaerobius. LZ-02 was 98.1% similar to S aureus subsp. aureus. LZ-03 was 99.0% similar to S. aureus subsp. anaerobius. LZ-04 was 98.8% similar to S. aureus subsp. aureus (Figure 7).

Whole cell fatty acid was analyzed following the standard protocol of MIDI [40]. The fatty acid profiles of LZ-01, LZ-02, LZ-03, LZ-04 and S. aureus type strains ATCC29213 and ATCC25923 [41,42] were presented in supplementary figure 3. The major fatty acids of the strains tested were $C_{15:0}$ iso, $C_{15:0}$ anteiso, $C_{16:0}$, $C_{17:0}$ iso, $C_{17:0}$ anteiso and $C_{18:0}$. This result was similar to the type strains of S. aureus confirming that all these four strains were closely related to S. aureus.

To further clarify the relationship between LZ-01 and S. aureus, we conducted Vitek phenotypic identification of strain LZ-01 and strains ATCC 25923 and ATCC 29213. The result showed that LZ-01 used 57 out of 64 of the same carbon sources with the two type strains. However, LZ-01 still displayed some major differences in comparison with the S. aureus type strains. LZ-01 produced α-glucosidase during growth while the other two type strains did not; LZ-01 could not utilize polybutylene and D-maltose while type strains could (Supplementary Figure 4). This data further confirmed that LZ-01 is closely related to S. aureus but still possesses significant differences from any known S. aureus strains.

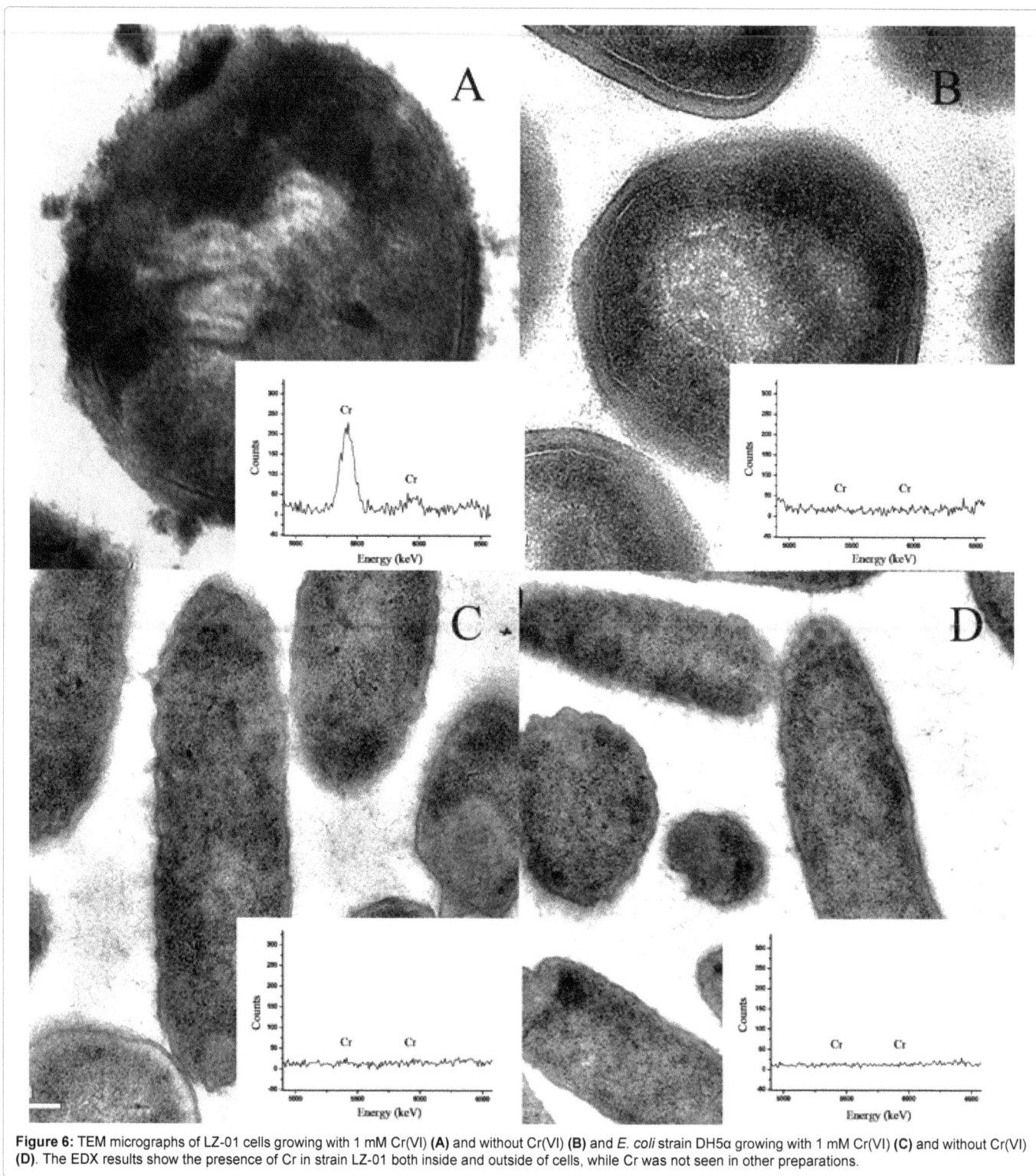

Figure 6: TEM micrographs of LZ-01 cells growing with 1 mM Cr(VI) **(A)** and without Cr(VI) **(B)** and *E. coli* strain DH5α growing with 1 mM Cr(VI) **(C)** and without Cr(VI) **(D)**. The EDX results show the presence of Cr in strain LZ-01 both inside and outside of cells, while Cr was not seen in other preparations.

Discussion

In this study, we screened surface sediment of Lanzhou reach of the Yellow River for Cr(VI)-reducing microbes. Thirteen hexavalent chromium-resistant strains were isolated. Four of them grew with up to 9 mM Cr(VI) and reduced Cr(VI) to Cr(III). All four isolates were closely related to *S. aureus*. Strain LZ-01 reduced Cr(VI) while growing aerobically and this activity appeared to be associated with both intracellular and extracellular activities. Cellular fatty acid and carbon source profiling suggested that LZ-01 is a novel subspecies of *S. aureus*, a result that was also supported by the transcriptome analysis.

Many Cr(VI)-reducing bacteria have been isolated from Cr(VI)-contaminated sites [30,31,43]. Many of these studies looked at

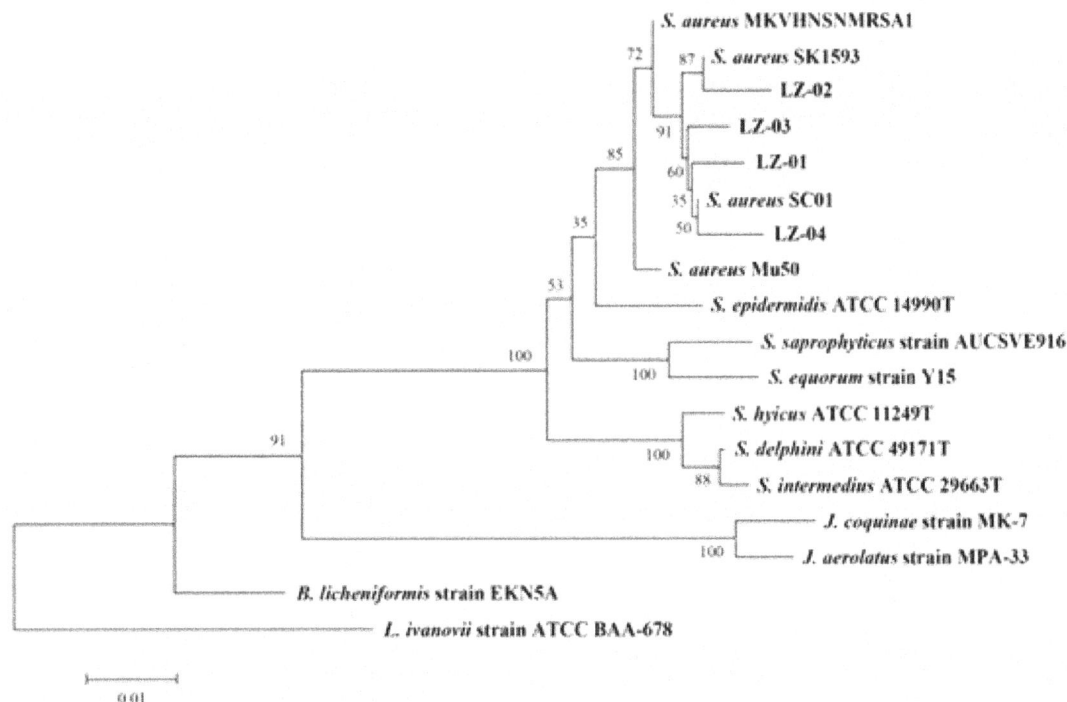

Figure 7: Phylogenetic tree based on 16S rRNA gene sequence showing the relationship between corresponding sequences of the genus *Staphylococcus* and our bacterial isolates LZ-01, LZ-02, LZ-03, LZ-04.

microbial Cr(VI)-reducing processes in anaerobic environment and found enzymes involved to be oxygen sensitive [35,36]. However, many Cr(VI)-contaminated aquatic environments are aerobic including Lanzhou reach. Anaerobic Cr(VI)-reducing bacteria may not be suitable for microbial remediation in such backgrounds. There are a few known strains that are able to reduce Cr(VI) aerobically, but their activity may not be as strong as anaerobic Cr(VI) reduction [44-47]. For example, *E. coli* produces a nitroreductase (NfsA) that has been shown to reduce Cr(VI) [45]. However, it is still sensitive to Cr(VI) and its ability to reduce Cr(VI) is limited (Figure 3).

Cr(VI) tolerance of strain LZ-01 is higher than any known aerobic Cr(VI) reducers and many anaerobic Cr(VI) reducers including *Desulfovibrio vulgaris* [35] and *Desulfovibrio desulfuricans* G20 [36]. Since LZ-01 was able to aerobically reduce Cr(VI) efficiently, it might be a good candidate for Cr(VI) remediation in Lanzhou reach.

Mechanisms of microbial Cr(VI) reduction can be divided into extracellular and intracellular reduction. In some previous studies, aerobic Cr(VI) reduction has been shown to occur intracellularly [45,46]. A cytoplasmic metal reductase in *Desulfovibrio desulfuricans* G20 reduces Cr(VI) using thioredoxin as electron donor [36]. However, *Desulfovibrio vulgaris* is known to reduce Cr(VI) in the periplasm with the involvement of cytochrome c [35]. In LZ-01, Cr(III) precipitates were detected both within and outside of cells (Figure 6) indicating that the reduction involved both intracellular and extracellular processes. Cd(II) is an inhibitor of thioredoxin and it repressed 42.6% of Cr(VI) reduction in LZ-01. NaN$_3$, a cytochrome oxidase inhibitor, inhibited only of 26.5% Cr(VI) reduction (Figure 4). These results suggest that these inhibitors lack specificity for the target proteins. But they also suggest that there might be two independent mechanisms of Cr(VI) reduction, an interpretation that is backed up by the results from TEM analysis.

Although *S. aureus* is frequently isolated from the environment [11,27-29,48], it has not previously been shown to reduce Cr(VI). A *S. aureus* plasmid pI258 has been shown to encoded a gene CadC that contains two distinct metal-binding sites for several metals [29], but not Cr(VI). A few other *Staphylococcus* strains have been reported to reduce Cr(VI). *S. epidermidis* L-02 can reduce Cr(VI) under anaerobic condition [11] and *Staphylococcus xylosus* can biosorb Cr(VI) and Cd(II) in single ion and binary solutions [48].

Strain LZ-01 is also resistant to As(V) and U(VI) (Table 1) but sensitive to Hg(II). Cr(VI), U(VI) and As(V) are all high valence oxidized metals existing in anions Cr$_2$O$_7$$^{2-}$, UO$_2$$^{2+}$ and AsO$_4$$^{3-}$ while Hg(II) is in the cation form Hg^{2+}. It is possible that the resistance mechanisms for Cr(VI), U(VI) and As(V) are similar in LZ-01. Several oxidoreductases have been shown to reduce both U(VI) and Cr(VI). e.g. ChrR and NfsA [45,47]. These proteins obtain electrons from NADH/NADPH and transfer two electrons to Cr(VI) or U(VI). Both ChrR and NfsA are cytoplasmic proteins and strain LZ-01 may have a similar mechanism for the reduction of these metals. *Desulfovibrio desulfuricans* G20 is also known to tolerate Cr(VI), U(VI) and As(V) with a common mechanism for both Cr(VI) and U(VI) reduction [36,49].

Metal-resistant microorganisms are often presented in contaminated soils [50-55]. These bacteria are already well adapted to the local environment and they will likely have a competitive advantage over allochthonous strains if introduced for bioremediation. Lanzhou reach of the Yellow River is heavily contaminated with metals, threatening downstream population with bioremediation occurring near the source of the metals. Hence, *S. aureus* may play an important role in Cr(VI) remediation at this site.

Acknowledgements

Funding for this work was supplied by grants from Lanzhou University, lzujbky-2011-32 and from NFSC 31200085, due to X, Li. We sincerely thank Dr. Chengxiang Fang from Wuhan University for helping on phylogenetic analysis on strain LZ-01 and Dr. Xiaoning Hu from Gansu CDC for helping on Vitek analysis.

References

1. Liding C, Bojie F (1996) Analysis of impact of human activity on landscape structure in yellow river delta-a case study of dongying region. Acta Ecologica Sinica 16: 337-344.

2. Gao H, Zhou L, Ma MQ, Chen XG, Hu ZD (2004) Composition and source of unknown organic pollutants in atmospheric particulates of the Xigu District, Lanzhou, People's Republic of China. Bull Environ Contam Toxicol 72: 923-930.

3. Liu C, Xu J, Liu C, Zhang P, Dai M (2009) Heavy metals in the surface sediments in Lanzhou Reach of Yellow River, China. Bull Environ Contam Toxicol 82: 26-30.

4. Zahoor A, Rehman A (2009) Isolation of Cr(VI) reducing bacteria from industrial effluents and their potential use in bioremediation of chromium containing wastewater. J Environ Sci (China) 21: 814-820.

5. Stearns DM (2000) Is chromium a trace essential metal? Biofactors 11: 149-162.

6. Hopkins LL Jr, Schwarz K (1964) Chromium (3) Binding to serum proteins, specifically siderophilin. Biochim Biophys Acta 90: 484-491.

7. Herring Wb, Leavell Bs, Paixo Lm, Yoe Jh (1960) Trace metals in human plasma and red blood cells. A study of magnesium, chromium, nickel, copper and zinc. I. Observations of normal subjects. Am J Clin Nutr 8: 846-854.

8. Macaskie LE, Dean AC (1989) Microbial metabolism, desolubilization, and deposition of heavy metals: metal uptake by immobilized cells and application to the detoxification of liquid wastes. Adv Biotechnol Processes 12: 159-201.

9. Abbas ZA, Steenari BM, Lindqvist O (2001) A study of Cr(VI) in ashes from fluidized bed combustion of municipal solid waste: leaching, secondary reactions and the applicability of some speciation methods. Waste Manag 21: 725-739.

10. Grevatt PC, Environmental Protection Agency (1998) Toxicological review of hexavalent chromium (CAS No. 18540-29-9) in support of summary informaation on the Integrated Risk Information System (IRIS). U.S. Environmental Protection Agency, Washington, D.C, USA.

11. Vatsouria A, Vainshtein M, Kuschk P, Wiessner A DK, Kaestner M, et al. (2005) Anaerobic co-reduction of chromate and nitrate by bacterial cultures of Staphylococcus epidermidis L-02. J Ind Microbiol Biotechnol 32: 409-414.

12. Rai LC, Gaur JP, Kumar HD (1981) Protective effects of certain environmental factors on the toxicity of zinc, mercury, and methylmercury to Chlorella vulgaris. Environ Res 25: 250-259.

13. Lee KY, Kim KW (2010) Heavy metal removal from shooting range soil by hybrid electrokinetics with bacteria and enhancing agents. Environ Sci Technol 44: 9482-9487.

14. Sano D, Myojo K, Omura T (2006) Heavy metal-binding proteins from metal-stimulated bacteria as a novel adsorbent for metal removal technology. Water Sci Technol 53: 221-226.

15. Cruz Viggi C, Pagnanelli F, Cibati A, Uccelletti D, Palleschi C, et al. (2010) Biotreatment and bioassessment of heavy metal removal by sulphate reducing bacteria in fixed bed reactors. Water Res 44: 151-158.

16. Mohanty M, Patra HK (2011) Attenuation of chromium toxicity by bioremediation technology. Rev Environ Contam Toxicol 210: 1-34.

17. Sanghi R, Sankararamakrishnan N, Dave BC (2009) Fungal bioremediation of chromates: conformational changes of biomass during sequestration, binding, and reduction of hexavalent chromium ions. J Hazard Mater 169: 1074-1080.

18. Seki H, Suzuki A, Maruyama H (2005) Biosorption of chromium(VI) and arsenic(V) onto methylated yeast biomass. J Colloid Interface Sci 281: 261-266.

19. Panda M, Bhowal A, Datta S (2011) Removal of hexavalent chromium by biosorption process in rotating packed bed. Environ Sci Technol 45: 8460-8466.

20. Zakaria ZA, Zakaria Z, Surif S, Ahmad WA (2007) Biological detoxification of

Cr(VI) using wood-husk immobilized Acinetobacter haemolyticus. J Hazard Mater 148: 164-171.

21. Tsibakhashvili NY, Kalabegishvili TL, Rcheulishvili AN, Murusidze IG, Rcheulishvili OA, et al. (2009) Decomposition of Cr(V)-diols to Cr(III) complexes by Arthrobacter oxydans. Microb Ecol 57: 360-366.

22. Elangovan R, Abhipsa S, Rohit B, Ligy P, Chandraraj K (2006) Reduction of Cr(VI) by a Bacillus sp. Biotechnol Lett 28: 247-252.

23. Goulhen F, Gloter A, Guyot F, Bruschi M (2006) Cr(VI) detoxification by Desulfovibrio vulgaris strain Hildenborough: microbe-metal interactions studies. Appl Microbiol Biotechnol 71: 892-897.

24. Gopalan R, Veeramani H (1994) Studies on microbial chromate reduction by Pseudomonas Sp. In aerobic continuous suspended growth cultures. Biotechnol Bioeng 43: 471-476.

25. Campos VL, Moraga R, Yánez J, Zaror CA, Mondaca MA (2005) Chromate reduction by Serratia marcescens isolated from tannery effluent. Bull Environ Contam Toxicol 75: 400-406.

26. He Z, Gao F, Sha T, Hu Y, He C (2009) Isolation and characterization of a Cr(VI)-reduction Ochrobactrum sp. strain CSCr-3 from chromium landfill. J Hazard Mater 163: 869-873.

27. Millar MR, Griffin N, Keyworth N (1987) Pattern of antibiotic and heavy-metal ion resistance in recent hospital isolates of Staphylococcus aureus. Epidemiol Infect 99: 343-347.

28. Ito T, Okuma K, Ma XX, Yuzawa H, Hiramatsu K (2003) Insights on antibiotic resistance of Staphylococcus aureus from its whole genome: genomic island SCC. Drug Resist Updat 6: 41-52.

29. Busenlehner LS, Weng TC, Penner-Hahn JE, Giedroc DP (2002) Elucidation of primary (alpha(3)N) and vestigial (alpha(5)) heavy metal-binding sites in Staphylococcus aureus pI258 CadC: evolutionary implications for metal ion selectivity of ArsR/SmtB metal sensor proteins. J Mol Biol 319: 685-701.

30. Zakaria ZA, Zakaria Z, Surif S, Ahmad WA (2007) Hexavalent chromium reduction by Acinetobacter haemolyticus isolated from heavy-metal contaminated wastewater. J Hazard Mater 146: 30-38.

31. Pattanapipitpaisal P, Brown NL, Macaskie LE (2001) Chromate reduction and 16S rRNA identification of bacteria isolated from a Cr(VI)-contaminated site. Appl Microbiol Biotechnol 57: 257-261.

32. Zakaria ZA, Zakaria Z, Surif S, Ahmad WA (2007) Hexavalent chromium reduction by Acinetobacter haemolyticus isolated from heavy-metal contaminated wastewater. J Hazard Mater 146: 30-38.

33. Wang YT, Shen H (1995) Bacterial reduction of hexavalent chromium. J Ind Microbiol 14: 159-163.

34. Wilson K (2001) Preparation of genomic DNA from bacteria. Curr Protoc Mol Biol.

35. Lovley DR, Phillips EJ (1994) Reduction of Chromate by Desulfovibrio vulgaris and Its c(3) Cytochrome. Appl Environ Microbiol 60: 726-728.

36. Li X, Krumholz LR (2009) Thioredoxin is involved in U(VI) and Cr(VI) reduction in Desulfovibrio desulfuricans G20. J Bacteriol 191: 4924-4933.

37. Denny MF, Hare MF, Atchison WD (1993) Methylmercury alters intrasynaptosomal concentrations of endogenous polyvalent cations. Toxicol Appl Pharmacol 122: 222-232.

38. Barak Y, Ackerley DF, Dodge CJ, Banwari L, Alex C, et al. (2006) Analysis of novel soluble chromate and uranyl reductases and generation of an improved enzyme by directed evolution. Appl Environ Microbiol 72: 7074-7082.

39. Barak Y, Nov Y, Ackerley DF, Matin A (2008) Enzyme improvement in the absence of structural knowledge: a novel statistical approach. ISME J 2: 171-179.

40. Müller KD, Husmann H, Nalik HP (1990) A new and rapid method for the assay of bacterial fatty acids using high resolution capillary gas chromatography and trimethylsulfonium hydroxide. Zentralbl Bakteriol 274: 174-82.

41. Sincoweay H, Miyagawa E, Kume T (1981) Cellular fatty acid composition in staphylococci isolated from bovine milk. Natl Inst Anim Health Q (Tokyo) 21: 14-20.

42. Asai S, Noda M, Yamamura M, Hozumi Y, Takase I, et al. (1993) Comparative study of the cellular fatty acids of methicillin-resistant and -susceptible Staphylococcus aureus. APMIS 101: 753-761.

43. He M, Li X, Guo L, Miller SJ, Rensing C, et al. (2010) Characterization and genomic analysis of chromate resistant and reducing Bacillus cereus strain SJ1. BMC Microbiol 10: 221.

44. Kwak YH, Lee DS, Kim HB (2003) Vibrio harveyi nitroreductase is also a chromate reductase. Appl Environ Microbiol 69: 4390-4395.

45. Ackerley DF, Gonzalez CF, Keyhan M, Blake R, Matin A et al. (2004) Mechanism of chromate reduction by the *Escherichia coli* protein, NfsA, and the role of different chromate reductases in minimizing oxidative stress during chromate reduction. Environ Microbiol 6: 851-860.

46. Campos J, Martinez-Pacheco M, Cervantes C (1995) Hexavalent-chromium reduction by a chromate-resistant *Bacillus sp.* strain. Antonie Van Leeuwenhoek 68: 203-208.

47. Cervantes C, Silver S (1992) Plasmid chromate resistance and chromate reduction. Plasmid 27: 65-71.

48. Ziagova M, Dimitriadis G, Aslanidou D, Papaioannou X, Litopoulou Tzannetaki E, et al. (2007) Comparative study of Cd(II) and Cr(VI) biosorption on *Staphylococcus xylosus* and *Pseudomonas sp.* in single and binary mixtures. Bioresour Technol 98: 2859-2865.

49. Li X, Krumholz LR (2007) Regulation of arsenate resistance in *Desulfovibrio*

desulfuricans G20 by an arsRBCC operon and an arsC gene. J Bacteriol 189: 3705-3711.

50. Kathiravan MN, Karthick R, Muthu N, Muthukumar K, Velan M (2010) Sonoassisted microbial reduction of chromium. Appl Biochem Biotechnol 160: 2000-2013.

51. Camargo FA, Bento FM, Okeke BC, Frankenberger WT (2003) Chromate reduction by chromium-resistant bacteria isolated from soils contaminated with dichromate. J Environ Qual 32: 1228-1233.

52. Singh SK, Tripathi VR, Jain RK, Vikram S, Garg SK (2010) An antibiotic, heavy metal resistant and halotolerant Bacillus cereus SIU1 and its thermoalkaline protease. Microb Cell Fact 9: 59.

53. Choudhary S, Sar P (2009) Characterization of a metal resistant *Pseudomonas sp.* isolated from uranium mine for its potential in heavy metal (Ni^{2+}, Co^{2+}, Cu^{2+}, and Cd^{2+}) sequestration. Bioresour Technol 100: 2482-2492.

54. Poli A, Romano I, Caliendo G, Nicolaus G, Orlando P, et al. (2006) *Geobacillus toebii subsp. decanicus subsp.* nov., a hydrocarbon-degrading, heavy metal resistant bacterium from hot compost. J Gen Appl Microbiol 52: 223-234.

55. Springael D, Diels L, Hooyberghs L, Kreps S, Mergeay M (1993) Construction and characterization of heavy metal-resistant haloaromatic-degrading Alcaligenes eutrophus strains. Appl Environ Microbiol 59: 334-339.

Stable Isotope Probing - A Tool for Assessing the Potential Activity and Stability of Hydrocarbonoclastic Communities in Contaminated Seawater

Petra J Sheppard[1]*, Eric M Adetutu[1], Alexandra Young[2], Mike Manefield[3], Paul D Morrison[1] and Andrew S Ball[1]

[1]RMIT University, School of Applied Science, Bundoora, Victoria 3083, Australia
[2]Flinders University, School of Biological Sciences, Bedford Park, SA, 5042, Australia
[3]Centre for Marine Bio innovation, University of New South Wales, Sydney, New South Wales, 2052, Australia

Abstract

To optimize bioremediation strategies, knowledge of which bacterial groups are actually degrading specific hydrocarbon fractions is required. In this study, we monitored the utilization rate of unlabeled (^{12}C) and labeled (^{13}C) substrates, benzene (0.559 µL L^{-1} h^{-1}) and hexadecane (0.330 µL L^{-1} h^{-1}) in seawater pre-exposed to hydrocarbons over 72 h in laboratory based microcosms. Microbial community analysis by RNA-SIP DGGE showed substantial differences between the banding pattern of ^{12}C and ^{13}C communities. Cluster analysis of the microbial community profiles showed that the labeled bacterial population was ~25% similar to the original community in the unlabeled microcosms. This suggested that only a subset of the original bacterial community appeared to have utilized the labeled substrates. Sequence analysis of 16S rRNA gene sequences revealed the presence of known hydrocarbon degraders including *Alcanivorax*, *Acinetobacter*, *Pseudomonas* and *Roseobacter*. The presences of a number of Firmicutes in both sets of mesocosms suggest that these species were able to utilize both benzene and hexadecane. This study highlights the benefits of incorporating RNA-SIP in remediation studies to enhance the understanding of microbial communities in contaminated seawater.

Keywords: Stable Isotope probing; Hexadecane; Benzene; DGGE; 16S rRNA

Introduction

Marine environments have a rich biodiversity which is essential for a functioning and stable ecosystem [1]. However, anthropogenic stressors such as habitat disturbances through drilling expeditions and hydrocarbon contamination threaten the biodiversity and their resilience to natural perturbations. Fortunately, microorganisms naturally play a role in carbon cycling and some have the capacity to degrade the hydrocarbons using them as a carbon and energy source [2]. Previously, more than 200 bacterial, algal and fungal genera have been identified as being able to metabolize hydrocarbons [3], indicating the ubiquity of hydrocarbon degrading capacity in the environment. Consequently, many investigations have been carried out on natural and enhanced microbial degradation of contaminant hydrocarbons in the marine environments [4,5].

A fundamental question that exists while assessing and characterizing any contaminated ecosystem is identifying which types of microorganisms are present. Assessment of the microbial communities by methods such as clone library' construction, genetic fingerprinting including denaturant gradient gel electrophoresis (DGGE), Temperature Gradient Gel Electrophoresis (TGGE), Terminal Restriction Fragment Length Polymorphism (T-RFLP) and metagenomics are approaches aimed to identify and survey the diversity of microbial communities' [6]. Whilst metagenomics (including metatranscriptomics) can provide greater information on microbial taxonomy and metabolism than other methods and can also be used for screening based on the expression of a selected phenotype [6], these methods do not provide direct information on how individual microorganisms relate to ecosystem functioning (e.g. hydrocarbon utilization). This ability of relating specific hydrocarbon degrading activities to specific groups in the microbial community can be difficult due to the high diversity and abundance of bacterial species in the marine environment. Therefore, one major challenge when studying microbial communities is that while there is abundant information on the diversity and potential function of hydrocarbon degrading marine microorganisms, there are gaps in our knowledge with regards to which hydrocarbon fractions are actually degraded by specific microbial groups. In order to overcome this limitation Stable Isotope Probing (SIP) can be used.

The incorporation of stable carbon isotopes in scientific experiments has shown promising results especially in studies for validating intrinsic bioremediation [7-9], examining contaminant migration and distribution [10] and those focusing on natural isotope abundance [11]. The use of SIP in natural abundance studies has demonstrated the application of this methodology to work *in situ*, which is advantageous for bioremediation studies. One of the advantages of using SIP in bioremediation studies is the ability to use this technique to target specific substrates and therefore specific subpopulations. Microbial utilization of substrates such as labeled carbon ^{13}C or nitrogen ^{15}N results in their incorporation into cellular biomarkers such as nucleic acids (e.g. RNA, DNA) [12]. Nucleic acids which contain the labeled substrate will be heavier than nucleic acids with unlabeled fractions. These different fractions can be separated by ultracentrifugation and fractionated into aliquots by displacing the gradient with water [12]. Furthermore, molecular identification of specifically labeled populations within the 'heavy' stable isotope enriched nucleic acids can be identified (through sequencing), corresponding to the microorganisms which were able to metabolise the labeled substrate of interest [13]. The use of RNA-SIP holds significant potential for exploring active populations as the use

*Corresponding author: Petra J Sheppard, RMIT University, School of Applied Science, Bundoora, Victoria 3083, Australia
E-mail: Petra.Sheppard@rmit.edu.au

of RNA achieve higher synthesis rates and requires reduced incubation times when compared to DNA-SIP [13].

The use of SIP has the potential to help in clarifying the identities of key microorganisms involved in degrading contaminants during bioremediation. SIP can aid in determining whether the hydrocarbon degrading microorganisms present during bioremediation are actually involved in contaminant degradation or merely a fraction of the microbial community that can tolerate the contaminant. In most investigations on microbial roles in bioremediation, the presence of hydrocarbonoclastic organisms is often associated with hydrocarbon degrading activities (especially when substantial degradation has occurred). This assumption is usually made with little or no information on the proportion of microbial groups that were directly involved in biodegradation. In addition there is limited information on microbial community dynamics during the bioremediation of a contaminated marine environment. Questions remain on which microbial groups are directly involved in the process, what phases of the degradation process are these groups active and the level of functional redundancy that exists within the community.

The aim of this study was to identify and compare the microbial communities of pre-exposed hydrocarbon contaminated seawater in closed marine systems using an RNA-SIP technique. In particular, this study focused on i) investigating and identifying the dominant microbial groups in hydrocarbon degradation through 16S rRNA analysis ii) determining if those prevalent microbial groups are actively involved in hydrocarbon mineralization using SIP, incorporating an aliphatic (hexadecane) and an aromatic substrate (benzene) and iii) understanding the dynamics of the active community using microbial community based tools such as range weighted richness (Rr), functional organization (Fo) and dynamics (Dy).

Materials and Methods

Substrate utilization microcosms

The ^{13}C hexadecane and ^{13}C benzene and their ^{12}C versions used for this study were obtained from Sigma- Aldrich (USA) with the ^{13}C labeled fractions having a chemical and isotope purity of 99 atom%. These hydrocarbons were selected as representatives of aromatic and aliphatic fractions commonly found in crude oil. Sterile glass serum bottles (500 mL) were each filled with seawater previously exposed (>27 weeks: Supplementary Table 1) to hydrocarbon contaminants (250 mL) (initially 1% w/v crude oil) and successfully subjected to bioremediation to completely eliminate the hydrocarbon contaminants. Prior to the start of this study, the pre-exposed seawater was subject to chemical analysis to validate the absence of hydrocarbon contaminants.

The microcosms were spiked with either hexadecane (^{13}C or ^{12}C) or benzene (^{13}C or ^{12}C) to a final concentration of approximately 65 μL L^{-1} and 30 μL L^{-1} respectively and immediately sealed with Teflon-coated butyl septa. The samples included i) pre-exposed seawater; ii) pre-exposed seawater with ^{12}C substrate (unlabeled benzene UB), iii) pre-exposed seawater with ^{12}C substrate (unlabeled hexadecane UH), iv) pre-exposed seawater with ^{13}C substrate (labeled benzene LB), v) pre-exposed seawater with ^{13}C labeled hexadecane LH) and vi) sterilized pre-exposed seawater (control). Replicate samples of treated and control samples were incubated over 72 h. Destructive sampling was carried out at times 0, 24, 48 and 72 h. Microcosms were incubated freestanding at room temperature with natural lighting until >40% degradation had occurred (~72 h). Autoclaved microcosms were used as controls throughout the experiment to ensure degradation was

by biotic processes. All standards and controls had the same liquid/headspace ratio and were held at the same temperature as the test samples.

Hydrocarbon substrate monitoring

Kitagawa tubes coupled with an aspirating pump AP-20 (Kitagawa, Japan) were used as per the manufacturer's protocol to determine the substrate concentration in each benzene microcosm. Briefly, benzene concentrations were measured from the headspace of each microcosm with a Kitagawa 118SC detection tube containing a chemical reagent calibrated for measuring 1-100 ppm benzene within a temperature range of 0-40°C. Hexadecane was also analyzed using Kitagawa tubes but with mixed results as most of the substrate concentration could not be recovered. Therefore, gas chromatography was then used for hexadecane analysis using a liquid extraction method. A standard calibration curve was constructed from hexadecane concentration dilutions (hexane), and the equation from the standard calibration curve was used in conjunction with the area under each chromatogram (correlated area) to determine the hexadecane concentration in each hexadecane-treated microcosms. The hexadecane samples were analyzed by mixing hexane (5 mL) in each microcosm (triplicate) and allowing the hexadecane to dissolve in the solvent. After separation, 1 mL of the solvent was extracted from each microcosm and sampled on a HP 6890 Gas Chromatogram (GC) system equipped with a 5973 Mass Spectrometry Detector (MSD). A programmed temperature split 20:1 injection was used. The injector temperature was initially set at 60°C, and then increased at a rate of 50°C/min to 270°C, where it was maintained. The capillary column used was an Agilent JW DB5 (30 m by 0.25 mm with 0.25 μm thickness), with helium as the carrier gas flowing at a rate of 1.9 mL/min in a constant flow mode.

RNA extraction

RNA was extracted from all treatments at each of the four different time points after the preparation of the microcosms, corresponding to time 0, 24, 48 and 72 h. The RNA was extracted using a phenol-chloroform extraction method [4]. Briefly, pelleted cells were resuspended in sodium phosphate buffer (SPB, 0.5 mL, pH 7.9) and phenol-chloroform-isoamyl alcohol (0.5 mL, 25:24:1) and then lysed by bead beating for 2×20 s (Mini Bead Beater K9, BioSpec). After centrifugation (12,000×g for 5 min at 4°C) the aqueous layer was removed and purified with phenol-chloroform-isoamyl alcohol (25:24:1). The RNA was further purified, precipitated, ethanol washed and resuspended in molecular grade water (30-50 μL). DNA was digested with RNase-free DNase (Promega, USA) in accordance with the manufacturer's protocol to remove residual DNA and purify the RNA. Test PCR was carried out using the appropriate primers to confirm the absence of DNA in the digested samples (detection of no amplicon). The extracted RNA was then subject to ultracentrifugation and fractionation.

Ultracentrifugation and fractionation

Equilibrium density gradient centrifugation and gradient fractionation were conducted in 1.99 g mL^{-1} cesium trifluoroacetate (GE Healthcare, UK) as per the Manefield et al. [12] protocol. Each gradient

Substrate*	Unlabelled (^{12}C) (μL L^{-1} hour^{-1})	% Degraded	Labeled (^{13}C) (μL L^{-1} hour^{-1})	% Degraded
Benzene	0.529 ± 0.154	72.0 ± 8.01	0.589 ± 0.175	80.0 ± 9.01
Hexadecane	0.300 ± 0.110	36.9 ± 3.51	0.361 ± 0.043	43.2 ± 2.07

Table 1: Substrate utilization rates and percentage degraded after 72 hours incubation.

was made up of 500-800 ng of total RNA. Hexadecane samples were sealed in polyallomer bell top Quickseal centrifuge tubes (13 x 32 mm), and spun in a TLA 100.3 rotor (Beckman Coulter) in an Optima TLX Ultracentrifuge (Beckman Coulter, USA) at 120,000 rpm at 20°C for 24 h; benzene samples were sealed in polyallomer Optiseal centrifuge tubes (13 x 51 mm) and spun in a NVT90 rotor (Beckman Coulter, USA) in an Optima L-100 XP ultracentrifuge (Beckman Coulter, USA) at 45,000 rpm at 20°C for 24 h. After centrifugation, gradients were fractionated from below by displacement with water using a syringe pump at a flow rate of 3.3 µL s^{-1}. The buoyant density of each gradient fraction was determined by weighing known volumes on a four-figure milligram balance. After the 'density location' of the fractions had been determined, fractions were subject to isopropanol precipitation and molecular analysis to further validate changes in the gradient density profile; separation of 'heavy' and 'light'. Thereafter, only the appropriate fractions (in triplicate) (based on their density location representative of 'heavy' and 'light' [14]) were used for molecular analysis.

16S rRNA amplification and denaturant gradient gel electrophoresis

Transcription of RNA into cDNA was performed with GoScript™ Reverse Transcriptase (Promega, USA) in accordance to the manufacturer's protocol using reverse primer 518R (5'ATTACCGCGGCTGCTGG-3') [15]. Two microliters of cDNA was used per 50 µL PCR mixture as previously described [16]. The primers used for PCR were 341F with GC clamp and 518R (10 pmol/µL). The amplified products were then separated on a 9% polyacrylamide gel with a 45-60% denaturant gradient. Electrophoresis was run at a constant voltage of 60 V for 18 h at 60°C in 1 X TAE running buffer. After electrophoresis, the gels were silver stained [17] and scanned with an Epson V700 scanner. Dominant bands (41, marked in Figure 1A and 1B) of interest were aseptically excised, purified and cleaned as previously described by Sheppard et al. [16]. Purified products were sent to the Australian Genome Research Facility (AGRF, Melbourne, Australia) and sequenced on an ABI Prism 3730. The nucleotide sequences were analysed using Sequencer software (Sequencer Version 5.0, Gene Codes Corporation, USA) and the consensus sequences were submitted to GenBank (see Table 2A and 2B for accession numbers). Homology searches were completed with the Basic Local Alignment Search Tool (BLAST) server of the National Centre for Biotechnology Information (NCBI) using a BLAST algorithm (http://www.ncbi.nlm.gov.library.vu.edu.au/BLAST/) for the comparison of a nucleotide query sequence against a nucleotide sequence database to determine putative identities of band sequences.

Data analysis

Digitised DGGE gel images were processed using Phoretix 1D advanced analysis package (Phoretix Ltd, UK). Banding patterns were analyzed with Phoretix to generate similarity profiles using the unweighted pair group method with mathematical averages (UPGMA) [15]. Pareto-Lorenz curves were used to estimate Functional Organization (Fo) and evenness within the microbial community [18]. Range weighted richness (Rr) were used to calculate the carrying capacity of the microbial community according to the equation Rr= (N^2×D$_g$) where N is the total number of bands multiplied by Dg which is the percentage of denaturing gradient of the sample analyzed [18]. Dynamics (Dy) is the number of species that on average come to significant dominance which was a concept based on the moving window analysis, the rate of change (Δt) parameter [18]. For each selected profile, the percentage similarities between each

time frame (T0-24 h, 24-48 h and 48-72 h) were obtained from the UPGMA dendrograms. The percentage change at each time-frame was calculated using the formula %change = 100-%similarity and (Δt) calculated as described by Marzorati et al. [18]. Statistical significance was determined using ANOVA and statistical significance accepted at P<0.05.

Results and Discussion

Substrate utilization and gradient evaluation

Rapid aerobic utilization of benzene was detected over a 72 h incubation period with a rate of 0.559 µL L^{-1} h^{-1} (Table 1) suggesting the presence of adapted hydrocarbon degraders. The rate of benzene utilization was higher than that observed for hexadecane (0.330 µL L^{-1} h^{-1}) (P<0.05). The higher rate of benzene utilization compared to hexadecane may be a result of differences in their chemical structure and solubility properties [19]. Solubility constraints with the labeled hexadecane (partition coefficients Log K$_{ow}$ 8.25 and Log K$_{ow}$ 2.13 for hexadecane and benzene respectively) may have resulted in some quantity of the substrate not being available for microbial activity but also as a result of the different microorganisms involved. Alternatively, it may be a result of an adaptive status of the microbial communities' given the pre exposure to weathered hydrocarbon [19]. As the samples were pre-exposed to hydrocarbon contamination; control readings taken prior to the start of the experiments confirmed that no background concentrations were present (no detectable hydrocarbon contaminant). The sterile controls showed no substantial difference in spiked hydrocarbons at the beginning and the end of the experiments (data not shown) confirming that degradation was by biotic processes.

After the incorporation of the labeled carbon atoms (substrate) into microbial nucleic acids as a result of biotic degradation, the RNA extracts were subjected to ultracentrifugation and a density gradient was formed (Supplementary Table 2). After validation of the separation of the 'heavy' and 'light' fractions, appropriate fractions were selected (in triplicate) at each of the four time points for both of the unlabeled and labeled samples (benzene and hexadecane). The average location of the gradient density selected corresponded to hexadecane (13×32 mm tube) labeled 1.834 (± 0.011) g mL^{-1}, hexadecane unlabeled 1.778 (± 0.013) g mL^{-1}, benzene (13×51 mm tube) labeled 1.798 (± 0.012) g mL^{-1} and benzene unlabelled 1.781 (± 0.013) g mL^{-1}. The location densities share similar results to those seen by Manefield et al. [14].

Labeled versus unlabeled fraction: bacterial community structure

The degradation results showed that the microbial communities contained active hydrocarbon degraders within each microcosm (Table 1). 16S rRNA gene analysis was therefore used to assess the microbial community structure and determine changes in its community composition in response to the addition of aliphatic and aromatic hydrocarbon substrates. The RNA-SIP DGGE results showed substantial differences between banding patterns from ^{12}C and ^{13}C bacterial communities.

Figure 1A shows the UPGMA dendrogram of bacterial community profiles obtained from microcosms supplemented with labeled and unlabeled benzene. The bacterial community in the unlabelled and labeled benzene microcosm each formed distinct clusters. However, there were substantial differences between the communities over the 72 h incubation period. For example, time 0-24 h communities had 0.55 and 0.48 similarity indices with 48-72 h having lower similarity indices of 0.38 and 0.34 for unlabeled and labeled microcosms

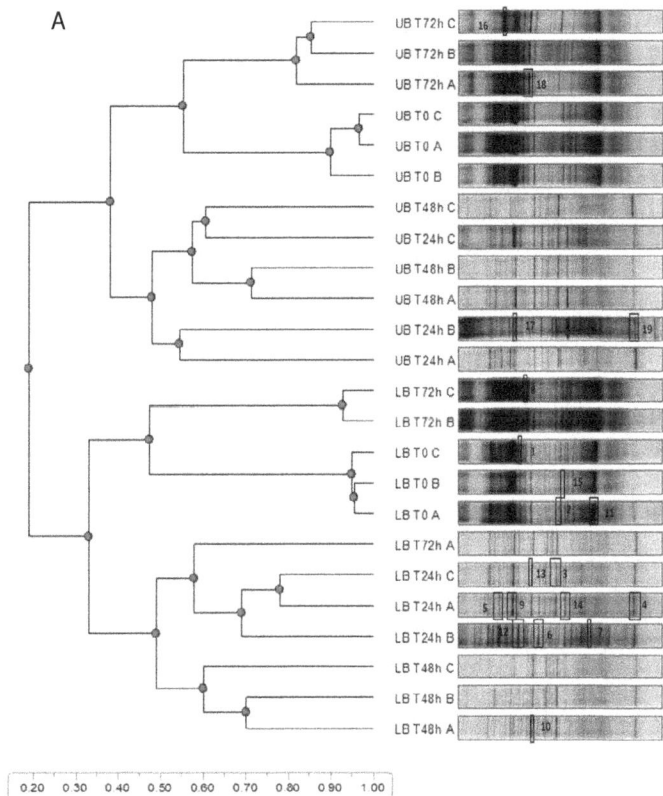

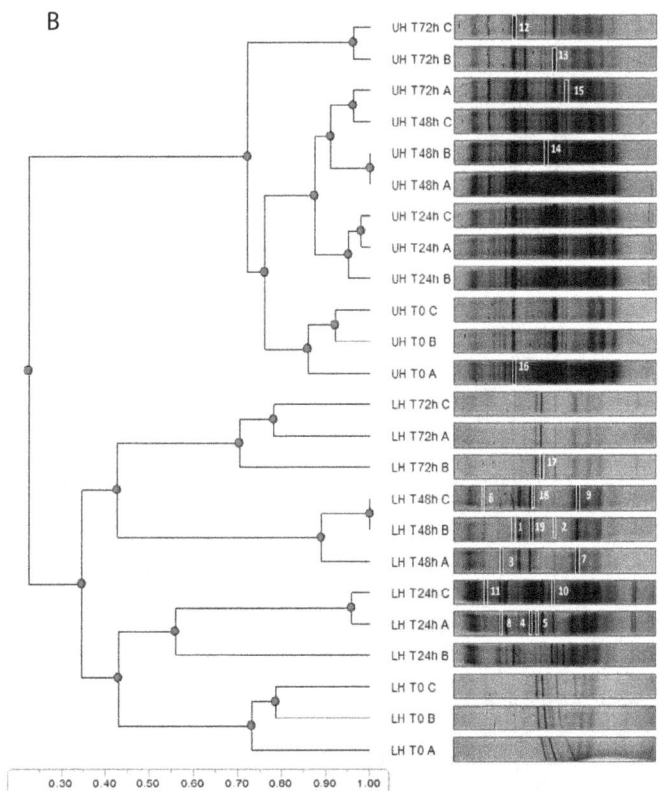

Figure 1: A & B Cluster analysis via UPGMA (Dice Sorensen's similarity index) of (A) benzene bacterial DGGE profiles and (B) hexadecane bacterial DGGE profiles. UB=unlabeled ^{12}C benzene, LB=labeled ^{13}C benzene, UH=unlabeled ^{12}C hexadecane, LH=labeled ^{13}C hexadecane, Txh=Time in hours, A-B-C=replicate samples, boxes represented excised bands and corresponding number.

Excised Band	Substrate	Division (or subdivision)	Nearest Taxon	Accession no.	Similarity (%)
1	LB	Firmicutes	*Bacillus pumilus*	JQ353824	99
2	LB	Gammaproteobacteria	*Stenotrophomonas maltophilia*	JX867287	100
3	LB	Gammaproteobacteria	*Stenotrophomonas sp*	JQ917790	99
4	LB	Firmicutes	*Lysinibacillus sphaericus*	JQ744637	99
5	LB	Actinobacteria	*Actinobacterium*	JN049471	99
6	LB	Firmicutes	*Bacillus sp*	HM026204	98
7	LB	Gammaproteobacteria	*Acinetobacter johnsonii*	JQ039983	100
8	LB	Gammaproteobacteria	*Pseudomonas putida*	JQ781586	100
9	LB	Betaproteobacteria	*Delftia sp*	HE716888	100
10	LB	Firmicutes	*Clostridiales* bacterium	GQ922846	97
11	LB	Actinobacteria	*Kineococcus radiotolerans*	CP000750	100
12	LB	Gammaproteobacteria	*Shewanella aquimarina*	FJ589034	100
13	LB	Firmicutes	*Lactobacillus helveticus*	CP003799	100
14	LB	Gammaproteobacteria	*Marinobacterium*	JQ670704	100
15	LB	Gammaproetobacteria	*Pseudidomarina sp*	GQ202579	98
16	UB	Epsilonproteobacteria	*Epsilonproteobacterium*	HQ607532	100
17	UB	Betaproteobacteria	*Massilia sp*	JQ014560	100
18	UB	Firmicutes	*Clostridium sp*	FR872934	99
19	UB	Actinobacteria	*Nocardioides islandensis*	NR044235	100

LB= Labeled benzene, UB-= unlabeled benzene

Table 2A: Summary of bacterial 16S rRNA partial sequences obtained from excised DGGE profiles (Figure 1A) of unlabeled and labeled benzene microcosms.

Excised Band	Substrate	Division (or subdivision)	Nearest Taxon	Accession no.	Similarity (%)
1	LH	Gammaproteobacteria	*Alcanivorax sp*	AB681379	100
2	LH	Alphaproteobacteria	*Roseobacter sp*	DQ659413	99
3	LH	Firmicutes	*Bacillus cereus*	JQ435695	98
4	LH	Alphaproteobacteria	*Brevundimonas sp*	JQ390020	100
5	LH	Firmicutes	*Bacillus sp*	AB715354	100
6	LH	Gammaproteobacteria	*Pseudomonas anguilliseptica*	JX177683	100
7	LH	Actinobacteria	*Rothia mucilaginosa*	NR044873	100
8	LH	Firmicutes	*Lysinibacillus sp*	JN224968	100
9	LH	Actinobacteria	*Rhodococcus aetherivorans*	JQ040007	100
10	LH	Alphaproteobacteria	*Methylobacterium sp*	GQ342553	100
11	LH	Bacteroidetes	*Chryseobacterium sp*	DQ673675	98
12	UH	Gammaproteobacteria	*Alcanivorax sp*	EU052761	99
13	UH	Gammaproteobacteria	*Roseobacter sp*	DQ659414	99
14	UH	Gammaproteobacteria	*Marinobacter sp*	JN106685	97
15	UH	-	Uncultured bacterium	JX255122	100
16	UH	Alphaproteobacteria	*Marivita sp*	HQ871858	97
17	LH	Gammaproteobacteria	*Enterobacteriaceae*	AB715425	99
18	LH	Betaproteobacteria	*Neisseria sp*	CQ900862	98
19	LH	Firmicutes	*Streptococcus sp*	JX861486	99
-*	NS	Alphaproteobacteria	*Ruegeria*	FN821687	92
-*	NS	Epsilonproteobacteria	Uncultured *Arcobacteria*	AY697901	99
-*	NS	Epsilonproteobacteria	*Arcobacteria*	FR870464	97

LH= Labeled hexadecane, UH= unlabeled hexadecane, NS= no substrate, - means no data, * means excised from seawater only.

Table 2B: Summary of bacterial 16S rRNA partial sequences obtained from excised DGGE profiles (Figure 1B) of unlabeled and labeled hexadecane microcosms.

respectively. There was also high replicate sample variability in some of the microcosms (Figure 1A). The microbial community in the labeled microcosms was also only ~20% similar to community in the unlabeled microcosms.

Figure 1B shows the UPGMA for hexadecane, which in contrast to benzene, indicates that the communities in replicate unlabeled fractions were grouped closely, ~0.73 on the Dice Sorensen's similarity index (irrespective of sampling time). These communities were also highly similar to the no substrate control (data not shown). However, the bacterial community in the labeled hexadecane microcosm showed a similar trend to the one observed in labeled benzene microcosms.

For example the community at day 0 was substantially different to that in 24 h (0.42 similarity index) and 48-72 h (0.35 similarity index) samples. The labeled fraction cluster was also only 0.23 similar to the unlabeled substrate cluster. This suggests that 20-25% (based on similarity coefficient) of the total community detected were involved in the breakdown of the benzene and hexadecane substrates at the time of sampling as well as under the investigated conditions (e.g. constant temperature).

However, this does not necessarily mean that this was the only fraction of the total microbial community that was capable of degrading the supplied substrate. These findings only suggest that in both benzene and hexadecane microcosm a fraction of the dominant microbial population in the unlabeled microcosms was involved in substrate utilization. This means that in the environment not all putative hydrocarbon degraders present in the community are involved in the degradation process.

The SIP-RNA-DGGE profile analysis also suggests that the community exposed to aromatic hydrocarbons was more sensitive than that to an aliphatic substrate. The DGGE profile was more reflective of the changes in the aromatic community (unlabeled) than the aliphatic community which had a stable community (unlabeled). Traditionally, DGGE profiles are known to reflect dominant microorganisms (top 1%) [6,15] and this dominance is assumed to have resulted from microbial activities and population. However, this study has shown that while DGGE gives a good picture of prevalent members in the community, not all the dominant microbial groups were involved in benzene and hexadecane utilization. Pre-exposure of seawater to hydrocarbon contaminants would usually have resulted in selection of hydrocarbon degrading microorganisms. The similarity cluster (Figure 1A and 1B) showed only 20-25% of this population compared to the total population in unlabeled microcosms was actually shown to have utilized the substrates. This suggests that only a subset of the dominant population in the original ^{12}C community (assumed to be predominantly hydrocarbonoclastic) is utilizing the ^{13}C substrate in the labeled microcosms. This could have been due the fact that the detected groups with ^{13}C incorporated into their DNA were the only groups capable of using the supplied substrate. Alternatively, this may simply be a reflection that the communities in these microcosms were functionally redundant, rather than a lack of hydrocarbon degrading capacities in the remaining 75-80% of the population [20].

In addition, the bacterial community in labeled and unlabeled benzene microcosms at 0 and 72 h (samples) were more similar to each other than to other samples collected at 24 or 48 h, unlike in hexadecane microcosms. The reason for this is not entirely clear. However, given that substantial degradation of benzene (up to 80%) had occurred by 72 h, the high similarities observed might have been due to the benzene stressed community returning to its original composition. This suggested that the bacterial community in the benzene supplemented microcosms were resilient. A key attribute of a resilient community is the ability of the altered community to return to its original composition after the substantial reduction or elimination of the stressor (benzene) [20]. By combining DGGE with RNA-SIP in this study, we were able to evaluate the dominant bacterial community and detect the fraction in this community that was actually degrading the specific hydrocarbon substrates. In the case of hexadecane, 16S rRNA-SIP was sensitive enough to pick differences which would have been undetectable using the conventional 16S rRNA based DGGE fingerprinting method. This was because the 16S rRNA-DGGE profile of the unlabeled microcosms showed very little community changes compared to the substantial changes observed in the labeled microcosm's profile (16S rRNA-SIP) over a 72 h period (Figure 1B).

Reliance on the 16S rRNA-DGGE profile alone could have led to the wrong conclusion that the community was largely stable when it was a dynamic and highly changing one. Therefore changes in bacterial community during bioremediation in closed contaminated marine environments are better assessed with 16S rRNA-SIP, although experiments longer than 72 h might be needed to validate this assumption providing that problems with cross feeding are properly addressed.

Microbial community dynamics

Given the assumed adaptation of the microbial community as a result of pre-exposure to contaminated seawater, we investigated the community dynamics during the experimental period. The diversity and structure of the bacterial communities were analysed using the Pareto-Lorenz curve to assess functional organization (Fo), windows moving analysis to assess dynamics (Dy) and the range weighted richness (Rr) to assess total number of species present.

To assess the functional organization (Fo) of the bacterial (benzene & hexadecane) community a Pareto-Lorenz curve distribution patterns were plotted of the 16S rRNA DGGE profiles. The Pareto-Lorenz curve is based on the number of bands and their intensities [21]. The functional organization of the bacterial communities in labeled and unlabeled benzene and hexadecane microcosms indicated communities with medium (45-60 %) functional organization (Figure 2A and 2B). The average functional organization indicates a community which can potentially deal with changing environmental conditions [18], such as the addition of a hydrocarbon substrate. This is logical as the samples were pre-exposed to hydrocarbon contamination which leads to adapted hydrocarbon degrading communities. Communities with medium functional organization are also assumed to be functionally redundant [18]. This appeared to be the case in this study as the UPGMA cluster analysis appears to show that only a fraction (20-25%; labeled community similarity to unlabeled community) of the total community was involved in benzene and hexadecane degradation (Figure 1).

The changes in a microbial community over time are reflected in the microbial community dynamics (Dy) analyses. This change over time is referred to as Δt. To evaluate the dynamics (Dy) of the 16S rRNA DGGE profiles, 24 hour-based deviations were established with moving window analysis (MWA). Δt values were very high (~62%) for the microbial communities in labeled and unlabeled benzene microcosms suggesting a high level of Dy (Supplementary Table 3). High (Dy) is reflective of a highly variable community (unstable) with broad dynamics with different species coming into dominance at different times [18]. While the Δt was also high (~59%) in labeled hexadecane communities, it was comparatively lower (~23%) in the unlabeled hexadecane community (Supplementary Table 2). This further validates the sensitivity of the RNA-SIP-DGGE methodology in picking up substantial changes in community dynamics which could have been missed by RNA-DGGE analyses which had suggested a comparatively more stable community and response to hexadecane introduction. All communities which had the incorporation of labeled benzene and hexadecane expressed high community dynamics and this was probably related to the pre-exposure (and presumably pre-adaptation) of the community to hydrocarbon contamination. This could have meant that members of the community had the capability to metabolize the introduced substrates and associated secondary and tertiary degradation products.

Similarly to Dy and Fo results, all samples showed similar results for weighted richness (Rr). The Rr value informs on the carrying capacity of an environment and whether the environment is habitable, adverse or exclusive [22]. High Rr value indicates a high microbial diversity (more bands on the DGGE profile) which correlates to an increase in the community carrying capacity and habitability. All samples produced high richness values, scoring values above 30%, therefore achieving high band numbers (diversity) on the DGGE analysis (data not shown).

Overall, these results indicate that all bacterial communities in the

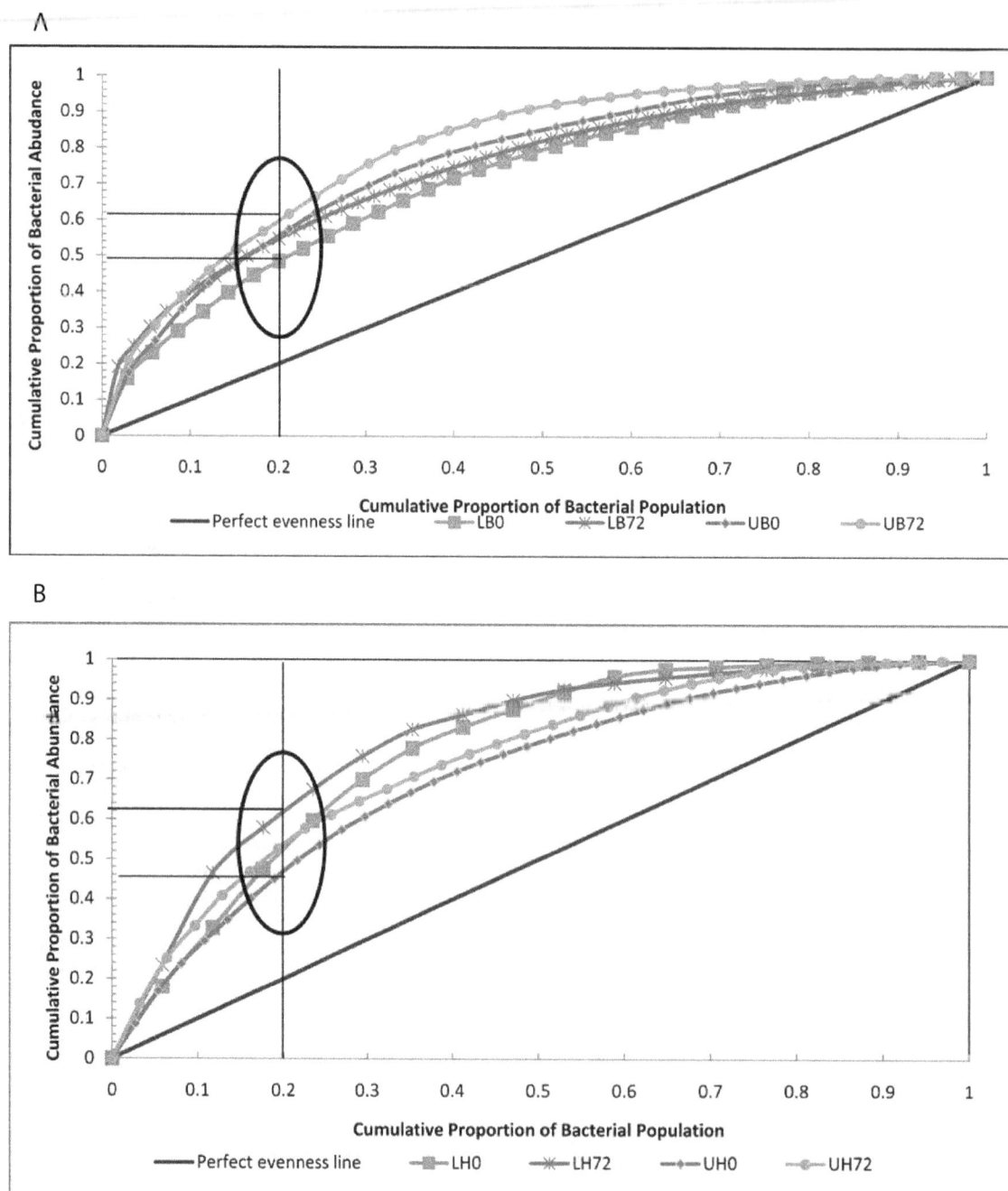

Figure 2: A & B Pareto-Lorenz distribution curves derived from 16S rRNA DGGE profiles of (A) benzene microcosm and (B) hexadecane microcosm. The vertical line is plotted at 0.2 x-axis in order to compare different Pareto values while the 45 degree diagonal line represents the perfect evenness of a community. Only the extreme curves are shown. UB=unlabeled benzene, LB=labeled benzene, UH=unlabeled hexadecane, LH=labeled hexadecane, 0 & 72=Time in hours.

microcosms reflected active hydrocarbon degrading microorganisms. Therefore sequence analysis was performed on the microcosm communities to identify the dominant species involved in the substrate utilization in order to validate this hypothesis.

Sequence analysis

Sequence analysis of the dominant 16S rRNA bands from the DGGE profiles revealed similarities to bacterial groups known to contain hydrocarbons degrading species including affiliations to Actinobacteria, Firmicutes and Proteobacteria [23].

Table 2A shows the sequence analysis for labeled benzene sequences which detected putative identities to hydrocarbon degrading species *Acinetobacter, Delftia, Pseudomonas putida, Shewanella,* and *Stenotrophomonas* [23]. Similarly, Table 2B shows the sequence analysis for labeled hexadecane microcosms. In contrast to the unlabeled and no substrate microcosm communities, the labeled hexadecane microcosms differed by the absence of the prevalent marine *Arcobacter*-related sequences. Additionally it differed by the presence of dominant bands whose sequences shared strong identities to known hydrocarbon degrading species including *Alcanivorax,*

Bacillus, Brevimonas, Chryseobacterium, Pseudomonas, Rhodococcus and *Roseobacter* [23]. In addition sequence analysis showed similarities in the species composition with in Firmicutes. The prevalent identification of *Bacillus* species in both substrate microcosms is not surprising in hydrocarbon degradation studies, as some *Bacillus* species are able to produce biosurfactants during hydrocarbon oxidation [24] thereby reducing the interfacial tension. Therefore this study suggests that members of this group played an important role in the degradation of the hydrocarbons in the closed marine system.

Conclusion

This study has further highlighted the benefits of RNA-SIP application in linking microorganism presence to degradation, by identifying species utilizing individual substrates by the incorporation of a labeled substrate. The utilization of individual hydrocarbons by a highly active and diverse community present in pre-exposed seawater within 72 h of incubation with benzene (unlabeled and labeled) was also successfully demonstrated. Microbial community analyses suggested that only 20-25% (based on similarity values) of the original community appeared to be involved in substrate utilization. We propose that this was likely due to a functional redundant and adapted microbial community rather than a lack of capacity by the remaining members of the community for hydrocarbon contaminant removal. Furthermore, the use of SIP based fingerprinting method was shown to be more sensitive to detecting changes in the community structure than conventional fingerprinting methods. Sequence analysis of each microcosm identified putative sequences of known hydrocarbon degrading microorganisms with similarities shown between the group Firmicutes. This study therefore shows how RNA-SIP application and the use of appropriate tools for community fingerprint analysis can enhance our understanding of microbial dynamics in remediation studies in contaminated seawater.

Acknowledgments

This research was supported by the Australian India Strategic Research Fund (BFO20032).

References

1. Paterson D (2005) Biodiversity and functionality of aquatic ecosystems. In: Barthott W, Porembski S, Linsenmair K (eds) Encyclopedia of Life Support Systems, Oxford, Canada.

2. Kostka JE, Prakash O, Overholt WA, Green SJ, Freyer G, et al. (2011) Hydrocarbon-degrading bacteria and the bacterial community response in gulf of Mexico beach sands impacted by the deepwater horizon oil spill. Appl Environ Microbiol 77: 7962-7974.

3. Yakimov MM, Timmis KN, Golyshin PN (2007) Obligate oil-degrading marine bacteria. Curr Opin Biotechnol 18: 257-266.

4. McKew BA, Coulon F, Osborn AM, Timmis KN, McGenity TJ (2007) Determining the identity and roles of oil-metabolizing marine bacteria from the Thames estuary, UK. Environ Microbiol 9: 165-176.

5. Vila J, María Nieto J, Mertens J, Springael D, Grifoll M (2010) Microbial community structure of a heavy fuel oil-degrading marine consortium: linking microbial dynamics with polycyclic aromatic hydrocarbon utilization. FEMS Microbiol Ecol 73: 349-362.

6. Rastogi G, Sani RK (2011) Molecular techniques to assess microbial community structure, function, and dynamics in the environment. Microbes and Microbial Technology: Agricultural and Environmental Applications. Springer Science + Business Media LLC, New York, USA.

7. Ahad JME, Sherwood Lollar B, Edwards EA, Slater GF, Sleep BE (2000) Carbon Isotope Fractionation during Anaerobic Biodegradation of Toluene: Implications for Intrinsic Bioremediation. Environ Sci Technol 34: 892-896.

8. Barth JA, Slater G, Schüth C, Bill M, Downey A, et al. (2002) Carbon isotope fractionation during aerobic biodegradation of trichloroethene by Burkholderia cepacia G4: a tool to map degradation mechanisms. Appl Environ Microbiol 68: 1728-1734.

9. Morasch B, Hunkeler D, Zopfi J, Temime B, Höhener P (2011) Intrinsic biodegradation potential of aromatic hydrocarbons in an alluvial aquifer--potentials and limits of signature metabolite analysis and two stable isotope-based techniques. Water Res 45: 4459-4469.

10. Hunkeler D, Chollet N, Pittet X, Aravena R, Cherry JA, et al. (2004) Effect of source variability and transport processes on carbon isotope ratios of TCE and PCE in two sandy aquifers. J Contam Hydrol 74: 265-282.

11. Gocht T, Barth JAC, Epp M, Jochmann M, Blessing M, et al. (2007) Indications for pedogenic formation of perylene in a terrestrial soil profile: Depth distribution and first results from stable carbon isotope ratios. Appl Geochem 22: 2652-2663.

12. Manefield M, Whiteley AS, Griffiths RI, Bailey MJ (2002) RNA stable isotope probing, a novel means of linking microbial community function to phylogeny. Appl Environ Microbiol 68: 5367-5373.

13. Simons KL, Sheppard PJ, Adetutu EM, Kadali K, Juhasz AL, et al. (2013) Carrier mounted bacterial consortium facilitates oil remediation in the marine environment. Bioresour Technol 134: 107-116.

14. Manefield M, Griffiths RI, Leigh MB, Fisher R, Whiteley AS (2005) Functional and compositional comparison of two activated sludge communities remediating coking effluent. Environ Microbiol 7: 715-722.

15. Muyzer G, de Waal EC, Uitterlinden AG (1993) Profiling of complex microbial populations by denaturing gradient gel electrophoresis analysis of polymerase chain reaction-amplified genes coding for 16S rRNA. Appl Environ Microbiol 59: 695-700.

16. Sheppard PJ, Adetutu EM, Makadia TH, Ball AS (2011) Microbial community and ecotoxicity analysis of bioremediated weathered hydrocarbon-contaminated soil. Soil Res 49: 261-269.

17. McCaig AE, Glover LA, Prosser JI (2001) Numerical analysis of grassland bacterial community structure under different land management regimens by using 16S ribosomal DNA sequence data and denaturing gradient gel electrophoresis banding patterns. Appl Environ Microbiol 67: 4554-4559.

18. Marzorati M, Wittebolle L, Boon N, Daffonchio D, Verstraete W (2008) How to get more out of molecular fingerprints: practical tools for microbial ecology. Environ Microbiol 10: 1571-1581.

19. Leahy JG, Colwell RR (1990) Microbial degradation of hydrocarbons in the environment. Microbiol Rev 54: 305-315.

20. Allison SD, Martiny JB (2008) Colloquium paper: resistance, resilience, and redundancy in microbial communities. Proc Natl Acad Sci U S A 105 Suppl 1: 11512-11519.

21. Wittebolle L, Van Vooren N, Verstraete W, Boon N (2009) High reproducibility of ammonia-oxidizing bacterial communities in parallel sequential batch reactors. J Appl Microbiol 107: 385-394.

22. Carballa M, Smits M, Etchebehere C, Boon N, Verstraete W (2011) Correlations between molecular and operational parameters in continuous lab-scale anaerobic reactors. Appl Microbiol Biotechnol 89: 303-314.

23. Prince RC, Gramain A, McGenity TJ (2010) Prokaryotic Hydrocarbon Degraders. In: Timmis KN (eds) Handbook of Hydrocarbon and Lipid Microbiology. Springer-Verlag, Berlin Heidelberg, Germany.

24. Kuyukina MS, Ivshina IB, Makarov SO, Litvinenko LV, Cunningham CJ, et al. (2005) Effect of biosurfactants on crude oil desorption and mobilization in a soil system. Environ Int 31: 155-161.

Eco-friendly and Cost-effective Use of Rice Straw in the Form of Fixed Bed Column to Remove Water Pollutants

Baljinder Singh*, Vasundhara Thakur, Garima Bhatia, Deepika Verma and Kashmir Singh

Department of Biotechnology, Panjab University, Chandigarh, India

Abstract

We investigated the removal of three pollutants: methylene blue (MB), phosphorus, and nickel (Ni [II]) from water by using modified rice straw powder (RSP) and fixed-bed column adsorption technique. The experiments were conducted in single and binary solutions to study the effects of initial pollutant concentration and column bed depth on adsorption. It was observed that the maximum adsorption capacity of RSP for MB, phosphorus, and Ni (II) was 21.99, 4.22, and 4 mg/g, respectively. In the MB-phosphorus binary solution, the presence of one pollutant did not affect the adsorption of other pollutants. In the Ni (II)-MB binary solution, exhaustion time significantly decreased for Ni (II) adsorption; however, it increased for MB adsorption. The adsorption mechanism was analysed by using the Adams-Bohart, Thomas, and Yoon and Nelson models for describing the column's dynamic behaviour. The results indicated that the Thomas model was very suitable for RSP column design.

Keywords: RSP; Fixed bed column; Pollutants; Adsorption

Introduction

Safe drinking water is a necessity for the humankind. Unfortunately, more than 1 billion people worldwide do not have access to safe drinking water; among these, 800 million live in rural areas [1]. Effluents of many industries generate multi-component wastewater containing metals, dyes, and phosphorus leading to the need for complicated treatment processes. The elimination of effluents containing these three types of pollutants needs immediate attention because of their potential toxicity and aesthetic problems. Presently, over more than 100,000 types of dyes with 700,000 tons of dyestuffs are generated annually; further, based on their structure, these can be categorized as anionic and cationic. Methylene blue (MB) (3, 7-bis (dimethylamino)-phenothiazin-5-iumchloride), a thiazine cationic dye, is used in dying paper cottons, wools, silk, wood, and temporary hair colorant [2,3]. Initially, MB was not considered hazardous; however, a recent study has indicated several harmful effects [4].

Phosphorus plays a critical role in the growth and development of living organisms. However, high concentrations of phosphorus released from industries into receiving water bodies lead to phosphorus pollution, which is known as eutrophication. Therefore, the removal of phosphorus from water is necessary to avoid water deterioration.

The conventional methods used for removal of these pollutants (MB and phosphorus) are coagulation and flocculation, biological oxidation, chemical precipitation, and activated carbon adsorption. These technologies are expensive and can result in the generation of toxic sludge that poses another serious problem. Therefore, there is a growing interest in using low cost and commercially available materials (agricultural wastes) for pollutant adsorption [5-7]. Several research studies have demonstrated adsorption of MB and phosphorus by using agro-waste materials [8-10].

Rice (*Oryza sativa* L.) is the world's largest cereal crop, and it produces large amounts of crop residues such as rice straw. Only about 20% of rice straw is used in the production of ethanol, paper, fertilizers, and fodders. Rice straw burning is a common post-harvest practice in many countries such as India, Egypt, Malaysia, Thailand, etc., which causes air pollution called the "Black Cloud" [11]. Rice straw mainly consists of cellulose, lignin, hemicellulose, and a small amount of mineral residues. It has many characteristics, which make it a potential adsorbent; it has binding sites that are able to remove metals from aqueous solutions.

Some efforts have been made to remove metals and dyes simultaneously [12]. Therefore, we have investigated the removal efficiencies for the two pollutants in co-pollutant solutions. The main objective of this study was to develop an innovative, cost-effective, and sustainable biosorbent for the removal of three pollutants (MB, phosphorus, and Ni [II]) using rice straw powder (RSP). The study was an extension of the previous studies where Ni (II) was removed using RSP [11]. We additionally investigated the adsorption efficiency of Ni (II) along with MB in single and binary solutions using RSP via column-based studies. The influential factors (for example, initial concentration, and bed depth) were investigated for process optimization. The adsorption mechanisms based on fixed columns were further analysed using the Adams-Bohart, Thomas, and Yoon and Nelson kinetic models for describing the dynamic behaviour of the column.

Materials and Methods

RSP

The rice straw was collected from a crop field area of Patiala, India. It was thoroughly washed with de-ionized water and heated in oven at 50°C for 90 h to remove all the moisture present in the material. Oven-dried straw was crushed, grounded, and sieved to the desired mesh size (300-500 μm) for use.

Modification of RSP

RSP was treated separately in two ways; first, with 0.1 M solution of sodium hydroxide (NaOH) and second, with solutions containing 0.1

---

*Corresponding author: Dr. Baljinder Singh, Assistant Professor, Department of Biotechnology, Panjab University, Chandigarh-160 014, India
E-mail: gilljwms2@gmail.com (or) sbaljinder@pu.ac.in

M NaOH and 0.25 M ferric chloride (FeCl$_3$). Each resulting mixture (RSP and treated solution) was taken in a beaker provided with a lid and stirred for 12 h. The resulting products were washed with deionized water and dried for 48 h at 60°C. These were named as treated RSP (TRSP) and iron loaded RSP (ILRSP), respectively.

Preparation of solutions

Stock solutions (1000 mg/L) of MB, phosphorus, and Ni (II) used in the study were prepared by dissolving an appropriate weight of dye, disodium hydrogen phosphate (Na$_2$HPO$_4$), and pure salt of NiSO$_4$ in the desired volume of de-ionized water, respectively. The working solutions for MB (100 and 200 mg/L), phosphorus (20 and 40 mg/L), and Ni (II) (75 ppm) were prepared using stock solutions. The pH of all the solutions was adjusted to 7.0 by using 1 M HCl and 0.1 M NaOH solutions.

Fixed-bed column

Continuous-flow sorption experiments were performed in a glass column designed with an internal diameter of 4 cm and length of 60 cm. At the bottom of the column, a 0.5-mm stainless sieve was attached followed by glass wool to ensure uniform liquid distribution (Figure 1).

Adsorption study with TRSP column: Column was packed with 1.5 g and 2 g of TRSP to obtain a particular bed height of the adsorbent (equal to 2, and 2.5 cm of bed depth, respectively) keeping the flow rate constant (10 mL/min). Five experiments were conducted. 1) In the first experiment, the varied influent (MB) concentration (100 and 200 mg/L) was pumped from the container to the fixed-bed column by a peristaltic pump. 2) In the second experiment, a fixed influent concentration of phosphate (20 mg/L) was loaded at a specific flow rate (10 mL/min) and bed depth (2 cm). 3) In the third experiment, binary solution with MB concentration of 100 mg/L and phosphate concentration of 20 mg/L were loaded at a specific flow rate (10 mL/min) and bed depth (2 cm). 4) In the fourth experiment, a fixed influent concentration of Ni (II) (75 mg/L) was loaded at a specific flow rate (10 mL/min) and bed depth (2 cm). 5) In the fifth experiment, binary solution with MB concentration of 100 mg/L and Ni (II) concentration of 75 mg/L were loaded at a specific flow rate (10 mL/min) and bed depth (2 cm). Effluents were collected at regular time intervals to determine the concentration of all the three pollutants. Flow to the column was continued until there was no further adsorption. The concentration of phosphorus was determined

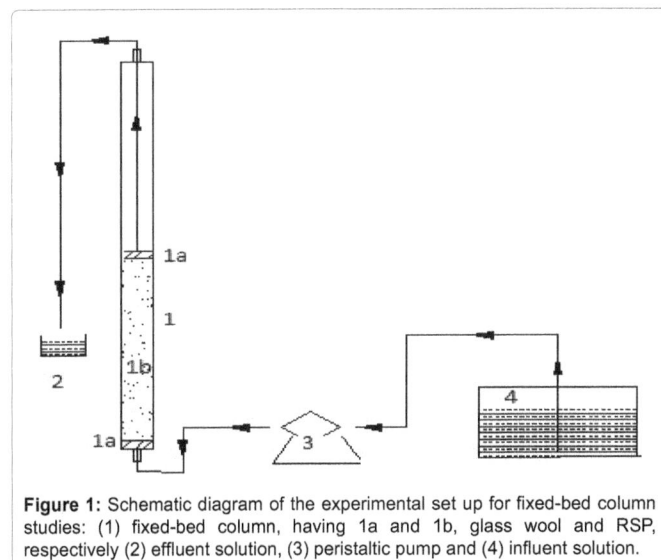

Figure 1: Schematic diagram of the experimental set up for fixed-bed column studies: (1) fixed-bed column, having 1a and 1b, glass wool and RSP, respectively (2) effluent solution, (3) peristaltic pump and (4) influent solution.

using modified molybdenum blue method [13]. Ni (II) concentration was analysed using atomic absorption spectrophotometry (AAS) using a GBC AVANTA GF 5000 model. MB was analysed at 667 nm by using UV-visible spectrophotometer (HITACHI U-2900). Analysis for Ni (II), phosphorus, and MB in binary solutions was performed by appropriate dilution.

Adsorption study with ILRSP column: A column was packed with a known amount of ILRSP (such as, 4 and 4.5 g) with different bed depths (such as, 2 and 2.5 cm). The first experiment was conducted using influent phosphate concentrations of 20 and 40 mg/L with a fixed flow rate (10 mL/min). In the second experiment, the influent concentration of MB (100 mg/L) was loaded at a fixed flow rate (10 mL/min) and bed depth (2 cm). In the third experiment, a binary solution with MB concentration of 100 mg/L and concentration of phosphate (20 mg/L) was loaded at a flow rate of 10 mL/min and a bed depth of 2 cm.

Analysis of fixed-bed column data

Various calculations related to the column data were performed as described in the previous study [11]. Breakthrough point is the point that effluent concentration (C$_t$) reaches about 0.1% of the influent concentration (C$_0$). The corresponding time is breakthrough time (t$_b$). When the effluent concentration reaches 95% of the influent concentration, it is exhaustion point, and the time is exhaustion time (t$_e$).

Breakthrough curves

A successful design of column adsorption process/method mainly requires prediction of the breakthrough curve for the effluent [11]. Typically, a breakthrough curve is a plot of C$_t$/C$_0$ as a function of time (t) or volume of effluent (V$_{eff}$, mL), where C$_t$ and C$_0$ (mg/L) are the concentration of the adsorbate in the effluent and influent, respectively. The breakthrough point is reached when the bed becomes saturated with the adsorbate. The breakthrough point is defined as the time when C$_t$ equals 5% of C$_0$. After reaching the breakthrough point the concentration of the adsorbate increases rapidly until it reaches the exhaustion point, where the column approaches complete saturation.

$$V_{eff} = Q\, t_{total} \qquad \text{eq. 1}$$

The value of the total mass of pollutants adsorbed, q$_{total}$ (mg):

$$q_{total} = Q/1000 \int_{t=0}^{t=total} C_{ad}\, dt \qquad \text{eq. 2}$$

Equilibrium pollutants uptake or maximum capacity of the column:

q$_{eq}$ (mg/g), in the column is calculated as the following:

$$q_{eq} = q_{total}/m \qquad \text{eq. 3}$$

where m is the dry weight of adsorbent in the column (g), that is, 2 g (constant for all concentrations).

Total amount of metal ion entering column (m$_{total}$) is calculated using the following equation:

$$m_{total} = C_0 Q t_{total}/1000 \quad [14] \qquad \text{eq. 4}$$

The removal percentage of pollutants can be obtained from Eq.

$$Y\% = q_{total}/m_{total} \times 100 \qquad \text{eq. 5}$$

The removal efficiency of pollutants was calculated by:

$$\text{Removal efficiency (\%)} = C_0 - C_t/C_0 \times 100 \qquad \text{eq. 6}$$

where C$_0$ and C$_t$ are the original and residual pollutants

Sample	Column	C_0 mg/L	Q mL/min	Z (cm)	t_{total} (min)	V_{eff} (mL)	m_{total} (mg)	q_{total} (mg)	q_{eq} (mg/g)	Y (%)
Methylene Blue	TRSP	200	10	2	60	600	120	116	29	96.67
		100	10	2	90	900	90	87.96	21.99	97.73
		100	10	2.5	110	1100	110	109	27.25	99.08
	ILRSP	100	10	2	30	300	30	15.9	3.975	53
Phosphate	ILRSP	40	10	2	50	500	20	18	4.5	90
		20	10	2	90	900	18	16.88	4.22	93.79
		20	10	2.5	100	1000	20	18.8	4.7	94
	TRSP	20	10	2	30	300	6	3.15	0.7878	52.5
Nickel	TRSP	75	10	2	50	500	37.5	16	4	42.66

Table 1: Parameters in fixed-bed column for MB, phosphorus, and Ni (II) adsorption in modified RSP. Co - the influent concentration, mg/L, **Q** - the volumetric flow rate, cm³/min, Z- the bed depth of the fix-bed, t_{total}- the total flow time, min, V_{eff} - the effluent volume, mL, m_{total} - total amount of pollutants sent to column, g, q_{total}- the total mass of pollutant adsorbed, mg, q_0 -equilibrium metal uptake or maximum capacity of the column, mg/g, Y-the removal per cent of Ni (II) ions, %.

concentrations in solution, respectively.

Modelling of the breakthrough curve

Predictive kinetic modelling of the effluent breakthrough curves is essential for the effective design of an adsorption column. This is because of the fact that the experimental evaluation of the performance of a column under different operating conditions is tedious and expensive. Several simple mathematical models have been developed for describing and analysing the lab-scale column studies for the purpose of industrial applications. Adams-Bohart, Thomas, and Yoon-Nelson models were performed to identify the best model for predicting the dynamic behaviour of the column.

Adams-Bohart model: Bohart and Adams [15] described the relationship between C_t/C_0 and t in a continuous system. It is used for describing the initial part of the breakthrough curve. The expression is as follows:

$$\ln(C_t/C_0)=k_{AB}C_0t-k_{AB}N_0(Z/U_0)$$

Where C_0 and C_t are the influent and effluent concentrations (mg/L), k_{AB} is the kinetic constant (L/mg/min), N_0 is the saturation concentration (mg/L), Z is the bed depth of the fix-bed column (cm), and U_0 is the superficial velocity (cm/min) defined as the ratio of the volumetric flow rate Q (cm³/min) to the cross-sectional area of the bed A (cm²).

Thomas model: The Thomas model [16] assumes plug flow behaviour in the bed. This is the most general and widely used model to describe the performance theory of the sorption process in fixed-bed column. The linearized form of this model can be described by the following expression:

$$\ln (C_0/C_t-1)=k_{Th}q_0m/Q-k_{Th}C_0t$$

where k_{Th} is the Thomas model constant (mL/min/mg), q_0 is the adsorption capacity (mg/g), and t stands for total flow time (min).

The Yoon-Nelson model: It was used to describe the column adsorption data. Use of this model could minimize the error resulting from the use of the Thomas model, especially at lower or higher time periods of the breakthrough curve. The Yoon-Nelson model [17] is based on the assumption that the rate of decrease in the probability of adsorption for each adsorbate molecule is proportional to the probability of adsorbate adsorption and the probability of adsorbate breakthrough on the adsorbent. The linearized Yoon-Nelson model for a single component system can be expressed as follows:

$$\ln(C_t/C_0-C_t)=k_{YN}t-\tau k_{YN}$$

where k_{YN} is the rate constant (min⁻¹), and τ is the time required for 50% adsorbate breakthrough (min).

Results and Discussion

TRSP fixed-bed column

Adsorption of MB: Experiments with columns packed with TRSP (bed depth, 2 cm) were conducted under the following conditions: fixed influent concentration of MB, 100 mg/L; flow rate, 10 mL/min; and bed depth, 2 cm. The results are shown in Table 1. The maximum adsorption capacity (q_{eq}) of RSP for MB was 21.99 mg/g, which is much higher than that of previous reported adsorbents in flask-level experiments, such as rice husk, wheat bran, guava seed, neem leaf powder, coconut coir, banana peel, spent rice biomass, and orange peel; however, it is less than some agricultural by-products such as pineapple stem, mango seed kernel, swede rape straw, pine cone, and coffee husk. These results have been thoroughly reviewed [18,19]. Moreover, the maximum adsorption capacity of MB was much higher than that of previous reported adsorbents such as cotton-alk (0.024 mg/g) [20]. Fourier transform infrared (FTIR) analysis of the biomass was conducted to identify possible interactions of pollutants and results of spectra are shown in Supplementary Figures 1 and 2.

Effect of MB concentration on % Removal efficiency: The effect of initial MB concentration on breakthrough curves is shown in Figure 2a. The breakdown time and exhaustion time increased with decreasing initial MB concentration. Change in the initial MB concentration had a significant effect on the breakthrough curve. It was obvious that the breakthrough curves became sharper while the breakthrough time and exhaustion points became shorter with increasing concentration of MB. This can be explained by the fact that RSP is more saturated at initial high MB concentration; a lower concentration gradient caused a slower transport owing to a decrease in the diffusion coefficient or mass transfer coefficient. These results are similar to that reported in previous studies [19,21].

Effect of bed depth on % removal efficiency of MB: The effect of bed depth (2.0 and 2.5 cm) on the adsorption of MB in columns was investigated under a given condition. Results showed that adsorption quantities and capacities in both the columns increased with increasing bed depth. The increase in MB uptake capacity with increasing contact time in the fixed bed column is likely to be because of an increase in the surface area, which provides more binding sites for column adsorption

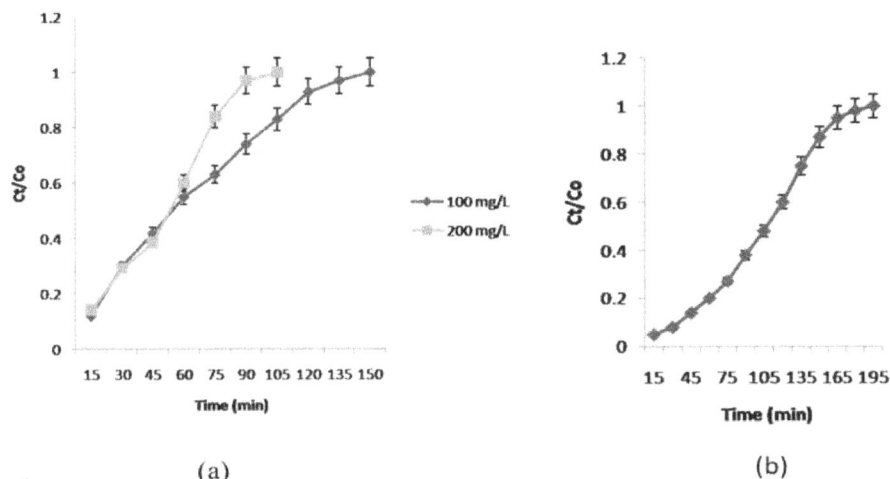

Figure 2: Breakthrough curve for MB for (a) bed depth 2 cm and (b) bed depth 2.5 cm. Data are presented as mean and standard error of three independent observations. Some error bars are not present because they are smaller than the diameter of the symbol.

[11]. Figure 2b shows that the % removal efficiency of MB in the column increases with the increase in the mass of adsorbent (rice straw). This is because of more adsorbent in the fixed-bed column, which provides more binding sites and larger surface area for the removal of phosphorus and MB.

Adsorption of phosphorus: The column was packed with TRSP, and a fixed influent concentration of phosphate (20 mg/L) was loaded at a specific flow rate (10 mL/min) and bed depth (2 cm). The results are shown in Table 1. Despite the differences in nature and chemical composition of MB and phosphorus, the results of the present study showed adsorption capacity (q_{eq}) 0.7875 mg/g of TRSP towards phosphorus. These results showed that TRSP was not much efficient for remediating phosphorus from aqueous solutions. This might be because of the available sites on the adsorbent for the dye that led to ionic interactions and further resulted in colour removal of dye.

Adsorption of MB and phosphorus in binary solution: Owing to the differences in the chemical properties of the two pollutants, the results for binary solutions remained the same as for a TRSP column (Table 1). The results showed that the presence of one pollutant does not affect the removal efficiency of another pollutant or collision between pollutants and active sites of RSP. This might be due to different binding sites for the two pollutants with differences in their chemical nature. Therefore, TRSP can be used to remove low levels of phosphorus in industrial effluents containing both MB and phosphorus.

Adsorption of MB and Ni (II) in binary solution: To study the behaviour of co-pollutants (that is, MB and Ni) in a fixed-bed column, binary solutions containing two pollutants were prepared as influent solution under given conditions, that is, for Ni (II), 2 cm bed depth, 10 mL/min flow rate, 75 ppm concentration; and for MB, 2 cm bed depth, and 10 mL/min flow rate. It was observed that for Ni (II) adsorption in binary solutions on the column, the exhaustion time decreased from 50 to 32 min, and both adsorption quantities (q_{total} from 16 to 10 mg) and capacities (q_{eq} from 4 to 2.5 mg/g) also decreased. This might be due to more favourable adsorption of MB on TRSP than Ni (II), which was derived from ionic interaction between MB and RSP in a packed column. In contrast, the exhaustion time increased from 90 to 105 min for MB adsorption from binary solution on columns, and both adsorption quantities and capacities were also enhanced. This might be due to the presence of π-π conjugated electrons in the complex, which

donates an electron to enhance the electron density to form stable complexes with the heavy metals [22].

ILRSP fixed-bed column

Effect of phosphorus concentration on % removal efficiency: The adsorption efficiency of columns was evaluated. Data are presented in Table 1. The maximum adsorption capacity of phosphorus was 4.22 mg/g (bed depth, 2 cm; with initial phosphorus concentration 20 mg/L). Meanwhile, the initial investigations showed that TRSP exhibited very low adsorption capacity (0.78 mg P/g (phosphorus per gram)). The presence of Fe (II) in the RSP played a decisive role in enhancing phosphorus capture. Treatment with FeCl$_3$ considerably improved phosphorus retention of RSP. It was found that ILRSP was 5.41 times more efficient than TRSP in remediating phosphorus from aqueous solutions. Despite the differences in nature and chemical composition between RSP and okara, the results of the present study fitted well with the findings reported by Ngugen et al. [9] in a flask-level study. Figure 3a shows that the % adsorption decreased with increase in initial concentration of phosphorus. However, the uptake capacity increased with an increase in initial concentration, which might be due to the availability of a greater number of phosphate ions in solution.

Effect of bed depth on % removal efficiency of phosphorus: The effect of contact time on the removal of phosphate ions by ILRSP was studied by varying contact time for an initial concentration of 20 mg/L. Relationship between the contact time and removal efficiency of phosphate ions is shown in Figure 3b at a constant bed depth (2 cm and 2.5 cm). It is demonstrated that as time increases, concentration of effluent decreases, and removal efficiency increases [14]. This may be due to the increase in surface area, which provides more binding sites for column adsorption [11]. The increase in phosphate uptake capacity of ILRSP was observed as the bed depth increased with the highest bed capacity. This is due to increased RSP mass at higher bed depth that would provide a larger surface area and more anion fixations towards the active binding sites for the adsorption process leading to an increase in the volume of the treated effluent. These results revealed that the bed depth of 2 cm provided optimum % removal efficiency.

Generally, biosorbents do not show significant affinity towards phosphate ions because of the lack of available binding sites [23]. Therefore, to enhance the sorption capacity of phosphorus, researchers

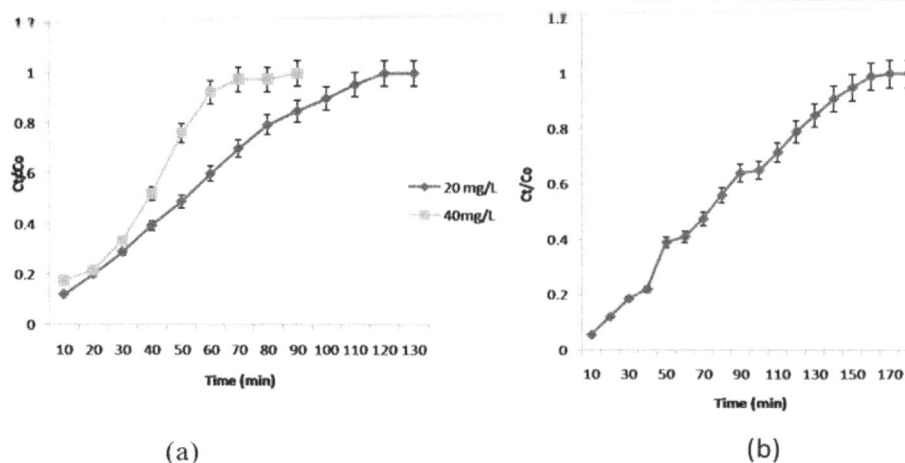

Figure 3: Breakthrough curve for Phosphorus for (a) bed depth 2 cm and (b) bed depth 2.5 cm. Data are presented as mean and standard error of three independent observations.

cationized biosorbents were impregnated with metal compounds [2,9]. The common metals used for this purpose include Fe (II, III), Zr (IV), La (III), Ce (III), and Zn (II). The base treatment of biosorbents is required for enhancing attachment of metals onto surface [8] because direct immersion of biosorbents into metal solutions usually does not have sufficient capacity for phosphate removal [23]. Many studies have confirmed the enhanced phosphorus removal of metal loaded biosorbents using agricultural by-products [6,8]. A summary of studies using metal loaded biosorbents for the decontamination of phosphorus was tabulated by Nguyen et al. [9].

Adsorption of MB in single solution and both MB and phosphorus in binary solution: Table 1 indicates that the maximum adsorption capacity (q_{eq}) of MB was 3.975 mg/g in columns packed with ILRSP at fixed influent concentration of MB (100 mg/L), and a bed depth 2 cm. It means that ILRSP column was 5.53 times less efficient than TRSP column for remediating MB from aqueous solutions. This might be because of available sites on the adsorbent towards dye that leads to ionic interactions, which further results in the colour removal of dye.

Owing to the differences in chemical properties of the two pollutants, the results of binary solution remained same for ILRSP column (Table 1). The results showed that the presence of one pollutant did not affect the removal efficiency of another pollutant.

Breakthrough curve modelling

Adams-Bohart model: The constants k_{AB} and N_0 were calculated from the linear plot of $\ln(C_t/C_0)$ against time (Figure 4a and 4b). Linear regression results and values of R^2 for kinetic models are presented in Table 2. The values of k_{AB} decreased with the increase in influent MB and phosphorus concentration; however, the values increased with increasing bed depth (Figure 3). It was indicated that the overall system kinetics was dominated by external mass transfer in the initial part of adsorption in the column [14]. Similar results were reported for phosphorus (Table 2). The value of R^2 was more than 0.9 with initial influent MB concentration of 200 mg/L and bed depth 2 cm, which indicates that Adams-Bohart model describes the column data very well for MB.

Thomas model: The values of k_{Th} and q_0 were determined from the linear plot of $\ln[(C_0/C_t-1)]$ against t (Figure 5a and 5b). The Thomas

model was applicable to the adsorption process, which indicates that the external and internal diffusions were not the limiting step [14,24]. It was observed that with an increase in the bed depth, the k_{Th} values decreased (Table 2 and Figure 4). The values of k_{Th} increased with increase in initial influent MB and phosphorus concentrations. Therefore, the higher influent concentration and bed depth would increase the adsorption of MB and phosphorus (single solution) using the TRSP and ILRSP columns, respectively. These model predictions agree well with the experimental results discussed earlier, which indicate that lower flow rates, higher influent concentrations, and higher bed heights enhance the adsorption ability of a fixed-bed column. Most values of R^2 are more than 0.9, which indicates that the Thomas model describes the column data very well for both MB and phosphate. The Thomas model has been shown to be more suitable for cases where the adsorption process is not controlled by mass transfer [11].

The Yoon-Nelson model: A linear plot of $\ln[C_t/(C_0-C_t)]$ against t determined the values of k_{YN} and τ from the intercept and slope of the plot (as shown in the Figure 6a and 6b). Different statistical parameters of the Yoon-Nelson model were calculated and shown in Table 2. The k_{YN} values decreased with increasing bed depth (Figure 5). Most values of R^2 are more than 0.5, which indicate that compared to the Thomas model, the Yoon-Nelson model is not an appropriate for the breakthrough curves.

Comparison of all three models Adams-Bohart, Thomas, Yoon-Nelson model: From Figures 4-6, Table 1, and supplementary data, it was observed that the Thomas model is an appropriate model to describe a fixed-bed system (TRSP column for MB and ILRSP column for phosphorus). However, in the case of the Adams-Bohart and Yoon-Nelson models, low correlation coefficients (R^2) were observed, which indicates that these models are not appropriate as predictors for the breakthrough curves. Thus, the Thomas model can be used to predict adsorption performance for the adsorption of both MB and phosphorus in a fixed-bed column.

Conclusion

In general, rice straw is an environment-friendly potential biosorbent for heavy metals. This work examined the efficiency of this sorbent for the removal of three important pollutants from single and binary solutions. The results indicate that several factors such as contact time, bed depth, and initial concentration affect the biosorption process.

Breakthrough curve models	Sample	C_0 (mg/L)	Q (ml/min)	Z (cm)	Parameters		
Adam-Bohart					k_{AB} (L/mg/min)	N_0 (mg/L)	R^2
	Methylene Blue	100	10	2	1.98×10^{-4}	2.78×10^3	0.811
		200	10	2	1.61×10^{-4}	4.04×10^3	0.919
		100	10	2.5	2.04×10^{-4}	2.85×10^3	0.894
	Phosphate	20	10	2	8.85×10^{-4}	0.65×10^2	0.859
		40	10	2	5.90×10^{-4}	0.99×10^2	0.879
		20	10	2.5	6.99×10^{-4}	0.78×10^2	0.81
Thomas					k_{TH} (mL/min/mg)	q_0 (mg/L)	R^2
	Methylene Blue	100	10	2	0.614×10^{-2}	9.665×10^2	0.9775
		200	10	2	0.499×10^{-2}	15.504×10^2	0.9528
		100	10	2.5	0.583×10^{-2}	16.217×10^2	0.9742
	Phosphate	20	10	2	2.36×10^{-2}	2.531×10^2	0.9949
		40	10	2	2.18×10^{-2}	3.333×10^2	0.9654
		20	10	2.5	1.95×10^{-2}	3.710×10^2	0.9473
Yoon-Nelson					k_{YN} (min^{-1})	τ (min)	R^2
	Methylene Blue	100	10	2	0.614	3.8661	0.5842
		200	10	2	0.998	3.0988	0.5126
		100	10	2.5	0.597	5.5384	0.5318
	Phosphate	20	10	2	0.472	5.0651	0.4873
		40	10	2	0.586	3.3948	0.5278
		20	10	2.5	0.391	7.4182	0.5953

Table 2: Parameters of Adam-Bohart model, Thomas model, and Yoon Nelson model under different condition. kAB - the kinetic constant, L/mg min, kTh - the Thomas model constant, mL/min mg, kYN - the rate constant, min^{-1}, No - the saturation concentration, mg/L.

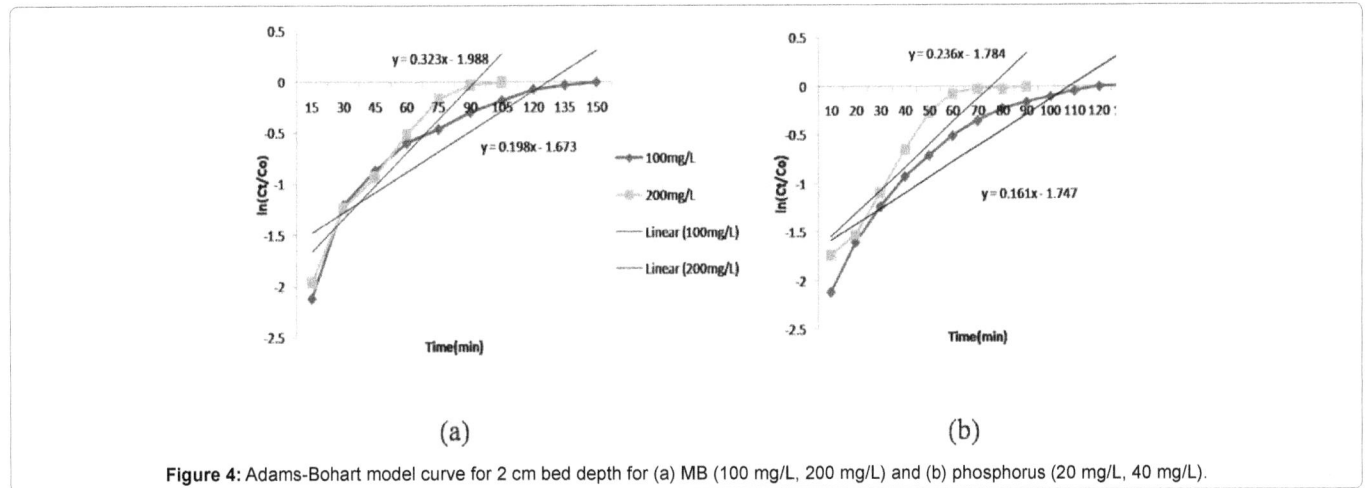

Figure 4: Adams-Bohart model curve for 2 cm bed depth for (a) MB (100 mg/L, 200 mg/L) and (b) phosphorus (20 mg/L, 40 mg/L).

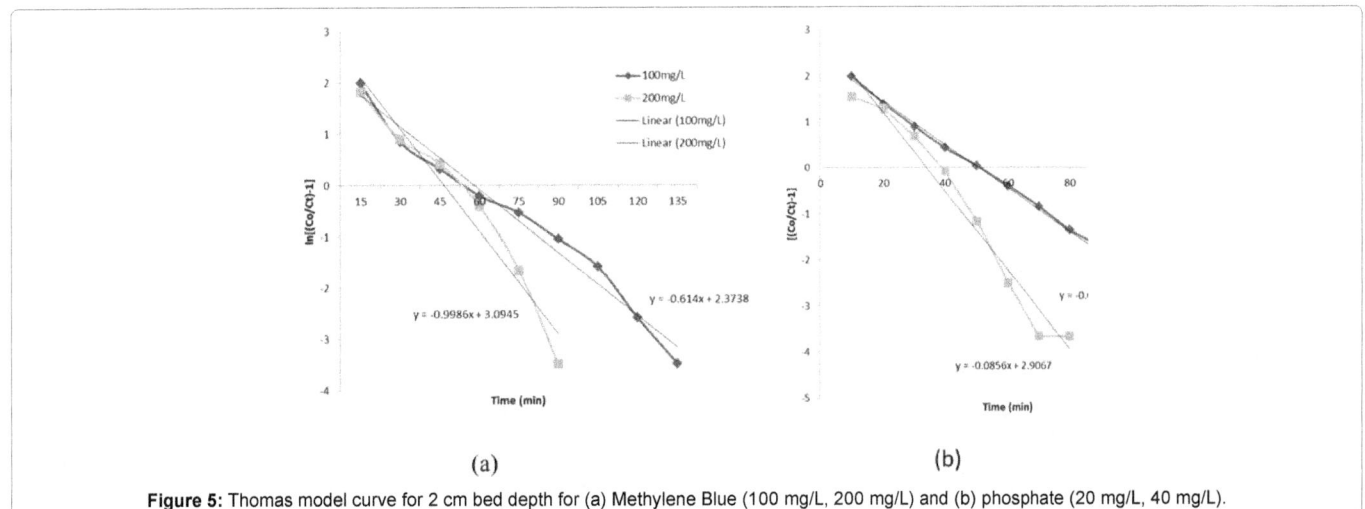

Figure 5: Thomas model curve for 2 cm bed depth for (a) Methylene Blue (100 mg/L, 200 mg/L) and (b) phosphate (20 mg/L, 40 mg/L).

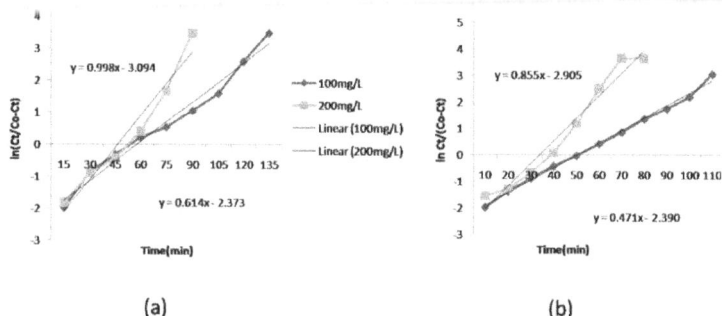

Figure 6: Yoon-Nelson model curve for 2 cm bed depth for (a) Methylene Blue (100 mg/L, 200 mg/L) and (b) phosphate (20 mg/L, 40 mg/L).

The physicochemical characteristics of wastewater from varying sources can be much more complex as compared to the aqueous metal solutions used in this study. Therefore, the effects of other components of wastewater on commercial pollutants' adsorption process should be determined. The two columns, TRSP and ILRSP, can be used in series to remove both these pollutants from the industrial effluents containing cationic dyes, heavy metals, and phosphorus. The saturated column was regenerated by 0.05 mol L^{-1} EDTA solution, and RSP could be reused in the removal of the three pollutants. Initial results showed the same removal efficiency after regeneration of column; however, further study under different conditions should be investigated. This work shows the possibility that rice straw may be a suitable material for the adsorption from aqueous solutions.

In this study, modified RSP was found to be an efficient biosorbent for sorption of MB, phosphorus, and Ni (II) in single and binary solutions. The adsorption capacity for a pollutant was increased with increasing bed height and initial dye concentration. It was found that Ni (II) adsorption was suppressed; however, MB uptake was enhanced in Ni (II)-MB binary solution in TRSP fixed-bed column. The column experimental data fitted well with the Thomas model; however, the Adams-Bohart and Yoon-Nelson models were found to be inapplicable.

Acknowledgements

The authors thank the Chairperson, Department of Biotechnology, Panjab University for providing necessary facilities.

References

1. Sekhon GS, Singh B (2013) Estimation of heavy metals in the groundwater of Patiala district of Punjab, India. Earth Resources 1: 1-4.

2. Han R, Wang Y, Zhao X, Wang Y, Xie F, et al. (2009) Adsorption of methylene blue by phoenix tree leaf powder in a fixed-bed column: experiments and prediction of breakthrough curves. Desalination 245: 284-297.

3. Li Y, Du Q, Liu T, Peng X, Wang J, et al. (2013) Comparative study of methylene blue dye adsorption onto activated carbon, graphene oxide, and carbon nanotubes. Chemical Engineering Research and Design 91: 361-368.

4. Chen L, Ramadan A, Lü L, Shao W, Luo F, et al. (2011) Biosorption of methylene blue from aqueous solution using lawny grass modified with citric acid. Journal of Chemical & Engineering Data 56: 3392-3399.

5. Benyoucef S, Amrani M (2011) Removal of phosphorus from aqueous solutions using chemically modified sawdust of Aleppo pine (*Pinus halepensis* Miller): kinetics and isotherm studies. The Environmentalist 31: 200-207.

6. Krishnan KA, Haridas A (2008) Removal of phosphate from aqueous solutions and sewage using natural and surface modified coir pith. Journal of Hazardous Materials, 152: 527-535.

7. Riahi K, Thayer BB, Mammou AB, Ammar AB, Jaafoura MH (2009) Biosorption characteristics of phosphates from aqueous solution onto Phoenix dactylifera L. date palm fibers. Journal of hazardous materials 170: 511-519.

8. Carvalho WS, Martins DF, Gomes FR, Leite IR, da Silva LG, et al. (2011) Phosphate adsorption on chemically modified sugarcane bagasse fibres. Biomass and Bioenergy 35: 3913-3919.

9. Nguyen TAH, Ngo HH, Guo WS, Zhang J, Liang S, et al. (2013) Feasibility of iron loaded 'okara' for biosorption of phosphorus in aqueous solutions. Bioresource Technology 150: 42-49.

10. Xu X, Gao BY, Yue QY, Zhong QQ (2010) Preparation of agricultural by-product based anion exchanger and its utilization for nitrate and phosphate removal. Bioresource Technology 101: 8558-8564.

11. Sharma R, Singh B (2013) Removal of Ni (II) ions from aqueous solutions using modified rice straw in a fixed bed column. Bioresource Technology 146: 519-524.

12. Deng JH, Zhang XR, Zeng GM, Gong JL, Niu QY, et al. (2013) Simultaneous removal of Cd (II) and ionic dyes from aqueous solution using magnetic graphene oxide nanocomposite as an adsorbent. Chemical Engineering Journal 226. 189-200.

13. Murphy JS, Riley J (1962) A modified single solution method for the determination of phosphate in natural waters. Analytica Chimica Acta 27: 31-36.

14. Chen S, Yue Q, Gao B, Li Q, Xu X, et al. (2012) Adsorption of hexavalent chromium from aqueous solution by modified corn stalk: A fixed-bed column study. Bioresource Technology 113: 114-120.

15. Bohart GS, Adams EQ (1920) Behaviour of charcoal towards chlorine. Journal of chemical Society 42: 523-529.

16. Thomas HC (1944) Heterogeneous ion exchange in a flowing system. Journal of the American Chemical Society 66: 1466-1664.

17. Yoon YH, Nelson JH (1984) Application of gas adsorption kinetics I. A theoretical model for respirator cartridge service life. The American Industrial Hygiene Association Journal 45: 509-516.

18. Mohammed MA, Shitu A, Ibrahim A (2014) Removal of methylene blue using low cost adsorbent: a review. Research Journal of Chemical Sciences 2231: 491-102.

19. Yagub MT, Sen TK, Afroze S, Ang HM (2014) Dye and its removal from aqueous solution by adsorption: a review. Advances in Colloid and Interface Science 209: 172-184.

20. Ding Z, Hu X, Zimmerman AR, Gao B (2014) Sorption and cosorption of lead (II) and methylene blue on chemically modified biomass. Bioresource Technology 167: 569-573.

21. Foo KY, Hameed BH (2012) Dynamic adsorption behaviour of methylene blue onto oil palm shell granular activated carbon prepared by microwave heating. Chemical Engineering Journal 203: 81-87.

22. Gong JL, Zhang YL, Jiang Y, Zeng GM, Cui ZH, et al. (2015) Continuous adsorption of Pb (II) and methylene blue by engineered graphite oxide coated sand in fixed-bed column. Applied Surface Science 330: 148-156.

23. Eberhardt TL, Min SH (2008) Biosorbents prepared from wood particles treated with anionic polymer and iron salt: Effect of particle size on phosphate adsorption. Bioresource Technology 99: 626-630.

24. Banerjee K, Ramesh ST, Gandhimathi R, Nidheesh PV, Bharathi KS (2012) A novel agricultural waste adsorbent, watermelon shell for the removal of copper from aqueous solutions. Iranian Journal of Energy & Environment 3: 143-156.

Evaluation of the Efficiency of Duckweed (*Lemna minor* L.) as a Phytoremediation Agent in Wastewater Treatment in Kashmir Himalayas

Irfana Showqi*, Farooq Ahmad Lone and Javeed Iqbal Ahmad Bhat

Division of Environmental Sciences, Sheri-Kashmir University of Agricultural Sciences and Technology of Kashmir, Shalimar, Jammu and Kashmir, India

Abstract

In the present study, the efficiency of duckweed (*Lemna minor* L.) as an effective natural biological tool in wastewater treatment was examined in an outdoor aquatic system. Duckweed plants were inoculated into wastewater and tap water systems for treatment over fifteen day's retention periods under local outdoor natural conditions. Water samples were taken below duckweed cover after fifteen days to assess the plant's efficiency in purifying waste water from different pollutants. For comparison, the plants were also grown in tubs containing tap water. The results show that concentrations of Nitrogen (N), Phosphorus (P), Potassium (K), Calcium (Ca), Magnesium (Mg), Copper (Cu), Zinc (Zn), Nickel (Ni), Chromium (Cr), Cadmium (Cd) and Lead (Pb) decreased by 93.4%, 99.9%, 93.9%, 98.5%, 91.9%, 85.0%, 95.0%, 90.0%, 99.8%, 99.5% and 95.0% respectively in waste water and subsequently these elements exhibited an increasing concentration in the plant body. Almost similar results were obtained when the plants were grown in tubs containing tap water. Biochemical parameters viz. chl-a, b, total-chl, carbohydrates and proteins as well as nutrient status of the macrophyte increased after the completion of the retention period both in waste and tap water. Results confirm that duckweeds can effectively be used for wastewater treatment systems.

Keywords: *Lemna minor*; Wastewater; Retention period

Introduction

Multiple environmental factors in association with anthropogenic activities have significantly altered our aquatic ecosystems. Over use of chemical fertilizers and intensification of industrial activities contaminate watercourses and water-bearing stratum with heavy metals and other pollutants. Toxic elements and heavy metals are an important category of pollutants and as such have major detrimental impacts on both human health as well as the health of terrestrial and aquatic ecosystems. The discharge of harmful chemical compounds into the aquatic environments disturbs the structure and functioning of the natural ecosystems. Restoration of water contaminated with potentially toxic metals and metalloids is of major global concern and thus there is a need to reduce their concentrations to a protective level in order to prevent eutrophication and other metallic enrichment. It is an established fact that in comparison to chemical treatments, the biological methods for the removal of contaminants from the environment are cheaper than the conventional remediation technologies [1]. In recent years, there has been a growing struggle to provide efficient, inexpensive, and environmentally friendly options for the remediation of trace elements and other contaminants in waste water. Moreover, aquatic plants are of special interest unlike the terrestrial plants, because they are capable of bio concentrating many nutrients and heavy metals in large quantities [2-4]. In Kashmir valley Anchar lake is one of the lakes severely suffering from nutrient enrichment and duck weed (*Lemna minor* L. belonging family Lemnaceae) is one of the common aquatic free floating macrophytes found in this water body. In order to evaluate the phytoremediation potential of duck weed a laboratory experiment was conducted in 2014-15 to analyze the nutrient and heavy metal absorption potential to assess its efficacy for waste water treatment.

Materials and Methods

Preparation steps

Waste water from a nearby ditch was transferred to the laboratory after filtering it to get rid of large suspended solids. The filtered water was immediately collected into four opaque tubs (as replicates) to prevent light entering except at the top [5]. Each tub was 50 cm long, 35 cm wide

and 25 cm deep and was filled with 10 L waste water. Duckweed (*Lemna minor* L.) plants were collected from a local ditch. The plants were cleaned by tap water followed by distilled water and were transferred to waste water systems for aquatic treatment. For comparison, similar set of tubs was also kept wherein duckweed was grown in tap water. The experiment was kept under outdoor local environmental conditions for fifteen days retention time.

Subsurface (under plant mat) water samples were collected in polyethylene bottles from all sides of each tank and then mixed after the completion of 15 days for chemical analysis. Similarly, plant samples were also collected from each tub for the analysis purposes.

Analysis of the water and plant samples

Macro, micro and heavy metal status in the water were carried out according to standard methods for examination of water and wastewater [6]. In plants except for the estimation of chlorophyll (a, b and total) where fresh samples were used, the plant samples were air dried and grounded in a grinder. The Photosynthetic pigments (Chlorophyll and Carotenoids were analyzed by following the methods of Strain et al. [7] and Duxbury and Yentsch [8] respectively. Similarly, carbohydrate and proteins were analyzed by following the methods of Dubios et al. [9] and Lowry et al. [10]. Dried plant was analysed for total N by Kjeldahl method [11], total P (Vanadatemolybdate method), Ca, Mg and K were determined using ammonium acetate method [12]. Cu, Zn, Fe, Mn, Ni, Cd, Cr and Pb were analyzed by using AAS (Atomic Absorption

*Corresponding author: Irfana Showqi, Division of Environmental Sciences, Sheri-Kashmir University of Agricultural Sciences and Technology of Kashmir, Shalimar, Jammu and Kashmir, India E-mail: irfanashowqi@gmail.com

Spectrophotometer). ELEMENTAS AAS4141-Electronics Corporation of India Limited).

Results and Discussion

Duckweed plant were inoculated into a wastewater system for aquatic treatment over 15 day's retention time periods to assess the plant's efficiency in improving macronutrients, micronutrients and heavy metal characteristics of wastewater and tap water. The wastewater and the plants that were used in the experiment were taken from the nearby ditch.

Macronutrients (N, P, K, Ca, Mg)

The results show that macronutrients such as nitrogen (N), phosphorus (P), potassium (K), calcium (Ca) and magnesium (Mg) in wastewater reached their minimum concentrations of 0.5, 0.001, 1.10, 0.4 and 2.6 mg L^{-1} respectively after growing *Lemna minor* for 15 days in it with a reduction percentage of 93.4%, 99.9%, 93.8%, 98.4% and 91.8% respectively (Table 1). However, these macronutrients have increased in the plant body during the retention period with an increasing percentage of 23.4%, 7.14%, 5.0%, 1.74% and 37.5% respectively (Table 2). Similar results have been observed in tap water that was used for comparison purposes where macronutrients had decreased in concentration after *Lemna minor* was grown in it for 15 days with a reduction percentage of 100%, 100%, 60.52%, 94.50% and 97.22% (Table 1) respectively and subsequently in plant body these nutrients have shown increasing percentage of 8.13%, 4.44%, 0.59%, 0.91% and 17.86% (Table 2) respectively during the experimental period. Fast growing duckweed in nutrient rich water is a highly efficient sink for both phosphorus and potassium; little of each, however, is required for rapid growth. In concurrence with the present findings, Verma and Suthar [13] have also reported that the nutrient load in wastewater reduced significantly by 83-89% of NO_3^- and 67-72% of total phosphorus. Pandey [14] reported that in the duckweed treatment system at Halisahar, nitrogen and phosphorus removals were in the range of 50-75% and 17-35% respectively. In the support of present findings Zaltauskaite et al. [15] found that *Lemna minor* efficiently removes N, P, K and also Ca and Mg from wastewaters compared to other treatment methods. The cause of reduction in nitrogen, phosphorus, potassium, calcium and magnesium was utilized by the *Lemna minor* as a nutrient as also reported out by Patel and Kanungo [16].

Micronutrients (Cu, Zn, Ni)

Micronutrients such as Copper (Cu), Zinc (Zn) and Nickel (Ni) reached their minimum concentrations of 0.6, 0.01 and 0.01 mg L^{-1} respectively after *Lemna minor* was grown in it for 15 days, with a

reduction percentage of 85.0%, 95.0% and 90.0% respectively (Table 1). On the other side, these micronutrient (Cu, Zn and Ni) show high concentration in the plant body during the retention time with an increasing percentage of 34.69%, 69.44%, and 66.67% respectively in the plant body (Table 2). Similar results have been observed in tap water that was used for comparison purposes where micronutrients zinc and copper had decreased in concentration after *Lemna minor* was grown in it for 15 days with a reduction percentage 100% and subsequently in plant body these nutrients have shown increasing percentage of 2.22% and 0.71% respectively (Table 2) however nickel has shown no change as it was not observed in tap water during initial analysis. In the line of present study Waffa et al. [17] reported that duckweed treatment system reduced zinc by 93.6% and copper by 100% during eight days retention period. Zaltauskaite et al. [15] also found that Zn was most efficiently removed, depending on the initial Zn concentration *Lemna minor* removed between 42.3-77.8% of Zn. They also found that after the 7 days of effluents exposure to *Lemna minor* treatment, the concentration nickel was reduced significantly.

The major cause of micronutrient reduction in waste water / tap water was utilization by the *Lemna minor* plant for body formation and development. Duckweeds require a number of macro and micronutrients for their normal growth. Nutrients are absorbed through all surfaces of the duckweed leaf. DWRP [18] reported that the highest growth rate for Lemnaceae under optimal laboratory conditions is about 0.66 generations per day that is equal to a doubling time of 16 hours. It has been reported that duckweeds generally double their mass in 16 hours to 2 days under optimal conditions which further helps in uptake of nutrients from the growing media for developing body tissue. As also reported by Korner and Vermaat [19] that the nutrients removed by duckweeds from growing media are mainly realized by newly grown tissue of the plants. Duckweed has a high mineral absorption capacity and can tolerate high organic loading as well as high concentrations of micronutrients.

Heavy metals (Cd, Cr and Pb)

The concentrations of heavy metals such as Cd, Cr and Pb reduced after *Lemna minor* was grown on wastewater for 15 days retention period. These metals have reached their minimum concentration of 0.001, 0.001 and 0.01 mg L^{-1} respectively with a reduction percentage of 99.5%, 99.8% and 95.0% respectively, however in plant body their concentration has increased during the study period with an increasing percentage of 92.06% 43.59% and 0.52% respectively. In case of tap water cadmium and lead were not observed during initial analysis however chromium had decreased in concentration from 0.20 ppm to 0.00 ppm with a reduction percentage of 100% (Table 1) and subsequently

Parameters (ppm)	Waste water			Tap water			$P \leq 0.05$
	Initial	Final	% Decrease	Initial	Final	% Decrease	
Nitrogen(N)	7.600	0.500	93.421	0.400	0.000	100.00	0.01
Phosphorus(P)	1.100	0.001	99.909	0.110	0.000	100.00	0.28
Potassium (K)	18.000	1.100	93.889	3.800	1.500	60.52	0.57
Calcium (Ca)	25.900	0.400	98.456	9.100	0.500	94.51	0.02
Magnesium (Mg)	32.000	2.600	91.875	3.600	0.100	97.22	0.06
Copper (Cu)	4.000	0.600	85.000	0.300	0.000	100.00	0.00
Zinc (Zn)	0.200	0.010	95.000	0.100	0.000	100.00	0.54
Nickel (Ni)	0.100	0.010	90.000	0.000	0.000	0.000	0.04
Cadmium (Cd)	0.200	0.001	99.500	0.000	0.000	0.000	0.02
Chromium (Cr)	0.600	0.001	99.833	0.200	0.000	100.00	0.38
Lead (Pb)	0.200	0.010	95.000	0.000	0.000	0.000	0.02

Table 1: Mean values of various parameters of waste water and tap water before and after treatment by *Lemna minor*.

Parameters	Waste water			Tap water			P≤ 0.05
	Initial	Final	%increase	Initial	Final	%increase	
N (%)	1.44	1.88	23.40	1.13	1.23	8.13	0.351
P (%)	0.52	0.56	7.14	0.43	0.45	4.44	0.06
K (%)	1.71	1.80	5.00	1.69	1.70	0.59	0.05
Ca (%)	1.13	1.15	1.74	1.09	1.10	0.91	0.04
Mg (%)	0.45	0.72	37.50	0.46	0.56	17.86	0.00
Cu (mg/kg)	24.85	38.05	34.69	25.08	25.26	0.71	0.11
Zn (mg/kg)	3.27	10.70	69.44	3.97	4.06	2.22	0.04
Ni (mg/kg)	0.02	0.06	66.67	0.01	0.01	0.00	0.15
Cr (mg/kg)	0.22	0.39	43.59	0.30	0.31	3.23	0.06
Cd (mg/kg)	0.05	0.63	92.06	0.05	0.05	0.00	0.55
Pb (mg/kg)	25.09	25.22	0.52	25.02	25.02	0.00	0.05
Carbohydrate (%)	15.28	20.01	23.64	15.33	17.54	15.19	0.08
Proteins (%)	12.23	17.61	30.55	12.15	13.20	7.95	0.04
Chl'a' (mg/g)	2.88	4.06	29.06	2.98	3.96	24.75	0.81
Chl'b' (mg/g)	1.13	2.01	43.78	1.40	1.93	27.46	0.59
Total chlorophyll (mg/g)	4.10	6.08	32.57	4.30	5.45	21.10	0.57
Carotenoids (mg/g)	1.48	1.80	17.78	1.49	1.59	6.29	0.08

Table 2: Mean values of various parameters in *Lemna minor* before and after growing them in Waste water and Tap water.

in plant body chromium have shown increasing percentage of 3.23%, whereas cadmium and lead have shown no change in their initial concentrations (Table 2). In concurrence with the present findings, Waffa [17] studied that duckweed aquatic treatment system performed 100% lead and 66.7% of cadmium removal from wastewater after 8 days treatment period. Similar results have also been found by Kara et al. [20] and concluded that duckweed *Lemna minor* L. take up Pb, Cu, Fe, Cd and Ni from contaminated solutions. Removal of heavy metals from wastewater was also studied by Zayed et al. [21] and the results confirm that duckweed (*Lemna minor* L.) is excellent accumulators of Cd, Se and Cu.

Biochemical parameters

The photosynthetic pigments chlorophyll 'a', chlorophyll 'b', total chlorophyll and carotenoids and other biochemical parameters such as total carbohydrates and total proteins, also reported an increase in their concentrations in *Lemna minor* during the experimental period.

The final result obtained after growing *Lemna minor* in the wastewater show that photosynthetic pigments such as Chl 'a' has increased in concentration from 2.88 mg/g to 4.06 mg/g exhibiting an increase of 29.06% ; Chl 'b' level has increased from 1.13 mg/g to 2.01 mg/g depicting an increase of 43.78%; total chlorophyll level has increased from 4.10 mg/g to 6.08 mg/g reporting an increase of 32.57% and carotenoid level has increased from 1.48 mg/g to 1.80 mg/g showing a increase of 17.78% in the tissues of *Lemna minor*. Also, other biochemical parameters such as total carbohydrate level has increased from 15.28% to 20.01% showing an increase of about 23.64% and total protein level have increased in concentration from 12.23% to 17.61% exhibiting an increase of about 30.55% in the tissues of *Lemna minor*. Similarly, the result obtained after growing *Lemna minor* in the tap water showed that photosynthetic pigments such as Chl 'a', Chl 'b', total chl-, carotenoids, carbohydrate and protein contents increased by 24.75%, 27.46%, 21.10% 6.29%, 15.19% and 7.95% respectively in the plant tissues of *Lemna minor*. Zaltauskaite et al. [15] have also found that wastewater treatment resulted in higher content of photosynthetic pigments in *L. minor* exposed to untreated wastewater. Similarly, Benerjee and Matai [22] have also reported concentration of carbohydrates in *Lemna minor* increased by 15.19% and 23.64% on dry weight basis when grown on tap water and wastewater respectively.

An earlier study on the biochemical composition of aquatic plants of Dal Lake in Kashmir valley revealed carbohydrates content in *Lemna minor* was 46.41-85.74% on dry weight basis [23]. However, the carbohydrate content observed during the present study remained within the findings of Mishra and Jha, Prasannakumari et al. and Mini [24-26]. Hammouda et al. [27] also found that protein content in duckweed increased when grown on wastewater and reached a maximum concentration of 47.1% with a maximum increase in all amino acids. Pandey, 2001 reported that duckweed had high nutrient value in the dried biomass; 20-31% protein, 0.5-2.2% fat, 0.008-0.01% vitamin C and 0.003-0.007% iron who recommended its use as a food supplement for fish, poultry, and cattle. The increase in concentration of biochemical parameters may be mainly due to increased absorption of macronutrient and micronutrients, which in turn increase the tissue proteins and carbohydrates.

Conclusion

Chemical analysis recorded that nutrients and heavy metals in wastewater and tap water were significantly reduced by *Lemna minor* over 15 days treatment. Therefore, this aquatic weed helps in a continuous gradual decrease in the nutrients and heavy metals from the wastewater. *Lemna minor* has been shown to be a potential scavenger of nutrients and heavy metals from wastewater and may be used in wastewater treatment systems. Further, there were no significant difference between duckweed treatment systems in waste water and tap water. Duckweeds efficiently remove nutrients and heavy metals from both the systems. Duckweed grows rapidly and is capable of nutrient uptake under a wide range of environmental conditions. Compared to most other aquatic plants, it is less sensitive to low temperatures, very high nutrient levels, pH fluctuations, pests and diseases.

It may be conclude that *Lemna minor* can be used as wastewater phytoremediation agent. Though protocols need to be developed and further studies are required to understand the mechanisms involved in the uptake of pollutants so that maximum potential can be utilized for use in phytoremediation technology.

References

1. Khan AG, Kuek C, Chaudhry TM, Khoo CS, Hayes WJ (2000) Role of Plants, Mycorrhizea and Phytocheleters in Heavy Metal Contaminated Land Remediation. Chemosphere 41: 197-207.

2. Keskinkan O, Goksu MZL, Basibuyuk M, Forster CF (2004) Heavy metal adsorption properties of a submerged aquatic plant (Ceratophyllum demersum). Bioresource Technology 92: 197-200.

3. Vardanyan LG, Ingole B (2006) Studies on heavy metal accumulation in aquatic macrophytes from Sevan (Armenia) and Carambolim (India) systems. Environmental International 32: 208-218.

4. Pratas J, Favas PJC, Paulo C, Rodrigues N, Prasad MNV (2012) Uranium accumulation by aquatic plants from uranium-contaminated water in Central Portugal. International Journal of Phytoremediation 14: 221-234.

5. Parr LB, Perkins RG, Mason CF (2002) Reduction in photosynthetic efficiency of Cladophora glomerata, induced by overlying canopies of Lemna spp. Water Research 36: 1735-1742.

6. American Public Health Association (1992) Standard Methods for the Examination of Water and Wastewater. Washington, USA.

7. Strain HH, Cope BT, Walter AS (1971) Analytical procedures for the isolation, identification, estimation and investigation of the chlorophylls. Methods in Enzymology, Academic Press, New York, USA 23: 452-457.

8. Duxbury AC, Yentsch CS (1956) Plankton pigment monographs. Journal of Marine Research 15: 91-101.

9. Dubois M, Gilles KA, Hamilton JK, Rebers PA, Smith F (1956) Colorimetric method for determination of sugars and related substances. Journal of Analytical Chemistry 28: 350-356.

10. Lowry OH, Rosenbrough NJ, Farr AL, Randale RJ (1951) Protein measurement with the Folin phenol reagent. Journal of Biology and Chemistry 193: 265-275.

11. Jackson ML (1974) Soil Chemical Analysis--Advanced Course. Published by the author, Madison, Wisconsin, p: 895.

12. Hesse PR (1971) A Text Book of Soil Chemical Analysis. John Nurray Williams Clowes and sons Ltd. London, UK, pp: 184-324.

13. Verma R, Suthar S (2015) Impact of density loads on performance of duckweed bioreactor: A potential system for synchronized wastewater treatment and energy biomass production. Environmental Progress & Sustainable Energy 34: 1596-1604.

14. Pandey M (2001) Duckweed Based Wastewater Treatment. Central Pollution Control Board New Delhi Entrepreneur. Invention Intelligence, pp: 9-10.

15. Zaltauskaite J, Sujetoviene G, Cypaite A, Auzbikaviciute A (2014) Lemna minor as a tool for wastewater toxicity assessment and pollutants removal agent. The 9th International Conference Environmental Engineering, pp: 22-23.

16. Patel DK, Kanungo VK (2010) Phytoremediation potential of duckweed (Lemna Minor L: A tiny aquatic plant) in the removal of pollutants from domestic wastewater with special reference to nutrients. The Bioscan 5: 355-358.

17. Wafaa AE, Gahiza I, Farid AE, Tarek T, Doaa H (2007) Assessment of the efficiency of duckweed (Lemna gibba) in wastewater treatment. International Journal of Agriculture & Biology 9: 681-687.

18. Duckweed Research Project (1997) Research programme for the development and testing of duckweed-based technologies. Dhaka, Bangladesh 6: 31.

19. Korner S, Vermaat JE (1998) The relative importance of (Lemna gibba L.), bacteria and algae for the nitrogen and phosphorus removal in duckweed-covered domestic waste water. Water Research 32: 3651-3661.

20. Kara YESI, Basaran DM, Kara I, Zeytunluoglu A, Genc H (2003) Bioaccumulation of nickel by aquatic macrophyta Lemna minor (Duckweed). International Journal of Agriculture and Biology 5: 281-283.

21. Zayed A (1998) Phytoaccumulation of trace elements by wetland plants. I. Duckweed. Journal of Environmental Quality 27: 715-721.

22. Banerjee A, Matai S (1990) Composition of Indian aquatic plants in relation to utilization as animal forage. Journal of Aquatic Plant Management 28: 69-73.

23. Pandit AK, Qadri MY (1986) Nutritive values of some aquatic life-forms of Kashmir. Environmental Conservation 13: 260-262.

24. Mishra PK, Jha SK (1996) Effect of water pollution on Biochemistry of Hydrophytes. Pollution Research 15: 411-412.

25. Prasannakumari AA, Arathy MS, Devi GT (2000) Bio-geochemical studies of a temple pond with reference to macroflora. Pollution Research 19: 623-631.

26. Mini (2003) Studies on bio organics and trace metals in the associated flora of a lotic ecosystem –Vamanapuram river. Ph.D thesis, University of Kerala, Kerala, India.

27. Hammouda O, Abdel-Hameed MS (1994) Response of phytoplankton populations to aquatic treatment by Lemna gibba. Folia Microbiologica 39: 420-427.

28

The Applicability of Electrical Current Based Treatment for the Remediation of Different Types of Polluted Soils Contaminated by Organic Compounds

E. C. Rada[1]* and I.A. Istrate[2]

[1]Department of Civil and Environmental Engineering, University of Trento, Via Mesiano 77, I-38050 Trento, Italy
[2]Department of Energy production and Use, University Politehnica of Bucharest, Splaiul Independentei 313, 060042, Bucharest, Romania

Abstract

Environmental pollution is considered a serious problem in Europe. It has been regulated by several directives with the aim of limiting the amount of pollutants that can reach the environment. Five years ago, on the 1st of January 2007, Romania became one of the European Union members. In the following years Romania had to comply with the EU directives in all the areas, but especially in the field of the contaminated sites. Presently the Romanian regulations regarding the contaminated sites are not very developed and must be enhanced.

In this paper, some results from a PhD research are presented. In the frame of a co-supervised doctorate, in 2007 an international scientific collaboration between the Politehnica University of Bucharest and the University of Trento, Italy, was signed in order to go on with the development of studies on remediation technologies for contaminated sites. The idea started from the need, especially, from the Romanian side, to identify alternative treatments for soils mainly polluted with petroleum products.

The main objective of this paper is to present some results regarding the effect of electrochemical treatment applied on the contaminated soil. Three types of matrices and two types of contamination were tested.

Keywords: Clay; Diesel; Electrochemical remediation; Hydrocarbons; PAH; Sand; Sediment

Introduction

During the last decades, the anthropogenic sources have contributed to organic compound penetration into the environment [1]. The existence of potential hazards that can arise from the addition of chemical substances to soils is not a matter of dispute. Indeed, there is a widespread knowledge that the alteration of the chemical conditions of a soil affects plant life and that substances deposited on land can later migrate to groundwater and surface water, entering the human food chain too. Soil contamination consists of many individual sites where the contamination occurred either in the past, or is still occurring. Soil contamination is in Europe a widespread problem of varying intensity and significance. Cleaning up all historically-contaminated sites, commonly of industrial origin, to background concentrations or levels suitable to all uses often is not viewed as technically or economically feasible [2]. As a result, clean-up strategies are designed to employ sustainable, long-term solutions, often using a risk-based approach to land management aimed at achieving the target "fitness for use", appropriate to the location. In the absence of specific EU legislation to address the clean-up of contaminated soil, Member States apply the "polluter pays" principle in clean-up programs. Public funds have been used in a number of Member States to finance remediation costs when necessary. Several economic activities are still causing soil pollution in Europe, particularly those related to inadequate waste disposal and leakages during industrial operations. It is expected that the implementation of preventive measures introduced by the legislation already in force would limit the inputs of contaminants into the soil in the coming years. As a consequence, most of the future management efforts will be concentrated on the clean-up of historical contamination. This is going to require large sums of public money which at present already account on average for 25% of the total remediation expenditure [3].

More than 80,000 sites have been cleaned up in the last 30 years in the countries where data on remediation are available. Although the range of polluting activities (and their relative importance as localized sources of soil contamination) may vary considerably across Europe, industrial and commercial activities as well as the treatment and disposal of waste are reported to be the most important sources. National reports indicate that heavy metals and mineral oil are the most frequent soil contaminants at investigated sites, while mineral oil and chlorinated hydrocarbons are the most frequent contaminants found in groundwater. A considerable share of remediation expenditure, about 35% on average, comes from public budgets. Although considerable efforts have been made yet, it will take decades to clean up a legacy of contamination [3]. In European Economic Area (EEA) member countries, it is estimated that potentially polluting activities have occurred at about three million sites.

In the present paper, a research regarding the treatment of diesel contaminated soils will be studied. The experimental research took place in the frame of a PhD research done both at the University of Trento, Italy as well as at University Politechnica of Bucharest, Romania. In both countries the organic contamination as a result of bad management from the petroleum industry is a well known problem. In Romania, in the last five years the authorities tried to adapt the national legislation accordingly to the European one. The research presented in

---

*Corresponding author: E. C. Rada, Department of Civil and Environmental Engineering, University of Trento, Via Mesiano 77, I-38050 Trento, Italy
E-mail: elena.rada@ing.unitn.it

this article will refer to two types of contamination and on three types of soils: sediments that have been affected by natural contamination, clay and sand that were artificially polluted by diesel. The method used for the trials were the electrochemical treatment that was included in the category of direct current technologies.

Electrochemical methods are attractive and may act as additional or alternative procedures because they are cost-effective and can be applied as *in situ* techniques [4].

Material and Methods

The electrochemical treatment

At first, DCT were used mainly for the remediation of metals, radio nuclides and polar inorganic pollutants from soil and groundwater. The process was called Electro kinetic Remediation. In recent years, several researches have been developed about DCT and their effectiveness in the removal of organic pollutants from soils and sediments. These studies seem to suggest that DCT can be effectively used for the mineralization of many organics, with lower energy expenditure, if compared to traditional electro kinetic remediation methods [5,6].

Electro Chemical Remediation Technologies (ECRTs) are phenomena related to colloid electrochemistry and belong to the class of Direct Current Technologies (DCTs) where DC electricity is passed between two electrodes. The primary distinctions between ECRTs and traditional electro kinetics are the operative mechanisms, energy input, nature of the direct current, and resulting outcome.

The study of the treatments based on the application of electrical current has been initiated since the 90's. They have been considered, at the beginning, as innovative methods for *in situ* restoration of hazardous waste sites. The principle of these technologies is the application of direct currents across the electrodes inserted in the soil to generate an electric field that will result in the mobilization and extraction of contaminants. Also, this type of treatment was used according to some researchers, like Akram Alshawabkeh to obtain biogeochemical modifications of polluted soils and slurries. Electroremediation or electro kinetic remediation is an *in situ* technology that consists of the controlled application of low intensity direct current through the soil between appropriately distributed electrodes [1,7,8].

The electrochemical remediation depends on several important factors, as it will be presented in the following

A) Soil chemistry or soil-contaminant interaction: The kinetics of the removal of contaminants is bound to adsorption phenomena, ion-exchange, buffering capacity.

- B) Water content: Inhomogeneous distribution of moisture and consolidation may take place during an Electro kinetic treatment.

- C) Soil structure: Clogging of the soil porous texture and blocking of the electro-osmotic flow may take place due to hydroxide formation (presence of heavy metals).

- D) Positioning of the electrodes and electrode structure: Solidity of the structure, easy workability, chemical stability, costs are the major actors. Silicon pig-iron, graphite, activated titanium are electrode materials of practical interest.

One of the most important advantages of the electro kinetic technique is its efficacy for the treatment of low hydraulic permeability soils, where other techniques as natural attenuation or traditional remediation efforts do not work. Empirical evidences also indicate that

the reaction rates are inversely proportional to grain size, such that this remediation technique is particularly effective in saturated low permeability soils (like clays and silts), which are often more difficult to treat with conventional chemical methods (such as chemical oxidation or soil flushing), because of their low permeability and their high sorption capacity. Usually, for *in situ* DCT applications, the current density is of the order of milliamperes per square centimeter (1mA/cm2) and the electric potential difference is on the order of a few Volts per centimeter across the electrodes placed in the ground (1V/cm) [8]. The electrodes can be made of different materials, as stainless steel or carbon, and they can be placed in the soil either in a vertical or horizontal array. The electrode distribution when electrochemical remediation is applied can vary from the bench scale to commercial installations as it is presented in figure 1.

There are several examples of application of DCTs to a real case of contamination in Europe and North America, but a deep knowledge of the phenomena ruling the remediation process has not been reached [9,10]. Therefore, at the moment the calibration of the remediation action is mainly based on empirical data and on the results of field preliminary tests.

Experimental setup description

During the basic experimental research two different types of setup were used to perform the bench tests. Each experimental setup includes the following elements:

- An electrochemical cell where the sample will be inserted.

- A pair of electrodes that have a shape as a square from stainless steel

- A direct current power supply that is connected to the electrodes with copper cables of about 2.5 mm in diameter.

- Two bottles for collecting the interstitial water that can be transported with electro osmotic flux in electrodes compartments (setup used for the tests performed on artificially contamination).

A representation of the experimental setup presented before can be seen in figure 2.

The characteristics of soils used for the research

For this research three kinds of soils were used as a material: kaolin (clay), sand and sediments. In the first two cases the contamination was artificial and for the last one was natural. In the following paragraphs the characteristics of these three types of soils will be described.

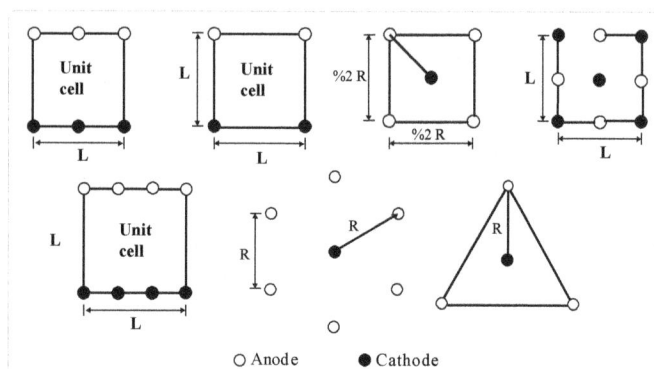

○ Anode   ● Cathode

Figure 1: Possible electrode distribution for the application of electrochemical remediation [11].

Clay – kaolin: Clay minerals are common weathering products of rocks and soils, characterized by small size particles with very high specific surfaces. They are composed of layer structures, named as phyllosilicates. Each phyllosilicate is composed by unit cells which bound to each other forming sheets and layers. In each clay particle, the sheets of tetrahedral (T) and octahedral (O) cells are stacked on each other to form layers. The silicon and aluminum layers are held together by share chemical bounds. The most common clay types are kaolinite, illite and bentonite, which is a mixture of different clay minerals, mainly composed of montmorillonite.

The type of clay used in the present research is the kaolinite known also as kaolin. The kaolin (figure 3) used in the experimental research is composed from particles with dimensions that range for 40% between 2μm and 75μm, for 60% less than 2 μm. The chemical composition of this kaolin is presented in table 1.

The kaolin that was used for this research has a Cation Exchange Capacity (CEC) of about 8.3 m_{eq}/100g and a pH of about 6. Because it was necessary to know what influence has the content of metals from the soil on the system efficiency, the quantity of iron and manganese in the kaolin was determined; 2794 mg/kg_{DM} of iron and 34 mg/kg_{DM} of manganese (DM= Dry Matter). The main characteristics of the kaolin used during this research, are presented in table 2.

Sand used for the research: The sand that was chosen for this research is dried silica sand, named VA-GA VG 12 (Figure 4). The sand was acquired by the Laboratory of Hydraulics of Faculty of Engineering of Trento from the company Va.Ga. S.r.l. (Pavia). This type of sand was chosen because, from all the available products, this one had the most uniform and fine granulometry. We looked a sand that was characterized by small particles because, as electrochemical remediation is more efficient for soils that are characterized by a very low hydraulic conductibility and a fine granulometry. The sand composition is presented in table 3.

The sand has undergone a series of treatments, done by the company Va.Ga. S.r.l., such as:

- Initially the inert was extracted and brought the screening and washing stages.

- After that the material was passed through a crushing installation, equipped with two mills, allowing the selection and the control of the material and allowing the recovery of the coarse parts of the extracted inert.

- At the end, the material arrived in a drying installation where it was inserted in a rotary oven at 1100°C; after that the sand was cooled down and cleaned.

For the mineralogical composition of the Vaga12 we had: main

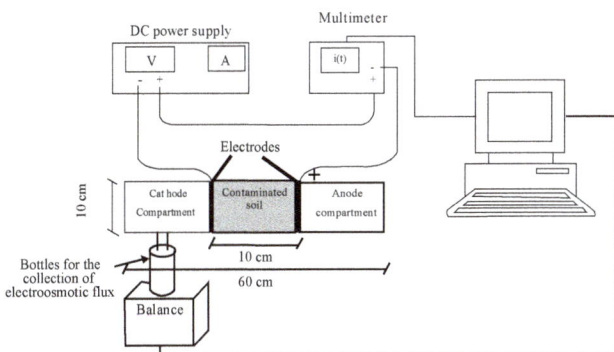

Figure 2: Schematic representation of the basic experimental setup used in this research.

Figure 3: Kaolin sample used during experimental research.

Compound	Chemical formula	Percentage presented in soil
Silicon	SiO_2	47%
Aluminum	Al_2O_3	38%
Others	-	15%

Table 1: Chemical composition of the kaolin used for the present research.

pH	CEC [m_{eq}/100g]	Fe [mg/kg_{ss}]	Mn [mg/kg_{ss}]	TOC [mg/kg_{ss}]
6.0	8.3	2794	34	4000

Table 2: Main chemical characteristics of the kaolin used for the research.

Figure 4: Sand sample used in experimental research.

Compound	Chemical formula	Percentages presented in soil
Silica	SiO_2	81.2%
Iron	Fe_2O_3	2.1%
Aluminum	Al_2O_3	6.6%
Calcium	CaO	1.2%
Magnesium	MgO	1.5%
Sodium	Na_2O	2.0%
Potassium	K_2O	2.1%

Table 3: Chemical composition of the sand after the treatments.

compound quartz – 61.8%; granite rocks – 16.5%; feldspar – 12.7%; the rest, about 9 %, formed from other types of minerals.

The granulometry analysis showed that the particles had a diameter that ranged between 0.01 mm to 0.20 mm, classified as sand. Vaga 12 had a Cation Exchange Capacity (CEC) of 0.7 m_{eq}/100g of soil, a very low value if we compare with CEC of kaolin (8.3 m_{eq}/100g).

The sand used for the experimentation was a dissolved material, monogranular, without particles of clay, without content of organic matter (Theory of Constraints (TOC) negligible amounts to 0.23g/kg_{DM}) and poor in nutrients (shortage of potassium, calcium and magnesium), because of these characteristics the sand had a low cation exchange capacity and consequently a low power buffer. The pH of this sandy soil was about 8, and from the analysis it resulted that the sand had an iron content of about 14425.5 mg/kg_{DM} and a manganese content of about 324.03 mg/kg_{DM}. All the main characteristics of the sand that was used during this research are presented in table 4.

Sediments: The contaminated sediments were collected from a canal in the town of Trento, which for several decades received industrial effluents polluted by organic and inorganic compounds. This canal is located in Trento-Nord, which can be found in the northern part of the town of Trento in the Italian region named Trentino Alto Adige. The canal, known as Roggia Campotrentino, was contaminated over the years mainly by ex- Carbochimica (42,700 m^2). Ex-Carbochimica sites are contaminated mainly by solvents, phenols and Polyaromatic Hydrocarbons (PAHs).

Several samples (total weight about 10 kg) of fine-grain silty sediments were collected from the first 30-40 cm layer at the bottom of the canal; these samples were then mixed together and mechanically stirred to produce an homogeneous sample.

Polycyclic Aromatic Hydrocarbons (PAHs) are found in nearly all soils. They are tightly bound to soil and sediment particles and have a reduced mobility. It is believed that PAHs in soils are a result of local or long-range air transport and subsequent deposition. Background PAHs levels are generally the highest in urban soils and adjoining near shore sediment due to the presence of large concentrations of anthropogenic activity in the urban corridors. More direct sources of PAHs in soils include sludge disposal from public wastewater treatment plants, automotive exhaust, leachate from bituminous coal storage sites, and use of contaminated soil compost and fertilizers. Soils directly adjacent to highways are susceptible to contamination from vehicle exhaust and wearing of tires and asphalt. Finally, the soils in or near landfill sites and industrial sites, such as creosote producing, wood-preserving, coking, and former gas manufacturing plants, all have the potential for high PAHs levels.

Analytical methods

In the present paragraph, the analytical methods used for treatment efficiency characterization will be described for pH, PAH and TOC determination from sediments, for Total Petroleum Hydrocarbon (TPH) and TOC from diesel contaminated soil.

In order to achieve a good resolution both for light and for heavy PAH species, light PAH concentrations in sediments samples were determined with analysis by Gas Chromatography (GC) and heavy PAH concentrations were detected by High Performance Liquid

Chromatography (HPLC). The setups were tested before analyzing the samples; external standards were used for HPLC calibration, while internal standards were used for GC calibration. The extraction efficiencies were about 85% both for HPLC and for GC.

As for light PAH detection by GC, the pollutants were at first extracted by sonication and solvent addition (HPLC grade acetonitrile, >99.99%). A 2 μL sample of solvent was then injected into the gas-chromatograph and analyzed using a Varian 4000 GC/MS. The following temperature program was used: 80°C for 1 min; isothermal from 80°C to 260°C at 10°C/min; isothermal from 260°C to 320°C at 20°C/min; 320°C were maintained for 3 minutes (trap temperature 180°C, marifold temperature 50°C, transfer time temperature 250°C).

As for heavy PAH detection, the samples were at first extracted by solvent addition (HPLC grade dichloromethane, >99.9%) and filtered on a 0.45 μm filter. The solvent was allowed to evaporate and then an acetonitrile solution (70% HPLC grade acetonitrile and 30% water) was added to the sample. A 100 μL sample of the obtained solution was injected into the HPLC and analyzed. The HPLC included: auto-sampler Gilson ASPEC XL (solid base extraction), Dionex P680 HPLC Pump, Dionex STH 585 Column Oven, HPLC detector Dionex UVD 340U (diode array). A Supelco-SIL LC-PAH column (520 mm × 4.6 mm i.d., 5μm particle size) was used for PAH detection. For the analysis, the acetonitrile concentration in the sample was maintained constant at 70% for 12 minutes, then it was increased with constant gradient up to 100% in 6.4 minutes; the 100% acetonitrile content was maintained for 1.6 min and then was decreased to 70% in 1 min. During the detection, the column temperature was maintained constant at 30°C and the eluent flux was set at a constant rate of 1.5 mL/min.

To detect the contaminant content in the sediment samples, the solid and liquid phases were extracted together; that allowed taking into account all PAHs in the sample. Only in the experiments which aimed at determining the PAH distribution in sediments and pore water, the solid and liquid phases were separated by filtration and analyzed separately.

BTEX (Benzene, Toluene, Ethylbenzene and Xylenes) concentrations in the untreated sediment samples were determined trough purge & trap extraction followed by gas chromatographic analysis with VARIAN 4000 GC/MS.

The TOC content was determined by IR analysis of thermal induced CO_2 with a TOC Analyzer Shimadzu TOC-V CSH, after heating the sample at 900°C with a Shimadzu Solid Sample Module.

The pH was taken in a sediment/water suspension using a pH-meter HI 99121 by Hanna Instruments, with HI 1292D electrode for soil pH measurement.

The TPH content was determined by gravimetric method after pollutant extraction by sonication and solvent addition. In order to perform the analysis, a soil sample (mass about 5 g) was carefully weighted and mixed with anhydrous sodium sulphate to eliminate soil humidity. The sample was extracted by sonication for 15 min, then 10 mL of solvent (HPLC grade nesane) were added to it. About 5 mL of the obtained solvent were collected and cleaned up on a Florisil tube. The solvent sample was transferred into a pre-weighted vial, it was allowed to evaporate and then the vial was weighted. The TPH content was determined from the difference between the initial and final weight of the vial.

pH	CEC [m_{eq}/100g]	Fe [mg/kg_{ss}]	Mn [mg/kg_{ss}]	TOC [mg/kg_{ss}]
8.0	0.7	14425.5	324.03	230

Table 4: Chemical parameters of the sand used in the experimental research.

Results

Artificial contamination – sand and clay

The experimental research performed during this project foresaw the application of electrooxidation on matrix of artificial contaminated soil with diesel. The diesel used was purchased from a petrol station in Trento. The choice of diesel as a contaminant of soil appeared from the need to represent the environmental pollution caused by oil spills and various oil products; indeed the most frequent sources of contamination are made up of losses of oil, which may arise from industrial activities, refineries and oil spills fuel. The diesel used in this project appeared light-colored, slightly amber and over time saw gradually reducing its sulphur content of up to 0.33%. During this research detailed analyses were done to determine the TOC and TPH of the diesel, because the removal efficiency of the remediation technique had to be evaluated in term of diminution of these two parameters (considered representative for the content of diesel in soil). The obtained values of TOC and TPH for diesel are presented in table 5.

The second type of contaminant analyzed during this present research is the polycyclic aromatic hydrocarbons (PAHs) that can be found in nearly all soils. They are tightly bound to soil and sediment particles and have a reduced mobility.

The tests were named ECR_ soil type that means Electro Chemical Remediation and the soil type used for the trial (K – kaolin, Sa – sand, Se – sediments).

The diesel fuel used in this study is commercially available and was purchased from a gasoline pump at a typical refuel station. To prepare the diesel contaminated soil samples, the soil was at first dried and then spiked with diesel fuel. One kilogram of dry soil was mixed with about 100 mL of diesel fuel, and then the sample was stirred with stainless steel spoons in a glass backer, to ensure the contaminants to be evenly distributed through the soil. After mixing, the sample was allowed to evaporate for about two weeks. Before the test, the spiked samples were saturated with demineralised water and allowed evaporating overnight at room temperature before being inserted in the experimental setup.

Two parameters were used to consider the contaminant content in the soil samples: TPH (total petroleum hydrocarbons), which refers to a family of many petroleum-based hydrocarbons, and TOC (total organic carbon), which represents the whole content of organic substances in the soil samples.

Tests on clay – kaolin: Test ECR_K, which aimed at investigating the variation of contaminant removal with test duration, was performed

with a constant voltage of 10 V (1 V/cm) for 4 weeks. The initial current flow across the soil specimen was 17 mA (the measurement was done at the beginning of the test, after 20 minutes). The results obtained during the monitoring, for pH, of the test ECR_K are presented in figure 5; in table 6 the final TOC, TPH concentrations are summarized and in figure 6 the removal percentages are compared.

An initial current of 2 mA was measured at the beginning of the experiment, while 1 mA was encountered at the steady state. Although the measured current was not very high, the current trend was once more characterized by a rapid decrease, and by some fluctuations during the period from the 1st to the 3rd day of the treatment, when some temporal rebounds of the current flowing were encountered.

TOC removal increased with the increase of the treatment time from 45% to 54 %, and for TPH the removal percentages are higher from 50 to 80%. The results are presented in figure 6.

The results obtained at the end of the test across the soil samples showed that pH decreased with the increase of distance from cathode, whilst TOC and TPH increased with the increase of the distance.

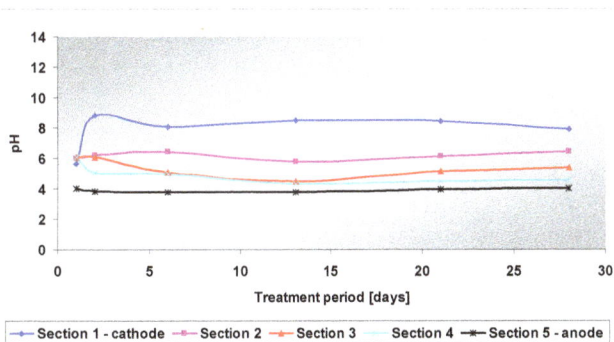

Figure 5: pH variation during test ECR_K – test performed on kaolin matrix.

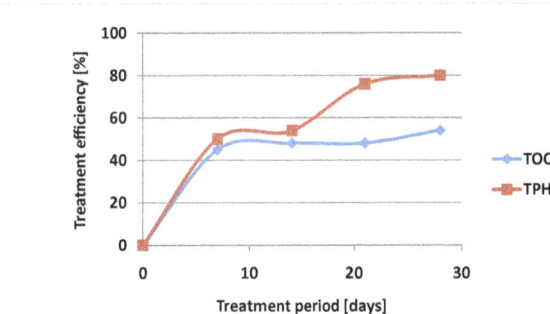

Figure 6: The removal percentages obtained during the test ECR_K – test performed on a kaolin matrix using a specific voltage of 1 V/cm for 28 days.

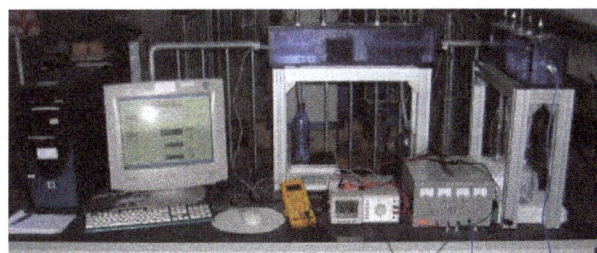

Figure 7: Complete experimental setup with all the instruments used for the research.

Contaminant	TOC [mg/L]	TPH [mg/L]
Diesel	639 231	847 898

Table 5: Values of TOC and TPH for the diesel used for the artificially contamination.

Elapsed time [d]	TOC [g/kg_{DM}]	TPH [mg/kg_{DM}]
0	208.1	130.977
7	113.6	64.966
14	109.2	60.338
21	107.8	30.873
28	95.5	26.566

Table 6: The removal percentages obtained during the test ECR_K – test performed on a kaolin matrix using a specific voltage of 1 V/cm for 28 days (DM=SS).

Tests on sand: The test ECR_Sa has been performed on a sample of sand Vaga 12 of about 1.8 kg, artificially contaminated by diesel fuel. The test was carried out in the experimental setup 1 (figure 7) where the sample had cubic shape (10 x 10 x 10 cm). During the test, a constant potential difference of 10 V was applied on the sample, which corresponds to a specific voltage of 1 V/cm, for a period of 28 days (table 7). The initial sample of soil had moisture equal to 23.9% and a pH of 8.88.

At the beginning of the test, the current was measured and presented a value of 3.98 mA. The behavior of the current was similar with the one observed with the tests performed with clay. This behavior was characterized by a modest increase of the current in the first hours of the tests. After that, a rapid decrease of the current in time was noticed after several hours from the beginning of the test. This decrease was observed until the current reached a constant value were remained until the end of the test (steady state value).

In this test, the following differences regarding the other tests performed with sand or clay were noticed:

- In the first days of the test, the current, after a first initial decreasing, started to increase and remained constant for some days; after this it started to decrease again until it reached a constant value of 0.5 mA.

- Decreasing of the current after some days from the beginning of the test was not as rapid as the one that occurred with clay.

- The steady state, where the value of the current was maintained during the time, was reached after almost 400 hours from the

beginning of the test (almost 17 days), while for the clay the phenomenon was reached after 30-50 hours.

- While the constant value of the current for sand was 0.5 mA, for the clay we had 1-9 mA.

Opposite to what was noticed with clay, in the sand case, the electroosmotic flux was formed after some days from the beginning of the sample (more precisely after 94 hours) as it can be seen in table 8. Moreover, the flux collected at the cathode increased for a maximum 8 days from the beginning of the test, after which stopped. This behavior is explainable because in between the third and the eight days of the test the current increased and reached a value a little bit higher than the initial one (4 mA) and it was maintained almost constant for a few days after which started again to decrease. The quantity of the electroosmotic flux was too small to be analyzed in the laboratory.

The pH profile was quite pronounced for this test (table 9). In the following the dynamics of pH change in time and in different sections of the sample (where section 1 is the section near the cathode and the section 5 is near the anode) is described:

- During the first days (1, 2, 3 days after the start of the trial) pH tended to increase (reaching values between 9.8 and 10.45) in section 1, 2, 3 and 4 while it was decreasing slightly near the anode (Section 5). The increase of pH can be explained by the fact that the sand had a very low buffer capacity (CSC equal to 0.7 m_{eq}/100g), and because of that the ground was unable to counter any change that occurred the soil. This behavior occurred when the current was gradually decreasing; the ground was losing ions (as H^+ ions) and was increasing its strength. Consequently, the soil tended to become basic near the anode where, because of the electrolysis, a reaction occurred causing an acid front (hydroxide ions underwent oxidation and gave electrons);

- After a week of testing, the pH values were decreasing compared with values measured in the early days of the test; the phenomenon could be caused by the fact that the acid front, formed at the anode, was beginning to move gradually to the cathode. An anomaly, not explainable, was found in section 4 where pH reached its maximum value of 9.77. It should be underlined that for a period of 5 days (from 3 to 8 days), the current increased and remained steady for a few days; this could have caused a stirring of charges in soil and therefore this could have led to a strange profile of pH in the soil sample;

- After two weeks of testing the pH values grew slightly respect to the values measured after 7 days in the range of 9.5 - 8.5; in particularly pH in sections 1, 2, 3, 4 showed similar values that decreased from the cathode to the anode. The increase in pH, although modest, could have been caused by the passage of the basic front flowing from the cathode to the anode with a smaller rate for the basic front;

- At the end of the test, the sample showed a pH profile equal to 9.05 at the cathode and 8 near the anode, but the pH values differed slightly compared to the initial terms of pH.

Regarding moisture, the initial value was 25.6% while at the end of the treatment it ranged from 22.7% close to anode, to 23.0% near the cathode (table 10). In general, it can be said that moisture remained constant and comparable with the initial value; this is also explained by the fact that the electroosmotic flux can be consider negligible because of its small quantity (1.3 mL).

Test	ECR_Sa	
Matrix	Fine sand	
Contaminant	Diesel	
Sample weight	1.8	[Kg]
Sample length	10	[cm]
Sample volume	1.0	[L]
Sample density	1.8	[kg/L]
Exposure time	28	[d]
Voltage (constant)	10	[V]
Specific voltage	1	[V/cm]

Table 7: Main characteristics of the test ECR_Sa – test performed on sand at a specific voltage of 1 V/cm.

Time [h]	Time [d]	Current [mA]	Electroosmotic flux [mL]
0	0.0	3.99	0.0
0.5	0.02	4.11	0.0
1	0.04	4.23	0.0
2	0.08	4.01	0.0
4	0.2	3.98	0.0
8	0.3	3.52	0.0
16	0.7	3.17	0.0
24	1	2.98	0.0
48	2	2.60	0.0
72	3	3.33	0.0
96	4	3.71	0.5
168	7	3.06	1.1
336	14	1.03	1.3
504	21	0.68	1.3
672	28	0.51	1.3

Table 8: The current and electro osmotic flux values monitored during test ECR_Sa.

The test ECR_Sa involved the application of electrooxidation to a homogeneous matrix composed of only sand for 4 weeks, using a specific voltage of 1 V/cm. Sampling was done after 3, 4, 7, 14, 21, 28 days, taking samples from the upper part of the cell. The values of TOC and TPH decreased in time. In the sample taken after 7 days it was noticed an increasing of the concentrations. This unusual trend was probably due to a possible migration of diesel contaminants that moved toward the top of the sample.

In this test, the removal percentages were modest (20% for the TOC and 27% for TPH – figure 8) and lower than the results obtained in the previous researches were the applied voltage was lower. So it can be noticed that an increasing of voltage does not always produce increasing pollutant removal.

For this test, it is able to trace the final concentration of the contaminant along the sample, as five samples were taken at different distances from the electrodes. The values of the profile of concentration are presented in figure 9.

In this figure, it is noticed the non-uniformity of contaminant removal along the sample. Indeed, the removal is more obvious near the electrodes, while in the middle of the sample it is observed that removal pollutant is almost imperceptible. Regarding ecotoxicity, values are presented in table 11. At the end of the test, a slight decrease in terms of mg/L for ecotoxicity parameter was noticed; that indicates a modest increase of toxicity for bacterial action. In this case the toxicity (both initial and final) of the diesel contaminated soil was much higher compared to the previous tests.

Tests on sediments – natural contamination

The contaminated sediments were collected from a canal in the town of Trento, which for several decades received industrial effluents polluted by organic and inorganic compounds. This canal is located in Trento-Nord, which can be found in the northern part of the town, in the Italian region Trentino Alto Adige. The canal, known as Roggia Campotrentino, was contaminated over the years mainly by a factory extended in an area of 42,700 m². This area was contaminated mainly by solvents, phenols and polyaromatic hydrocarbons (PAHs).

Several samples (total weight about 10 kg) of fine-grain silty sediments were collected from the first 30 - 40 cm layer at the bottom of the canal; these samples were then mixed together and mechanically stirred to produce an homogeneous sample.

In order to investigate the effectiveness of the electrochemical treatment on PAH contaminated sediments and to evaluate the best, two types of tests were conducted using setup of Figure 10.

- Exposure time of four weeks, with a voltage of 10 V – test ECR_Se1.

- Exposure time of four weeks, with a voltage of 20 V – test ECR_Se2.

For these tests the parameters that have been monitored were: the current, pH, PAHs and TOC, after a treatment period of 7, 14, 21 and 28 days. The values for the first two parameters are presented in table 12 and in figure 11. The difference for the value of the initial current (50 mA for test 2 respect to 6.5 mA for test 1) is due to the fact that in test ECR_Se2 the specific voltage was double respect to the one for test ECR_Se1.

At first, the collected sediments were characterized and analyzed for BTEX (i.e. aromatic hydrocarbons: benzene, toluene, ethylbenzene, xylene), PAHs, pH and total organic matter, represented by the value of TOC. Both organic pollutants and natural organic matter occurred in the sediment samples, which proved to be contaminated by PAHs, but not by BTEX, whose presence was detected just in traces.

The initial total PAH concentration in sediment samples for tests ECR_Se1 and ECR_Se2 was about 1032.8 mg/kg$_{DM}$ (light PAHs about 893.4 mg/kg$_{DM}$, heavy PAHs about 139.4 mg/kg$_{DM}$) and 90% degradation was required to meet the remediation goals. The initial PAH concentrations in the samples may vary because the pollutant content of the sediments was not homogeneous. The TOC content was 88.5 g/kg$_{DM}$, the sediment pH was about 9. The sediments also showed a significant metal content, with a total iron concentration of 36350 mg/kg$_{DM}$ and a manganese content of 522 mg/kg$_{DM}$. The results are summarized in figure 12.

It can be seen that the removal percentages increased with the increase of the exposure time for both tests. Also, the applied voltage did not influence very much the results because for example the removal percentages obtained for 10 V were higher than the ones for 20 V. Anyway the target of these tests was to obtain a removal percentage higher than 90%. So, we can say that the proposed target was reached.

The influence of the treatment period on the removal percentages obtained at the end of the tests is more pronounced for the PAH removal than for TOC removal. The application of only electrochemical

The distance from the cathode [cm]	Time [d]								
	0	1	2	3	4	7	14	21	28
1	8.88	10.45	10.36	10.35	10.11	9.23	9.73	9.64	9.05
3	8.88	10.26	10.00	9.98	9.97	8.77	9.70	9.22	8.78
5	8.88	10.21	9.98	9.95	9.09	8.88	9.60	8.89	8.65
7	8.88	10.14	9.89	9.82	9.76	9.77	9.50	8.79	8.67
9	8.88	7.96	8.72	8.5	8.63	7.75	8.55	8.00	8.00

Table 9: pH profile of the sample used for test ECR_Sa.

The distance from the cathode [cm]	Initial humidity	Final humidity
1	23.9%	23.0%
3	23.9%	23.7%
5	23.9%	23.2%
7	23.9%	23.0%
9	23.9%	22.7%

Table 10: Humidity distribution in soil sample.

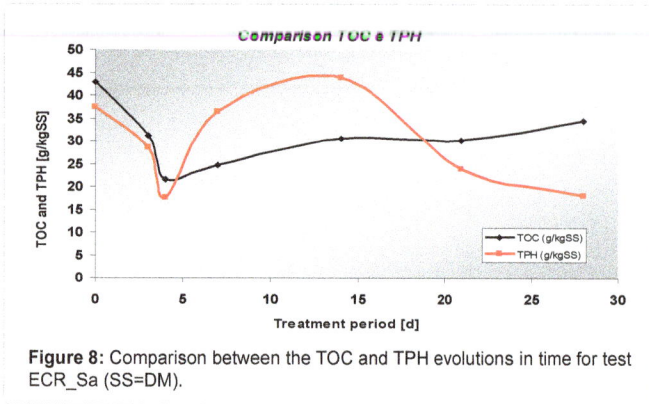

Figure 8: Comparison between the TOC and TPH evolutions in time for test ECR_Sa (SS=DM).

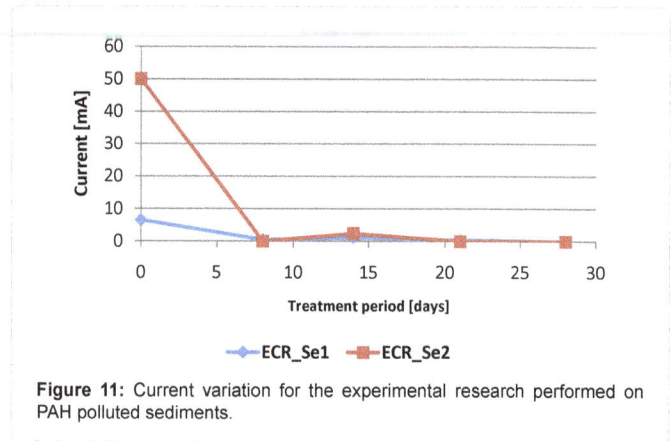

Figure 9: The distribution of TPH concentration along the sample, before and after the treatment (SS=DM).

Time [d]	Ecotoxicity [mg/l]
0	5.8
28	1.8

Table 11: Ecotoxicity values at the beginning and end of the test ECR_Sa.

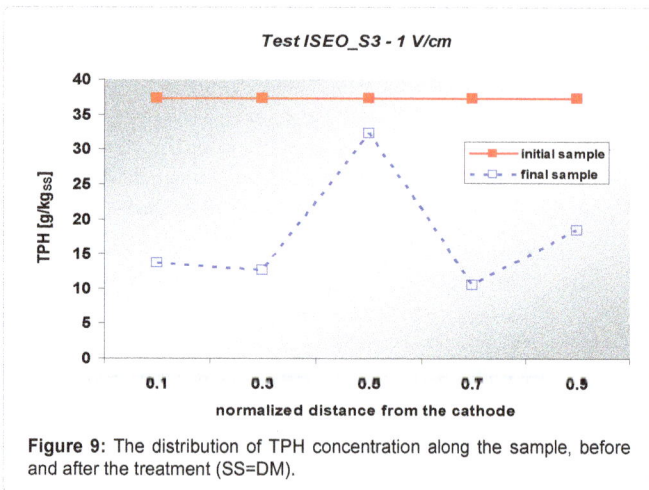

Figure 10: Setup used for the experiments with PAH contaminated sediments.

oxidation showed to be able to achieve a very good PAH removal (above 90%) after a four week treatment. The applied voltage seemed to have a limited influence on the efficiency of the remediation action, good results being achieved with specific voltages as low as 1 V/cm, with low energy expenditures, while the remediation efficiency proved to increase significantly with the process duration.

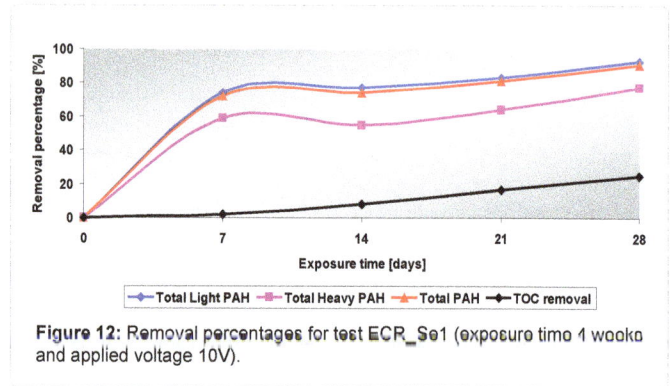

Figure 11: Current variation for the experimental research performed on PAH polluted sediments.

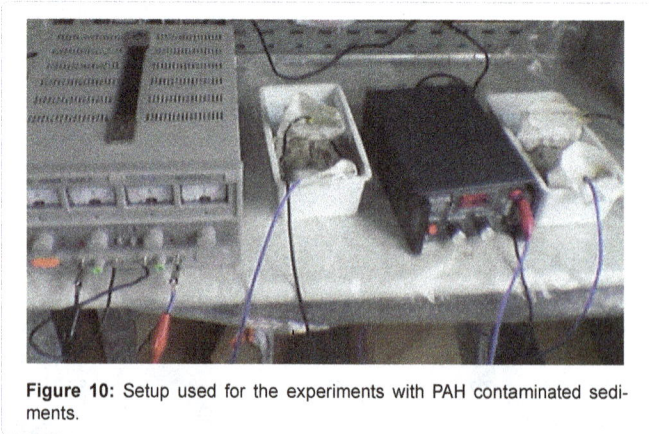

Figure 12: Removal percentages for test ECR_Se1 (exposure time 4 weeks and applied voltage 10V).

Conclusions

The implementation of the electrochemical remediation system in the field is relatively simple. Its design and operation for successful remediation is cumbersome due to complex dynamic electrochemical transport, transfer, and transformation processes that occur under applied electric potential [7]. The efficacy of electrochemical remediation depends strongly on contaminated media characteristics such as buffer capacity, mineralogy and organic matter content, among others. If geochemistry, soil-contaminants interaction, and subsurface heterogeneity are well understood, the electrochemical remediation systems can be engineered to achieve remediation in an effective and economic manner [5,7].

This experimental research aimed at evaluating the effectiveness of electrooxidation for the removal of organic pollutants from artificial contaminated kaolin and sand and from natural contaminated sediments. During this experimental research, tests were performed on three types of matrix: sediments, silty clay (mainly composed by kaolin) and sand. Several tests were performed with one-dimensional setup for lab experiments with specific voltages of 1 V/cm for 28 days for artificial contamination and 1 V/cm and 2 V/cm for natural contamination with the same treatment period. For sand the conclusion is that only the electrochemical treatment is not efficient if applied alone, but for kaolin and sediments the results are quite encouraging (more than 80% for kaolin and more than 90% for sediments). In the case of sediments it was noticed that a higher specific voltage does not mean also significant differences at the end of the test respect to the smaller specific voltage tested.

| | Day | Voltage [V] | Current [mA] | pH | | | | |
				out 1	out 2	out 3	out 4	out 5
ECR_Se1	0	10	6.5	not measured				
	8	10	0.5	9.87	8.8	8.84	8.6	8.7
	14	10	1.1	8.98	8.93	8.66	8.25	7.68
	21	10	0.35	8.71	8.24	8.25	8.75	8.83
	28	10	0.009	8.26	8.98	9.52	9.24	9.07
ECR_Se2	0	20	50	not measured				
	8	20	0.003	10.06	8.63	8.53	8.52	8.54
	14	20	2.4	9.18	8.48	8.4	8.57	8.27
	21	20	0.005	9.36	8.11	8.7	8.45	8.29
	28	20	0.001	8.54	8.15	8.21	8.24	8.02

Table 12: The monitored data for current flow and pH variation during test ECR_Se1 and ECR_Se2.

Acknowledgments

This work was partially supported under the Sectorial Operational Programme Human Resources Development 2007-2013 of the Romanian Ministry of Labour, Family and Social Protection through the Financial Agreement POSDRU/89/1.5/S/62557.

References

1. Pazos M, Rosales E, Alcántara T, Gómez J, Sanromán MA (2010) Decontamination of soils using PAHs by electroremediation: A review. J Hazard Mater 177: 1-11.

2. U. S. Environmental Protection Agency (2009) How To Evaluate Alternative Cleanup Technologies For Underground Storage Tank Sites: A Guide For Corrective Action Plan Reviewers (EPA 510-B-94-003; EPA 510-B-95-007 and EPA 510-R-04-002).

3. Environmental European Agency (2007) Progress in management of contaminated sites.

4. Cong Y, Ye Q, Wu Z (2005), Electrokinetic behavior of chlorinated phenols in soil and their electrochemical degradation. Process Safety and Environmental Protection 83: 178-183.

5. Acar YB, Gale RJ, Alshawabkeh AN, Marks RE, Puppala S, et al. (1995) Electrokinetic remediation: basics and technology status. J Hazard Mater 40: 117-137.

6. Acar YB, Alshawabkeh AN (1993) Principles of electrokinetic remediation. Environ Sci Technol 27: 2638-2647.

7. Reddy RK, Claudio C (2009) Electrochemical remediation technologies for polluted soils, sediments and groundwaters, Wiley.

8. U.S. AEC (2000) In-situ electrokinetic remediation of metal contaminated soils - technology status report. U.S. Army Environmental Center, Report Number: SFIM-AEC-ET-CR-99022.

9. Andreottola G, Ferrarese E, Saroj DP, Oprea IA (2007) Feasibility study of electro-oxidation of landfill clay barrier contaminated by leachate, XI International Waste Management and Landfill Symposium, S. Margherita di Pula (Cagliari), Sardinia, Italy.

10. Reddy KR, Ala PR, Sharma S, Kumar SN (2006), Enhanced electrokinetic remediation of contaminated manufactured gas plant soil. Engineering Geology 85: 132-146.

11. De Battisti A (2008) Electrochemical remediation, Seminar "Dall'Emergenza delle bonifiche ad una gestione consapevole del territorio", Provincia Autonoma di Trento.

29

The Role of Microorganisms in Distillery Wastewater Treatment

Terefe Tafese Bezuneh*

College of Natural Science, Arbaminch University, Arbaminch, Ethiopia

Abstract

Distilleries are one of the most polluting industries generating large volume of wastewater having a serious environmental concern. Distillery effluent is characterized by dark brown color, acidic pH, high temperature, low dissolved oxygen (DO), high biochemical oxygen demand (BOD) and chemical oxygen demand (COD). Distillery wastewater disposed onto the environment prior to treatment is hazardous and leads to soil and water pollution. The dark brown color of distillery effluent causes reduction of sunlight penetration, decreased photosynthetic activity and dissolved oxygen concentration in rivers, lakes and lagoons, hence becomes detrimental to aquatic life. It also causes reduction in soil alkalinity and inhibition of seed germination. Different physicochemical and biological methods have been investigated for the treatment of distillery effluents. In recent years, increasing attentions has been directed towards biological wastewater treatment methods. Bioremediation of wastewater using microorganisms is efficient and cost effective method. Microorganisms as bacteria, fungi, and algae have been shown to exhibit bioremediation activities mainly due to their production of complex and non-specific enzymatic systems capable of degrading various forms of pollutants from wastewater. The main concern of the present review is also to explore the role of microorganisms in wastewater treatment disposed from distilleries. Further, the mechanisms of color removal by fungi, bacteria and algae have also been incorporated.

Keywords: Microbial; Physicochemical; Treatment; Wastewater

Introduction

Distilleries can be categorized among the most polluting industries generating large volume of wastewater known as spent wash [1]. Distilleries generate wastewater at various stages in the manufacturing process as distillation, condenser cooling, fermenter cooling, fermentation and washing stages. Larger amount of the effluent is produced at distillation and condenser cooling stages [2]. The characteristics of the wastewater generated depend on the feed stock used. Distilleries are agro-based industries, which utilize agricultural products as sugar cane juice, sugar cane molasses, sugar beet molasses, corn, wheat, cassava, rice, barley as raw materials [3,4].

Distillery effluent is characterized by its acidic pH (4-5), dark brown color, high temperature (53-100°C), low dissolved oxygen (DO), high biochemical oxygen demand (BOD) (40,000-50,000 mg/L) and chemical oxygen demand (COD) (80,000-100,000 mg/L) [5]. Apart from the high BOD and COD load, distillery effluent also contains significant amount of phenols (7,202 mg/L), chlorides (7,997 mg/L), sulphates (1,100 mg/L), nitrates, phosphates (1,625 mg/L) and heavy metals [6,7]. The dark brown color of the effluent is mainly due to the formation of polymer melanoidin by a non-enzymatic browning reaction called Maillard reaction [8]. Melanoidins are highly recalcitrant and have antioxidant properties which make them toxic to many microorganisms [9].

Distillery wastewater disposed to the environment prior to treatment is hazardous and can be a major source of soil and water pollution. It induces toxic substances into water bodies as rivers, lakes and lagoons which adversely affect aquatic plants and animals. The highly colored nature of the effluent also leads to the reduction of sunlight penetration in rivers, lakes or lagoons which, in turn, reduces oxygenation of the water by photosynthesis and hence becomes detrimental to aquatic life [10]. Disposal of distillery wastewater into soil is equally harmful, as it reduces soil alkalinity and manganese availability [11]. It also imparts high concentration of heavy metals viz., copper, nickel, silver, cadmium, iron and mercury which are capable of inhibiting seed germination and seeding growth [12,13]. According to various reports, application of distillery effluent for irrigation without proper monitoring might result in reduction of soil fertility by suppressing the activity of soil microorganisms as nitrogen fixing bacteria *rhizobium* and *azotobacter* [14,15].

Distillery effluent must be treated before it is disposed into the environment which helps to minimize the adverse effect posed by the effluent. There have been several treatment technologies explored so far for the reduction of the pollutants from distillery wastewater. Treatment methods can vary based on the chemical composition of the effluent as well as economic viability of the technology. Generally, wastewater treatment methods can be categorized as physical, chemical and biological methods. Physicochemical treatment methods such as adsorption, sedimentation, screening, coagulation, pH adjustment, reverse osmosis, ultrafiltration, flotation, oxidation, electrolysis, membrane filtration and evaporation have been used for treatment of distillery effluents.

Physicochemical methods of wastewater treatment have so many drawbacks such as consumption of chemicals, high cost, large amount of sludge left after treatment, and possible formation of harmful by-products [16]. As a result of this, in recent years, biological wastewater treatment using microorganisms has attracted the attention of researchers all over the world. Microbial degradation and decolorization of distillery effluents have been found as cost effective and environmental friendly alternative to physicochemical methods. Various types of microorganisms as bacteria, fungi, and algae have been reported for their potential in degradation and decolorization of various

*Corresponding author: Terefe Tafese Bezuneh, College of Natural Science, Arbaminch University, Arbaminch, Ethiopia
E-mail: teretafe@gmail.com (or) teretafe2007@gmail.com

industrial effluents including that of distilleries. Hence, this review discusses and summarizes the role of microorganisms in degradation and decolorization of distillery wastewater. Moreover, the mechanisms of microbial degradation of melanoidin by fungal, bacterial and algal systems are also discussed.

The Role of Microorganisms in Effluent Treatment

Microorganisms play a key role in bioremediation process and have been proven as an efficient, low cost and environmental friendly alternative to physicochemical methods. Several microbial species including fungi, bacteria and algae have been studied for their capacity to degrade and decolorize toxic chemical pollutants present in various industrial wastewater including distilleries. Free or immobilized cells have been studied widely for bioremediation of distillery wastewater. Immobilizing microorganism on inert support material including alginate, polyacrylamide, agar, polystyrene, and polyurethane is more advantageous compared to that of free-cell. Some of the advantages include compact physical structure of carrier pellets, high biomass retention, reusability of culture and easier separation process [17,18]. The potential of microorganisms in distillery wastewater treatment is highly dependent on the type of chemical composition of wastewater, nutrient, pH, temperature, oxygen and inoculum size [19,20].

Bacterial treatment

A wide variety of bacterial cultures as *Pseudomonas aeruginosa*, *Pseudomonas putida*, *Lactobacillus plantarum*, *Bacillus circulans*, *Bacillus megaterium*, *Bacillus firmus*, *Bacillus thuringiensis*, *Bacillus cereus*, *Lactobacillus hilgardii*, *Lactobacillus coryniformis*, *Xanthomonas fragariae* have been reported for their activity in degradation and decolorization of pollutants from distillery effluents. Table 1 presents some of the bacterial cultures involved in biodegradation of distillery wastewater.

A wide variety of aerobic or anaerobic bacterial strains have been involved in bioremediation of distillery wastewater. However, a large number of bacterial species including *Bacillus* sp., *Pseudomonas* sp., *Alcaligenes* sp. and acetogenic bacteria operates effectively under aerobic conditions. Tiwari et al. [21] isolated thermotolerant bacterial culture comprised of *Bacillus subtilis*, *Bacillus cereus* and *Pseudomonas aeruginosa* from soil contaminated with distillery wastewater. Among which *Bacillus subtilis* showed maximum decolorization 85% at 45°C in the presence of little amount of carbon (0.1%, w/v) and nitrogen sources (0.1%, w/v) within a very short incubation period 24 hr. *Bacillus cereus* and *Pseudomonas* sp. showed 73 and 69% decolorization, respectively under optimum conditions. *Bacillus subtilis* showed best thermotolerance ability and could tolerate 35-50°C without affecting exponential growth phase. According to reports from different investigations the genus *Bacillus* showed the highest bioremediation efficiency compared to other bacterial cultures. Various strains of *Bacillus* sp. showed an average decolorization 75-81%, COD 80-85% and BOD 85-95% removal efficiency whereas other species as *Alcaligenes* sp., *Pseudomonas* sp. and acetogenic bacteria removed color by about 50-78%, COD 62-76% and BOD 70-82% under optimum conditions [1,6,22-27].

Aerobic bacterial strains are very effective in bioremediating distillery effluents under aerobic conditions. However, those bacterial strains are not economical due to high energy consumption for aeration thus; it was very difficult to apply those bacterial strains on an industrial scale. Considering this problem it is important to isolate bacterial strains that can degrade distillery wastewater under anaerobic condition. Anaerobic bacterial strains are advantageous than that of aerobic strains due to low energy consumption hence, minimizes cost of wastewater treatment. Ohmomo et al. [28] reported the first bacterial strain *Lactobacillus hilgardii* W-NS capable of decolorizing molasses melanoidins under anaerobic condition. This bacterial strain decolorized about 28% of molasses melanoidin under

Wastewater Composition	Microorganism	Treatment Process	Initial COD Load (g/L)	COD Removal Efficiency (%)	Treatment Time (days)	Ref
Color (dark brown), pH (2-4), COD (146.38 g/L), TDS (8.9 g/L), sulphate (1.1 g/L), phosphorus (5.1 g/L), free chlorine (5.84 g/L)	*Pseudomonas sp.*	A pure culture of the isolate was transferred in medium with 10% diluted spent wash and incubated at room temperature.	146.4	63	3	[6]
pH (7.6), COD (12.1 g/L), BOD (6.88 g/L), TS (17.8 g/L), nitrogen (0.98 g/L), phosphorus (0.38 g/L)	*B. cereus*	Each bacterial isolate was immobilized on sodium alginate and degradation of anaerobically digested distillery wastewater was carried in batch experiment.	30	81	2	[1]
	X. fragariae			76		
	B. megaterium			76		
Color (brown), pH (3.85), BOD (40 g/L), COD (100 g/L), TDS (38.4 g/L), TSS (105 g/L), TVA (2.8 g/L)	*B. circulans*	Bioremediation of molasses spent wash was carried in medium supplemented with glucose, yeast extract, KH_2PO_4 and $MgSO_4.7H_2O$ and kept for incubation for 15 days.	100	80.8	15	[22]
	B. megaterium			80.9		
	B. firmus			78.9		
BOD (52.5 g/L), COD (112 g/L), TS (2.1 g/L), TDS (1.985 g/L), TSS (0.07 g/L), sulphate (6.26 g/L), phosphate (0.0024 g/L), phenol (0.584 g/L)	*Mixed culture of: B. thuringiensis, B. brevis, Bacillus sp. (MTCC6506)*	The decolorization of Sucrose-Glutamate-Acid (SGA) was carried out by the mixed bacterial culture supplemented with 15% glucose under shaking flask conditions (150 rpm) at pH 7.0 and 37°C.	112	63.39	1	[47]
pH (7.5), COD (20.6 g/L), TS (29.6 g/L), TN (1.7 g/L), Total phosphorus (0.1 g/L)	*Lactobacillus L-2*	The isolated bacteria was supplemented with 10 g/L glucose and bio-remediate 12.5% diluted anaerobically digested spent wash.	20.6	57	7	[10]
pH (7.5–8), BOD (8–10 g/L), COD (45-52 g/L), TS (72.5 g/L), TSS (40.7 g/L), phosphates (1.625 g/L), sulphates (3.875 g/L), chlorides (8 g/L), phenols (7.2 g/L)	A bacterial consortium of: *P. aerugenosa* A01, *S. maltophia, P. microbilis*	The degrading and decolorizing of anaerobically treated distillery wastewater was studied using isolated bacterial consortium.	45	51	3	[7]
pH (8.2), BOD (3 g/L), COD (54 g/L), phosphorous as P (0.07 g/L), TS (45 g/L)	Acetogenic bacteria of strain No. BP103	Decolorization of molasses wastewater was carried in replacement culture system.	54	70.9	7	[65]

Note: Total Solid (TS); Biological Oxygen Demand (BOD); Chemical Oxygen Demand (COD); Total Volatile Acid (TVA); Total Suspended Solid (TSS); Total Nitrogen (TN)

Table 1: Bacterial Cultures Employed for Treatment of Distillery Effluent.

optimum condition. Another facultative anaerobic bacterial culture L-2 belonging to *Lactobacillus* showed similar decolorization of 91% for 12.5% (v/v) diluted digested spent wash in 7 days of incubation. Along with decolorization this bacterial culture also removed 56.2% COD [10]. Nakajima et al. [29] observed decolorization yield of 35.5% using bacterial strain MD-32 within 20 days of cultivation under both thermophilic and anaerobic conditions. The COD and color removal efficiency of anaerobic bacterial strains is lower than that of aerobic bacteria. Hence, it is important to isolate bacterial strains capable of degrading and decolorizing toxic chemical pollutants under anaerobic conditions.

Fungal treatment

In recent years, several fungal strains have been investigated for their ability to degrade and decolorize distillery wastewater. Table 2 presents some of the fungal cultures involved in bioremediation of distillery wastewater. One of the most studied fungi having high molasses wastewater bioremediation activity belongs to the genera of *Aspergillus*. Miranda et al. [30] studied color elimination from anaerobic-aerobically treated beet molasses spent wash using *Aspergillus niger*. The fungal culture showed COD and color removal yield of about 65 and 75%, respectively when supplemented with 10 g/L sucrose, 1.8 g/L NH_4NO_3, 1 g/L KH_2PO_4 and 0.5 g/L $MgSO_4 \cdot 7H_2O$ with an initial pH of 5. In the culture with the optimal nutrient concentration 83% of the total color removed was eliminated biologically and 17% by adsorption on the mycelium. Ohomomo et al. [31] used mycelia of a thermophilic strain *Aspergillus fumigatus* G-2-6 for batch and continuous decolorization of melanoidin solution. This strain decolorized about 75 and 70% of a molasses melanoidin solution under batch and continuous culture, respectively when the strain was cultivated on a glycerol-peptone medium at 45°C within 3 days. At the same time, about 51% of the chemical oxygen demand and 56% of the total organic carbon in the initial solution were removed. Later on they observed similar decolorization yield of 75% using autoclaved mycelium of *Aspergillus oryzae* No. Y-2-32 when it was cultivated at 35°C for 4 days on glycerol-peptone medium with shaking. The main melanoidin decolorization activity of this strain was due to the adsorption of melanoidin to mycelia. This fungal strain adsorbed lower weight fractions of melanoidin and

degree of adsorption was influenced by the kind of sugars used for cultivation [28].

White-rot fungi are among most widely exploited microorganisms because of their capacity in bioremediation of toxic compounds. They produce various forms of complex and non-specific intracellular and extracellular enzymatic system including laccases, manganese peroxidases, lignin peroxidase, and sugar oxidase involved in the degradation of various toxic pollutants [32,33]. The most widely studied white-rot fungal species in bioremediation of distillery wastewater are *Phanerochaete sp.*, *Flavodon sp.*, *Coriolus sp.* and *Trametes* sp. Among white-rot fungi the highest melanoidin decolorization in a range of 80-82% have been reported for *Coriolus* sp. No. 20 [34,35], *Coriolus versicolor* Ps4a [36], *Trametes versicolor* and *Trametes hirsuta* [37] under optimum conditions. Along with decolorization and COD removal, white rot-fungi are also effective in removing phenolic compounds from distillery wastewater. It has been reported that *Trametes pubescens* MB 89 and *Phanerochaete chrysospotium* can remove 80 and 63% total phenolic compounds from wastewater [38,39]. In another report *Flavodon flavus* removed benzo(a)pyrene a polycyclic aromatic hydrocarbons (PAHs) 68% from molasses spent wash within 5 days [18].

Algal treatment

Microalgae are unicellular microorganisms that are known for their capacity in bioadsorption and biodegradation of toxic chemical pollutants as phenols, heavy metals, pesticides, polycyclic aromatic hydrocarbons (PAHs), xenobiotics and melanoidins from wastewater [40,41]. Utilizing microalgae for bioremediation purpose is advantageous compared to that of bacterial and fungal systems in many ways. The first advantage is that, microalgae has a great potential in utilizing contaminants as ammonium, nitrate and phosphate as a nutrient hence minimizes the amount of externally added nutrient in case of fungi and bacteria. Secondly fungi and bacteria require optimum condition for growth and bioremediation activity whereas microalgae can grow rapidly and adapt harsh conditions. Thirdly microalgae produces valuable products as ethanol, methane, livestock feed, or it can also be used as organic fertilizer due to its high N:P ratio [42]. Hence, utilizing microalgae for bioremediation purpose

Wastewater Composition	Microorganism	reatment Process	Initial COD Load (g/L)	COD Removal Efficiency (%)	Treatment Time (days)	Ref
Color (greenish dark brown), pH (7.2), BOD (5-6.5 g/L), COD (34.8 g/L), TDS (4.5-4.62 g/L), sulphates (0.16 g/L), free chlorine (0.8 g/L)	C. cladosporioides	The fungus was used in a batch experiment to decolorize 100 mL of 15 % diluted ADSW cultivation medium supplemented with carbon and nitrogen source.	34.8	62.5	10	[64]
pH (5.2), COD (80.5 g/L), TS (109 g/L), TSS (3.6 g/L), sulfates (5 g/L), Total phenols (0.45 g/L)	Penicillium sp. P. decumbens	Aerobic degradation of beet molasses alcoholic fermentation wastewater diluted to 50%.	52.1 50.7	57	5	[61]
Color (brown), pH (3.85), BOD (40 g/L), COD (100 g/L), TDS (38.4 g/L), TSS (105.2 g/L), TVA (2.8 g/L)	A. fumigatus	Bioremediation of molasses spent wash was carried in medium supplemented with glucose, yeast extract, KH_2PO_4 and $MgSO_4 \cdot 7H_2O$ and with fungi and kept for incubation for 15 days.	100	84.0	15	[22]
	A. terreus			84.0		
Color (dark-brown), pH (4.1), COD (0.06 g/L), TS (52.4 g/L), SS (12.8 g/L)	Trametes sp. I-62	20% (v/v) of distillery effluent was added to the culture medium and incubated for 7 days at 28°C under sterile condition.	0.06	61.7	7	[32]
NR	P. chrysosporium	The fungus was immobilized on different support materials, such as polyurethane foam (PUF) and scouring web (SW), in rotating biological contactor (RBC).	NR	48	17	[39]
Color (dark brown), pH (4.3), BOD (42 g/L), COD (80 g/L)	F. flavus	The isolated fungi was immobilized on polyurethane foam cube and decolorized 10% diluted molasses wastewater.	80	50	5	[18]
Total phenols (0.54 g/L), pH (3.9), COD (25.5 g/L)	T. pubescens MB 89	The isolated fungus was used in flask cultures and a bubble lift bioreactor to treat 10% diluted wastewater.	25.5	79	15	[33]

Note: Total solid (TS); Biological Oxygen Demand (BOD); Chemical Oxygen Demand (COD); Total Volatile Acid (TVA); Total Suspended Solid (TSS); Anaerobically Digested Spent Wash (ADSW); Suspended Solids; (SS); Not Recorded (NR)

Table 2: Fungal Cultures Employed for Treatment of Distillery Effluent.

is advantageous compared to fungal and bacterial system. Various species of microalgae as *Chlorella vulgaris, Oscillatoria boryana, Chlorella pyrenoidosa, Chlorella sorokiniana, Coenochloris pyrenoidosa, Nostoc muscorum, Neochloris oleoabundans, Phormidium valderianum, Chlorella zofingiensis,* and *Chlorella ellipsoidea* have been used in bioremediation of wastewater.

The green microalgae belonged to the genera of *Chlorella* have been studied most widely due to its capacity of bioremediating toxic chemicals pollutants. Valderrama et al. [43] carried out research to develop a procedure for treatment of recalcitrant wastewater from ethanol and citric acid production using first the microalga *Chlorella vulgaris* followed by the macrophyte *Lemna minuscula*. In the first stage of treatment, *Chlorella vulgaris* resulted in a reduction of ammonium ion 71.6%, phosphorus 28% and chemical oxygen demand 61% from 10% diluted wastewater within 4 days of treatment. Travieso et al. [44] evaluated the performance of a laboratory-scale microalgae pond for secondary treatment of distillery wastewater previously digested in an anaerobic fixed bed reactor using *Chlorella vulgaris* SR/2. *Chlorella vulgaris* SR/2 removed volatile suspended solids (VSS) 78.8%, total solids (TS) 60.6%, total suspended solids (TSS) 53.4%, chemical oxygen demand (COD) 83.2% and biochemical oxygen demand (BOD) 88.0% from the effluent. More recently, Solovchenko et al. [45] investigated phycoremediation of alcohol distillery wastewater with a novel *Chlorella sorokiniana* strain isolated from White Sea. This algal strain showed maximum reduction in chemical oxygen demand (COD) 92.5%, nitrate 95%, phosphate 77% and sulfate 35% within four days. Another marine cyanobacterium *Oscillatoria boryana* decolorized pure melanoidin pigment (0.1%) and crude pigment in the distillery effluent (5%) by about 75% and 60%, respectively, within 30 days of treatment [46].

Mixed culture treatment

Several researchers studied the efficiency of mixed culture microorganisms for degradation and decolorization of distillery wastewater. The mixed microbial cultures exhibited increase in mineralization of effluents over that showed by individual cultures. This might be due to the enhanced effect of coordinated metabolic interactions present in mixed community [47,48]. Bharagava et al. [49] observed enhanced growth, enzyme production and melanoidin degradation by mixed bacterial culture compared to axenic bacterial culture. In that report a mixed consortium comprised of *Bacillus licheniformis, Bacillus* sp. and *Alcaligenes* sp. showed melanoidin decolorization of about 73.79 and 69.83% for synthetic and natural melanoidins whereas axenic cultures decolorized 65.88, 62.56 and 66.10% synthetic and 52.69, 48.92 and 59.64% natural melanoidins, respectively. In another report, a mixed bacterial culture comprised of *Bacillus thuringiensis, Bacillus brevis* and *Bacillus* sp. exhibited two-to four fold increase in melanoidin decolorisation over that showed by any individual *Bacillus* isolate [47]. Pant and Adholeya [48] developed a novel fungal consortium comprised of *Penicillium pinophilum, Alternaria gaisen, Aspergillus flavus, Fusarium verticillioides, Aspergillus niger* and *Pleurotus florida* for decolorization of distillery effluent using agricultural residues as a growth substrate. The fungal consortium was run on a bioreactor with undiluted distillery effluent for 40 days. In the first 14 days, 61.5% color and 65.4% COD removal was achieved.

Mechanism of Melanoidin Decolorization by Microorganisms

The mechanism of microbial degradation of melanoidin is difficult to inspect since its chemical structure is yet to be fully discovered. However, several studies revealed the role of microorganisms in degradation and decolorization of melanoidin from distillery wastewater. Different mechanisms of color removal from distillery effluent have been inspected. Melanoidin removal by microorganisms can take place through enzymatic degradation, utilizing the pigment as a carbon and nitrogen source, flocculation by microbially secreted substances, and adsorption onto the surface of living (resting) and dead (autoclaved) cells [6,48,50,51]. Various forms of intracellular and extracellular enzymes as laccases, manganese peroxidases, lignin peroxidase, sugar oxidases such as sorbose oxidase have been reported to show melanoidin degradation activity [37,52].

The melanoidin decolorization mechanism by adsorption onto the surface of both living (resting) and dead (autoclaved) cells have been reported for *Rhizoctonia* sp. D-90 [53], *Aspergillus fumigatus* G-2-6 [31], *Aspergillus oryzae* Y-2-32 [28], *Coriolus* No.20 [35], *Coriolus versicolor* Ps4a [36,54], acetogenic bacteria BP103 [27] and *Lactobacillus plantarum* No. PV71-1861 [55]. The melanoidin decolorization mechanism involved adsorption onto mycelium first, then incorporated into the cell and then decomposed by intracellular enzyme which require active oxygen molecules and sugar in reaction mixture [35,36,54]. Watanabe et al. [34] purified enzymes from *Coriolus* sp. No.20 which was identified as sorbose oxidase and involved in melanoidin decolorization activity. It was suggested that melanoidins were decolorized by the active oxygen such as hydrogen peroxide (H_2O_2) species produced by the enzymatic oxidation reaction with sugar oxidase (L-sorbose oxidase and Glucose oxidase) in the presence of sugar such as glucose, maltose, sucrose, lactose, sorbose, galactose and xylose as a substrate [34,56].

Microorganisms particularly white-rot fungi produces various forms of nonspecific extracellular enzyme including H_2O_2, lactase and oxidases namely lignin peroxidases (LiP) and manganese peroxidase (MnP). Lignin peroxidase (LiP) catalyzes the oxidative degradation of lignin by H_2O_2. Both lignin peroxidases (LiP) and manganese peroxidase (MnP) oxidizes a variety of substrates including Mn^{2+}, phenolic and non-phenolic compounds and different types of dyes [57]. LiP and MnP differ in their catalytic mechanism where the former catalyzes one-electron oxidation of phenolic and non-phenolic compounds by H_2O_2, promoting production of the corresponding free radicals while the latter catalyzes H_2O_2 dependent oxidation of Mn(II) to Mn(III) and the oxidized Mn(III) then catalyzes one-electron oxidation of phenolic and non-phenolic compounds by H_2O_2, promoting production of the corresponding free radicals [49,58]. The free radicals generated by the two mechanisms are responsible for the degradation of wide variety of pollutants including melanoidins. Production of H_2O_2, laccase, manganese-dependent peroxidase (MnP) and lignin peroxidase (LiP) have been reported in several fungal, bacterial and algal species as *Bacillus licheniformis, Alcaligenes* sp., *Penicillium pinophilum, Alternaria gaisen, Coriolus hirsutus, Emericella nidulans, Flavodon flavus, Oscillatoria boryana* BDU 92181 and *Neurospora intermedia* [18,46,49,56,59,60].

Conclusion

Microorganisms as fungi, bacteria and algae plays key role in bioremediating toxic pollutants from distillery wastewater for safe disposal. There are many reposts showing the activity of microorganism in biodegrading and decolorizing distillery wastewater. However, large number of reports on fungal, bacterial and algal treatment has been limited to laboratory-scale experiments. The application of the process to full-scale was still inconvenient due to lack of stability, nutrient supplement, long growth cycle, spore formation, loss of extracellular enzymes and lack of an appropriate reactor system. Thus, application of the process to field scale would need further research. Moreover, it is also found necessary to isolate, characterize and genetically improve microbes for better bioremediation yield [61-65].

Acknowledgements

Dedicated to my wife Hirut Mamo and my daughter Obse Terefe.

References

1. Jain N, Minocha AK, Verma CL (2002) Degradation of predigested distillery effluent by isolated bacterial strains. Indian J Exp Biol 40: 101-105.

2. Satyawali Y, Balakrishnan M (2008) Wastewater treatment in molasses-based alcohol distilleries for COD and color removal: A review. J Environ Manage 86: 481-497.

3. Wilkie AC, Riedesel KJ, Owens JM (2000) Stillage characterization and anaerobic treatment of ethanol stillage from conventional and cellulosic feedstock. Biomass and Bioenerg 19: 63-102.

4. Kawa-Rygielska J, Chmielewska J, Pląskowska E (2007) Effect of raw material quality on fermentation activity of distillery yeast. Polish Journal of Food and Nutrition Science 57: 275-279.

5. Kumar GS, Gupta SK, Singh G (2007) Biodegradation of distillery spent wash in anaerobic hybrid reactor. Water research 41: 721-730.

6. Chavan MN, Kulkarni MV, Zope VP, Mahulikar PP (2006) Microbial degradation of melanoidins in distillery spent wash by an indigenous isolate. Indian J Biotechnol 5: 416-421.

7. Mohana S, Desai C, Madamwar D (2007) Biodegradation and decolorization of anaerobically treated distillery spent wash by a novel bacterial consortium. Bioresource Technol 98: 333-339.

8. Martins SI, Van Boekel MA (2005) A kinetic model for the glucose/glycine Maillard reaction pathways. Food Chem 00: 257-260.

9. Kitts DD, Wu CH, Stich HF, Powrie WD (1993) Effect of glucose glycine millard reaction products on bacterial and mammalian cells mutagenesis. J Agr Food Chem 41: 2353-2358.

10. Kumar V, Wati L, FitzGibbon FJ, Nigam P, Banat IM, et al. (1997) Bioremediation and decolourisation of anaerobically digested distillery spent wash. Biotechnol Lett 19: 311-313.

11. Agarwal CS, Pandey GS (1994) Soil pollution by spent wash discharge:depletion of manganese (II) and impairment of its oxidation. J Environ Biol 15: 49-53.

12. Pandey SN, Nautiyal BD, Sharma CP (2008) Pollution level in distillery effluent and its phytotoxic effect on seed germination and early growth of maize and rice. J Environ Biol 29: 267-270.

13. Ramana S, Biswas AK, Kundu S, Saha JK, Yadava RBR (2002) Effect of distillery effluent on seed germination in some vegetable crops. Bioresource Technol 82: 273-275.

14. Ilic SZ, Mirecki N, Trajkovic R, Kapoulas N, Milenkovic L, et al. (2015) Effect of Pb on germination of different seed and His translocation in bean seed tissues during sprouting. Fresen Environ Bull 24: 670- 675.

15. Srinivas J, Purushotham AV, Murali K (2013) The effects of heavy metals on seed germination and plant growth on Coccinia, Mentha and Trigonella plant seeds in timmapuram, EG District, Andhra Pradesh, India. International Research Journal of Environmental Science 2: 20-24.

16. Boopathy MA, Senthilkumar SNS (2014) Media optimization for the decolorization of distillery spent wash by biological treatment using Pseudomonas fluorescence. International Journal of Innovations in Engineering and Technology 4: 1-8.

17. Sankaran K, Pisharody L, Narayanan GS, Premalatha M (2015) Bacterial assisted treatment of anaerobically digested distillery wastewater. Royal Society of Chemistry Advances 5: 70977-70984.

18. Raghukumar C, Mohandass C, Kamat S, Shailaja MS (2004) Simultaneous detoxification and decolorization of molasses spent wash by the immobilized white-rot fungus Flavodon flavus isolated from a marine habitat. Enzyme and Microbial Technology 35: 197-202.

19. Gay M, Cerf O, Davey K R (1996) Significance of pre-incubation temperature and inoculum concentration onsubsequent growth of Listeria monocytogenes at 14°C. Journal of Applied Bacteriology 81: 433-438.

20. Surega P, Krzywonos M (2015) Screening of medium components and process parameters for sugar beet molasses vinasse decolorization by Lactobacillus plantarum using placket-burman experimental design. Polish Journal of Environmental Study 24: 683-688.

21. Tiwari S, Gaur R, Rai P, Tripathi A (2012) Decolorization of distillery effluent by thermotolerant Bacillus subtilis. American Journal of Applied Science 9: 798-806.

22. Kumar MA, Abdullah SS, Kumar PS, Dheeba B, Mathumitha C (2008) Comparitive study on potentiality of bacteria and fungi in bioremediation and decolorization of molasses spent wash. Journal of Pure and Applied Microbiology 2: 393-400.

23. Singh KD, Sharma S, Dwivedi A, Pandey P, Thakur RL, et al. (2007) Microbial decolorization and bioremediation of melanoidin containing molasses spent wash. Journal of Environmental Biology 28: 675-677.

24. Sankaran K, Divakar S, Nagarajan SM, Vadanasundari V (2011) Analysis on biodegradation and color reduction of distillery effluent spent wash. International Journal of Recent Scientific Research 2: 4-9.

25. Rani A, Saharan BS (2009) Optimization of cultural conditions for anaerobically treated distillery effluent bioremediation by an isolate Pseudomonas putida SAG$_{45}$. Journal of Applied and Natural Science 1: 132-137.

26. Dahiya J, Singh D, Nigam P (2001) Decolorization of synthetic and spentwash melanoidins using the white-rot fungus Phanerochaete chrysosporium JAG-40. Bioresource Technol 78: 95-98.

27. Sirianuntapiboon S, Phothilangka P, Ohmomo S (2004) Decolorization of molasses wastewater by a strain No. BP103 of acetogenic bacteria. Bioresource Technol 92: 31-39.

28. Ohmoho S, Kainuma M, Kamimura K, Sirianuntapiboon S, Aoshima I, et al. (1988) Adsorption of melanoidin to the mycelia of Aspergillus oryzae Y-2-32. Agr Biol Chem Tokyo 52: 381-386.

29. Nakajima-Kambe T, Shimomura M, Nomura N, Chanpornpong T, Nakahara T (1999) Decolorization of molasses wastewater by Bacillus sp. under thermophilic and anaerobic conditions. Journal of Bioscience and Bioengineering 87: 119-121.

30. Miranda MP, Gonzalez G, Cristobal NS, Nito CH (1996) Color elimination from molasses wastewater by Aspergillus niger. Bioresource Technol 57: 229-235.

31. Ohmoho S, Kaneko Y, Sirianuntapiboon, Somchai P, Atthasampunna, et al. (1987) Decolorization of molasses wastewater by a thermophilic strain, Aspergillus Fumigatus G-2-6. Agr Biol Chem Tokyo 51: 3339-3346.

32. Gonzalez T, Terron MC, Yague S, Zapico E, Galletti GC, et al. (2000) Pyrolysis/gas chromatography/mass spectrometry monitoring of fungal-biotreated distillery wastewater using Trametes sp. I-62 (CECT 20197). Rapid Commun Mass Sp 14: 1417-1424.

33. Strong PJ, Burgess JE (2007) Bioremediation of a wine distillery wastewater using white-rot fungi and the subsequent production of laccase. Water Sci Technol 56: 179-186.

34. Watanabe Y, Sugi R, Tanaka Y, Hayashida S (1982) Enzyamatic decolorization of melanoidin by Coriolus sp. No. 20. Agr Biol Chem Tokyo 46: 1623-1630.

35. Sirianuntapiboon S, Chairattanawan K (1998) Some properties of Coriolus sp. No. 20 for removal of color substances from molasses wastewater. Thammasat International Journal of Science and Technology 3: 74-79.

36. Aoshima I, Tozawa Y, Ohmoho S, Ueda K (1985) Production of decolorizing activity from molasses pigment by Cloriolus versicolor P$_s$4$_a$. Agr Biol Chem Tokyo 49: 2041-2045.

37. Couto SR, Sanroman MA, Gubitz GM (2005) Influence of redox mediators and metal ions on synthetic acid dye decolorization by crude laccase from Trametes hirsuta. Chemosphere 58: 417-422.

38. Benito GG, Miranda MP, Santos DR (1997) Decolorization of wastewater from an alcoholic fermentation process with Trametes versicolor. Bioresource Technol 61: 33-37.

39. Guimarãesa C, Porto P, Oliveira R, Mota M (2005) Continuous decolorization of a sugar refinery wastewater in a modified rotating biological contactor with Phanerochaete chrysosporium immobilized on polyurethane foam disks. Process Biochem 40: 535-540.

40. Shashirekha S, Uma L, Subramanian G (1997) Phenol degradation by the marine cyanobacterium Phormidium valderianum BDU 30501. J Ind Microbiol Biot 19: 130-133.

41. Maynard HE, Ouki SK, Williams SC (1999) Tertiary lagoons: A review of removal mechanisms and performance. Water Res 33: 1-13.

42. Mata TM, Martins AA, Caetano NS (2010) Microalgae for biodiesel production and other applications: A review. Renewable and Sustainable Energy Reviews 14: 217-232.

43. Valderrama LT, Del Campo CM, Rodriguez CM, de-Bashan LE, Bashan Y (2002) Treatment of recalcitrant wastewater from ethanol and citric acid production using the microalga *Chlorella vulgaris* and the macrophyte *Lemna minuscule*. Water Res 36: 4185-4192.

44. Travieso L, Benítez F, Sánchez E, Borja R, León M, et al. (2008) Performance of a laboratory-scale microalgae pond for secondary treatment of distillery wastewaters. Chem Biochem Eng Q 22: 467-473.

45. Solovchenko A, Pogosyan S, Chivkunova O, Selyakh I, Semenova L, et al. (2014) Phycoremediation of alcohol distillery wastewater with a novel *Chlorella sorokiniana* strain cultivated in a photobioreactor monitored on-line via chlorophyll fluorescence. Algal Research 6: 234-241.

46. Kalavathi DF, Uma L, Subramanian G (2001) Degradation and metabolization of the pigment melanoidin in distillery effluent by the marine cyanobacterium Oscillatoria boryana BDU 92181. Enzyme Microb Tech 29: 246-251.

47. Kumar P, Chandra R (2006) Decolorisation and detoxification of synthetic molasses melanoidins by individual and mixed cultures of Bacillus spp. Bioresource Technol 97: 2096-2102.

48. Pant D, Adholeya A (2010) Development of a novel fungal consortium for the treatment of molasses distillery wastewater. Environmentalist 30: 178-182.

49. Bharagava RN, Chandra R, Rai V (2009) Isolation and characterization of aerobic bacteria capable of the degradation of synthetic and natural melanoidins from distillery effluent. World J Microb Biot 25: 737-744.

50. Chandra R, Bharagava RN, Rai V, Singh SK (2009) Characterization of sucrose-glutamic acid Maillard products (SGMPs) degrading bacteria and their metabolites. Bioresource Technol 100: 6665-6668.

51. Kumar V, Wati L, Nigam P, Banat IM, Yadav BS, et al. (1998) Decolorization and biodegradation of anaerobically digested sugarcane molasses spent wash effluent from biomethanation plants by white-rot fungi. Process Biochem 33: 83-88.

52. Freitas AC, Ferreira F, Costa AM, Pereira R, Antunes SC (2009) Biological treatment of the effluent from a bleached kraft pulp mill using basidiomycete and zygomycete fungi. Sci Total Environ 407: 3282-3289.

53. Sirianuntapiboon S, Sihanonth P, Somchai P, Atthasampunna P, Hayashida (1995) An absorption mechanism for the decolorization of melanoidin by

Rhizoctonia sp. D-90. Biosci Biotech Bioch 59: 1185-1995.

54. Ohmomo S, Aoshima I, Tozawa Y, Sakurada N, Ueda K (1985) Purification and some properties of melanoidin decolorizing enzymes and P-4 from mycelia of *Coriolus versicolor* Ps4a. Agr Biol Chem Tokyo 49: 2047-2053.

55. Tondee T, Sirianuntapiboon S (2008) Decolorization of molasses wastewater by *Lactobacillus plantarum* No. PV71-1861. Bioresource Technol 99: 6258-6265.

56. Pant D, Adholeya A (2007) Identification, ligninolytic enzyme activity and decolorization potential of two fungi isolated from a distillery effluent contaminated site. Water Air Soil Poll 183: 165-176.

57. Perez-Boada M, Ruiz-Duenas FJ, Pogni R, Basosi R, Choinowski T, et al. (2005) versatile peroxidase oxidation of high redox potential aromatic compounds: site-directed mutagenesis, spectroscopic and crystallographic investigation of three long-range electron transfer pathways. Journal of Molecular Biology 354: 385-402.

58. Miyata N, Mori T, Iwahori K, Fujita M (2000) Microbial decolorization of melanoidin-containing wastewaters: combined use of activated sludge and the fungus *Coriolus hirsutus*. Journal of Bioscience and Bioengineering 89: 145-150.

59. Kaushik G, Thakur IS (2009) Isolation of fungi and optimization of process parameters for decolorization of distillery mill effluent. World J Microb Biot 25: 955-964.

60. Miyata N, Iwahori K, Fujita M (1998) Manganese-independent and dependent decolorization of melanoidin by extracellular hydrogen peroxide and peroxidases from *Coriolus hirsutus* pellets. J Ferment Bioeng 85: 550-553.

61. Jemenez AM, Borja R, Martin A (2003) Aerobic-anaerobic biodegradation of beet molasses alcoholic fermentation wastewater. Process Biochemistry 38: 1275-1284.

62. Krzywonos M, Seruga P (2012) Decolorization of sugar beet molasses vinasse, a high strength distillery wastewater, by lactic acid bacteria. Pol J Environ Stud 21: 943-948.

63. Murakami M, Ikenouchi M (1997) The biological CO_2 fixation and utilization project by rite (2)-screening and breeding of microalgae with high capability in fixing CO_2. Energ Convers Manage 38: 493-497.

64. Ravikumar R, Vasanthi NS, Saravanan K (2011) Single factorial experimental design for decolorizing anaerobically treated distillery spent wash using *Cladosporium cladosporioides*. Int J Environ Sci Te 8: 97-106.

65. Sirianuntapiboon S, Prasertsong K (2008) Treatment of molasses wastewater by acetogenic bacteria BP103 in sequencing batch reactor (SBR) system. Bioresource Technol 99: 1806-1815.

30

Heavy Metal Removal from Wastewater Using Low Cost Adsorbents

Ashutosh Tripathi* and Manju Rawat Ranjan

Amity Institute of Environmental Sciences, Amity University, Noida-125, Gautam Buddha Nagar, U.P, India

Abstract

With the onset of industrialization mankind has witnessed various environmental issues in the society. This industrialization has not only brought development and prosperity but eventually disturbed the ecosystem. One of the impacts is visible, in form of water pollution. In the present study heavy metal contamination of water bodies has been discussed. Effluents from large number of industries viz., electroplating, leather, tannery, textile, pigment & dyes, paint, wood processing, petroleum refining, photographic film production etc., contains significant amount of heavy metals in their wastewater. The conventional methods of treatment of heavy metal contamination includes chemical precipitation, chemical oxidation, ion exchange, membrane separation, reverse osmosis, electro dialysis etc. These methods are costly, energy intensive and often associated with generation of toxic byproducts. Thus, the adsorption has been investigated as a cost effective method of removal of heavy metals from wastewater. In the present study various low cost adsorbent has been reviewed as an abatement of heavy metal pollution from wastewater. These adsorbent includes materials of natural origin like zeolites, clay, peat moss and chitin are found to be an effective agent for removal of toxic heavy metals like Pb, Cd, Zn, Cu, Ni, Hg, Cr etc. Apart from these various agricultural wastes like rice husk, neem bark, black gram, waste tea; Turkish coffee, walnut shell etc. were also established as a potent adsorbent for heavy metal removal. Beside that low cost industrial by products like fly ash, blast furnace sludge, waste slurry, lignin, iron (III) hydroxide and red mud, coffee husks, Areca waste, tea factory waste, sugar beet pulp, battery industry waste, sea nodule residue and grape stalk wastes have been explored for their technical feasibility to remove toxic heavy metals from contaminated water.

Keywords: Agricultural waste; Heavy metal; Low cost adsorbent; Wastewater; Toxicity

Introduction

Water pollution caused due to addition of heavy metals resulting from the industrial activities is increasing tremendously and is a matter of global concern. Mining, mineral processing and metallurgical operations are generating effluents containing heavy metals. The heavy metals present in the wastewater is persistent and non degradable in nature. Moreover, they are soluble in aquatic environment and thus can be easily absorbed by living cells. Thus, by entering the food chain, they can be bioaccumulated and biomagnified in higher trophic levels also. The heavy metals, if absorbed above the permissible labels, could lead to serious health disorders. In light of the facts, treatment of heavy metals containing industrial effluent becomes quite necessary before being discharged into the environment. The scientists and environmental engineers are therefore facing a tough task of cost effective treatment of wastewater containing heavy metals. The conventional methods for heavy metal removal from wastewater includes chemical precipitation, chemical oxidation, ion exchange, membrane separation, reverse osmosis, electro dialysis etc. These methods are not very effective, are costly and require high energy input. They are associated with generation of toxic sludge, disposal of which renders it expensive and non ecofriendly in nature. In the recent past, number of approaches has been investigated for safe and economical treatment of heavy metal laden wastewater. Adsorption has emerged out to be better alternative treatment methods. It is said to be effective and economical because of its relatively low cost. Authors have claimed adsorption to be easiest, safest and most cost-effective methods for the treatment of waste effluents containing heavy metals [1,2]. The key benefit of adsorption method for heavy metal removal is less initial as well as operation cost, unproblematic design and less requirement of control systems [3]. Generally the heavy metals are present in the wastewater at low concentrations and adsorption is suitable even when the metal ions are present at concentrations as low as 1 mg/L. This makes adsorption an economical and favorable technology for heavy metal removal from wastewater. The adsorbent may be of mineral, organic or biological origin. It could be zeolites, industrial byproducts, agricultural waste, biomass and polymeric material. One of the conventional adsorbent, activated carbon has been extensively used in many applications. However, the high cost effectiveness of activation processes limits its usage in wastewater treatment processes. The present research activity aims toward contributing in the search for cost effective or low cost adsorbents of natural origin and their applicability in recovery as well as removal of heavy metals from the industrial wastewater.

Industrial Wastewater and Heavy Metals

Heavy metals are commonly released in the wastewater from various industries. Electroplating and surface treatment practices leads to creation of considerable quantities of wastewaters containing heavy metals (such as cadmium, zinc, lead, chromium, nickel, copper, vanadium, platinum, silver and titanium). Apart from this wastewater from leather, tannery, textile, pigment & dyes, paint, wood processing, petroleum refining industries and photographic film production contains significant amount of heavy metals. These heavy metal ions are toxic to both human beings and animals. The toxic metals cause physical discomfort and sometimes life threatening illness and irreversible damage to vital body system [4]. The metals get bioaccumulated in the auatic environment and tend to biomagnified along the food chain.

*Corresponding author: Ashutosh Tripathi, Amity Institute of Environmental Sciences, Amity University, Noida-125, Gautam Buddha Nagar, (U.P.-201303),India
E-mail: tripathiashutos@gmail.com; atripathi1@amity.edu

Thus, the organisms at higher trophic level are more susceptible to be affected by their toxicity. There are 20 metals which are almost persistant and cannot be degraded or destroyed. Mercury (Hg), lead (Pb), cadmium (Cd), chromium (Cr [VI]), Zinc (Zn), Arsenic (As), Nickel (Ni) etc., are toxic heavy metals from ecotoxicoligal point of view. The table below shows Maximum Contaminant Level (MCL) standards for some heavy metals established by USEPA [5]. These heavy metals can lead to serious effects such as stunted growth, damage to vital organs, damage to brain, cancer and in some cases death also. Health hazard related to heavy metal toxicity are not new. Human diseases like minamata, itai itai, fluorosis, Arsenicosis etc. are due to heavy metal ingestion above permissible levels. Treating the industrial effluents contaminated with heavy metals within the industrial premises before being discharged is efficient way to remove heavy metals rather than treating high volumes of wastewater in a general sewage treatment plant. Thus it is advantageous to develop separate handling modus operandi for removal of heavy metals from the industrial effluents. The current work focuses on study of natural coagulants as an effective and economical alternative treatment process for heavy metals removal from industrial wastewater. (Table 1)

Adsorption

As discussed earlier, adsorption has emerged out as effective, economical and ecofriendly treatment technique. It is a process potent enough to fulfill water reuse obligation and high effluent standards in the industries. Adsorption is basically a mass transfer process by which a substance is transferred from the liquid phase to the surface of a solid, and becomes bound by physical and/or chemical interactions [5]. It is a partition process in which few components of the liquid phase are relocated to the surface of the solid adsorbents. All adsorption methods are reliant on solid-liquid equilibrium and on mass transfer rates. The adsorption procedure can be batch, semi-batch and continuous. At molecular level, adsorption is mainly due to attractive interfaces between a surface and the group being absorbed. Depending upon the types of intermolecular attractive forces adsorption could be of following types:

Physical adsorption

It is a general incident and occurs in any solid/liquid or solid/gas system. Physical adsorption is a process in which binding of adsorbate on the adsorbent surface is caused by van der Waals forces of attraction. The electronic structure of the atom or molecule is hardly disturbed

upon physical adsorption. Van der Waals forces originate from the interactions between induced, permanent or transient electric dipoles. Physical adsorption can only be observed in the environment of low temperature and under appropriate conditions, gas phase molecules can form multilayer adsorption. Commercial adsorbents utilize physical adsorption for its surface binding.

Chemical adsorption

It is a kind of adsorption which involves a chemical reaction between the adsorbent and the adsorbate. The strong interaction between the adsorbate and the substrate surface creates new types of electronic bonds (Covalent, Ionic). Chemical adsorption is also referred as activated adsorption. The adsorbate can form a monolayer. It is utilized in catalytic operations.

In general, the main steps involved in adsorption of pollutants on solid adsorbent are:

Transport of the pollutant from bulk solution to external surface of the adsorbent.

Internal mass transfer by pore diffusion from outer surface of adsorbent to the inner surface of porous structure.

Adsorption of adsorbate on the active sites of the pores of adsorbent.

The overall rate of adsorption is decided by either film formation or intra particle diffusion or both as the last step of adsorption are rapid as compared to the remaining two steps.

Low Cost Adsorbents

The removal of heavy metals by using low cost adsorbent is found to be more encouraging in extended terms as there are several materials existing locally and profusely such as natural materials, agricultural wastes or industrial by-products which can be utilized as low-cost adsorbents [6]. To be commercially viable, an adsorbent should have high selectivity to facilitate quick separations, favorable transport and kinetic characteristics, thermal and chemical stability, mechanical strength, resistance to fouling, regeneration capacity and low solubility in the liquid in contact. Adsorption process has several advantages over the conventional methods of heavy metal removal. Some of the gains of adsorption process are: (I) Economical, (II) metal selectivity, (III) Regenerative, (IV) Absence of toxic sludge generation (V) metal recovery and most importantly (VI) effective. Various low cost adsorbent derived from various natural as well as anthropogenic sources have been implemented for treatment of waste water contaminated with heavy metals. The adsorbents mostly used are agricultural waste, industrial byproducts, natural materials or modified biopolymers.

Adsorption by Natural Materials

Zeolites

They are naturally occurring crystalline alumino silicates consisting of a skeleton of tetrahedral molecules, connected with each other by mutual oxygen atoms. Ion exchanging capacities of zeolites make them a suitable candidate for removal of heavy metals. Adsorption in zeolites is in fact a choosy and reversible packing of crystal cages, so surface area is not a significant aspect. Zeolites consist of a wide variety of species such as clinoptilolite and chabazite. Among the different zeolites, clinoptilolite has been extensively studied and was shown to have high selectivity for metals like Pb (II), Cd (II), Zn (II)

Heavy metal	Toxicity	MCL (mg/L)
Arsenic (As)	Skin manifestations, visceral cancers, vascular disease	0.050
Cadmium (Cd)	Kidney damage, renal disorder, human carcinogen	0.01
Chromium (Cr)	Headache, diarrhoea, nausea, vomiting, carcinogenic	0.05
Copper (Cu)	Liver damage, Wilson disease, Insomnia	0.25
Nickel (Ni)	Dermatitis, nausea, chronic asthma, coughing, human carcinogen	0.20
Zinc (Zn)	Depression, lethargy, neurological signs and increased thirst	0.80
Lead (Pb)	Damage the fetal brain, diseases of kidney, circulatory system and nervous system	0.006
Mercury (Hg)	Rheumatoid arthritis and disease of kidneys, circulatory and nervous system	0.00003

Table 1: The MCL standards for the most hazardous heavy metals [5].

and cu (II). Several zeolites are modified during the past few years to increase their efficiency. Clinoptilolite was found to be more effectively removing heavy metals owing to its ion exchange capability, followed by pretreatment [5,7].

Clay

There are three main groups of clays: kaolinite, montmorillonite-smectite, and mica. The montmorillonite has the highest cation exchange capacity and its recent market price is found to be 20 times cheaper as compared to activated carbon. Their heavy metals removal capacity is less as compared to zeolites but their easy availability and economical properties give back their less efficiency. Efficiency for heavy metal removal by clay could be improved by modifying them to clay-polymer composites [8-10].

Peat moss

Abundant in nature and has a very high organic content. Its large surface area ($\geq$ 200 m^2/g) and high porosity makes it an effective agent for heavy metal removal from wastewater. It was observed that peat moss plays an important role in treatment of metal-bearing industrial effluents such as Cu^{2+}, Cd^{2+}, Zn^{2+} and Ni^{2+} [11]. The adsorption capacity of sphagnum peat moss was found to be 132 mg of Cr^{6+}/g at a pH range of 1.5-3.0. The most striking benefit of this adsorbent in treatment is the easiness of the system, low cost, and the capability to acknowledge a wide variation of effluent composition [12].

Chitin: It is the second most abundant natural biopolymer followed by cellulose. Chitin is a long-chain polymer of a N-acetylglucosamine, a derivative of glucose. It is the main component of the cell walls of fungi, the exoskeletons of arthropods such as crustaceans (e.g., crabs, lobsters and shrimps) and insects, the radulas of mollusks, and the beaks and internal shells of cephalopods, including squid and octopuses. It has been used for removal of several heavy metals in the past. Currently, chitosan, which is produced by alkaline N-deacetylation of chitin, is drawing an increased amount of research interest for its heavy metal removal capability due to chelating property. It can be made by treating shrimp and other crustacean shells with the alkali sodium hydroxide. Chitosan has been used for treatment of Hg^{2+}, Cu^{2+}, Ni^{2+}, Zn^{2+}, Cr^{6+}, Cd^{2+} and Pb^{2+}.

Adsorption by Agricultural Wastes

Use of agricultural byproducts as adsorbents for heavy metal removal from industrial waste water has been increasing nowadays. Most of the studies were focused on plant wastes such as rice husk and neem bark [13,14], Black gram husk [15], Waste tea, Turkish coffee, Walnut shell [16] etc. Some more adsorbents like papaya wood [17], maize leaf [18], teak leaf powder [19], coraindrum sativum [20], lalang (Imperata cylindrica) leaf powder [21], peanut hull pellets [22], sago waste [23], saltbush (Atriplex canescens) leaves [24,25], tree fern [26-28], grape stalk wastes [29], etc. are also studied in detail. The benefits of using agricultural wastes for wastewater treatment include easy technique, needs modest processing, superior adsorption ability, selective adsorption of heavy metal ions, economical, easy availability and easy regeneration. On the other hand, the use of untreated agricultural wastes as adsorbents can also fetch a number of problems such as small adsorption ability, elevated chemical oxygen demand (COD) and biological chemical demand (BOD) as well as total organic carbon (TOC) due to discharge of soluble organic compounds contained in the plant materials [30,31]. The increase of the COD, BOD

and TOC can cause diminution of dissolved oxygen (DO) content in water and can make threats to the aquatic life. Consequently, plant wastes require to be modified or treated ahead of being applied for the cleansing of heavy metals.

New products such as jackfruit, rice husk, pecan shells, hazenut shell, maize cob or husk are also used for adsorbent for heavy metal elimination after chemical modifications. Chemically modified agricultural wastes have been found to have enhanced chelating efficiency. Wheat bran, a by-product of wheat milling industries proved to be a good adsorbent for removal of many types of heavy metal ions which eventually results in better efficiency of adsorption of copper ions as reported by O zer et al. [32]. Orange peel has been used for Ni (II) removal from simulated wastewater [33]. Similarly, Adsorption of divalent heavy metal ions particularly Cu2+, Zn2+, Co2+, Ni2+ and Pb2+ onto acid and alkali treated banana and orange peels was performed by Annadurai et al. in 2002 [34]. Activated Coconut shell carbon powder (ACSCP) and Activated charcoal powder (ACP) is used as adsorbent for removal of Lead from electrochemical industry effluent [35]. There are several other examples of chemically modified agricultural wastes also. Moreover, factors like pH, temperature, contact period, initial concentration of metal, agitation rate, dosage of adsorbent etc. affects the adsorption capacity [36].

Adsorption by Industrial Wastes

Various industrial wastes have also got adsorption capacity and can be used for adsorbing heavy metals from wastewater. These industrial wastes are produced as a by-product and are used rarely for any purpose. The by-product nature renders it to be easily available and very economical also. These industrial wastes are found to have good application as adsorbent. Adsorptive capacity of these wastes could be increased followed by slight processing. Industrial by-products such as fly ash [37,38], blast furnace sludge [39,40], waste slurry, lignin-a black liquor waste of paper industry [41,42,43], iron (III) hydroxide [44,45] and red mud [46,47] have been explored for their technical feasibility to remove toxic heavy metals from contaminated water. Other industrial wastes, coffee husks [48], Areca waste [49], tea factory waste [50], sugar beet pulp [51], waste pomace of olive oil factory waste [52], battery industry waste, waste biogas residual slurry [53], sea nodule residue [54] and grape stalk wastes [29] have been utilized as low-cost adsorbents for the removal of toxic heavy metals from wastewater. Several adsorbents have been used for adsorption of Zinc from waste water. Some of the highest adsorption capacities reported for Zn2+ are 168 mg/g powdered waste sludge, 128.8 mg/g dried marine green macroalgae, 73.2mg/g lignin, 55.82mg/g cassava waste, and 52.91mg/g bentonite [55].

Conclusion

The recent worldwide trend to achieve higher environmental standards favors the usage of low cost systems for treatment of effluents. In the meantime various low cost adsorbent derived from agricultural waste or natural products have been extensively investigated for heavy metal removal from contaminated wastewater. It has been found that after chemical or thermal modifications, agricultural waste exhibited tremendous heavy metal removal capability. Concentration of adsorbate, extent of surface modification and adsorbent characteristics are the factors responsible for metal adsorption capability. Cost effectiveness and technical applicability are the two important key factors for selecting effective low cost adsorbent for heavy metal removal.

Acknowledgement

Authors would like to acknowledge the motivation and facilities provided by Amity University Uttar Pradesh to carry out the present work.

References

1. Shah BA, Shah AV, Singh RR (2009) Sorption isotherms and kinetics of chromium uptake from wastewater using natural sorbent material. International Journal of Environmental Science and Technology 6: 77-90.

2. Rahmani K, Mahvi AH, Vaezi F, Mesdaghinia AR, Nabizade R, et al (2009) Bioremoval of lead by use of waste activated sludge. International Journal of Environmental Research, 3: 471-476.

3. Acar FN, Eren Z (2006) Removal of Cu (II) ions by activated poplar sawdust (Samsun Clone) from aqueous solutions. J Hazard Mater B 137: 909–914.

4. Malik A (2004) Metal bioremediation through growing cells. Environmental International, 30: 261-278.

5. Babel S, Kurniawan TA (2003) Various treatment technologies to remove arsenic and mercury from contaminated groundwater: an overview. In: Proceedings of the First International Symposium on Southeast Asian Water Environment, Bangkok, Thailand, 24-25 October: 433-440.

6. Siti Nur AA, Mohd Halim SI, Lias Kamal Md, Shamsul Izhar (2013) Adsorption Process of Heavy Metals by Low-Cost Adsorbent: A Review. World Applied Sciences Journal 28: 1518-1530.

7. Bose P, Bose MA, Kumar S (2002) Critical evaluation of treatment strategies involving adsorption and chelation for wastewater containing copper, zinc, and cyanide. Adv Environ Res 7: 179-195.

8. Vengris T, Binkiene R, Sveikauskaite A (2001) Nickel, copper, and zinc removal from wastewater by a modified clay sorbent. Appl Clay Sci 18:183-90.

9. Solenera M, Tunalib S, O¨zcan ,A S, O¨zcanc A, Gedikbey T (2008) Adsorption characteristics of lead (II) ions onto the clay/ poly(methoxyethyl)acrylamide (PMEA) composite from aqueous solutions. Desalination 223: 308-322.

10. Abu-Eishah SI (2008) Removal of Zn, Cd, and Pb ions from water by Sarooj clay. Appl Clay Sci 42 : 201-205

11. Gosset T, Trancart JL, Thevenot DR (1986) Batch Metal removal by peat Kinetics and thermodynamics. Water Res 20: 21-26.

12. Sharma DC, Forster CF (1993) Removal of Hexavalent Chromium using Sphagnum moss peat. Water Res 27: 1201-1208.

13. El-Said AG, Badawy NA, Garamon SE (2012) Adsorption of Cadmium (II) and Mercury (II) onto Natural Adsorbent Rice Husk Ash (RHA) from Aqueous Solutions: Study in Single and Binary System, International Journal of Chemistry 2012: 58-68.

14. Bhattacharya AK, Mandal SN, Das SK (2006) Adsorption of Zn(II) from aqueous solution by using different adsorbents. Chem Eng J 123: 43-51.

15. Saeed A, Iqbal M (2003) Bioremoval of cadmium from aqueous solution by black gram husk (Cicer arientinum). Water Res 37: 3472-3480.

16. Orhan Y, Büyükgüngör H (1993) The removal of heavy metals by using agricultural wastes. Water Sci Technol 28(2): 247-255.

17. Saeed A, Iqbal M, Akhtar MW (2005) Removal and recovery of heavy metals rom aqueous solution using papaya wood as a new biosorbent. Sep. Purif Technol 45: 25-31.

18. Babarinde NAA, Oyebamiji Babalola J, Adebowale Sanni R (2006) Biosorption of lead ions from aqueous solution by maize leaf. Int. J Phys Science 1: 23-26.

19. King P, Srivinas P, Prasanna Kumar Y, Prasad VSRK (2006) Sorption of copper (II) ion from aqueous solution by Tectona grandis l.f. (teak leaves powder). J Hazard Mater B136: 560-566.

20. Karunasagar D, Balarama Krishna MV, Rao SV, Arunachalam J (2005) Removal of preconcentration of inorganic and methyl mercury from aqueous media using a sorbent prepared from the plant Coriandrum sativum. J. Hazard Mater B 118: 133-139.

21. Hanafiah MAK, Ngah WSW, Zakaria H, Ibrahim SC (2007) Batch study of liquid-phase adsorption of lead ions using Lalang (Imperata cylindrica) leaf powder. J Biol Sci 7: 222-230.

22. Johnson PD, Watson MA, Brown J, Jefcoat IA (2002) Peanut hull pellets as a single use sorbent for the capture of Cu (II) from wastewater. Waste Manage 22: 471-480.

23. Quek SY, Wase DAJ, Forster CF (1998) The use of sago waste for the sorption of lead and copper. Water SA 24: 251-256.

24. Sawalha MF, Peralta-Videa JR, Romero-Gonza´lez J, Duarte-Gardea M, Gardea-Torresdey JL (2007) Thermodynamic and isotherm studies of the biosorption of Cu(II), Pb(II), and Zn(II) by leaves of saltbush(Atriplex canescens). J Chem. Thermodyn 39: 488-492.

25. Sawalha MF, Peralta-Videa JR, Romero-Gonza´lez J, Gardea-Torresdey JL (2007) Biosorption of Cd (II), Cr (III), and Cr(VI) by saltbush (Atriplex canescens) biomass: thermodynamic and isotherm studies. J Colloid Interface Sci 300: 100-104.

26. Ho YS, Wang CC (2004) Pseudo-isotherms for the sorption of cadmium ion onto tree fern. Process Biochem. 39: 759-763.

27. Ho YS, Chiu WT, Hsu CS, Huang CT (2004) Sorption of lead ions from aqueous solution using tree fern as a sorbent. Hydrometallurgy. 73: 55-61.

28. Ho YS (2003) Removal of copper ions from aqueous solution by tree fern. Water Res 37: 2323-2330.

29. Villaescusa I, Fiol N, Martínez N, Miralles N, Poch J, et al. (2004) Removal of copper and nickel ions from aqueous solutions by grape stalks wastes. Water Research. 38: 992-1002.

30. Gaballah I, Goy D, Allain E, Kilbertus G, Thauront J (1997) Recovery of copper through decontamination of synthetic solutions using modified barks. Met Metall Trans.B 28: 13-23.

31. Nakajima A, Sakaguchi T (1990) Recovery and removal of uranium by using plant wastes. Biomass 21: 55-63.

32. Ozer A, Ozer D (2004) The adsorption of copper (II) ions onto dehydrated wheat bran (DWB): determination of equilibrium and thermodynamic parameters. Process Biochem 39: 2183-2191.

33. Ajmal M, Rao RAK, Ahmad R, Ahmad J (2000) Adsorption studies on Citrus reticulata (fruit peel of orange) removal and recovery of Ni (II) from electroplating wastewater. J Hazard Mater 79: 117-31.

34. Annadurai G, Juang HS, Lee DJ (2002) Adsorption of heavy metal from water using banana and orange peels. Water Sci Technol 47: 185-190.

35. Nishigandha JB, Suryavanshi AA, Tirthakar SN (2015) Removal of heavy metal lead (pb) from electrochemical industry waste water using low cost adsorbent. Int J Research in Engineering and Technology 04: 731-733.

36. Parmar M, Thakur Lokendra Singh (2013) Heavy metal Cu, Ni and Zn: toxicity, health hazards and their removal techniques by low cost adsorbents: a short overview. Int J Plant Animal & Env. Science 3: 143-147.

37. Bayat B (2002) Combined removal of zinc (II) and cadmium (II) from aqueous solutions by adsorption onto high-calcium Turkish Fly Ash. Water Air Soil Pollut. 136: 69-92.

38. Wang SB, Li L, Zhu ZH (2007) Solid-state conversion of fly ash to effective adsorbents for Cu removal from wastewater. J Hazard. Mater 139: 254-259.

39. Dimitrova SV (1996) Metal sorption on blast-furnace slag. Water Res. 30: 228-232.

40. Srivastava S, Gupta V, Mohan D (1997) Removal of Lead and Chromium by Activated Slag-A Blast-Furnace Waste. J Environ Eng, 123: 461-468.

41. Suhas, Carrott PJM, Ribeiro Carrott MML (2007) Lignin-from natural adsorbent to activated carbon: A review Bioresour Technol 98: 2301-2312.

42. Srivastava SK, Singh AK, Sharma A (1994) Studies on the uptake of lead and zinc by lignin obtained from black liquor-a paper industry waste material. Environ Technol 15: 353-361.

43. Demirbas A (2004) Adsorption of lead and cadmium ions in aqueous solutions into modified lignin from alkali glucerol delignification. J of hazard Mater, 109: 221-226.

44. Namasivayam C, Ranganathan K (1993) Waste Fe (III)/Cr (III) hydroxide as adsorbent for the removal of Cr (VI) from aqueous solution and chromium plating industry wastewater. Environ Pollut. 82: 255-261.

45. Namasivayam C, Ranganathan K (1998) Effect of organic ligands on the removal of Pb(II), Ni(II), and Cd(II) by waste Fe(III)/Cr(III) hydroxide. Water Res 32: 969-971.

46. Gupta VK, Ali Imran (2002) Adsorbents for water treatment: Low cost

alternatives to carbon, Encyclopaedia of surface and colloid science, (edited by Arthur Hubbard), Marcel Dekker, New York, USA Vol. 1: 136-166.

47. Altundogan HS, Altundogan S, Tumen F, Bildik M (2000) Arsenic removal from aqueous solutions by adsorption on red mud. Waste Manage. 20:761-767.

48. Oliveira WE, Franca AS, Oliveira LS, Rocha SD (2008) Untreated coffee husks as biosorbents for the removal of heavy metals from aqueous solutions. J Hazard Mater. 152: 1073-81.

49. Zheng W, Li XM, Wang F, Yang Q, Deng P, et al. (2008) Adsorption removal of cadmium and copper from aqueous solution by areca-a food waste. J Hazard Mater 157: 490-495.

50. Malkoc E, Nuhoglu Y (2007) Potential of tea factory waste for chromium (VI) removal from aqueous solutions: Thermodynamic and kinetic studies. Sep Purific Technol 54: 291-297.

51. Pehlivan E, Cetin S, Yanik BH (2006) Equilibrium studies for the sorption of zinc and copper from aqueous solutions using sugar beet pulp and fly ash. J Hazard Mater 135: 193-199.

52. Malkoc E, Nuhoglu Y, Dundar M (2006) Adsorption of chromium (VI) on pomace-An olive oil industry waste: Batch and column studies. J Hazard Mater 138: 142-151.

53. Namasivayam C, Yamuna RT (1995) Waste biogas residual slurry as an adsorbent for the removal of Pb(II) from aqueous solution and radiator manufacturing industry wastewater. Bioresour Technol 52: 125-131.

54. Agrawal A, Sahu KK, Pandey BD (2004) Removal of zinc from aqueous solution using sea nodules residue, Colloids and Surfaces A: Physicochem. Eng Aspects 237: 133-140.

55. Zwain Haider M, Vakili Mohammadtaghi, Dahlan Irvan (2014) Waste Material Adsorbents for Zinc Removal from Wastewater: A Comprehensive Review. Int. J Chem Engg. 2014: 1-13.

31

Potential of Microbial Inoculated Water Hyacinth Amended Thermophilic Composting and Vermicomposting in Biodegradation of Agro-Industrial Waste

Alkesh Patidar*, Richa Gupta and Archana Tiwari

School of Biotechnology, Rajiv Gandhi Proudyogiki Vishwavidhyalaya, State Technological University of Madhya Pradesh, Airport Bypass Road, Bhopal, India

Abstract

Potential of *Streptomyces viridosporus*, *Aspergillus niger* and *Moraxella osloensis* combination in thermophilic composting followed vermicomposting of jatropha seed cake with 2:1 ratio of Water Hyacinth (WH) and Cow Dung (CD) was tested. Significant decrement in MC, TOC, and C: N ratio and increment in temperature, TP was observed. TKN, pH and TK were first increases in composting and then decreased in vermin composting. Most stable and matured vermin compost was obtained in S1 substrate in terms of C: N ratio below 20 (19.50 ± 1.10). Maximum earthworm number (47 ± 6), maximum net biomass worm⁻¹ (150 ± 15.6) obtained in S3 substrate and maximum cocoons (23 ± 2) were counted S1 substrate. 20% (w/w) vermin compost amended soil induced the growth in root length, shoot length and 100% GI while >20% (w/w) was inhibitory for plant growth. This approach decreases time period of degradation and produce good quality vermin compost if mixed up to 20%.

Keywords: Jatropa seed cake; Vermicompost; *Perionyx excavates*; Inoculation

Introduction

India has huge biomass of agro-industrial processing waste viz. food processing, pulp and paper, rice husk, cotton dust, jatropa seed cake, viticulture and winery waste, sugarcane trash, guar gum industrial waste, but the capability of these organic resources is not fully utilized, and disposal of these wastes through incineration, open dumping and land filling, raises serious human health risk and environmental issue [1]. For future biodiesel production from the non-food biodiesel crop in southern Asia and Africa, Jatropa curcas is cited as one of the best candidates [2]. But the processing of jatropa seed produce 0.68 million tons of seed cake (at the rate of 65% oil recovery with 3 tons/ha oil yield), which is quite a significant amount. Jatropha seed cake cannot be used as animal feed because of its toxic properties. It has been reported to possess phorbol esters, a potential toxic compound to animals. Toxicity of *J. curcas* seeds has been studied extensively in different animals like goat, sheep, mice, rats and fish. Decrease in glucose level, lack of appetite, diarrhoea, dehydration and other hemorrhagic effects in different organs were the common observations when animals were fed on phorbol ester containing feed [3-5]. Generation of huge amount of biomass also pose serious threats to environment.

Due to high nitrogen content, jatropa seed cake can be processed into organic fertilizers via microbial thermophillic composting which is a technique that has been used to address the issues of environmental pollution, non-reliance on chemical fertilizers, sustainable natural soil fertility, and minimizing the development of new dumps and landfills [6]. In recent times, interest in the use of a closely-related technique, known as vermin composting (using earthworms to breakdown organic materials) has increased [7]. Industrial wastes that are rich in organic matter and free from toxic substances or ions could be suitable substrates for vermin composting [8]. A combine approach of composting and subsequent vermicomposting brings the advantages of both process and minimizes the adverse impact of waste on the environment. The product obtained with desirable characteristics and at a faster rate than either of the individual processes [9]. Through composting the required temperature for ensuring satisfactory pathogen kill would be achieved and product passes EPA rules for pathogen reduction. Through vermin composting, the rate of decomposition of organic materials increased and microbially active casts would be achieved.

Although, the microbial community naturally present in wastes usually carries out the process satisfactorily, the inoculation of residues with lingocellulolytic microorganisms is a strategy that could potentially enhance the way the process takes place or the properties of the final product. Inoculation with bacteria and fungi which can break down ligno-cellulolytic material has been reported to be effective in composting [10]. Inoculation increases cellulase activity, promote biodegradation of organic matter and accelerate the composting process [11,12] showed that mixed inoculation with complex microorganisms and ligno-cellulolytic microorganisms during composting of municipal solid waste had a clear advantage over inoculation with complex microorganisms or lingo-cellulolytic microorganisms alone.

Water hyacinth (*Eichhornia crassipes* (Mart) Solms – Laubach; family: Pontederiaceae) has been listed as most troublesome weed in aquatic systems. One of the easiest ways to utilize water hyacinth is to subject it to composting. Water hyacinth compost has been shown to have positive effect on plant growth. Research conducted by Umsakul et al. [13] showed that agro-industrial waste should be mixed with water hyacinth for composting to facilitate the development of thermophilic phase during the composting and to improve the product.

This present study was conducted to investigate the efficiency of water hyacinth and mixed microbial inoculation in biodegradation of industrial waste (Jatropa seed cake) during integrated composting

---

*Corresponding author: Alkesh Patidar, School of Biotechnology, Rajiv Gandhi Proudyogiki Vishwavidhyalaya, State Technological University of Madhya Pradesh, Airport Bypass Road, Bhopal (M.P.), India, E-mail: pati.ak9@gmail.com

and vermin composting technology. The objective of this study was to evaluate physical, chemical and biological parameters to determine the stability and maturity of composts prepared by mixing water hyacinth and cow dung in agro-industrial wastes.

Materials and Methods

Microbial source

The fungal strains *A. niger, S. viridosporus* and the bacterial strain *M. osloensis* were procured from IMTECH, Chandigarh, India. The fungi *A. Niger* and *S. viridosporus* were maintained by subculturing them on czepek yeast extract agar and trypton yeast extract respectively whereas bacteria was subcultured on tryptone yeast extract medium.

Industrial wastes (IW)

Jatropa seed cake (IW) was obtained from biodiesel plant at RGPV Bhopal.

Water hyacinth, cow dung and earthworm

Water Hyacinth (WH) was collected from the Lake situated in Bhopal. Cow manure (CD) was procured from a nearby poultry farm. Earthworms, *Perionyx excavatus*, used in this study came from the experimental farm of the Agronomic.

Experimental set-up

Different proportion of six feed mixture of industrial waste (Jatropa seed cake) was mixed in 2:1 ratio of CD and WH including one with CD and WH only were established (Table 1). Five kilogram of mixture (dry weight) was put in each rectangular pit (1 m length×1 m width×1 m height). A combination of pure cultures of *A. niger* and *S. viridosporus* (500 g mycelium per ton substrate) and *M. osloensis* (50 ml/kg substrate having 106 cells per ml) were inoculated in each pit on first day of composting. These mixtures were turned over manually every 24 h for 20 days in order to eliminate volatile substances potentially toxic to animals. The substrate with different treatments was pre-decomposed, in triplicates, for 20 days.

Subsequent vermicomposting

After 20 days, 20 adult individuals of *Perionyx excavatus* were introduced into each pit and then subjected to vermin composting for 20 days. The moisture content was maintained at 70 ± 10% of water holding capacity by periodic sprinkling of an adequate quantity of water. A plastic pipe with upper cap was installed in the middle of each bed and brought out of the ground through the tied mouth of the bag, in order to measure the temperature during composting. The temperature was measured daily by placing the thermometer through the plastic pipe. The earthworms and cocoons were removed manually and the vermin compost was chemically analyzed.

Compost sampling

Homogenized samples of the feed material were drawn at 0, 20 and

Substrate	Cow dung: Water hyacinth (2:1) (Kg)	Jatropa seed cake (Kg)
S1	4 (80%)	1 (20%)
S2	3 (60%)	2 (40%)
S3	2 (40%)	3 (60%)
S4	1 (20%)	4 (80%)
S5	0 (0%)	5 (100%)
Control	5 (100%)	0 (0%)

Table 1: Substrates composition.

40 days from each pit. The zero days refers to the time of initial mixing of the industrial wastes with cow dung and water hyacinth before preliminary decomposition.

Compost analysis

The pH of the compost was determined in distilled water with 1:10 (w/v) compost: water ratio. The moisture content was measured by drying to constant weight in an oven at 105°C. Changes in the temperature of the compost pile were recorded using mercury thermometer. The mean of the pile temperature at two monitoring points is recorded. Total Kjeldahl Nitrogen (TKN) and Total Organic Carbon (TOC) of the pre-decomposed bio inoculated residue and the vermicompost were measured with the Micro-Kjeldahl method [14] and Walkely and Black [15] Rapid titration method (1934). Total phosphorus (TP) was determined spectrophotometrically [16]. Total potassium (TK) was detected by the flame emission technique by Flame photometer [17].

Germination assay

Impact of most stable and matured vermin compost in terms of C: N ratio below 20 was tested on germinated seeds of cress plant. The amendments of vermin compost with garden soil were carried out at different concentrations (0, 20, 40, 60 80 and 100%). A control of garden soil was also used. Different concentrations of vermicompost amended soils (2.5 kg) were filled in different plastic pots. The soaked and sterile 30 cress (*L. sativum* L.) seeds were evenly sown in each pot to a depth of 2 inch. Prior to sowing, the cress seeds were sterilized in 3% formalin for 5 min to avoid any fungal contamination, washed thrice with double distilled water and soaked overnight in water. Experiments were conducted in triplicate with a parallel set of control. All the pots were watered daily till seed germination. All the pots were kept in a natural condition with an average durable temperature (25–35°C) and low humidity (50–55%). The germination index (GI) was calculated by the method reported by Fuentes et al. [18]. Subsequently, the plants were harvested after 5 days of sowing and repeatedly washed with tap water to remove any attached particles. Furthermore, these were rinsed with 10 mmol CaCl solution along with washing with deionised water. Root length, shoots length was measured manually using an inch-scale immediately after harvesting.

Statistical analysis

Results are the means of three replicates. One way analysis of variance (ANOVA) was done using the SIGMASTAT programme. The objective of statistical analysis was to determine any significant differences among the parameters analyzed in different treatments during the composting and vermin composting process.

Results and Discussion

Physico-chemical parameter

The physico-chemical characteristics of industrial waste, water hyacinth and cow dung are given in Table 2.

Changes in Temperature and Percent moisture content during composting are given in Figures 1 and 2. An initial temperature in the range of 26-28°C was noted in all five substrates and in control. Final temperature of 54°C in S2 and 54°C in control was recorded at 20th day of composting. It has been previously reported that the increase of temperature in the early phase of composting may be the result of thermophilic phase of composting. The temperature rise as a

Parameter	Jatropa seed cake	Water hyacinth	Cow dung
pH	6.8 ± 0.20	7.2 ± 0.20	8.7 ± 0.20
Total organic carbon (TOC)	35.64 ± 0.21	42.6 ± 0.10	39.2 ± 0.21
Total Kjeldahl nitrogen (TKN)	0.84 ± 0.03	0.86 ± 0.05	0.63 ± 0.02
C: N ratio	42.42 ± 1.12	49.3 ± 1.01	62.2 ± 1.05
Total phosphorus (TP)	0.50 ± 0.02	0.82 ± 0.13	0.59 ± 0.14
Total potassium (TK)	0.57 ± 0.10	0.82 ± 0.05	0.62 ± 0.12

Table 2: Physico-chemical characteristics of the initial raw materials used in experiment.

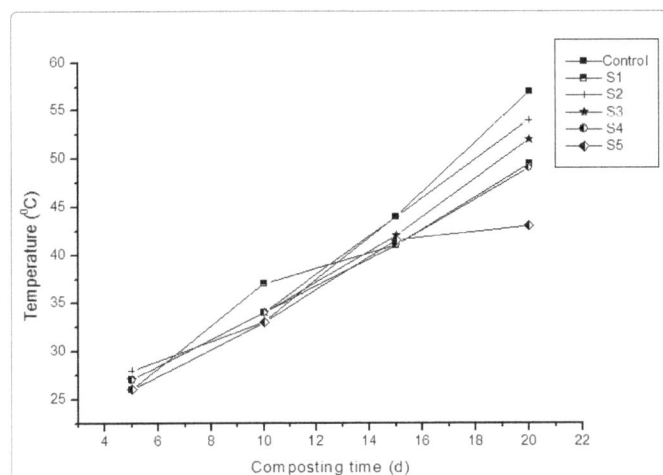

Figure 1: Changes in temperature during jatropa seed cake composting (20 days).

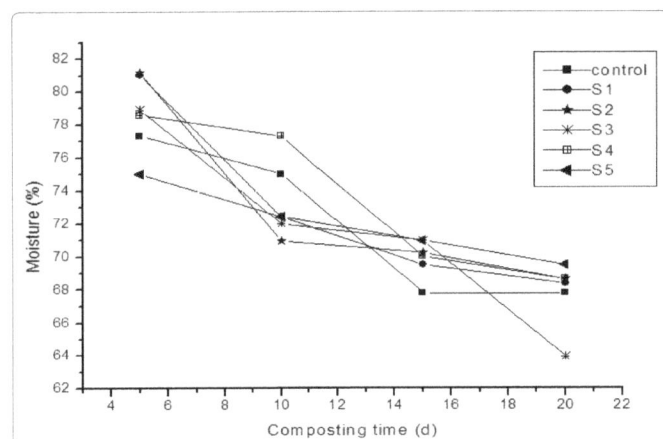

Figure 2: Changes in % moisture content during jatropa seed cake composting (20days).

consequence of the rapid breakdown of organic matter, mostly readily available substrate rich in carbon, by microbial metabolism [19].

Moisture content is an important factor in degradation because it influences the structural and thermal properties of the material, as well as the rate of degradation and metabolic process of microorganisms. Initial moisture content was in the range of 81.15-75 on 5th day of composting. Final moisture content was decreased and in the range of 63.93-69.49 in all the treatments. Lowest percentage was noted in S3 treatment. After composting of jatropa seed cake pH was increased in all the treatments and after vermin composting on 40th day it was decreased (Figure 3). Increase in pH during composting during

composting was due to mineralization of the nitrogen and phosphorus into nitrites/nitrates and orthophosphates. The decrease in pH during vermicomposting may be due to CO_2 and organic acids produced by microbial metabolism [20].

Total Organic Carbon (TOC) decreased significantly in all treatments during composting and subsequent vermin composting, extremely in S1 and S2 which contained lower concentrations of industrial wastes. It may be due to the mineralization of organic matter [21]. During composting maximum carbon loss of 11.84% was noted in S1 treatment and least was 1.98% in S5 treatment (Table 3). This decrement was due to combined potential of added microbes and water hyacinth both initiate the thermophilic phase of composting. During vermin composting maximum carbon loss of 7.9% was noted in S2 and least was 1.92 in S3. It was the action of Earthworms which break and homogenizes the ingested material through muscular action of their foregut and, also adds mucus and enzymes in ingested material, this increase the surface area for microbial action. Thus action of water hyacinth earthworms and microorganisms bring TOC loss from the substrates in the form of CO_2 [22].

Percentage loss in carbon during composting was fast as compared to that of the vermin composting and it was due to higher initial N concentration, which might have increased the microbial activity in composting phase [23].

During composting phase significant increment in TKN at 20th day was noted. Maximum percentage of TKN increment was noted in S1 treatment which receives the 80% of 2:1 ratio of cow dung and water hyacinth (Table 3). This increment might be due to potential of water hyacinth which facilitate and initiate the development of thermophilic phase during the composting process [13] and the inoculated microbes especially *moraxella osloensis* which fix the nitrogen present in the substrate. During subsequent vermin composting nitrogen content was decreased in all the treatments possibly due to ammonification, NH_3 volatilization and denitrification as reported by Bernal et al. [24].

C: N ratio-index of compost maturity

The C: N ratio of a substrate material reflects the organic waste mineralization and stabilization during the process of composting and vermin composting. The loss of carbon as CO_2 through microbial respiration and simultaneously addition of nitrogen by worms in the form of Mucus and nitrogenous excretory material lower the C: N ratio of the substrate [25]. C: N ratio is considered as a parameter to

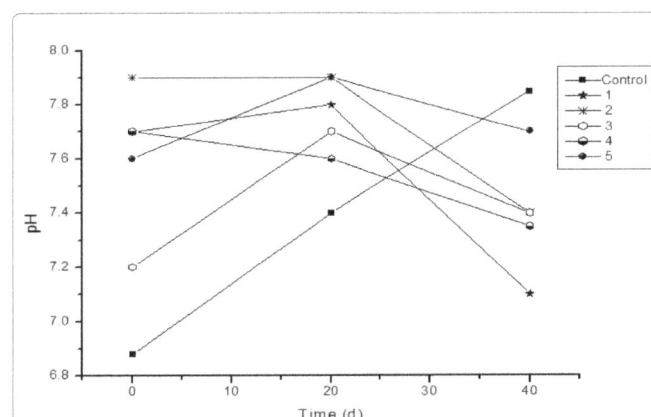

Figure 3: Changes in pH during jatropa seed cake composting (20 days) and subsequent vermicomposting (20 days).

establish the degree of maturity of compost and its agronomic quality [26]. Decline of C: N ratio to less than 20 indicates an advanced degree of organic matter stabilization and reflects a satisfactory degree of maturity of organic wastes [27]. Microbial composting and vermin composting of jatropa seed cake resulted in a loss of carbon because of mineralization of organic waste (Figure 4). Initial C: N ratio was in the range of 37.89-65.47 at zero days in different substrates. The initial C: N ratio was more in those feed mixtures which had higher percentage of jatropa seed cake. After composting and vermincompostubg C: N ratio was in the range of 18.59-37.82 and 18.46-39.11 respectively. Maximum decrement in C: N ratio was observed in S1 substrate comparable to control (P<0.001). The decomposition of the waste during vermicomposting was slow compared to that of the composting

and might have been due to higher initial N concentration, which might have increased the microbial activity in the beginning, thus decreasing the C/N ratio [23].

Percentage of total phosphorus was increased significantly during composting and vermicomposting in all the treatments with respect to control (P<0.001). The overall percent increase in the TP content was maximum in the substrate S3 (1.71 ± 0.10), and minimum percent increase was in the substrate S5 (0.65 ± 0.20) (Table 4). Yadav and Garg [28] noticed 1.5-2-fold increase in total phosphorus from industrial sludge vermicomposting. Lee [29] states that when organic material passes through the gut of earthworm then insoluable phosphorus is converted to soluable form by phosphatase produced within the

Substrate	0 days		20 days		40 days	
	TOC	TKN	TOC	TKN	TOC	TKN
Control	36.00 ± 0.12*	0.95 ± 0.21	27.74 ± 0.01*	1.49 ± 0.10*	25.66 ± 0.20*	1.39 ± 0.01*
S1	45.50 ± 0.21*	0.92 ± 0.01	33.66 ± 0.15*	1.57 ± 0.12	27.70 ± 0.15*	1.42 ± 0.02
S2	53.40 ± 0.25*	0.93 ± 0.03	49.40 ± 0.18*	1.45 ± 0.10	41.50 ± 0.01*	1.42 ± 0.01
S3	55.00 ± 0.10*	0.84 ± 0.02	53.40 ± 0.20*	1.48 ± 0.10	51.48 ± 0.01*	1.36 ± 0.10
S4	47.52 ± 0.10*	0.86 ± 0.01	45.50 ± 0.05*	1.42 ± 0.20	43.56 ± 0.05*	1.32 ± 0.01
S5	35.64 ± 0.21	0.84 ± 0.03	33.66 ± 0.31*	0.89 ± 0.10*	31.68 ± 0.24*	0.81 ± 0.02*

All values are mean (n=3) ± S.D. and given in percentage.
*Significant (P<0.001).

Table 3: Percentage of total organic carbon (TOC), total Kjeldahl nitrogen (TKN) in different substrate composition during microbial composting (0–20 days) and subsequent vermicomposting (20 days).

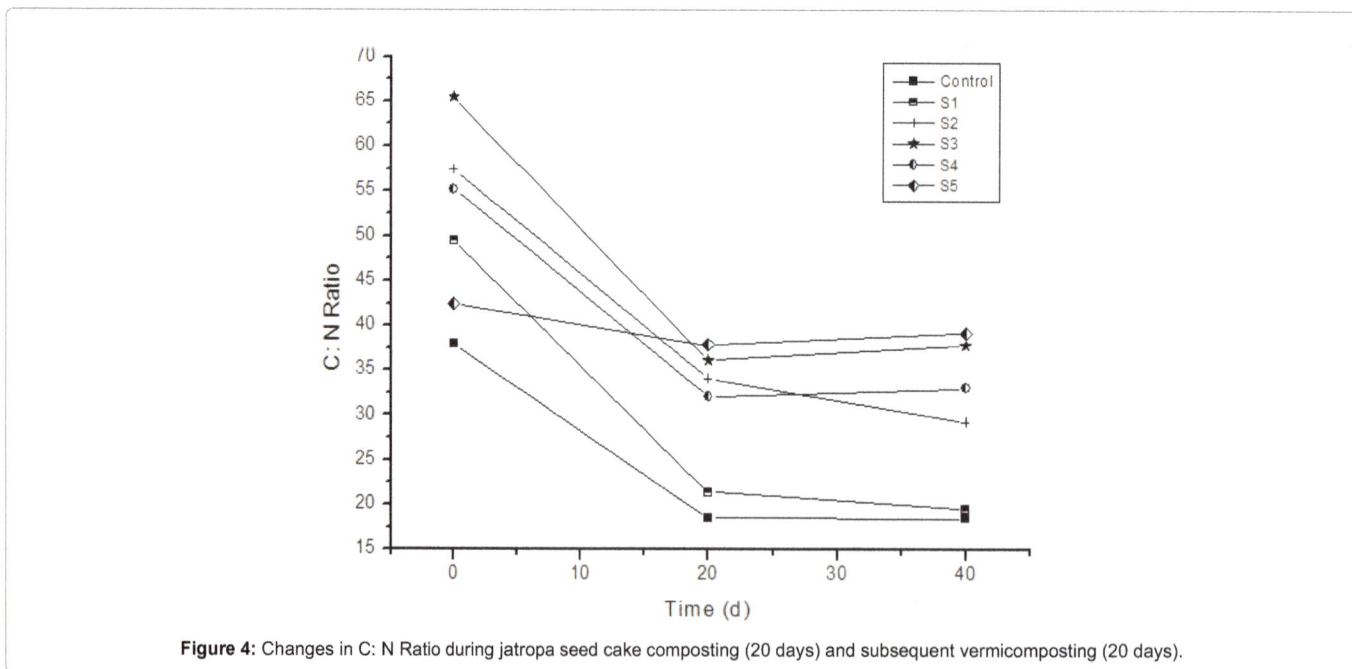

Figure 4: Changes in C: N Ratio during jatropa seed cake composting (20 days) and subsequent vermicomposting (20 days).

Substrate	0 days	20 days	40 days
Control	37.89 ± 1.10*	18.59 ± 1.20*	18.46 ± 1.20*
S1	49.45 ± 1.10*	21.43 ± 1.02*	19.50 ± 1.10
S2	57.41 ± 1.20*	34.06 ± 1.12*	29.22 ± 1.20*
S3	65.47 ± 1.21*	36.08 ± 1.02*	37.85 ± 1.15*
S4	55.25 ± 1.02*	32.04 ± 1.12*	33.00 ± 1.10*
S5	42.42 ± 1.12*	37.82 ± 1.10*	39.11 ± 1.10*

All values are mean (n=3) ± S.D. and given in percentage.
*Significant (P<0.001).

Table 4: Changes in C/N ratio during composting and vermicomposting.

earthworm and secreted by inoculated microbes which is easily available to plants.

Data in Table 4 reported that total potassium during microbial composting increase significantly in all the treatments as well as in control (P<0.001). Maximum increment was noted in S3 substrate (1.57 ± 0.15) and minimum was in S1 substrate (0.72 ± 0.10). However, we observed a marginal decrease in potassium during vermicomposting. Only S1 substrate showed slight increment in TK. This decrement in potassium was supported the result of Kumar et al. [30] who reported lower K content in sugar-cane waste by-products after vermicomposting. There are contradictory reports regarding the potassium content in vermicompost obtained from different substrates. Hait and Tare [31] have reported higher potassium content in the sewage sludge vermicompost.

This difference in potassium content can be attributed to the differences in the chemical nature of substrate. This might be also due to leaching of potassium by excess water that drained through the feeds. Benitez et al. [32] have reported that the leachate collected during vermicomposting process had higher potassium concentrations. It has been suggested that earthworm processed residue contains higher concentration of exchangeable K due to enhanced microbial activity during vermicomposting, which consequently enhances the rate of mineralization [25].

Total no. of earthworm, cocoons and biomass

The different pattern of earthworm number and cocoons production depending on the percentage of substrate in the feed mixtures was observed after vermicomposting of jatropa seed cake (Table 5). Maximum earthworm number was observed in substrate S3 and maximum cocoons production was noted in S1. The addition of industrial waste up to 40% in cow dung and water hyacinth was found significant in cocoon production, but concentration above this level was not able to support the cocoon production. The increase in earthworms' growth may also be attributed to a low C: N ratio of the microbial composted easily consumable water hyacinth added substrate [8]. Fayolle et al. [33] suggested that food source play an important role on cocoon production rates by worms.

The maximum net biomass production by *Perionyx excavatus* was observed in substrate S2 (150 ± 15.6 mg earthworm^{-1}) and lowest in S5 substrate (53 ± 15.2 mg earthworm^{-1}). Increasing percentage of jatropa seed cake in the feed mixtures resulted in a decrease in biomass of earthworm. It is concluded from the results that addition of 40% jatropa seed cake to the water hyacinth and cow dung is acceptable during the vermicomposting in terms of growth and reproductive success of the earthworms.

Biological parameter of vermicompost analysis

The evaluation of industrial wastes toxicity by chemical characterisation and biological testing is extremely important for screening the suitability of vermicompost for land application. To evaluate the biological parameter of vermicompost, the amendment of most stable and matured compost on the basis of C: N ratio less than 20, substrate S1 with garden soil were carried out at different concentrations (20, 40, 60, 80 and 100%). In general, percentage of seed germination decreased with increase in vermicompost concentrations from S1a to S1e (Table 6). In present study, it was observed that up to 20% (w/w) concentration of vermicompost has no inhibitory effect on seed germination while at higher concentration (>40%), decrease in percent germination was recorded. This decrease in percent germination except in control and 20% vermicompost was due to phytotoxic effect produced by combination of several factors, rather than one [34]. These factors include the presence of heavy metals, ammonia, salts, and low molecular weight organic acids [18] all of which have been shown to have inhibitory effects.

The vegetative growth parameters of cress plant (shoot length and root length) at all tested concentrations showed significant inhibitory effect (except 20% vermicompost amended soil) versus the control (Table 7). Maximum root length was noted in S1a trial (3.6 ± 0.12). Similarly maximum shoot length was observed in S1a trial (5.6 ± 0.14) more than the control. As the concentration of vermicompost increased root and shoot length was decreased. It has been reported that high vermicompost content was suppressive for plant growth

Substrate	0 days		20 days		40 days	
	TP	TK	TP	TK	TP	TK
Control	0.85 ± 0.10*	0.90 ± 0.20*	0.98 ± 0.10	1.70 ± 0.15*	1.25 ± 0.10	1.67 ± 0.20*
S1	0.70 ± 0.10	0.42 ± 0.18*	0.88 ± 0.10	0.72 ± 0.10*	1.65 ± 0.10	0.78 ± 0.10*
S 2	0.67 ± 0.01	0.80 ± 0.15	0.90 ± 0.10	1.57 ± 0.15*	1.70 ± 0.20	1.47 ± 0.25
S3	0.66 ± 0.02	0.77 ± 0.15	0.93 ± 0.20	1.65 ± 0.20	1.71 ± 0.10	1.57 ± 0.15
S4	0.61 ± 0.10*	0.70 ± 0.10	0.67 ± 0.15	1.35 ± 0.10	0.83 ± 0.15	1.46 ± 0.15
S5	0.50 ± 0.02*	0.60 ± 0.10	0.52 ± 0.20	1.10 ± 0.20*	0.65 ± 0.20*	1.17 ± 0.16*

All values are mean (*n*=3) ± S.D. and given in percentage.
*Significant (P<0.001).

Table 5: Percentage of total phosphorus (TP) and total potassium (TK).

Substrate	Initial No. of earthworms	Total earthworm	Cocoons no.	Mean Biomass worm^{-1} (mg)		Net biomass worm^{-1} (mg)
				Initial	Final	
Control	20	31 ± 4*	11 ± 4*	235 ± 2.5*	375 ± 12*	140 ± 11.2*
S1	20	28 ± 5	23 ± 2*	238 ± 1.7	368 ± 12	130 ± 5.6
S2	20	42 ± 4*	18 ± 4	240 ± 3.6	390 ± 14	150 ± 15.6
S3	20	47 ± 6*	07 ± 1	230 ± 1.7	325 ± 11*	95 ± 2.4*
S4	20	31 ± 2	04 ± 3	230 ± 1.6	284 ± 4.3*	54 ± 4.9*
S5	20	29 ± 5	03 ± 2	225 ± 1.7*	278 ± 11*	53 ± 15.2*

All values are mean (*n*=3) ± S.D.
*Significant (P<0.001).

Table 6: Changes in growth, cocoon production and net biomass of earthworms.

Vermicompost concentration (%)	GI (%)	Root length (cm)	Shoot length (cm)
Control (0)	100	3.6 ± 0.30	4.2 ± 0.10*
S1a (20)	100	3.6 ± 0.12	5.6 ± 0.14*
S1b (40)	96	3.2 ± 0.21	5.4 ± 0.13*
S1c (60)	70	3.2 ± 0.20	4.3 ± 0.10*
S1d (80)	70	3.0 ± 0.10	3.8 ± 0.12*
S1e (100)	60	2.6 ± 0.21*	3.2 ± 0.10*

All values are mean (n=3) ± S.D.
*Significant (P<0.001).

Table 7: Effect of vermicompost S1-amended soil on germination of cress seed.

hormone (s) (auxin and gibberline) which are responsible for the growth and development of plants [35]. The reduction in plant growth at high concentrations of vermicompost might be due to the entrance of the metal into the protoplasm resulting in the loss of intermediatory metabolites which are essential for further growth and development of plants. These findings are very much in accordance with the earlier reported results [36].

Conclusion

The chemical analyses of the compost produced by microbial inoculated composting and vermicomposting of jatropa seed cake admixed with water hyacinth and cow dung, point towards its patentability. On the basis of C: N ratio below 20, earthworm biomass, growth and reproduction, results demonstrated that after the addition of jatropa seed cake (20%) in 2:1 ratio of water hyacinth, cow dung, and inoculated microbes, *S. viridosporus, A.niger* and *M. osolensis*, it can be used as a raw material. Germination index using cress seeds showed that concentration above 20% of vermicompost had inhibitory effect and cause phytotoxicity.

Acknowledgements

Author would like to thank the Indian institute of soil science for help with the compost analyzing.

References

1. Raj D, Antil RS (2011) Evaluation of maturity and stability parameters of composts prepared from agro-industrial wastes. Bioresour Technol 102: 2868-2873.

2. Sachs G (2010) Commodity prices and volatility: old answers to new questions. Goldman Sachs Global Economics, Commodities and Strategy Research, Global Economics Paper No: 194.

3. Adam SE (1974) Toxic effects of Jatropha curcas in mice. Toxicology 2: 67-76.

4. Adam SE, Magzoub M (1975) Toxicity of Jatropha cucurcas for goats. Toxicology 4: 347-354.

5. Makkar HP, Becker K (1999) Nutritional studies on rats and fish (carp Cyprinus carpio) fed diets containing unheated and heated Jatropha curcas meal of a non-toxic provenance. Plant Foods Hum Nutr 53: 183-192.

6. Papadimitriou EK, Balis C (1996) Comparative study of parameters to evaluate and monitor the rate of a composting process. Compost Sci Utilization 4: 52-61.

7. Logsdon G (1994) Worldwide progress in vermicomposting. Biocycle 35: 63-65.

8. Ndegwa PM, Thompson SA (2000) Effects of C-to-N ratio on vermicomposting of biosolids. Bioresour Technol 75: 7-12.

9. Ndegwa PM, Thompson SA (2001) Integrating composting and vermicomposting in the treatment and bioconversion of biosolids. Bioresour Technol 76: 107-112.

10. Nair J, Okamitsu K (2010) Microbial inoculants for small scale composting of putrescible kitchen wastes. Waste Manag 30: 977-982.

11. Ghaffari S, Sepahi AA, Razavi MR, Malekzadeh F, Haydarian H (2011)

Effectiveness of inoculation with isolated *Anoxybacillus* sp MGA110 on municipal solid waste composting process. Afr J Microbiol Res 5: 6373-6378.

12. Wei Z, Xi B, Zhao Y, Wang S, Liu H, et al. (2007) Effect of inoculating microbes in municipal solid waste composting on characteristics of humic acid. Chemosphere 68: 368-374.

13. Umsakul K, Dissara Y, Srimuang N (2010) Chemical, physical and microbiological changes during composting of the water hyacinth. Pak J Biol Sci 13: 985-992.

14. Shaw J, Beadle LC (1949) A simplified ultramicro Kjeldahl method for the estimation of protein and total nitrogen in fluid samples of less than 1-0 mu 1. J Exp Biol 26: 15-23.

15. Walkely JA, Black JA (1934) Estimation of organic carbon by the chronic acid titration method. Soil Sci 37: 29-31.

16. Fiske CH, Subbarow Y (1925) The colorimetric determination of phosphorus. J Biol Chem 66: 375-400.

17. Pearson RW (1952) Potassium – supplying power of eight Alabama Soils. Soil Sci 74: 301-310.

18. Fuentes A, Lloréns M, Sáez J, Aguilar MI, Ortuño JF, et al. (2004) Phytotoxicity and heavy metals speciation of stabilised sewage sludges. J Hazard Mater 108: 161-169.

19. Kutzner HJ (2000) Microbiology of composting technology. In: Rehm HJ, Reed G (eds) Biotechnology Environmental Processes II, John Wiley & Sons, USA.

20. Elvira C, Sampedro L, Benítez E, Nogales R (1998) Vermicomposting of sludges from paper mill and dairy industries with *Eisenia andrei*: A pilot-scale study. Bioresour Technol 63: 205-211.

21. Kaushik P, Garg VK (2003) Vermicomposting of mixed solid textile mill sludge and cow dung with the epigeic earthworm Eisenia foetida. Bioresour Technol 90: 311-316.

22. Edwards CA (1988) Breakdown of animal, vegetable and industrial organic wastes by earthworms. Earthworms in Waste and in Environment management. SPB Academic Publishing, The Netherlands.

23. Eiland F, Klamer M, Lind AM, Leth M, Bååth E (2001) Influence of Initial C/N Ratio on Chemical and Microbial Composition during Long Term Composting of Straw. Microb Ecol 41: 272-280.

24. Bernal MP, Navarro AF, Roig A, Cegarra J, Garcia D (2006) Carbon and nitrogen transformation during composting of sorghum bagasse. Biol Fertil Soils 22: 141-148.

25. Suthar S (2008) Bioconversion of post harvest crop residues and cattle shed manure into value-added products using earthworm Eudrilus eugeniae Kinberg. Ecological Engineering 32: 206-214.

26. Singh D, Suthar S (2012) Vermicomposting of herbal pharmaceutical industry solid wastes. Ecological Engineering 39: 1-6.

27. Senesi N (1989) Composted materials as organic fertilizers. Sci Total Environ 81/82: 521-524.

28. Yadav A, Garg VK (2009) Feasibility of nutrient recovery from industrial sludge by vermicomposting technology. J Hazard Mater 168: 262-268.

29. Lee KE (1992) Some trends and opportunities in earthworm research or: Darwin's children--the future of our discipline. Soil Biol Biochem 24: 1765-1771.

30. Kumar R, Verma D, Singh BL, Kumar U, Shweta (2010) Composting of sugar-cane waste by-products through treatment with microorganisms and subsequent vermicomposting. Bioresour Technol 101: 6707-6711.

31. Hait S, Tare V (2011) Vermistabilization of primary sewage sludge. Bioresour Technol 102: 2812-2820.

32. Benitez E, Nogales R, Elvira C, Masciandaro G, Ceccanti B (1999) Enzyme activities as indicators of the stabilization of sewage sludge composting with *Eisenia foetida*. Bioresour Technol 68: 297-303.

33. Fayolle L, Michaud H, Cluzeau D, Stawiecki J (1997) Influence of temperature and food source on the life cycle of the earthworm *Dendrobaena veneta* (Oligochaeta). Soil Biol Biochem 29: 747-750.

34. Hoekstra NJ, Bosker T, Lantinga EA (2002) Effects of cattle dung from farms with different feeding strategies on germination and initial root growth of cress (*Lepidium sativum L.*). Agric Ecosys Environ 93: 189-196.

Chitosan– A Low Cost Adsorbent for Electroplating Waste Water Treatment

Bhavani K[1], Roshan Ara Begum E[1], Selvakumar S[2] and Shenbagarathai R[1]*

[1]PG and Research Department of Zoology and Biotechnology, Lady Doak College, Madurai-625 002, Tamil Nadu, India
[2]Department of Environmental Studies, School of Energy Environment and Natural Resources, Madurai Kamaraj University, Madurai-625 021, Tamil Nadu, India

Abstract

In the wake of rapid industrialization, urbanization and population explosion, the basic needs of life viz. air, water and land are continuously being polluted over the period of the time. Effluent waste from Electroplating industries considerably pollutes the environment. It is of major concern as the discharged heavy metals such as Cr^{6+}, Zn^{2+}, Cu^{2+}, Pb^{2+}, Cd^{2+} etc. are harmful to the ecosystem. Although various techniques are used for the removal of heavy metals. Bioadsorption process is one of the mainly adopted methods to recycle and reuses the wastewater. Hence the present study focused in assessing the usage of chitosan, a marine biopolymer extracted from the locally available shells of crab *Portunus pelagicus* as an adsorbent. The ability of chitosan to adsorb heavy metals such as Cr^{6+}, Zn^{2+}, Cu^{2+}, Pb^{2+}, Cd^{2+} from the electroplating industry waste water effluents was investigated at different pH. It was found that maximum adsorption of heavy metals occurred at a pH range between 5 to 7. The study results indicate that the cost incurred towards extraction was less than that the cost incurred by the commercial treatment facilities. The -OH and $-NH_2$ functional groups in chitosan facilitate the adsorbent function and make it an ideal adsorbent for the treatment of wastewater. This study brings to light that chitosan is function as an economically useful adsorbent for the treatment of electro plating effluent containing heavy metals.

Keywords: Electroplating; Waste water; Bioadsorption; Chitosan; Heavy metals

Introduction

In the course of human development, industrialization has made possible higher standards of living in our modern Society. Such "progress" has created increased problems with waste from processing operations and their ultimate disposal creating water pollution. There are 400 registered electroplating units in the state of Tamilnadu and are located in Thiruvallur, Kancheepuram, Krishnagiri, Madurai and Coimbatore districts. These units are mainly operating in a small scale sector and 90% of them are exploiting zinc, nickel and chromium plating. In Madurai district alone 54 electroplating units are functional. While these industries economically benefit and support to the local living community it produce hazardous electroplating wastes. Moreover, these units are not only spaced inadequately and small in size, but they do not contain waste water treatment plants. Hence, the waste water effluents containing a large amount of metals and chemicals which are disposed into the mainstream water resources. These wastes contain heavy metals which are very harmful because of their non-biodegradable nature and long biological half-lives. These metals tend to accumulate in different body parts of the living organisms that cause various diseases and disorders. Hence it is imperative to control and eliminate the hazardous waste generated by the electroplating industries. Over the decade, a variety of methodologies and technologies have been developed for waste minimization and elimination from these industrial wastes [1].

A wide range of physical and chemical process such as precipitation, adsorption, oxidation, coagulation, flocculation and membrane filtration processes are available for the removal of these metals from waste waters. Among them, adsorption techniques have been proved to be an excellent method to treat waste water. It is now recognized that adsorption using low cost adsorbents is an effective and economic method for the removal of heavy metals in waste water. However, low cost adsorbents with high adsorption capacities are still under development to minimize disposal problems. Much attention has recently been focused on biopolymers as bioadsorbents which are naturally produced by all living organism. The adsorbents may be of mineral, organic or biological origin, activated carbons zeolites or low-cost adsorbents (industrial by-products, agricultural wastes, biomass and polymeric materials). In the recent years, numerous studies have aimed to develop cheaper and more effective natural polymers adsorbents. Among these biopolymers, polysaccharides such as chitin and starch and their derivatives chitosan, represent an interesting and attractive alternative class of adsorbents. Their particular structure, physico-chemical characteristics, chemical stability, high reactivity and excellent selectivity towards aromatic compounds and metals, are due to the presence of chemical reactive groups (hydroxyl, acetamido or amino functions) in polymer chains. Chitosan is extracted from crustacean shell waste produced by the sea food industry. Crab shell waste is an environmental pollutant with significant health hazards. The most frequent method employed for its disposal is burning, which becomes environmentally costly, due to its low burning capacity. In such a scenario, conversion of crab shell waste to chitosan, a commercially valuable product with myriad of uses, could serve as an effective mode of shell remediation and more importantly enable waste water treatment for the electroplating units.

The shell fish industry produces about 60,000 tons of wastes. The disposal of such enormous amount of waste has become a serious environmental concern. Although these wastes are biodegradable but the rate of degradation of a large amount of waste generated per processing operation is comparatively slow. The immediate solution

*Corresponding author: Shenbagarathai R, PG and Research Department of Zoology and Biotechnology, Lady Doak College, Madurai-625 002, Tamil Nadu, India, E-mail: shenbagarathai@rediff.com

of this problem seems to recycling of the crustacean shells generated and extraction of commercially viable polysaccharide such as chitosan. It carries strong cationic charges at and below pH 6.5 and strong anionic charges above this pH. Therefore it has strong affinity for ions because it comprises of sequenced amino groups (-NH$_2$) and hydroxyl groups (–OH) and thus it is used in many applications in waste water treatment. Hence, this study explored the possibility of using chitosan as bioadsorbent to remove the heavy metals present in the effluent of the electroplating industries.

Materials and Methods

Extraction of chitosan and sample preparation

Crab shells were collected from the Ramaeshwaram, Tamilnadu, India. The shells were washed and dried under hot air oven at 60°C for 24 hours to remove the viscera and tissues. The dried samples were blended and sieved. Twenty grams of shell powder was used for further analysis. The sample was then deproteinized with 1N NaOH at 80°C for 24 hours with constant stirring. It was then washed with distilled water, followed by demineralization with 2N hydrochloric acid. The sample was then washed again with distilled water so as to remove the acidic residues [2].

Electro plating waste water was collected from local electro plating industry near Dindugal district, Tamilnadu. Collected waste water samples were stored in the sterile bottles that has been pre-washed with 10% nitric acid and then thoroughly rinsed with de-ionized water.

Deacetylation of chitin into chitosan

Chitin was deacetylated with 40% NaOH at 110°C, for 6 hours with constant stirring. Then 10% acetic acid was added to the sample and stored for 12 hours at room temperature. The dissolved sample was reprecipitated by adding 40% NaOH (pH 10). The sample was washed by distilled water until a pH of 6.5 was achieved. Then it was centrifuged at 10,000 rpm for 10 minutes and freeze dried subsequently.

Physico-chemical characterization of chitosan

Fourier Transform – Infra Red spectroscopy: Domszy and Roberts [3] FT-IR spectroscopy was used for the qualitative analysis of organic chemicals. The spectral location of their IR absorptions could be used to determine the general structure of compounds. FT-IR spectroscopy (Bio-Rad FTIS-40 model, USA) was used to compare and confirm the chemical conformation of standard chitosan with that of extracted chitosan. The spectra of chitosan samples were obtained with a frequency range of λ=400-4000 cm^{-1}.

Yield: Chitosan yield was determined by comparing weight measurements of the raw material and of the chitosan obtained after treatment.

Ash content: Approximately 3-5 g of chitosan samples were accurately weighed into clean, dry, pre-weighed porcelain crucibles and charred over a Bunsen burner. The charred samples were heated in a muffle furnace and maintained at 550 ± 1°C until a gray ash was obtained. Crucibles were subsequently cooled in desiccators and weighed. Ash content was calculated as percent ratio of the mass of the ash obtained after ignition to that of the original material.

The ash content was calculated using the following formula:

$$\% \text{ Ash} = \frac{\text{weight of the residue } (g)}{\text{weight of the sample } (g)} \times 100$$

Degree of deacetylation: Domszy et al. [3] the degree of deacetylation (DD) of the chitosan was calculated using the baseline by Domszy and Roberts.

The computation equation for the baseline is given below:

DD=100–[(A1655/A3450) × 100/1.33]

Where A1655 and A3450 were the absorbance at 1655 cm^{-1} of the amide-I band as a measure of the N-acetyl group content and 3450 cm^{-1} of the hydroxyl band as an internal standard to correct for film thickness. The factor 1.33 denoted the value of the ratio of A1655/A3450 for fully N-acetylated chitosan.

Moisture Content: Moisture content of the chitosan was analysed by the gravimetric method. The water mass was determined by drying the sample to constant weight and then measuring the sample weight before and after drying. The water weight was the difference between the weight of the wet and oven dried samples.

$$\text{Moisture content } (\%) = \frac{\text{Wet Weight } (g) - \text{dry Weight } (g)}{\text{Wet Weight } (g)} \times 100$$

Adsorption studies

Determination of optimum pH for adsorption: Tewari et al. [4] found optimum pH for adsorption of heavy metals (Cr^{6+}, Zn^{2+}, Cu^{2+}, Pb^{2+}, Cd^{2+}) by chitosan was determined experimentally. 50 mL of electro plating effluent sample was dispensed in 7 conical flasks. The initial pH of the sample was adjusted in range of 2-9. Subsequently, 50 mg of chitosan powder was added to each flask. The flasks were kept for 24 hours in agitation using shaker, maintained at 25°C at 100 rpm. After 24 hours, flask content was filtered. The filtrate were analysed for heavy metal concentrations (Cr^{6+}, Zn^{2+}, Cu^{2+}, Pb^{2+}, Cd^{2+}). Samples of the Cr^{6+} were analysed using UV Visible Spectrophotometer (U -2000, Hitachi) at 540 nm, according to the 1,5-diphenyl-carbazide method. The respective samples of the solution of Zn^{2+}, Cu^{2+}, Pb^{2+}, Cd^{2+} were analysed by Atomic adsorption Spectrophotometer. The amount of metallic ion adsorbed by the chitosan was calculated using the equation

q-(C$_0$-C$_e$)v/w

C$_o$ - Initial Concentration of Heavy Metals (Mg/L)

C$_e$ - Final Concentration of Heavy Metals (Mg/L)

V- The Volume of Aqueous Solution (L)

W- The Mass of Chitosan (g)

q- The adsorbate concentration or sorption capacity of chitosan (mg/g)

Effect of adsorbent dose: The dependence of Cr^{6+}, Zn^{2+}, Cu^{2+}, Pb^{2+}, Cd^{2+} adsorption on dose was studied by varying the amount of adsorbents from 1 to 6 g/L, by keeping the other parameters (pH, contact Time) constant. 50 mL of electro plating effluent sample was dispensed in 6 conical flasks with different concentration of chitosan from 1 g/L to 6 g/L. Flasks were agitated at 150 rpm at different time intervals. The solution was filtered and the filtrate were analysed by atomic adsorption spectrophotometer Zn^{2+}, Cu^{2+}, Pb^{2+}, Cd^{2+} and spectrophotometer for Cr^{6+} at 540 nm [5].

Effect of contact time for adsorption: The optimum time for adsorption of heavy metals (Cr^{6+}, Zn^{2+}, Cu^{2+}, Pb^{2+}, Cd^{2+}) by chitosan was determined experimentally. 50 mL of electro plating effluent sample was dispensed in 22 conical flasks. Contact time for treatment was varied from 40 min to 460 min. The pH of the solution was adjusted to 5. Flasks were agitated at 150 rpm at different time interval. The solution were filtered and the filtrate was analysed to atomic adsorption Zn^{2+},

Cu²⁺, Pb²⁺, Cd²⁺ and spectrophotometer for Cr⁶⁺. In order to investigate the controlling mechanism of the adsorption processes, pseudo first-order and pseudo second-order kinetics were applied in the data [6].

Desorption studies

Determination of optimum pH for desorption study: Vieira et al. [7] After carrying out the adsorption process, the chitosan particles adsorbed with heavy metal were filtered and let a few minutes to get dry. 50 mg of dried chitosan particles dissolved with 50 mL distilled water. The pH of these solutions was adjusted from 2 to 8 with 0.1N NaOH and 0.1 M HCl. The flasks were kept for 24 hours in agitation using shaker, maintained at 25°C at 100 rpm. After 24 hours, flask content was filtered. The filtrate were analyzed for heavy metal concentrations (Cr⁶⁺, Zn²⁺, Cu²⁺, Pb²⁺, Cd²⁺). Samples of the Cr⁶⁺ were analyzed using UV Visible Spectrophotometer (Model U -2000, Hitachi) at 540 nm, according to the 1,5-diphenyl-carbazide method. The respective samples of the solution of Zn²⁺, Cu²⁺, Pb²⁺, Cd²⁺ were analyzed by Atomic adsorption Spectrophotometer. The optimum pH for desorption were ascertained from maximum metal concentration in solution.

Results and Discussion

Extraction of chitosan

Chitosan, the deacetylated form of chitin has found wide applications due to its versatility, non-toxicity, biodegradability, biocompatibility and bio-renewability. Keeping the importance of chitin and chitosan in mind, an attempt was made to extract chitosan from the exoskeleton of Crab (*Portunus pelagicus*). The yield of chitosan obtained after deacetylation of chitin was found to be 59% (Table 1). These result were in agreement with Ref. [8] reports, where the yield of chitosan from freshwater crab (*Potamon potamios*) and blue swimmer crab (*Portunus pelagicus)* was 60%. The various parameters studied after the extraction and the values obtained are tabulated in Table 1.

Degree of deacetylation

Deacetylation degree of chitosan is the most important quality parameter in this study. The degree of deacetylation showed that the percentage of acetyl groups removed from the chitin to produce chitosan. High deacetylation degree showed that the acetyl groups contained in the chitosan is low. The deacetylated chitosan enhances the interaction between the ions and hydrogen bond. Degree of deacetylation of chitosan obtained in this study ranged from 76.26-91.60%, with an average overall deacetylation degree was 82%.

Ash content

The ash content in chitosan is an important parameter that affects its solubility, viscosity and also other important characteristics. The extracted chitosan had an ash content of less than 1% when compared to commercially available chitosan.

FTIR spectroscopy

The structure of the extracted chitosan was confirmed by FTIR analysis in this study, the IR spectra of the isolated sample of chitosan from crab shells was analyzed and compared with the IR spectrum of standard chitosan. The isolated fractions gave IR spectra similar to that of the standard chitosan (Figures 1 and 2). The results indicated a strong similarity between both the standard and extracted compound. The FT-IR spectrum of the standard chitosan showed eight major peaks at the ranges of 837, 896, 1253, 1335, 1599, 1640, 2878 and 3374 cm⁻¹, whereas

S No	Parameters studied	*P. pelagicus*
1	Yield	59%
2	pH	6.8
3	Ash content	0.4%
4	Degree of deacetylation	83%

Table 1: Physico-Chemical characterization of chitosan.

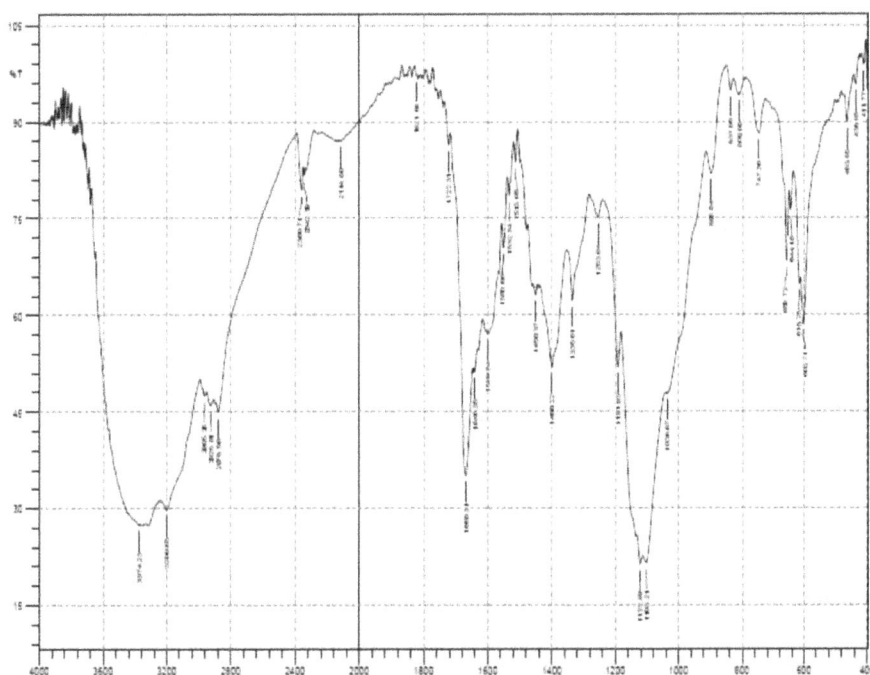

Figure 1: FT-IR spectrum of standard chitosan.

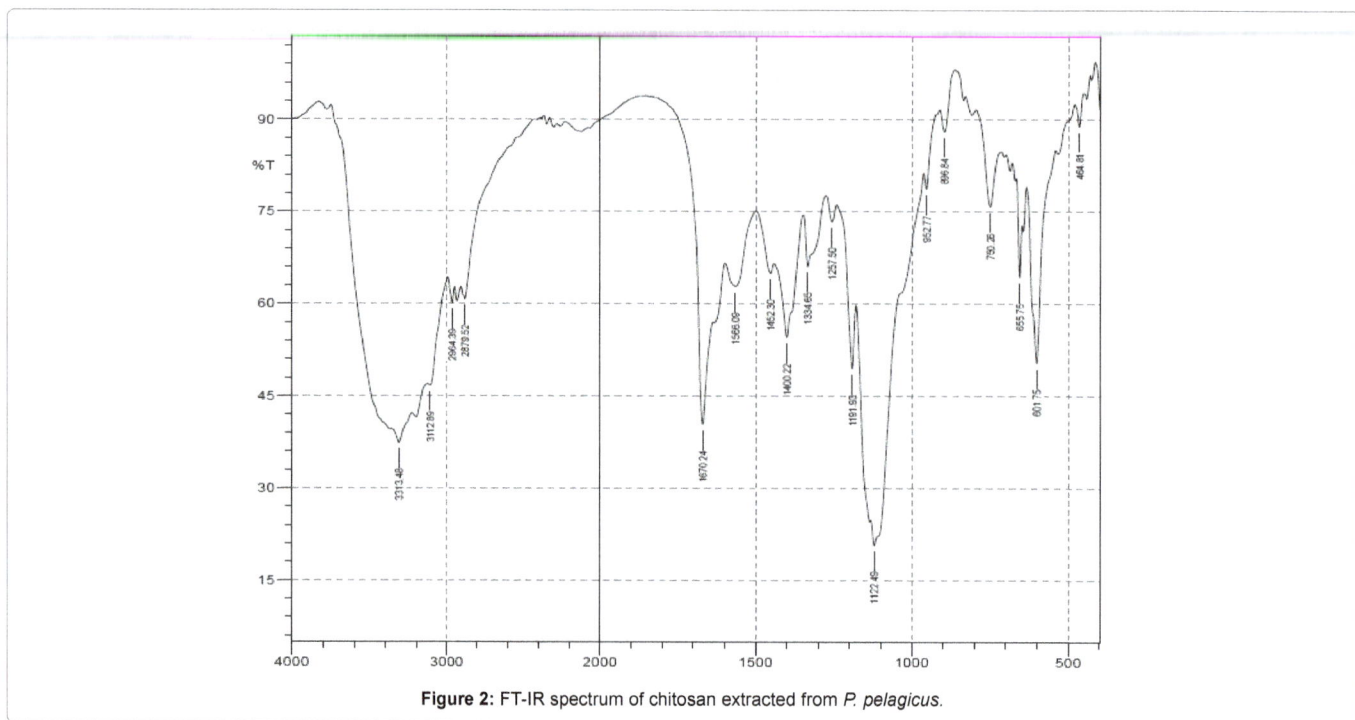

Figure 2: FT-IR spectrum of chitosan extracted from *P. pelagicus*.

the Chitosan sample from shells of *P. pelagicus* matched with that of seven major peaks of standard chitosan. The absorbance bands of 891, 897, 1026, 1259, 1422, 1587, 3377 indicate the HPO_4^{2-}, (NH) amide III, PO_4^{-3}, PO_3^{-4}, OH group (monomer), ($-NH_2$) Amide II, structural unit respectively.

Batch adsorption experiments

Effect of pH: pH plays an important role in the adsorption of metal ions as it affects both the dissociation degree of the functional group from the adsorbent surface and the solubility of metal ions. The electroplating wastewater contains 2.64, 2.75, 3, 2.96, 2.9 mg/L of Cr^{6+}, Zn^{2+}, Cu^{2+}, Pb^{2+}, Cd^{2+} respectively. The effect of pH on adsorption of heavy metals (Cr^{6+}, Zn^{2+}, Cu^{2+}, Pb^{2+}, Cd^{2+}) by chitosan was studied over a pH range of 2 to 9 at 25°C, at an agitation speed of 150 rpm. From the study, the adsorption capacity of chitosan for heavy metals like Cr^{6+}, Zn^{2+}, Cu^{2+}, Pb^{2+}, Cd^{2+}) was found to be 70%, 78%, 81%, 82%, 79% respectively, at a pH 5, while adsorption was lowest at pH 2 (Figure 3). This result was in accordance with Ref. [4], who reported that adsorption of heavy metals reaches a peak at a pH range of 5 to 8. At

nearly neutral pH (4 to 5) the cationic heavy metals exist as free ions and get adsorbed on to chitosan. Whereas at higher pH, adsorption performance is improved due to decrease in H^+ ion that causes decrease in both binding sites and the electrostatic repulsion. Furthermore, at a lower pH, H^+ ion concentration is high, so the protonation of amino group induces an electrostatic repulsion of metallic ions.

Effect of contact time: It has been observed that at a constant concentration of metal ions and fixed amount of adsorbent, the adsorption efficiency increases with increasing the contact time up to a certain level and then it reaches the equilibrium. Figure 4 shows that

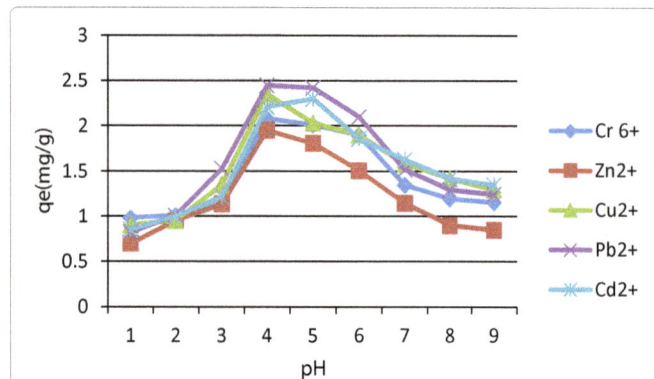

Figure 4: Pseudo Second order Kinetics for Cr^{6+}, Zn^{2+}, Cu^{2+}, Pb^{2+}, Cd^{2+}, Pb^{2+}, Zn^{2+}.

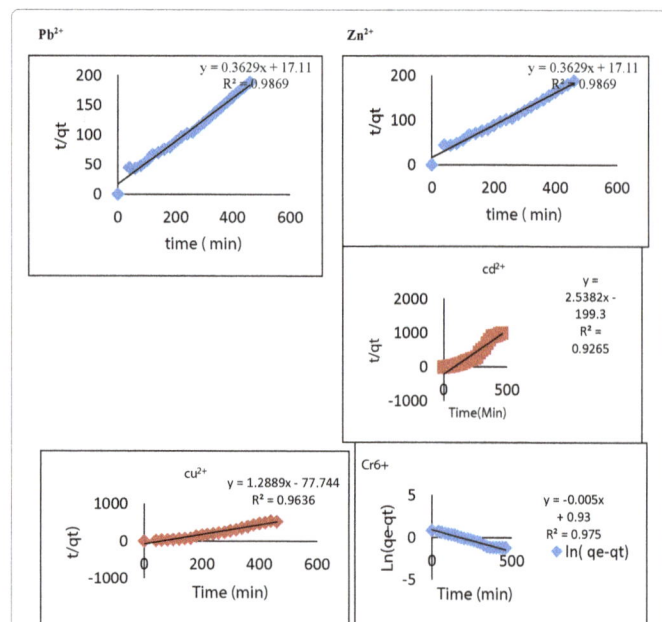

Figure 3: Effect of pH on adsorption capacity of chitosan for Cr^{6+}, Zn^{2+}, Cu^{2+}, Pb^{2+}, Cd^{2+}.

adsorption rate first increased rapidly as the contact time increases, but after reaching the optimum time of about 340 min there is no significant increase. The effect may be due to the saturation of adsorption sites with metal ions on the solid particle. The optimum contact time for Cr^{6+}, Zn^{2+}, Cu^{2+}, Pb^{2+}, Cd^{2+} was 320 min. The slight decrease in adsorption after optimum contact time may be due to the breakage of newly formed weak adsorption bonds due to constant shaking. In order to investigate the controlling mechanism of adsorption processes such as mass transfer and chemical reaction, the pseudo first order and pseudo second order equations are applied to model the kinetics of Cr^{6+}, Zn^{2+}, Cu^{2+}, Pb^{2+}, Cd^{2+} adsorption onto chitosan. The pseudo first-order kinetic model did not adequately describe the adsorption results with a low correlation coefficient for the Zn^{2+}, Cu^{2+}, Pb^{2+}, Cd^{2+} (0.8) except Cr^{6+} (0.975). However, the pseudo second-order kinetic model provided a comparable correlation for the adsorption of Zn^{2+}, Cu^{2+}, Pb^{2+} ions in contrast to the pseudo first-order model (0.99). The pseudo first-order rate model has been widely used for sorption of metals, which is widely used for reversible reactions with an equilibrium being established between liquid and solid phases. In many cases, the pseudo first order does not fit well to the whole range of contact time and is generally applicable over the initial stage of adsorption process.

Effect of adsorbent dose: The effect of the adsorbent dose was studied at room temperature by varying the sorbent amounts from 1 to 6 g. For all these runs, initial concentration of the metal ions was fixed. Figure 5 shows the adsorption of Cr^{6+}, Zn^{2+}, Cu^{2+}, Pb^{2+}, Cd^{2+} ions increase rapidly with increase in the amount of adsorbent due to greater availability of the surface area at higher concentration of the adsorbent. The significant increase in uptake was observed when the dose was increased from 1 to 4 g. Any further addition of the adsorbent beyond this did not cause any significant change in the adsorption. This might be due to overlapping of adsorption sites as a result of overcrowding of the adsorbent particles. These results indicate that removal efficiency is directly related to the number of available adsorption sites. Once equilibrium is attained, there is no effect on adsorption efficiency thus proving that even at a lower concentration of 4 g/L of the adsorbent chitosan was enough for removal of heavy metals.

Determination of optimum pH for desorption study

The nature of the adsorption process and the recovery of metal ions. Regeneration of the adsorbent material is of crucial importance in economic development [9]. Regeneration must produce small volume of metal concentrates suitable for metal recovery process without damaging the capacity of the adsorbent. The effect of pH on desorption of the heavy metals by chitosan over a pH range 2 to 8, at room temperature with an agitation speed of 150 rpm is illustrated

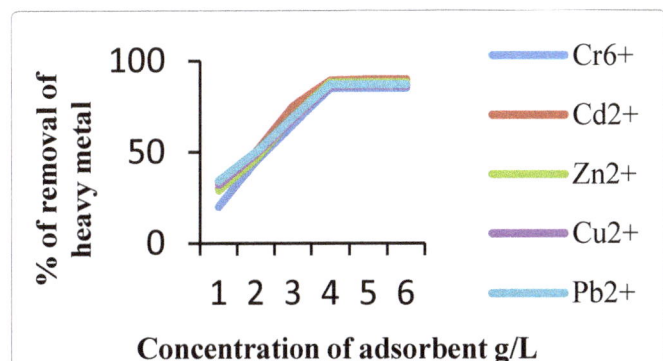

Figure 5: Effect of adsorbent dose for adsorption of Cr^{6+}, Zn^{2+}, Cu^{2+}, Pb^{2+}, Cd^{2+}.

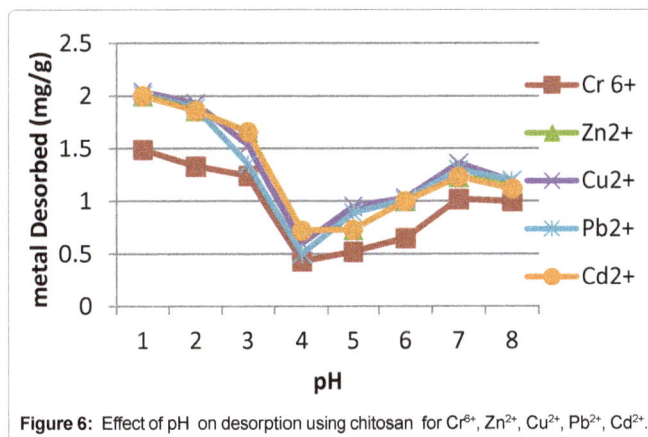

Figure 6: Effect of pH on desorption using chitosan for Cr^{6+}, Zn^{2+}, Cu^{2+}, Pb^{2+}, Cd^{2+}.

Cost of Commercial chitosan (100 g)	Rs. 2500
Chemical used for extraction of chitosan	
NaOH (500 ml)	Rs.120
HCl (500 ml)	Rs.98
Acetic acid (500 ml)	Rs.200
Crab powder (100 gms)	Free of cost
Cost of extracted chitosan	Rs.700

Table 2: Comparative cost for extracted chitosan and commercial Chitosan.

in (Figure 6). Maximum desorption occurs at pH 2. The percentage of desorption of Cr^{6+}, Zn^{2+}, Cu^{2+}, Pb^{2+}, Cd^2 was found to be 70%, 78%, 81%, 82%, 79% at pH 2. The percentage of desorption was lowest at pH 5(25%) [10]. Observed that qualitative decline in the metal uptake capacity of chitin and chitosan, for an extended series of metals when the pH was reduced from neutral to acetic. Desorption study results indicate that when abundant protons compete for the adsorption site with metals at low pH, it releases the metal ions adsorbed on chitosan particle into the suspending medium, causing desorption. Hence, higher amount of metals desorbed from chitosan at lower pH values. The adsorption-desorption cycle results clearly demonstrate that the regeneration and subsequent use of the chitosan would enhance the removal of heavy metals from wastewater.

Thus, from the above study it is suggested that chitosan can be used as low cost adsorbent for electroplating effluent treatment. An adsorbent is considered as low cost if it requires minimal processing and is abundant in nature. It has been reported by Ref. [11], that the expense of individual adsorbent varies depending on the degree of processing required and local availability. In this study, Chitosan is extracted from the shells of *P. pelagicus* which is considered as a waste and is abundantly available in Rameshwaram Coast, and hence the production of which requires minimal processing and the raw materials used for chitosan extraction is freely available. Table 2 shows the cost effectiveness of the extracted chitosan [12-17]. Thus, the use of low-cost adsorbents like chitosan may also contribute to the sustainability of the surrounding environment.

Conclusion

Chitosan extracted from the locally available crab shell was utilised as an adsorbent for the removal of heavy metals from electroplating waste water. The optimum pH necessary for removing these heavy metals in solution was found to range between 5-7. Metal removal increased with an increase in pH, as there is a decrease in competition between proton and metal cation for some functional group along with a decrease in positive surface charge, resulting in a lower electrostatic repulsion between surface and metal ions. Pseudo-second order model showed the best fit to the experimental data. Thus, this study shows that the use of crab shell chitosan for heavy metal removal appears to be technically feasible, and eco-friendly with very high efficacy. Further studies would reveal the concentrations of the adsorbent required to resolve the COD and BOD level in the effluent after treatment.

Acknowledgements

Authors would like to thank DBT – BIF lab, Lady Doak College, Madurai. Supported by Ministry of Science and Technology, Department of Biotechnology, Bioinformatics Division, New Delhi (Sanc. No. BT/B I/25/001/2006).

References

1. Muzzarelli RAA (1977) Enzymatic Synthesis of Chitin & Chitosan. In Chitin. Pergamon Press, Oxford, pp: 5-17.

2. Takiguchi Y (1991) Preparation of chitosan and partially deacetylated chitin. Gihodou Shupan Kaisha Japan 2: 9-17.

3. Domszy JG, Roberts G (1985) Evaluation of infrared spectroscopic techniques for analyzing chitosan. Macromol Chem 186: 1671-1677.

4. Tewari K, Promod AK, Mishra RP, Sinha S, Shrivastav RPS (1989) Removal of toxic metals from electroplating industries (effect of pH on removal by adsorption). Indian Journal of Environmental Health 31: 120-124.

5. Orumwense FFO (1996) Removal of lead from water by adsorption on a kaolinitic clay. J Chem Tech Biotech 65: 363-369.

6. Annadurai G, Ling LY, Lee JF (2008) Adsorption of reactive dye from an aqueous solution by chitosan: isotherm, kinetic and thermodynamic analysis. J Hazard Mater 152: 337-346.

7. Vieira RS, Beppu MM (2006) Dynamic and static adsorption and desorption of Hg(II) ions on chitosan membranes and spheres. Water Res 40: 1726-1734.

8. Tureli C, Çelik M, Erdem U (2000) Comparison of Meat Composition and Yield of Blue Crab (Callinectes sapidus RATHBUN 1896) And Sand Crab (Portunus pelagicus LINNE, 1758) Caught in Iskenderun Bay, North-East Mediterranean. Turk J Vet Anim Sci 24: 195-203.

9. Tsezos M, Volesky B (1981) Biosorption of Uranium and Thorium. Biotechnol Bioeng 23: 583-604.

10. Muzzarelli RA, Tubertini O (1969) Chitin and chitosan as chromatographic supports and adsorbents for collection of metal ions from organic and aqueous solutions and sea-water. Talanta 16: 1571-1577.

11. Chui VMD, Mok KW, Ng CY, Luong BP, Ma KK (1996) Removal and recovery of copper (II), chromium (III), and nickel (II) from solutions using crude shrimp chitin packed in small columns. Environ Int 22: 463-468.

12. Wan Ngah WS, Kamari A, Koay YJ (2004) Equilibrium and kinetics studies of adsorption of copper (II) on chitosan and chitosan/PVA beads. Int J Biol Macromol 34: 155-161.

13. Tsezos M, Volesky B (1981) Biosorption of Uranium and Thorium. Biotechnol Bioeng 23: 583-604.

14. Kofuji K, Qian CJ, Nishimura M, Sugiyama I, Murata Y, et al. (2005) Relationship between physic chemical Characteristics and functional properties of chitosan. Eur Polym 411: 2784 -2791.

15. Jeon C, Holl WH (2004) Application of the surface complexation model to heavy metal sorption equilibria onto aminated chitosan. Hydrometallurgy 71: 421-428.

16. Sarkar M, Acharya PK, Bhattacharya B (2003) Modeling the adsorption kinetics of some priority organic pollutants in water from diffusion and activation energy parameters. J Colloid Interface Sci 266: 28-32.

17. Su-Hsia L, Juang J (2002) Heavy metal removal from water by sorption using surfactant-modified montmorillonite Hazard. Journal of Hazardous Materials 92: 315-326.

Permissions

List of Contributors

Melissa Paola Mezzari
Biotechnology and Sciences Program, West University of Santa Catarina, Videira, SC 89560-000, Brazil

Jean Michel Prandini
Department of Chemical Engineering, Federal University of Santa Catarina, Florianópolis, SC 88040-900, Brazil

Jalusa Deon Kich and Márcio Luís Busi da Silva
EMBRAPA Swine and Poultry, Concórdia, Brazil

Lamia Ayed, Nadia Chammam, Nedra Asses and Moktar Hamdi
Carthage University, National Institute of Applied Science and Technology (INSAT). Laboratory of Microbial Ecology and Technology, Department of Biological and Chemical Engineering, B.P .676,1080 Tunis, Tunisia

Sandip R Sabale and Bhaskar V Tamhankar
P.G. Department of Chemistry, Jaysinpur College, Jaysingpur-416101, M.S., India

B S Mohite
Department of Chemistry, Shivaji University, Kolhapur-416004, M.S., India

Meena M Dongare
Department of Botany, Shivaji University, Kolhapur-416004, M.S., India

Natasha Nageswaran
Department of Microbiology and Fermentation Technology, Jacob School of Biotechnology & Bioengineering, SHIATS Allahabad-211007, Uttar Pradesh, India

P.W. Ramteke
Department of Biological Sciences, Jacob School of Biotechnology & Bioengineering, SHIATS Allahabad-211007, Uttar Pradesh, India

O.P. Verma and Avantika Pandey
Department of Molecular & Cellular Engineering, Jacob School of Biotechnology & Bioengineering, SHIATS Allahabad-211007, Uttar Pradesh, India

Peter Firbas
Rivate Researcher, SICRIS Id. No.11784, Private Laboratory for Plant Cytogenetics, Ljubljanska c. 74. SI-1230 Domžale, Slovenia

Tomaž Amon
Researcher, Gorazdova ul. 3. SI–1000 Ljubljana, Slovenia

David Nzioka Mutua, Eliud Nyaga Mwaniki Njagi, George Orinda and Geoffry Obondi
Department of Biochemistry and Biotechnology, School of Pure and Applied Science, Kenyatta University, PO Box 43844, 00100, Nairobi, Kenya

Frank Kansiime
Department of Biochemistry, Institute of Environment and Natural Resources, Makerere University, PO Box 7062, Kampala, Uganda

Joseph Kyambadde, John Omara, Robinson Odong and Hellen Butungi
Department of Biochemistry, Faculty of Science, Makerere University, PO Box 7062, Kampala, Uganda

D A Turner, and J V Goodpaster
Department of Chemistry and Chemical Biology, Forensic and Investigative Sciences Program, Indiana University Purdue University Indianapolis (IUPUI), Indianapolis, USA

J. Pichtel
Department of Natural Resources and Environmental Management, Ball State University, Muncie, USA

Y Rodenas
Department of Biology, Ball State University, Muncie, USA
Department of Natural Resources and Environmental Management, Ball State University, Muncie, USA

J McKillip
Department of Biology, Ball State University, Muncie, USA

Gopan Mukkulath and Santosh G. Thampi
Department of civil engineering, National Institute of Technology Calicut, Calicut-673 601, Kerala, India

Sibi G
Department of Biotechnology, Indian Academy Degree College, Centre for Research and Post Graduate Studies, Bangalore, India

R. Karthikeyan
Biological and Agricultural Engineering, Texas A&M University, College Station, TX 77845-2117, USA

S.L.L. Hutchinson
Biological and Agricultural Engineering, Kansas State University, Manhattan, KS 66502, USA

L. E. Erickson
Chemical Engineering, Kansas State University, Manhattan, KS 66502, USA

Kiraye M, John W and Gabriel K
Department of Chemistry, Makerere University, Kampala, Uganda

Shweta Saraswat
National Institute of Science, Communication and Information Resources (CSIR) 14, India

Korrapati Narasimhulu
Department of Biotechnology, National Institute of Technology, Warangal, India

Y. Pydi Setty
Department of Chemical Engineering, National Institute of Technology, Warangal, India

Philip Antwi, Jianzheng Li and En Shi
State Key Laboratory of Urban Water Resource and Environment, School of Municipal and Environmental Engineering, Harbin Institute of Technology, 73 Huanghe Road, Harbin 150090, PR China

Portia Opoku Boadi
School of Management, Harbin Institute of Technology, 92 West Dazhi Street, Nan Gang District, Harbin 150001, PR China

Frederick Ayivi
Department of Geography, University of North Carolina, 237 Graham building, 1009 Spring Garden St, Greensboro, NC27412, USA

Mousumi Saha and Bidyut Bandhophadhyay
Department of Biotechnology, Vidyasagar University, West Bengal, India

Agniswar Sarkar
Department of Biotechnology (Recognized by DBT, Govt. of India), The University of Burdwan, Burdwan, West Bengal, India

Hindumathy CK and Gayathri V
Department of Biotechynology, Vinayaka Mission University, Salem, Tamil Nadu, India

Rajdeo Kumar, Nisha Yadav and Laxmi Rawat
Forest Ecology and Environment Division, Forest Research Institute, Dehradun, India

Ashish Chauhan and Manish Kumar Goyal
National Institute of Pharmaceutical Education and Research, Mohali, India

M. B. Adewole and Y. I. Bulu
Institute of Ecology and Environmental Studies, Obafemi Awolowo University, Ile-Ife, Nigeria

Mohamed Ateia and Chihiro Yoshimura
Department of Civil Engineering, Tokyo Institute of Technology, 2-12-1-M1-4 Ookayama, Tokyo 152-8552 Japan

Mahmoud Nasr
Department of Sanitary Engineering, Faculty of Engineering, Alexandria University, PO Box 21544, Alexandria, Egypt

Danah Khazaal Al-Ameeri and Mohammad Al Sarawi
Earth and Environmental Science, College of Science, Kuwait University, Kuwait

Quainoo AK and Konadu A
Department of Biotechnology, Faculty of Agriculture, University for Development Studies, Tamale, Ghana

Kumi M
Water Research Institute, Tamale, Ghana

Alaa El Din Mahmoud and Manal Fawzy
Environmental Sciences Department, Faculty of Science, Alexandria University, Alexandria, Egypt

Prabhakar R. Pawar
Veer Wajekar Arts, Science and Commerce College, Mahalan Vibhag, Phunde, MH, India

Zhang X, Zhengsheng Yu, Yong Chen, Pu Liu and Xiangkai Li
MOE Key Laboratory of Cell Activities and Stress Adaptations, School of Life Sciences, Lanzhou University, Lanzhou, Gansu, 730000, P. R. China

Krumholz LR
Department of Botany and Microbiology and Institute for Energy and the Environment, University of Oklahoma, Norman, Oklahoma, 73019, U.S.A

Petra J Sheppard, Eric M Adetutu, Paul D Morrison and Andrew S Ball
RMIT University, School of Applied Science, Bundoora, Victoria 3083, Australia

Alexandra Young
Flinders University, School of Biological Sciences, Bedford Park, SA, 5042, Australia

Mike Manefield
Centre for Marine Bio innovation, University of New South Wales, Sydney, New South Wales, 2052, Australia

Baljinder Singh, Vasundhara Thakur, Garima Bhatia, Deepika Verma and Kashmir Singh
Department of Biotechnology, Panjab University, Chandigarh, India

Irfana Showqi, Farooq Ahmad Lone and Javeed Iqbal Ahmad Bhat
Division of Environmental Sciences, Sheri-Kashmir University of Agricultural Sciences and Technology of Kashmir, Shalimar, Jammu and Kashmir, India

E. C. Rada
Department of Civil and Environmental Engineering, University of Trento, Via Mesiano 77, I-38050 Trento, Italy

I.A. Istrate
Department of Energy production and Use, University Politehnica of Bucharest, Splaiul Independentei 313, 060042, Bucharest, Romania

Terefe Tafese Bezuneh
College of Natural Science, Arbaminch University, Arbaminch, Ethiopia

Ashutosh Tripathi and Manju Rawat Ranjan
Amity Institute of Environmental Sciences, Amity University, Noida-125, Gautam Buddha Nagar, U.P, India

Alkesh Patidar, Richa Gupta and Archana Tiwari
School of Biotechnology, Rajiv Gandhi Proudyogiki Vishwavidhyalaya, State Technological University of Madhya Pradesh, Airport Bypass Road, Bhopal, India

Bhavani K, Roshan Ara Begum E and Shenbagarathai R
PG and Research Department of Zoology and Biotechnology, Lady Doak College, Madurai-625 002, Tamil Nadu, India

Selvakumar S
Department of Environmental Studies, School of Energy Environment and Natural Resources, Madurai Kamaraj University, Madurai-625 021, Tamil Nadu, India

Index

A

Aas, 15-16, 18-19, 106, 171, 177

Allium Metaphase Test, 28-29

Ammonia Oxidizing Bacteria, 88

Anaerobic Sludge, 79-80, 86-87, 104

Antibiotic Susceptibility Test, 23-24

Antibiotic-resistant Bacteria, 1

Arsenic Biosorption Potential, 51

Arsenic Concentration, 51

Arsenic Removal, 50, 53, 56-57, 200

Aspergillus Niger, 5-14, 75, 77-78, 143, 192, 194, 201

B

Bacteria, 1-4, 15-16, 22-23, 25-28, 33, 37, 39-44, 57, 59, 64-65, 70, 87-97, 99-104, 114, 119, 128, 137, 153, 155, 158-161, 169, 180, 190, 192-195, 201-202

Biofilters, 46, 49

Biogas Yield, 79-80, 83-85

Biological Treatment, 5, 13, 33-34, 44, 56, 111-114, 194

Bioremediation, 3, 15, 23, 26, 37, 42, 45, 67, 72, 74, 89, 93, 97, 99, 104, 116, 119, 128, 130, 136, 143, 152, 160, 163, 167, 169, 190, 194, 199

Bioremediation Rate, 64-67

Biosorption, 5, 9-16, 22, 50-53, 55-57, 75-78, 115, 137-139, 141-143, 153, 155, 160-161, 174, 176, 199, 213

C

Calopogonium Mucunoides, 105, 108-109

Chlorpyrifos, 93-94, 97-98, 128

Chromosome Damage, 28, 30-31

City Abattoir, 33-34

Coir Geotextiles, 46-49

Constructed Wetland, 28, 30, 114-115

Contaminated Soils, 39, 41, 56, 59, 71, 93-94, 96-97, 99, 104, 116, 118-119, 128, 159, 181, 189

Crude Oil, 64, 67, 105-106, 108-109, 163, 169

D

Decolorization, 5-9, 11, 13, 15, 22, 190, 192-195

Degradation, 9, 11, 13, 15, 18, 22, 39-46, 49, 58-63, 68, 71, 88, 93-102, 105, 108, 112, 119, 121, 128, 150, 162-164, 167, 169, 187, 190, 195, 201, 203, 208

Disc-diffusion, 23

Dried Biomass, 50, 53, 55-56, 179

E

East Kolkata Wetland, 88-90

Environmental Pollutants, 28, 67, 69, 104, 118

F

Fresh Water Microalgae, 50

G

Genotoxicity, 28-32

H

Heavy Metal Ions, 15-16, 18-19, 22, 25, 50, 56, 196, 198

Hexavalent Chromium, 75, 143, 152, 155, 160, 199

Hila Gene, 1

Hydrocarbons, 39, 43, 63-67, 99, 104-105, 109-110, 116, 118-123, 127-128, 152, 162-164, 167-169, 181, 184-185, 187, 192

I

Ignitable Liquids, 39, 44

Immobilization, 75-76, 78, 113

In-situ Treatment, 111, 113-114

Industrialization, 99, 196, 208

Isotherm, 51, 53, 57, 75-77, 176, 199, 213

L

Lake Albert, 64-67

Lake Water, 15, 18, 113

Living Biomass, 11, 51, 53, 55-56

M

Malodorants, 58

Meat Processing, 33-34, 36-37

Microalgae, 1-3, 50-51, 53, 55-57, 192-193, 195

Microbial Degradation, 39-44, 63, 71, 93, 97, 99, 108, 112, 119, 162, 169, 190, 193-194

Minimum Inhibitory Concentration, 23-24, 154

Modeling, 79, 86-87, 114, 142-143, 213

Multi-drug Resistance, 23

Multiple Non-linear Regression, 80

N

Nitrite Oxidizing Bacteria, 33, 88, 90

Non-living Biomass, 51

Non-rhizospheric, 93-95

O

Optimization, 5, 7, 13, 97, 104, 137, 139-140, 142-143, 170, 194-195

Organic Fertilizer, 88, 105-109, 192

P

Pandorina, 50-53, 55-56

Parallel Control Test, 29

Patent Analysis, 69

Pentachlorophenol, 71, 99-104

Permittivity, 46-49

Pesticide, 15-20, 22, 61, 64, 88, 93-99, 104, 130

Pesticides, 15-20, 22, 28, 94-95, 99, 104, 118-119, 128, 192

Petroleum Hydrocarbons, 64-67, 105, 109-110, 118-123, 127-128, 185

Phytoextraction, 69-70, 118, 121

Phytoremediation, 56-57, 69-71, 74, 88, 91, 93, 95, 105, 114, 116, 118-122, 127-130, 136, 177, 179-180

Plasmid Curing, 23-25

Potato Starch, 79-80, 84-86

Pseudomonas Aeruginosa, 3, 64-67, 91, 104

R

Remediation, 15, 19, 44, 56, 58, 63, 65, 67, 72, 78, 99, 104, 111-115, 117-119, 121, 128, 130, 136, 153, 159, 162, 169, 177, 179, 181-183, 185, 189, 208

Removal Mechanism, 111

Rhizoremediation, 116, 118-119, 127-128

Rhizospheric, 93-96

Root Meristems Cell, 30

S

Salmonella Enterica Serovar Typhimurium, 1-3

Scenedesmus, 1-4, 50-53, 55-57

Scenedesmus Spp., 1-3

Sequencing Batch Reactors, 33-34, 37

Soil Chemistry, 39, 182

Soils, 39-45, 56, 58-59, 61-63, 71-72, 93-99, 104-105, 109, 116-119, 121, 124-126, 128-130, 136, 153, 159, 161, 181-185, 189, 202, 206

Sulfur Compounds, 58, 63

Surface Water, 28, 31, 111-115, 117, 181

Swine Wastewater, 1-3

T

Tannase, 5-7, 11-13

Tensile Strength, 46-49

Tertiary Butyl Mercaptan, 58, 62-63

Topw, 5-12

Trace Metals, 16, 56, 116, 119-121, 123, 127, 160, 180

U

Uasb, 79-80, 83-87, 115

W

Wastewater, 1-9, 13, 28, 31, 33-38, 46, 49, 58, 63, 69, 75, 80, 88, 92, 115, 122, 131, 137, 143, 152, 160, 170, 176-180, 184, 190, 200, 208, 211-212

Water Quality, 15, 29-30, 34, 83, 88-89, 112, 114-115, 128, 136, 151